Advances in Intelligent Systems and Computing

Volume 923

Series editor

Janusz Kacprzyk, Systems Research Institute, Polish Academy of Sciences, Warsaw, Poland

The series "Advances in Intelligent Systems and Computing" contains publications on theory, applications, and design methods of Intelligent Systems and Intelligent Computing. Virtually all disciplines such as engineering, natural sciences, computer and information science, ICT, economics, business, e-commerce, environment, healthcare, life science are covered. The list of topics spans all the areas of modern intelligent systems and computing such as: computational intelligence, soft computing including neural networks, fuzzy systems, evolutionary computing and the fusion of these paradigms, social intelligence, ambient intelligence, computational neuroscience, artificial life, virtual worlds and society, cognitive science and systems, Perception and Vision, DNA and immune based systems, self-organizing and adaptive systems, e-Learning and teaching, human-centered and human-centric computing, recommender systems, intelligent control, robotics and mechatronics including human-machine teaming, knowledge-based paradigms, learning paradigms, machine ethics, intelligent data analysis, knowledge management, intelligent agents, intelligent decision making and support, intelligent network security, trust management, interactive entertainment, Web intelligence and multimedia.

The publications within "Advances in Intelligent Systems and Computing" are primarily proceedings of important conferences, symposia and congresses. They cover significant recent developments in the field, both of a foundational and applicable character. An important characteristic feature of the series is the short publication time and world-wide distribution. This permits a rapid and broad dissemination of research results.

** Indexing: The books of this series are submitted to ISI Proceedings, EI-Compendex, DBLP, SCOPUS, Google Scholar and Springerlink **

Advisory Editors

Nikhil R. Pal, Indian Statistical Institute, Kolkata, India

Rafael Bello Perez, Faculty of Mathematics, Physics and Computing, Universidad Central de Las Villas, Santa Clara, Cuba

Emilio S. Corchado, University of Salamanca, Salamanca, Spain

Hani Hagras, Electronic Engineering, University of Essex, Colchester, UK

László T. Kóczy, Department of Automation, Széchenyi István University, Gyor, Hungary

Vladik Kreinovich, Department of Computer Science, University of Texas at El Paso, EL PASO, TX, USA

Chin-Teng Lin, Department of Electrical Engineering, National Chiao Tung University, Hsinchu, Taiwan

Jie Lu, Faculty of Engineering and Information Technology, University of Technology Sydney, Sydney, NSW, Australia

Patricia Melin, Graduate Program of Computer Science, Tijuana Institute of Technology, Tijuana, Mexico

Nadia Nedjah, Department of Electronics Engineering, University of Rio de Janeiro, Rio de Janeiro, Brazil

Ngoc Thanh Nguyen, Faculty of Computer Science and Management, Wrocław University of Technology, Wrocław, Poland

Jun Wang, Department of Mechanical and Automation Engineering, The Chinese University of Hong Kong, Shatin, Hong Kong

More information about this series at http://www.springer.com/series/11156

Ana Maria Madureira · Ajith Abraham ·
Niketa Gandhi · Maria Leonilde Varela
Editors

Hybrid Intelligent Systems

18th International Conference on Hybrid
Intelligent Systems (HIS 2018) Held in Porto,
Portugal, December 13–15, 2018

 Springer

Editors
Ana Maria Madureira
School of Engineering
Instituto Superior de Engenharia (ISEP/IPP)
Porto, Portugal

Ajith Abraham
Machine Intelligence Research Labs
(MIR Labs)
Auburn, WA, USA

Niketa Gandhi
Machine Intelligence Research Labs
Auburn, WA, USA

Maria Leonilde Varela🆔
Department of Production and Systems
University of Minho
Guimarães, Portugal

ISSN 2194-5357 ISSN 2194-5365 (electronic)
Advances in Intelligent Systems and Computing
ISBN 978-3-030-14346-6 ISBN 978-3-030-14347-3 (eBook)
https://doi.org/10.1007/978-3-030-14347-3

Library of Congress Control Number: 2019933215

This Springer imprint is published by the registered company Springer Nature Switzerland AG
The registered company address is: Gewerbestrasse 11, 6330 Cham, Switzerland

Preface

Welcome to Porto, Portugal, and to the 18th International Conference on Hybrid Intelligent Systems (HIS 2018); the 10th International Conference on Soft Computing and Pattern Recognition (SoCPaR 2018); and the 13th International Conference on Information Assurance and Security (IAS 2018) held at Instituto Superior de Engenharia do Porto (ISEP) during December 13–15, 2018.

Hybridization of intelligent systems is a promising research field of modern artificial/computational intelligence concerned with the development of the next generation of intelligent systems. A fundamental stimulus to the investigations of hybrid intelligent systems (HISs) is the awareness in the academic communities that combined approaches will be necessary if the remaining tough problems in computational intelligence are to be solved. Recently, hybrid intelligent systems are getting popular due to their capabilities in handling several real-world complexities involving imprecision, uncertainty, and vagueness. HIS 2018 builds on the success of HIS 2017, which was held in Delhi, India, during December 14–16, 2017. HIS 2018 received submissions from 30 countries, and each paper was reviewed by at least five reviewers in a standard peer review process. Based on the recommendation by five independent referees, finally 56 papers were accepted for the conference (acceptance rate of 40%).

Conference proceedings are published by Springer Verlag, Advances in Intelligent Systems and Computing Series. Many people have collaborated and worked hard to produce this year successful HIS 2018 conference. First and foremost, we would like to thank all the authors for submitting their papers to the conference and for their presentations and discussions during the conference. Our thanks to Program Committee members and reviewers, who carried out the most difficult work by carefully evaluating the submitted papers. We are grateful to our three plenary speakers:

* *Petia Georgieva, University of Aveiro, Portugal*
* *J. A. Tenreiro Machado, Polytechnic of Porto, Portugal*
* *Henrique M. Dinis Santos, University of Minho, Portugal*

Our special thanks to the Springer Publication team for the wonderful support for the publication of this proceedings. Enjoy reading!

Ana Maria Madureira
Ajith Abraham
General Chairs

Maria Leonilde Varela
Oscar Castillo
Simone Ludwig
Program Chairs

Contents

A Machine Learning Approach to Contact Databases' Importation for Spam Prevention

Duarte Coelho[1](✉) ⓘ, Ana Madureira[2]ⓘ, Ivo Pereira[1]ⓘ, and Bruno Cunha[2]ⓘ

[1] E-goi, Matosinhos, Portugal
{dcoelho,ipereira}@e-goi.com
[2] Interdisciplinary Studies Research Center (ISRC), ISEP/IPP, Porto, Portugal
{amd,bmaca}@isep.ipp.pt

Abstract. This paper aims to provide a solution to a problem shared by online marketing platforms. Many of these platforms are exploited by spammers to ease their job of distributing spam. This can lead to platforms domains being black-listed by ISP's, which translates to lower deliverability rates and consequently lower profits. Normally, platforms try to counter the problem by using rule-based systems, which require high-maintenance and are not easily editable. Additionally, since analysis occurs when a contact database is imported, the regular approach of judging messages' contents directly is not an effective solution, as those do not yet exist. The proposed solution, a machine-learning based system for the classification of contact database's importations, tries to surpass these aforementioned systems by making use of the capabilities introduced by machine-learning technologies, namely, reliability in regards to classification and ease of maintenance. Preliminary results show the legitimacy of this approach, since various algorithms can be successfully applied to it. The most proficient of the ones applied being Ada-boost and Random-forest.

Keywords: Machine-learning · Spam · E-marketing

1 Introduction

In recent years, the act of traditional marketing has been gradually losing relevance to the concept of e-marketing, the process of advertising and selling products and services on the internet [7].

At its core, e-marketing remains the same as traditional marketing, organizations must meet their customers needs. More than that, it could be said that the internet was the final nail in the power shift from organizations to users. By giving full control to the user over what he wanted to do, marketers could not keep persons captive while watching advertisements, i.e. it generated a great degree of awareness [7].

ⓒ Springer Nature Switzerland AG 2020
A. M. Madureira et al. (Eds.): HIS 2018, AISC 923, pp. 1–10, 2020.
https://doi.org/10.1007/978-3-030-14347-3_1

The use of e-marketing became an integral part of any modern organization, and in order to ease the process of creating, managing and distributing campaigns over different channels, organizations like E-goi were created [3].

E-goi is an organization which owns a Software as a Service multichannel marketing automation platform. As the definition implies, it employs an array of communication channels in order to bring marketing campaigns to fruition. These channels are: e-mail, SMS, voice, smart SMS, push notifications, and, in the near future, Facebook Messenger.

E-goi directs its business model, i.e. marketing automation solutions, to all types of organizations, from micro enterprises to large multinationals, while also not forgetting small and medium businesses.

Currently, E-goi has nearly three hundred thousand users (checked on February 2018) from all over the world. Over 250 new account sign-ups occur each day (on average) and hundreds of thousands of people log into its platform every hour. These users then use E-goi to send more than 20 million e-mails daily. All of this so that they can more easily and reliably transmit to their clients the news and information they really need to know about their business.

Each of the new accounts created daily import their Contact Databases (CDBs), which can contain over a million subscribers each, in order to create their campaigns. This all happens while existing accounts continue to, frequently, import even more contacts.

Presently, the organization uses a rule based system in their CDB importation judging system. This system takes into account various aspects of the user's information (such as account type, location, etc.) and the contents of the database to be imported to reach a conclusion about whether the database should or not be imported, in order to avoid various types of malicious behaviours (namely, spamming the e-mails that were loaded). It differs and complements a regular spam classification tasks because it acts before messages are created and uses information and statistics inherent to the user and its importation.

Internet Service Providers (ISPs) (term which will be used in an interchangeable fashion with E-mail Service Provider (ESP) throughout this paper, since ISPs are, in most cases, ESPs), are strongly opposed this type of activity (spamming). It is not uncommon for e-mail addresses or even domains associated with this type of practices to get marked as a target whose e-mails should be blocked or even fully deleted. This means that the organization should be very careful when approaching this point, as one wrongly classified/detected spammer could be responsible for a block to the e-mails sent by many other users.

Due to that, the presence of a CDB importation analysis system is a necessity. However, the type of rule's based system, which is currently in use, is extensive and not easily editable in case a change needs to be made to the way the importations are judged. It may even miss key characteristics that would make for good indicators of the user's intent to realize malicious actions.

As such, the main objective tackled in this paper is the creation of a system that can be more easily editable that the current one, while also being both equally or more efficient and accurate, and also taking into account more, possibly relevant, criteria. In order to organize it in a timely fashion, first a couple of previous works relating to the issues at hand will be presented, followed by the process of developing a proposed solution for the problem and then evaluating said solution.

2   Previous Work

The idea of using machine learning for either filtering or blocking spam is not new. Through the last decade, partially due to the rise of spam in prominence, works in this area have become common. Normally, spam classification problems focus on the judging the contents of messages and not so much in the classification of databases. That being the case, two previous works will be presented ([8] and [1]), one focusing on the comparison of algorithms in "regular" spam detection, and another focusing on algorithms performance in binary classification (since we want to find if a CDB is that of a spammer or not).

In the paper, *A comparative study for content-based dynamic spam classification using four machine learning algorithms* [8], the authors proposed the execution of an empirical evaluation of four different machine learning algorithms, relatively to their spam classification capabilities. The encompassed algorithms included: one based in a Naïve Bayes approach; an Artificial Neural-network; a SVM approach; and a Relevance Vector Machine (RVM) approach.

The various approaches were evaluated based on different data-sets and feature sizes in terms of: accuracy, the percentage of e-mail correctly classified by the algorithm; spam precision, the ratio of spam e-mail correctly classified from all e-mail classified as spam; and spam recall, the proportion between the e-mails which the algorithm managed to classify as spam and the true number of spam e-mails present in the testing set [8].

From the results depicted in that paper, it was possible to conclude that: the NN classifier is unsuited to be used by itself in the task, as it is the one that consistently got the lowest results; both the Support Vector Machine (SVM) and RVM classifiers seemed to slightly outperform the NB classifier; comparatively to SVM, RVM provided similar results with less relevance vectors and faster testing time, although the learning procedure was slower overall.

One work presenting a comparison between the performance of supervised learning algorithms in regards to binary problem scenarios is *An Empirical Comparison of Supervised Learning Algorithms Using Different Performance Metrics* [1]. In this work the authors used various data-sets and performance metrics in order to compare seven supervised learning algorithms based on different approaches. These approaches were based on: K Nearest Neighbours (KNN), Artificial Neural Networks (ANN), Decision Tree (DT), Bagged Decision Tree (BAG-DT), Random Forest (RF), Boosted Decision Tree (BST-DT), Boosted Stumps (BST-STMP), SVM.

These algorithms were then evaluated according to three different metric groups (each containing three metrics), these were: threshold metrics, ordering/rank metrics and probability metrics. Additionally, this comparison between algorithms occurred using seven different binary classification problems (each referring to a different data-set whose sources can be checked at [1]).

The results of the performed comparisons were then normalized and averaged over the nine testing metrics (which included accuracy, area under the respective ROC curve, average precision, and others). The results obtained from the various performed tests allowed diverse conclusions to be reached. ANN's, SVM's and BAG-DT were the best algorithms regarding their effectiveness; SVM's performance was probably a result of the low dimensionality of the problems used, although when taking into account probability metrics (that is, *"metrics that are uniquely minimized, in expectation, when the predicted value for each case coincides with the true underlying probability of that case being positive"* [1]) BST-DT perform poorly overall, if one takes into account only threshold and rank metrics, this algorithm was the best performing one out of all; KNN performed well if attributes are scaled, by their gain ratio, for instance. However, it was not as competitive as other choices; single DT's did not perform well comparatively to other methods regardless of type; lastly, BST-STMP's did not perform nearly as well as BST-DT's.

3   Solution

The first step to find a solution is to understand the problem. In this case some contact database importations are being performed by spammers which then use those lists to produce spam e-mail campaigns. So, how can this be prevented? The answer is straightforward, information. If enough information is presented about a subject then it can be understood. However, due to the context of the problem, information can only be extracted regarding a CDB's characteristics and not a campaign's content. Various parameters can be extracted from a single CDB. In a global perspective, factors such as the overall size of an importation are important, as spammers' importations will probably differ in size in comparison to those made by both singular users as well as corporate users. If a more sharp and singular outlook is used, then characteristics obtained from the e-mails present in a CDB also become important, e.g. a regular CDB shouldn't have any spam-trap addresses in its constitution, while that of a spammer's might.

3.1   Feature Selection

Many different features can be extracted from a CDB importation. The important thing to keep in mind, however, is that not every single one has the same degree of contribution to the final decision. This becomes clear if we consider, for example, two features, the first one representing the percentage of e-mails that is similar to a known domain but not an actual one (e.g. example@gmail.com), and the second one representing the percentage of spam-traps in the imported CDB. By definition a spam-trap classifies any domain that tries to send messages to it as a possible spammer, which means that even a low percentage of these e-mails in a CDB can carry more weight than a greater percentage similar e-mails, in which case the worst outcome is that a bounce will occur.

The problem now becomes how to distinguish these relevant features in order to train the algorithm which will classify each CDB importation attempt. A good starting point would be the concept of sensibility analysis. Sensitivity analysis in normally defined as: *"The study of how uncertainty in the output of a model (numerical or otherwise) can be apportioned to different sources of uncertainty in the model input"* [6]. However this definition is not of much use without knowing what a model means in this context. Simply put a model is a representation of a given system or problem made through a mathematical approach (e.g. a simple body mass index prediction model could be accomplished by computing the formula $BMI = Weight/Height^2$). Knowing this we can then say that sensibility analysis can be summed up as an activity that aims to investigate the degree to which a variable used in a model (e.g. the height in the BMI example) can affect that models results (i.e. the BMI itself).

Although various ways of conducting this type of analysis exist, the most straightforward one is direct observation of data, namely, through the use of scatter-plots and other such methods. Evidently, the presence of a possible pattern in a scatter-plot is usually a good sign of the sensitivity between a variable and a model's output. The more clear the pattern is, the more sensible the output is to that variable. For instance, in Fig. 1, Y is increasingly more sensitive to Z_1, Z_2, Z_3 and Z_4, that is the case since it is clearly possible to see that a pattern gradually becomes more clear in each case's plot [6].

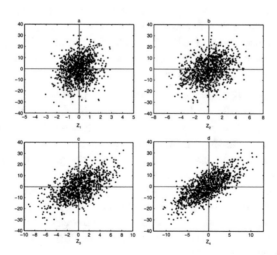

Fig. 1. Scatter-plot of output variable Y versus input variables Z_1, Z_2, Z_3 and Z_4 [6]

This method may be adequate if the relations between the various input feature and the generated outputs are distinct, however if the relation between them does not seem to show a well defined pattern, then using it a basis for decision may prove difficult. A plausible way of countering this problem would be to compute the correlation coefficients of input features versus their expected outputs. This would allow for a solid and easy to understand hierarchy to be formed between the input features, as, by definition, correlation coefficients are measured in a $[-1, 1]$ scale. However, if this method is to be used one should consider the various assumptions that any correlation coefficient method carries, as well as what it stands for. E.g. the use of Pearson's correlation coefficient is not adequate to measure relations which are not linear, this is the case since Pearson's correlation coefficient is a measure of the strength of the linear relationship between two variables, on the other hand, Spearman's coefficient assesses how well an arbitrary monotonic function can describe a relationship between two variables, without making any assumptions about their frequency distribution [5].

3.2   Classifier Implementation/Tuning

After having identified the most relevant features for a classifier's training, there is nothing of extreme relevance stopping the implementation of possible prototypes. In this case, four different prototypes were implemented while structuring the system around a modular structure which would allow for extras if necessary. These four classifier prototypes were based in the following algorithms: Naïve Bayes, Support Vector Machines, Random Forest and Adaptive Boost (a.k.a. Ada-boost).

Although no immediate problem to a classifier's training exists when data is available and features to use have been identified. If the classes for which the classifier is being trained are heavily unbalanced, it may begin to pay undue attention to the majority class while ignoring the minority class, which is clearly not good for a classification solution. Normally, three different approaches exist to solve this problem, each of which presents disadvantages:

1. Under-sampling - Namely cutting the number of training instances of the majority class in order to balance the data-set and, in that way, train the classifier. This process means cutting factually real data, which is not good since in machine-learning data is the be all and end all. Discarding data without thought can be dangerous as some niche group of a given class may be entirely excluded.
2. Over-sampling - The generation of minority class cases based on the minority class instances which do exist. This process too, may be dangerous, as the generated data, no matter how cautiously generated, will not be equivalent to that of real cases. This means that the possibility of training the classifier in a wrongful way will exist.
3. Training weights - This means attributing a different weight to the classes for which the classifier is being trained. This method allows for a compromise. While a kind of "instance repetition" exists by attributing a greater weight than one to each minority class instance, the fact that it allows for all the majority class instances to be used, albeit with a lower weight than one, is a great benefit.

Having decided which algorithms should be used for the classifier's prototypes, as well as the way to deal with the problem of class imbalance (training weights). The next step to take is to find which hyper-parameters would be ideal in order to train the classifiers. However, to do that, it is necessary to understand which classifier metric should be optimized. The performance of a classifier in an unbalanced classification problem can normally be determined by the recall and precision of its various classes. For clarification sake, precision can be defined as the proportion of classifications of a given class which was actually correct, while recall expresses the proportion of actual class instances which were identified correctly. Taking into account the problem's context, it is possible to conclude that the most important metric to optimize is the recall of the spammer class. This is the case since if a misclassification of a regular case occurs there is no major repercussion, while if a misclassification of a spam importation were to occur it can lead to the black-listing of e-mails/addresses.

In order to perform this optimization/tuning process regarding the hyper-parameters of the classifier prototypes, the chosen approach to be use was nested cross-validation. Nested cross-validation estimates the generalization error of the underlying model and its hyper-parameter search. It does this to counter the fact that model selection without nested cross-validation uses the same data to tune model parameters and evaluate model performance, which may lead information to "leak" to the model causing over-fitting. It operates by using a series of train/validation/test set splits. In an inner loop the score is approximately maximized by fitting a model to each training set, and then directly maximized in selecting hyper-parameters over the validation set. Then, using a outer loop, generalization error is estimated by averaging test set scores over several data-set splits.

By applying training weights to the prototype classifiers' training phase and optimizing their hyper-parameters through the use of nested cross-validation, all four prototypes were implemented [2].

4   Results

Now that the prototypes have been implemented, it is necessary to evaluate how well they perform comparatively to each other. For clarity, during this chapter positive class will refer to spam importations while negative class will refer to "regular" importations. By comparing some intermediate results obtained through the development phase, it was possible to establish a rough performance order of: Naïve-Bayes < SVM < Random-forest < Ada-boost. However, this order is based on single runs used for hyper-parameter optimization, as such a better comparison method is necessary. As such, it was decided that three different factors would be compared between prototypes: training time, precision/recall and area under the ROC curve.

To verify the training time differences between classifiers, ten different runs for training sets with a varying numbers of instances were executed for each classifier. A timer was started when each training session began and stopped when it finished. In this way, a set of points related to each classifier's necessary training time for a certain number instances was created. By applying a generalized linear model to each group of observations, the following plot was created (Fig. 2):

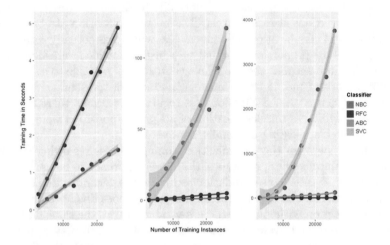

Fig. 2. Classifier's training time(s) per number of training instances

By itself, this plot does not constitute any kind of statistical evidence, hence an appropriate test is necessary. In this case, while taking into account the data's characteristics a two-sample sign test was used. By applying this test to each pair of classifiers, the following Table 1 was formed:

Table 1. *p-values* for two-sample sign test between each classifier's populations

	NBC-RFC	NBC-ABC	NBC-SVC	RFC-ABC	RFC-SVC	ABC-SVC
p-value	0.002	0.002	0.002	0.002	0.002	0.002

Since for each combination of classifier's populations, the resulting p-value was less than the α for a confidence level of 95%, the null hypothesis is rejected. This means that the true median difference between populations is not equal to zero, which indicates that they are distinct. Which corroborates the order previously established trough the aforementioned plot regarding which classifier was less affected by the number of training instances.

To test if significant differences existed regarding the precision/recall of the various classifiers, the first step taken was to collect data regarding the classification provided by each algorithm. To do that, the various algorithms were repeatedly exposed to a fifteen-fold cross-validation scheme where a Stratified-KFold approach was used in order to obtain various balanced train/test splits (which were formed simultaneously). The resulting confusion matrices of each cross-validation cycle were then summed and stored. In this way, the process was repeated until enough data could be stored so as to allow for a quasi-normal distribution to occur according to the central limit theorem. The number of observations collected for each classifier was forty two, in which twenty one targeted the positive class and the rest the negative class. This data was then used in a Mann–Whitney U test. The performed test allowed to conclude the following:

- Negative Class' Precision: $NBC < RFC < ABC$
- Negative Class' Recall: $NBC < ABC < RFC$
- Positive Class' Precision: $NBC < ABC < RFC$
- Positive Class' Recall: $NBC < RFC < ABC$.

Finally, in order to test the area under the ROC curve of the various classifiers, the first step taken was to extract data targeting the classification probabilities generated by the same train/test split in each of their cases. This train/test split was based on a data-set with 26240 entries, and its division followed a ratio of 70% training data to 30% testing data, meaning that the classification probabilities were based in 7872 instances.

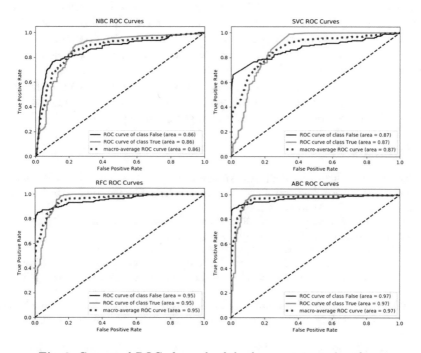

Fig. 3. Generated ROCs for each of the four prototype classifiers

Once that was concluded the next step taken was to use that data to plot the ROC for each individual classifier and calculate their area under the curve (seen in Fig. 3). The computed values for each area were: 0.86 for the Naïve Bayes classifier, 0.87 for the SVM classifier, 0.95 for the Random-forest classifier, and 0.97 for the Ada-boost classifier.

As is plain to see, if we make a direct comparison without further analysis then the order established between algorithms, in regards to Area Under the Curve (AUC), would be: Naïve Bayes < Support Vector Machine < Random Forest < Ada-Boost. This ordering, however, may not be correct, as the differences between each of the ROC curves could be caused by a random set of factors. To solve this problem the methodology described in [4] was employed, allowing to conclude with a confidence level of 95%, that the order which correctly expressed the relation between the area under the curve for the various ROC curves was: Naïve Bayes = SVM < Random-forest < Ada-boost.

5   Conclusion

Through the process described above, a significant problem of online marketing platforms was identified, and a possible solution to it was implemented. By using machine-learning capabilities, the developed solution was able to obtain comparable success to the currently employed rules-based approach to spam prevention in contact database importations. Additionally, various prototypes were developed allowing for multiple approaches to the same problem to exist and comparison to one-another to occur, which in turn benefits further optimization of the system. This comparison allowed to assert that, for the problem at hand, the best performing algorithms were the ones based in Ada-boost, Random-forest, SVMs and Naïve Bayes, in that order. This performance order is not immutable, since as the system evolves more efficient prototypes may be created or the existent prototypes may suffer changes in their proficiency.

References

1. Caruana, R., Niculescu-Mizil, A.: An empirical comparison of supervised learning algorithms. In: Proceedings of the 23rd International Conference on Machine Learning, pp. 161–168. ACM (2006)
2. Coelho, D.: Intelligent analysis of contact databases' importation. Master's thesis. Instituto Superior de Engenharia do Porto (2018)
3. Gonçalves, M.: E-goi (2018). https://www.e-goi.pt/
4. Hanley, J.A., McNeil, B.J.: A method of comparing the areas under receiver operating characteristic curves derived from the same cases. Radiology **148**(3), 839–843 (1983)
5. Hauke, J., Kossowski, T.: Comparison of values of Pearson's and Spearman's correlation coefficients on the same sets of data. Quaestiones Geographicae **30**(2), 87–93 (2011)
6. Saltelli, A., Ratto, M., Andres, T., Campolongo, F., Cariboni, J., Gatelli, D., Saisana, M., Tarantola, S.: Global Sensitivity Analysis: The Primer. Wiley, Hoboken (2008)
7. Strauss, J., et al.: E-Marketing. Routledge, Abingdon (2016)
8. Yu, B., Xu, Z.B.: A comparative study for content-based dynamic spam classification using four machine learning algorithms. Knowl.-Based Syst. **21**(4), 355–362 (2008)

Post-processing of Wind-Speed Forecasts Using the Extended Perfect Prog Method with Polynomial Neural Networks to Elicit PDE Models

Ladislav Zjavka[✉], Stanislav Mišák, and Lukáš Prokop

ENET Centre, VŠB-Technical University of Ostrava, Ostrava, Czech Republic
ladislav.zjavka@vsb.cz

Abstract. Anomalies in local weather cause inaccuracies in daily predictions using meso-scale numerical models. Statistical methods using historical data can adapt the forecasts to specific local conditions. Differential polynomial network is a recent machine learning technique used to develop post-processing models. It decomposes and substitutes for the general linear Partial Differential Equation being able to describe the local atmospheric dynamics which is too complex to be modelled by standard soft-computing. The complete derivative formula is decomposed, using a multi-layer polynomial network structure, into specific sub-PDE solutions of the unknown node sum functions. The sum PDE models, using a polynomial PDE substitution based on Operational Calculus, represent spatial data relations between the relevant meteorological inputs->output quantities. The proposed forecasts post-processing is based on the 2-stage approach of the Perfect Prog method used routinely in meteorology. The original procedure is extended with initial estimations of the optimal numbers of training days whose latest data observations are used to elicit daily prediction models in the 1st stage. Determination of the optimal models initialization time allows for improvements in the middle-term numerical forecasts of wind speed in prevailing more or less settled weather. In the 2nd stage the correction model is applied to forecasts of the training input variables to calculate 24-h prediction series of the target wind speed at the corresponding time.

Keywords: Polynomial neural network ·
General Partial Differential Equation ·
Operational calculus polynomial substitution · Perfect prog post-processing

1 Introduction

Continual fluctuations in wind speed arise from complex chaotic interaction processes included in the large-scale air circulation primarily induced by global pressure and temperature differences. Numerical Weather Prediction (NWP) systems solve sets of primitive differential equations, simplifying the atmosphere behavior to the ideal gas flow, starting from the initial conditions to predict the time-change of a variable in each central grid cell with respect to its neighbors. The NWP models cannot reliably recognize conditions near the ground as they simulate multi-layer physical processes in

A. M. Madureira et al. (Eds.): HIS 2018, AISC 923, pp. 11–21, 2020.
https://doi.org/10.1007/978-3-030-14347-3_2

atmosphere to forecast large-scale upper air patterns. On the contrary statistical autonomous models, developed from historical data series, can correctly represent local weather anomalies in the surface level although their predictions are usually worthless beyond a few hours [3]. The biases of NWP models are induced due to the physical parameterization and computation limitations. Additional NWP errors can arise from the systems inability to account for physical processes at a scale smaller than the grid used in numerical equations [1].

Two main post-processing methods, called Model Output Statistics (MOS) and Perfect Prog (PP), are compared in Table 1. MOS typically derive a set of linear equations to describe relationships between local observation data and the forecasts at a certain time. MOS detail specific surface feature effects to eliminate systematic forecast errors [2]. PP models the local dynamical relations between inputs and output observations without using forecasted data [4]. These skills are unrelated to NWP models outputs so can secondary improve their representation of chaotic atmospheric processes [6]. PP and MOS use a constant size of data sets to develop prediction models which do not allow for improvements in numerical forecasts as these are corrected by NWP post-processing utilities nowadays, using mostly the MOS approach.

Table 1. MOS and PP conveniences (blue) and drawbacks (red)

Characteristics	MOS	Perfect Prog (PP)
Relations between	Forecasts and observations	Inputs->output observations
Reduction of	NWP models biases	Non-systematic NWP errors
NWP model resolution	Non-increased	Increased
Catching weather	Less-sensitive	More-sensitive
NWP model dependent	Yes	No
Longer time-horizon	NWP error sensitive	NWP error non-sensitive
Prediction horizon	Middle-term	Short-term
Forecasting	converges to climatology	rare events
Data set size	Constant (months, years)	Constant (months)

Artificial Neural Networks (ANN) are simple 1 or 2-layer structures which cannot model complex patterns in local weather described by a mass of data. ANN require data pre-processing to significantly reduce the number of input variables, which leads usually to the models over-simplification. The number of parameters in polynomial regression grows exponentially with the number of variables, contrary to ANN. Polynomial Neural Networks (PNN) decompose the general connections between inputs and output variables expressed by the Kolmogorov-Gabor polynomial (1).

$$Y = a_0 + \sum_{i=1}^{n} a_i x_i + \sum_{i=1}^{n}\sum_{j=1}^{n} a_{ij} x_i x_j + \sum_{i=1}^{n}\sum_{j=1}^{n}\sum_{k=1}^{n} a_{ijk} x_i x_j x_k + \ldots \tag{1}$$

n - number of input variables x_i $a_i, a_{ij}, a_{ijk}, \ldots$ - polynomial parameters

Group Method of Data Handling (GMDH) evolves a multi-layer PNN structure in successive steps, adding one layer at a time to calculate its node parameters and select the best ones to be applied in the next layer. PNN nodes decompose the complexity of a system into a number of simpler relationships, each described by low order polynomial transfer functions (2) for every pair of input variables x_i, x_j [5].

$$y = a_0 + a_1 x_i + a_2 x_j + a_3 x_i x_j + a_4 x_i^2 + a_5 x_j^2$$

x_i, x_j - *input variables of polynomial neuron nodes*

(2)

Differential polynomial neural network (D-PNN) is a novel type of neural network which extends the multi-layer PNN structure to define the 2nd order sub-PDEs in its nodes which decompose the general Partial Differential Equation (PDE). It uses a polynomial PDE substitution of Operational Calculus to form pure rational functions which represent the Laplace transforms of the unknown partial node functions. The inverse Laplace transformation is applied to the function images to solve the node sub-PDEs whose sum models the searched complete n-variable separable function. D-PNN expands the general derivative formula into a number of node specific sub-PDEs, i.e. neurons in this context, analogous to the GMDH decomposition of the general connection polynomial (1).

The proposed extended forecast correction procedure uses spatial inputs->output observations from the last few days to elicit the PDE models which post-process daily NWP model forecasts of the input variables to calculate the target output prediction series, analogous to PP [4]. It additionally estimates initial time-intervals of the last days whose data samples are used to adapt the models parameters. The optimal daily training parameters compensate for the weather dynamics and inaccuracies of processed forecasts [8]. The D-PNN models represent the current spatial inputs->output weather data relations which are supposed to be actual in the prediction time-horizon to allow improvements in 24-h local forecasts of regional NWP models [7].

2   Post-processing of NWP Data Using Extended Perfect Prog

The proposed method is a 2-stage process analogous to the PP approach. In the 1st stage, the training is made on real observations to derive the PDE model (Fig. 1, left) which is applied to the forecasts to calculate 24-h output predictions (Fig. 1, right) at the corresponding time in the 2nd PP stage. The models formed with data from the last days and applied in the prediction day with similar data patterns give usually more accurate predictions than models developed with large data sets including various weather conditions. The 1st PP stage is extended into 2 steps:

1. The optimal number of training days, used to elicit the prediction model, is initially estimated by an assistant test model.
2. Parameters of the prediction model are adapted for inputs->output data samples to process forecasts of the training input variables.

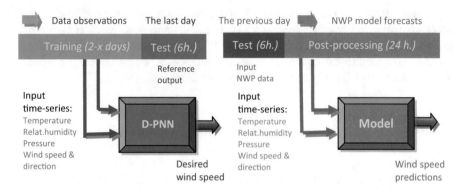

Fig. 1. D-PNN is trained with spatial data observations (blue-left) over the last few days to form daily prediction PDE models that process forecast series of the input variables (red-right)

The assistant model is an additional D-PNN test model formed initially before the 1st PP step with observations from increasing number of the previous days *2, 3, ..., x*. It processes NWP data from the previous day to compare regularly the output with the latest 6-h observations to detect a change in the atmosphere state in the previous days, which determines the training period length. The optimal number of data samples, giving the minimal test error, is used in the daily prediction model formation.

The D-PNN models can post-process and improve forecasts in more or less settled weather which tends to prevail and whose patterns do not change fundamentally within short-time intervals. The model derived from the previous last days is not actual in the case of a significant overnight change in the atmosphere state because of different spatial data relations. If the assistant model cannot obtain an acceptable test error, defined in consideration of failed forecast revisions in the previous days, the post-processing model is flawed and should not be applied to the current NWP data.

3   A Polynomial PDE Substitution Using Operational Calculus

D-PNN defines and substitutes for the general linear PDE (3) whose exact form is unknown and which can describe uncertain dynamic systems. It decomposes the n-variable general PDE into 2nd order specific sub-PDEs in the PNN nodes to model the unknown functions u_k whose sum gives the complete *n*-variable model *u* (3).

$$a + bu + \sum_{i=1}^{n} c_i \frac{\partial u}{\partial x_i} + \sum_{i=1}^{n}\sum_{j=1}^{n} d_{ij} \frac{\partial^2 u}{\partial x_i \partial x_j} + \ldots = 0 \qquad u = \sum_{k=1}^{\infty} u_k \qquad (3)$$

$u(x_1, x_2, \ldots, x_n)$ - *unknown separable function of* $n -$ *input variables*

$a, b, c_i, d_{ij}, \ldots$ - *weights of terms* $\qquad\qquad u_i$ - *partial functions*

Considering 2-input variables in the PNN nodes, the derivatives of the 2nd order PDE (4) correspond to variables of the GMDH polynomial (2). This type of the PDE is most often used to model physical or natural systems non-linearities.

$$\left(\sum \frac{\partial u_k}{\partial x_1}, \sum \frac{\partial u_k}{\partial x_2}, \sum \frac{\partial^2 u_k}{\partial x_1^2}, \sum \frac{\partial^2 u_k}{\partial x_1 \partial x_2}, \sum \frac{\partial^2 u_k}{\partial x_2^2} \right) \quad (4)$$

u_k - *node partial sum functions of an unknown separable function u*

The polynomial conversion of the 2nd order PDE (4) using procedures of Operational Calculus is based on the proposition of the Laplace transform (L-transform) of the function n^{th} derivatives in consideration of the initial conditions (5).

$$L\{f^{(n)}(t)\} = p^n F(p) - \sum_{k=1}^{n} p^{n-i} f_{0+}^{(i-1)} \qquad L\{f(t)\} = F(p) \quad (5)$$

$f(t), f'(t), \ldots, f^{(n)}(t)$ - *originals continuous in* $<0+, \infty>$ p, t - *complex and real variables*

This polynomial substitution for the $f(t)$ function n^{th} derivatives in an Ordinary Differential Equation (ODE) leads to algebraic equations from which the L-transform image $F(p)$ of an unknown function $f(t)$ is separated in the form of a pure rational function (6). These fractions represent the L-transforms $F(p)$, expressed with the complex number p, so that the inverse L-transformation is applied to them to obtain the original functions $f(t)$ of a real variable t (6) described by the ODE.

$$F(p) = \frac{P(p)}{Q(p)} = \sum_{k=1}^{n} \frac{P(\alpha_k)}{Q_k(\alpha_k)} \frac{1}{p - \alpha_k} \qquad f(t) = \sum_{k=1}^{n} \frac{P(\alpha_k)}{Q_k(\alpha_k)} e^{\alpha_k \cdot t} \quad (6)$$

α_k - *simple real roots of the multinomial* $Q(p)$ $F(p)$ - *L−transform image*

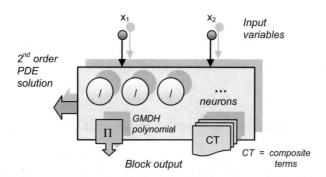

Fig. 2. A block of derivative neurons - 2nd order sub-PDE solutions in the PNN nodes

The inverse L-transformation is analogously applied to the selected neurons, i.e. rational terms (6) produced in D-PNN node blocks (Fig. 2), to substitute for the specific 2nd order sub-PDEs (7) and obtain the originals of node functions u_k whose sum gives the model of the unknown separable output function u (3). Each block contains a single output polynomial (2) which is used to form neurons which can be included in the total network output sum of a general PDE solution.

$$F\left(x_1, x_2, u, \frac{\partial u}{\partial x_1}, \frac{\partial u}{\partial x_2}, \frac{\partial^2 u}{\partial x_1^2}, \frac{\partial^2 u}{\partial x_1 \partial x_2}, \frac{\partial^2 u}{\partial x_2^2}\right) = 0 \tag{7}$$

where $F(x_1, x_2, u, p, q, r, s, t)$ is a function of 8 variables

While using 2 input variables in the PNN nodes the 2nd order PDE can be expressed in the equality of 8 variables (7), including derivative terms formed with respect to variables corresponding to the GMDH polynomial (2).

$$y_1 = w_1 \frac{a_0 + a_1 x_1 + a_2 x_2 + a_3 x_1 x_2 + a_4 sig(x_1^2) + a_5 sig(x_2^2)}{b_0 + b_1 x_1} \cdot e^\varphi \tag{8}$$

$$y_3 = w_3 \frac{a_0 + a_1 x_1 + a_2 x_2 + a_3 x_1 x_2 + a_4 sig(x_1^2) + a_5 sig(x_2^2)}{b_0 + b_1 x_2 + b_2 sig(x_2^2)} \cdot e^\varphi \tag{9}$$

$$y_5 = w_5 \frac{a_0 + a_1 x_1 + a_2 x_2 + a_3 x_1 x_2 + a_4 sig(x_1^2) + a_5 sig(x_2^2)}{b_0 + b_1 x_1 + b_2 x_{12} + b_3 x_1 x_2} \cdot e^\varphi \tag{10}$$

$\varphi = arctg(x_1/x_2)$ - *phase representation of 2 input variables* x_1, x_2

a_i, b_i - *polynomial parameters* w_i - *weights*   *sig - sigmoidal transformation*

Each D-PNN block can form 5 simple derivative neurons, in respect of single x_1, x_2 (8) squared x_1^2, x_2^2 (9) and combination $x_1 x_2$ (10) derivative variables, which can solve specific 2nd order sub-PDEs in the PNN nodes (7). The Root Mean Squared Error (RMSE) is calculated in each iteration step of training and testing to select and optimize the parameters (11).

$$RMSE = \sqrt{\frac{\sum_{i=1}^{M} (Y_i^d - Y_i)^2}{M}} \to min \tag{11}$$

Y_i - *produced and* Y_i^d - *desired D − PNN output for* i^{th} *training vector of M − data samples*

4   Backward Selective Differential Polynomial Network

Multi-layer networks form composite functions (Fig. 2). The blocks of the 2nd and next hidden layers can produce additional Composite Terms (CT) equivalent to the neurons. The CTs substitute for the node sub-PDEs with respect to variables of back-connected blocks of the previous layers according to the composite function (12) partial derivation rules (13).

$$F(x_1, x_2, \ldots, x_n) = f(z_1, zz_2, \ldots, z_m) = f(\phi_1(X), \phi_2(X), \ldots, \phi_m(X)) \tag{12}$$

$$\frac{\partial F}{\partial x_k} = \sum_{i=1}^{m} \frac{\partial f(z_1, z_2, \ldots, z_m)}{\partial z_i} \cdot \frac{\partial \phi_i(X)}{\partial x_k} \quad k = 1, \ldots, n \tag{13}$$

The 2rd layer blocks can form and select one of the neurons or additional 10 CTs using applicable neurons of the 1st layer 2 blocks in the products (13) for composite sub-PDEs (14). The 3rd layer blocks can select from additional 10 + 20 CTs using neurons of 2 and 4 back-connected blocks in the previous 2nd and 1st layers (20), etc. The number of possible block CTs doubles along with each joined preceding layer (Fig. 3). The L-transform image $F(p)$ is expressed in the complex form, so the phase of the complex representation of 2-variables is applied in the inverse L-transformation.

$$y_{2p} = w_{2p} \cdot \frac{a_0 + a_1 x_{11} + a_2 x_{13} + a_3 x_{11} x_{13} + a_4 x_{11}^2 + a_5 x_{13}^2}{x_{11}} \cdot e^{-\varphi_{21}} \cdot \frac{b_0 + b_1 x_1 + b_2 x_1^2}{x_{11}} \cdot e^{\varphi_{11}} \tag{14}$$

$$y_{3p} = w_{3p} \cdot \frac{a_0 + a_1 x_{21} + a_2 x_{22} + a_3 x_{21} x_{22} + a_4 x_{21}^2 + a_5 x_{22}^2}{x_{21}} \cdot e^{-\varphi_{31}} \cdot c_{21} \cdot \frac{b_0 + b_1 x_2}{x_{11}} \cdot e^{\varphi_{11}}$$

y_{kp} - p^{th} *Composite Term* (CT) *output* $\varphi_{21} = arctg(x_{11}/x_{13})$ $\varphi_{31} = arctg(x_{21}/x_{22})$ (15)
c_{kl} - *complex representation of the* l^{th} *block inputs* x_i, x_j *in the* k^{th} *layer*

The CTs include the L-transformed fraction of the external function sub-PDE (left) in the starting block and the selected leaf neuron (right) for the internal function from a block in preceding layers (14). The complex representation of 2-inputs (16) of blocks in the CTs inter-connected layers can substitute for the node internal function PDEs (15). The pure rational fractions (6) correspond to the amplitude r (radius) in Eulers's notation of a complex number c (16) in polar coordinates.

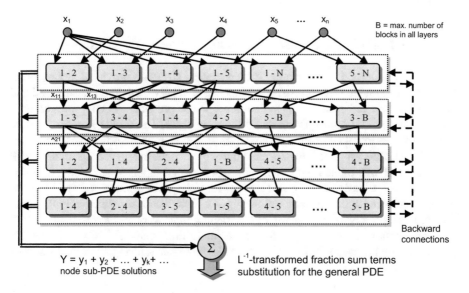

Fig. 3. D-PNN selects from possible 2-variable combination blocks in each hidden layer

$$c = \underbrace{x_1}_{Re} + i \cdot \underbrace{x_2}_{Im} = \sqrt{x_1^2 + x_2^2} \cdot e^{i \cdot \arctan\left(\frac{x_2}{x_1}\right)} = r \cdot e^{i \cdot \varphi} = r \cdot (\cos \varphi + i \cdot \sin \varphi) \quad (16)$$

Recursive algorithms can efficiently perform the back-production and output calculation of the CTs in the D-PNN tree-like structure (Fig. 3). All the D-PNN blocks in each layer form equivalent neurons and CTs, i.e. node specific sub-PDE solutions, from which only one can be selected to be included in the total output sum. N-variable D-PNN selects from the possible 2-combination node blocks in each layer (analogous to GMDH) as the number of the input combination couples grows exponentially in each next layer. The D-PNN complete PNN structure is initialized with an estimated number of layers and the node blocks, which can agree with the number of n-input variables. A convergent combination of selected applicable neurons and CTs can form a general PDE solution. The D-PNN total output Y is the arithmetic mean of active neurons + CTs output values to simplify the parameters adaptation (17).

$$Y = \frac{1}{k} \sum_{i=1}^{k} y_i \quad k = the \ number \ of \ active \ neurons + CTs \ (node \ sub-\mathrm{PDEs}) \quad (17)$$

Two processes of the optimal D-PNN structure formation and sub-PDEs definition, i.e. the selection of 2-input blocks and neurons + CTs combinations, are followed by the pre-optimization of the polynomial parameters and term weights using the Gradient Steepest Descent (GSD) method. This iteration algorithm skips from the actual to the next block, one by one, to minimize the training RMSE in consideration of a continual test using the External complement of GMDH [5].

5   Spatial NWP Data Post-processing Using PDE Models

National Oceanic and Atmospheric Administration (NOAA) provides, among other services, free tabular 24-h forecasts of temperature, relative humidity, wind speed and direction at selected localities [9]. The atmospheric pressure tendency prognosis is missing but the WU hourly forecasts [10] can supply it. The Global Forecast System (GFS) provides initial and boundary conditions for the regional North American Meso-scale (NAM) forecast system to generate hourly predictions every 6th hour. In the 1st PP stage, D-PNN was trained with hourly averaged spatial inputs->output data observations from 3 + 1 bordering + central stations for the optimal periods of 2 to 11 days to model wind speed at the central location (Fig. 4). Next, the daily prediction PDE model post-processes the NOAA forecasts of the training input variables to calculate 1–24-h output prediction series, corresponding to the inputs time, in the 2nd PP stage (Sect. 2). The results are presented for initial NAM model forecasts issued at 00 of the local time (Figs. 5 and 6).

Fig. 4. Observation and NWP data from the central forecasted and 3 bordering weather stations

The D-PNN prediction models can be formed in consideration of the optimal daily training errors to terminate the parameters adaptation and avoid the over-fitting. The initial assistant models can additionally detect this 2nd training parameter, along with the optimal lengths of training periods, according to the minimal test errors.

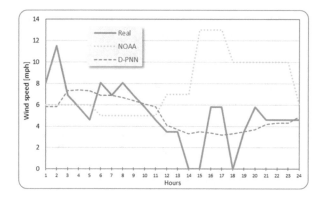

Fig. 5. 25.10.2015, Great Falls - RMSE: NOAA = 5.25 and D-PNN = 2.09 mph

The D-PNN models can demonstrate large prediction errors in sporadic days of an overnight break change in the current weather as they represent conditions in the last days that do not correspond to patterns in the forecasted day (Fig. 6). Larger testing errors of the initial assistant models can indicate this weather change. If the effect is not apparent from the last 6-h test, i.e. a passing atmospheric front is not evident until the prediction hours, it may be detected by a computing or comparative analysis between forecast and observation data patterns in the last days (in consideration of the previous failed revisions). NOAA provides free historical weather data archives [11], shared also by WU [12], and current 2-day hourly tabular observations [13]. Figure 7 shows the daily average prediction RMSEs of the original NOAA NAM and D-PNN models in a week period (the x-axis represents the real data).

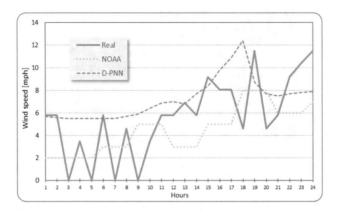

Fig. 6. 28.10.2015, Great Falls - RMSE: NOAA = 3.15 and D-PNN = 3.30 mph

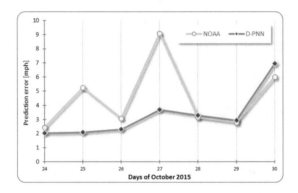

Fig. 7. 1-week average wind-speed prediction RMSE: NOAA = 4.54 and D-PNN = 3.32 mph

6 Conclusions

D-PNN uses latest inputs->output data observations from the estimated optimal training periods to develop the PP models which post-process 24-h forecast series of the trained input variables (supplying unknown real data). Some sorts of settled weather allow for improvements in the performance of middle-scale NWP models as the post-processing PDE models detail the current ground conditions which are supposed to be actual in the prediction day. D-PNN develops the PNN structure to decompose the n-variable general PDE into node specific 2nd order sub-PDE solutions, using the polynomial substitution based on OC. Its PDE sum models can properly reflect the dynamic uncertainty in local atmospheric dynamic processes which are too complex to be represented comprehensively by conventional computing techniques.

Acknowledgement. This paper was supported by the following projects: LO1404: Sustainable Development of ENET Centre; CZ.1.05/2.1.00/19.0389 Development of the ENET Centre Research Infrastructure; SP2018/58 and SP2018/78 Student Grant Competition and TACR TS777701, Czech Republic.

References

1. Durai, V.R., Bhradwaj, R.: Evaluation of statistical bias correction methods for numerical weather prediction model forecasts of maximum and minimum temperatures. Nat. Hazards **73**, 1229–1254 (2014)
2. Klein, W., Glahn, H.: Forecasting local weather by means of model output statistics. Bull. Am. Meteorol. Soc. **55**, 1217–1227 (1974)
3. Lei, M., Shiyan, L., Chuanwen, J., Hongling, L., Yan, Z.: A review on the forecasting of wind speed and generated power. Renew. Sustain. Energy Rev. **13**, 915–920 (2009)
4. Marzban, C., Sandgathe, S., Kalnay, E.: MOS, perfect prog, and reanalysis. Mon. Weather Rev. **134**, 657–663 (2006)
5. Nikolaev, N.Y., Iba, H.: Adaptive Learning of Polynomial Networks. Genetic and Evolutionary Computation. Springer, New York (2006)
6. Vislocky, R.L., Young, G.S.: The use of perfect prog forecasts to improve model output statistics forecasts of precipitation probability. Weather Forecast. **4**, 202–209 (1989)
7. Zjavka, L.: Wind speed forecast correction models using polynomial neural networks. Renew. Energy **83**, 998–1006 (2015)
8. Zjavka, L.: Numerical weather prediction revisions using the locally trained differential polynomial network. Expert Syst. Appl. **44**, 265–274 (2016)
9. NAM forecasts. http://forecast.weather.gov/MapClick.php?lat=46.60683&lon=-111.9828333&lg=english&&FcstType=digital
10. WU forecasts. www.wunderground.com/hourly/us/mt/great-falls?cm_ven=localwx_hour
11. NOAA National Climatic Data Center archives. www.ncdc.noaa.gov/orders/qclcd/
12. WU data. www.wunderground.com/history/airport/KGTF/2015/10/3/DailyHistory.html
13. NOAA data observations. www.wrh.noaa.gov/mesowest/getobext.php?wfo=tfx&sid=KGTF

Classifying and Grouping Narratives with Convolutional Neural Networks, PCA and t-SNE

Manoela Kohler[1(✉)], Leonardo Sondermann[2], Leonardo Forero[2], and Marco Aurelio Pacheco[1]

[1] Pontifical Catholic University of Rio de Janeiro, Rio de Janeiro, Brazil
{manoela,marco}@ele.puc-rio.br
[2] State University of Rio de Janeiro, Rio de Janeiro, Brazil
leosonder@gmail.com, mendonza@ele.puc-rio.br

Abstract. Each week, the Consumer Financial Protection Bureau (CFPB) receives thousands of consumer complaints about financial products and services. These complaints must be forwarded to the responsible company and posted on the site after 15 days or when the company responds to the complaint, whichever comes first. Published complaints and solutions help consumers solve their problems and also serve as a repository of help for other consumers to avoid or solve problems on their own. Every complaint provides information about the problems people are having, helping them to identify inappropriate practices and allowing them to stop before they become major problems. Culminating in better results for consumers and a better financial market for everyone. Each of the complaints contains information on submission date, company to send the complaint, complaint narrative, among others. However, complaints do not have information on the department to which it should be forwarded. Therefore, in this work, the three approaches to analyze each complaint are: (i) convolutional neural network (CNN) to classify the narratives; (ii) principal components analysis (PCA); and (iii) t-distributed stochastic neighbor embedding (t-SNE) to create a three-dimensional embedding for clustering. Embedding from scratch, Pre-trained Word Vectors (word2Vec) and Global Vectors (GloVe) vectors are used and compared in six different CNNs modeling. The results increase the evidence that pre-trained word vectors is important and that convolutional neural networks and t-SNE can perform remarkably well on real text classification data.

Keywords: Deep neural network · Convolutional neural network · t-SNE · PCA · Classification · Clustering · Natural language processing · Word2vec · Glove

1 Introduction

Deep Learning (DL) is an emerging topic within the field of Artificial Intelligence as a subcategory of machine learning that addresses inference tasks with the use of neural networks with multiple layers to improve results. DL is fast becoming one of the most studied and sought-after fields within the modern computer science.

© Springer Nature Switzerland AG 2020
A. M. Madureira et al. (Eds.): HIS 2018, AISC 923, pp. 22–30, 2020.
https://doi.org/10.1007/978-3-030-14347-3_3

Even though single hidden layer neural networks are universal approximators [1], to determine the number of nodes needed in one hidden layer is difficult. Therefore, adding more layers (apart from increasing computational complexity to the training and testing phases), allows for more easy representation of the interactions within the input data, as well as allows for more abstract features to be learned and used as input into the next hidden layer [2].

In recent years, deep learning has been used in image classification, text detection and recognition, time series forecasting, regression problems and so on.

Deep Learning models trains computer models so that it can process natural language. The model relates terms and words to infer meaning in large amounts of data. Text classification is significant for Natural Language Processing (NLP) systems, where there has been an enormous amount of research on sentence classification tasks, specifically on sentiment analysis. NLP systems classically treat words as discrete, atomic symbols where the model leverages a small amount of information regarding the relationship between the individual symbols [3]. Traditional methods usually use bag-of-words [4] or n-gram [5] approaches. Recently, it has become more common to use DNNs in NLP applications, where much of the work involves learning word representations through neural language models [6, 7] and then performing a composition over the learned word vectors for classification. These approaches have led to new methods for solving the data sparsity problem. Consequently, several neural network-based methods for learning word representations – like DNN, Recurrent Neural Networks (RNN) and CNN – followed these approaches [3].

The objective of this work is to train a DNN and a CNN with a database of consumer complaints, and thus to be able to classify each sentence according to the type of complaint. This way, we can streamline the process, and reduce time taken to the company to respond the consumer. PCA and t-SNE are evaluated as two methods to classify these sentences, and results are visualized in a 3D graph.

As 97% of the complaints sent by consumers receive a timely response, by submitting a complaint, consumers can be heard by financial firms, get help with their own problems, and help others avoid similar ones. This work will help to identify more quickly and efficiently inappropriate practices and stop them before they become big problems. The result will be better communication for consumers and a better financial market for all.

The paper is organized as follows: Sect. 2 presents the description on convolutional neural networks, PCA and t-SNE, and explains the proposed modelling to handle narratives. Sections 3 presents details on the dataset and word embeddings. Results and discussions are presented in Sect. 4 and Sect. 5 concludes the work.

2 Deep Learning Models in NLP

2.1 Convolutional Neural Networks

CNNs utilize layers with convolving filters that are applied to local features [8]. Originally invented for computer vision, CNN models have subsequently been shown to be effective natural language processing where much of the work has involved learning word vector representations through neural language models [9–11] and performing composition over the learned word vectors for classification [12]. In the context of artificial intelligence and machine learning, a convolutional neural network has recently achieved state-of-the-art performance in a variety of pattern recognition tasks. Given that CNN is able to select objects with the next human level, questions arise about the differences between the computer and a human view.

A model based on Kim Yoon's CNN [6] for text classification was used. In order to compare results, in the present work, six CNNs were trained. Not only learning embeddings from scratch (one-hot vectors), but also on top of word vectors obtained from: (i) word2vec, an unsupervised neural language model trained by Mikolov [11] in 100 billion Google News words (publicly available[1]); and (ii) GloVe, a model proposed by a Stanford research group [13] which tries to generate the vector representation of words by using similarity between words as an invariant. Both models learning from word vectors were modelled as a static and a non-static CNN (static and dynamic word vector, respectively). The sixth model is a two-channel CNN with static and dynamic GloVe vectors, i.e., one channel is adjusted during training and the other is not.

The first layers (embedding) map the words into a representation of low-dimensional vectors. The next layer performs convolutions on the embedded word vectors using various filter sizes (128 filters of sizes 2, 3, 4). Then, the max-pool layer groups the result of the convolutional layer (one layer for each filter size) into a long feature vector, and finally, comes the dropout regularization, and the classification is a result of the softmax layer.

2.2   PCA

PCA is one of the oldest and widely used methods for dimensionality reduction in data science [14]. The goal of PCA is to reduce the dimensionality of a data set, i.e., extract low-dimensional subspaces from high-dimensional data, while preserving as much variability as possible [15]. Over the past decades, thanks to its simple, nonparametric nature, PCA has been used as a descriptive and adaptive exploratory method on numerical data of various types [14].

In this work, PCA is used to create three-dimensional embeddings to explore linear relationships (e.g., word analogies).

2.3   T-SNE

Recently, Stochastic Neighbor Embedding (SNE) [16] and its extensions have drawn many researchers' attention to perform the dimensionality reduction and visualization task. SNE converts the high-dimensional Euclidean distances between data points into conditional probability distribution related to Gaussian which represents the pairwise similarity, then requires the low-dimensional data to retain the same probability distribution.

---

[1] https://www.consumerfinance.gov/data-research/consumer-complaints/.

The t-SNE [17, 18] is an efficient nonlinear dimensionality reduction approach for embedding the high-dimensional data into 2D or 3D data points [19]. t-SNE extended SNE in two aspects, (i) it substitutes the joint probability distribution for conditional probability distribution, forming the symmetric SNE which leads to simpler gradients of cost function; (ii) it employs Student t-distribution with one degree of freedom to model pairwise similarity in low-dimensional space to alleviate the "crowding problem" [20].

In this work, t-SNE is used to create three-dimensional embeddings to assess the data overall structure and to highlight separation between word categories.

2.4 Hyperparameters and Training

In order to define a stopping criterion, to prevent overfitting and to evaluate the generalization capacity of the trained model, the original database is randomly divided into training (60%), validation (20%) and test (20%) sets and training of each run took 500 epochs.

CNN: rectified linear units, filter windows of 2, 3 and 4 with 900 feature maps each, dropout rate of 0.5, and mini-batch size of 64 are used. Softmax activation function is applied to the output of the penultimate fully-connected layer. These values were chosen empirically.

3 Dataset and Word Embedding

The database used is composed of a narrative, taken from the body of emails received by the department of consumer financial protection[2], and each of these narratives or complaints are duly classified among 11 possible categories. The list below shows all possible classes, as well as the complaints frequency of occurrences in each of them:

1. Debt collection 17552
2. Mortgage 14919
3. Credit reporting 12526
4. Credit card 7929
5. Bank account or service 5711
6. Consumer Loan 3678
7. Student loan 2128
8. Prepaid card 861
9. Payday loan 726
10. Money transfers 666
11. Other financial service 110.

3.1 Data Preprocessing

Data preprocessing aims to eliminate incomplete, noisy and inconsistent data [21]. Preprocessing helps in maximizing classifier performance, therefore, the narratives classification is expected to benefit from this process.

The dataset contains 66806 complaints classified among 11 possible categories. All text contained in these records is cleaned before being used to train the proposed model.

In this step, the removal of stopwords, stemming of words, as well as removal of special characters, numbers, white space, punctuation, and everything else that does not help the model to perform the classification or grouping task is performed.

3.2   Embeddings from Scratch

The simpler approach learns embeddings from scratch: a vocabulary index is built and each word is mapped to an integer between 0 and the vocabulary size. Each narrative becomes then a vector of integers [22].

3.3   Pre-trained Word Vectors

- *word2vec*: Initializing word vectors with those obtained from an unsupervised neural language model is a popular method [12, 23]. We use the publicly available word2vec vectors – with dimensionality of 300 – that were trained on 100 billion words from Google News. Words not present in the set of pre-trained words are initialized randomly.
- *glove*: GloVe is a new global log bilinear regression model that combines the advantages of the two major model families in the literature: global matrix factorization and local context window methods. The model produces a vector space with meaningful substructure. It also outperforms related models on similarity tasks and named entity recognition [13].

4   Results and Discussion

To solve the presented problem and for purposes of methodology comparison, several approaches was tested using CNN, PCA and t-SNE. All tests were run using Intel AI DevCloud nodes having the following configuration (Table 1):

Table 1. Intel cluster configuration

Intel AI DevCloud	Cluster's compute nodes configuration
	Feature description
Family	Intel® Xeon® Processor
Architecture	Skylake architecture (Intel Xeon Scalable Processors Family)
Model	Intel® Xeon® Gold 6128 CPU
Memory Size	Ram memory: 96 GiB
Networking	Gigabit Ethernet Interconnect

Mean results for all 10 experiments from all six implementations CNNs follow (Table 2):

Table 2. CNNs results

CNN	Validation		Test	
	Accuracy	Loss	Accuracy	Loss
Embeddings from scratch	0.959	0.0406	0.775	0.3056
Non-trainable word2vec	0.966	0.0325	0.815	0.2654
Trainable word2vec	0.968	0.0348	0.82	0.2558
Non-trainable GloVe	0.97	0.0826	0.832	0.2381
Trainable GloVe	**0.973**	**0.0761**	**0.85**	**0.2053**
Two-channel GloVe + word2vec	0.836	0.1655	0.841	0.224

The best-trained model, the Trainable Glove model, resulted in a validation accuracy of 97.3%, and a test accuracy of 85%, which can be considered a good result due to the complexity of the problem.

In order to compare performance with Xeon Scalable Processor and Tensorflow compiled with the Intel MKL optimizations, all tests were also run in a different hardware with default Deep Learning libraries (no optimization). The hardware used is described in Table 3.

Table 3. Hardware for comparison

	Computer configuration
	Feature description
Family	Intel® Core i7 Processor
Model	Intel® Core i7 4770 CPU
Memory size	Ram memory: 32 GiB

Performance comparisons are shown in tables below. Table 4 presents the median of samples per second, number of CPU cores in each architecture, speed up in throughput (throughput on Xeon CPU/throughput on i7 CPU: measures the relative performance of two systems processing the same problem) and efficiency – speed up per CPU core: speed up is normalized by number of CPU cores, which make the tests more comparable.

Table 4. CNN training and hardware comparison

CNN	Xeon – 6 cores			i7–8 cores		
	Samples/sec	Speed up	Efficiency	Samples/sec	Speed up	Efficiency
Embeddings from scratch	360.04	8.866x	147.76%	40.61	1.000x	12.5%
Non-trainable word2vec	483.26	6.187x	103.12%	78.11	1.000x	12.5%
Trainable word2vec	325.5	5.362x	89.36%	60.71	1.000x	12.5%
Non-trainable GloVe	432.61	5.570x	92.83%	77.67	1.000x	12.5%
Trainable GloVe	284.74	5.288x	88.13%	53.85	1.000x	12.5%
Two-channel GloVe + word2vec	224.25	6.934x	115.57%	32.34	1.000x	12.5%

The clustering approach was evaluated using PCA and t-SNE. To simplify the problem and for reasons of data limitation imposed by the graphical tool of the tensorboard[3] – maximum of ten thousand records, only five classes of the original eleven were evaluated: Debt collection, Mortgage, Credit reporting, Bank account or service, and Student loan.

To plot the 3D PCA plot, the transformed data of the three first components were chosen. Figure 1 shows the result of the transformation. We can see that PCA is capable of grouping well three of the five classes. PCA treats all words as individual points – not as analogy pairs – and simply captures the direction of the largest variation among all words. In many cases, the intergroup variance can be larger than the variance between the two concepts, resulting in an embedding that does not preserve the analogy relationship. Therefore, it is a not so good approach to the problem in question.

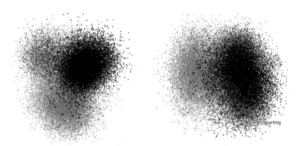

Fig. 1. 3D visualization of PCA: first three principal components.

Figure 2 below shows the clustering made with t-SNE analysis to provide a high-level overview of the embedding space. It is possible to notice in the 3D graph that the approach was able to perfectly discriminate the five classes present in the simplified database. The relationship between colors and classes is shown in the list below:

- Debt collection: blue
- Mortgage: orange

- Credit reporting: red
- Bank account or service: pink
- Student loan: purple.

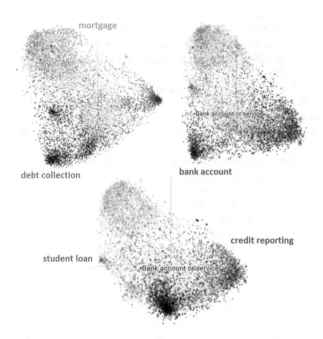

Fig. 2. 3D visualization of t-SNE transformed data in three different angles.

5   Conclusion

In the present work, we have describe several approaches to investigated plain text classification problem with convolutional neural networks – learning embeddings from scratch, on top of word2vec and GloVe vectors, the data was also evaluated as clustering problem using PCA and t-SNE. These results increase the evidence that pretrained word vectors is an important ingredient in deep learning for NLP. Furthermore, convolutional neural networks and t-SNE can perform remarkably well on real text classification data.

Acknowledgment. This research was supported by Intel Corporation. The authors gratefully acknowledge funding and use of Intel's cloud services.

To PUC-Rio and Intel Corporation, without which this work could not have been accomplished.

References

1. Haykin, S.S.: Neural Networks: A Comprehensive Foundation. Prentice Hall, Upper Saddle River (1998)
2. Goodfellow, I., Bengio, Y., Courville, A.: Deep Learning. MIT Press, Cambridge (2016)
3. Hassan, A., Mahmood, A.: Convolutional recurrent deep learning model for sentence classification. IEEE Access **6**, 13949–13957 (2018)
4. Collobert, R., Weston, J.: A unified architecture for natural language processing. In: Proceedings of the 25th International Conference on Machine Learning, ICML 2008, pp. 160–167 (2008)
5. Joachims, T.: Text categorization with support vector machines: learning with many relevant features (2005)
6. Kim, Y.: Convolutional neural networks for sentence classification, August 2014
7. Mikolov, T., Chen, K., Corrado, G., Dean, J.: Efficient estimation of word representations in vector space (2013)
8. Lecun, Y., Bottou, L., Bengio, Y., Haffner, P.: Gradient-based learning applied to document recognition. Proc. IEEE **86**(11), 2278–2324 (1998)
9. Bengio, Y., Ducharme, R., Vincent, P., Jauvin, C., Ca, J.U., Kandola, J., Hofmann, T., Poggio, T., Shawe-Taylor, J.: A neural probabilistic language model (2003)
10. Yih, W.-T., Toutanova, K., Platt, J.C., Meek, C.: Learning discriminative projections for text similarity measures. Association for Computational Linguistics (2011)
11. Mikolov, T., Sutskever, I., Chen, K., Corrado, G., Dean, J.: Distributed representations of words and phrases and their compositionality (2013)
12. Collobert, R., Weston, J., Com, J., Karlen, M., Kavukcuoglu, K., Kuksa, P.: Natural language processing (almost) from scratch (2011)
13. Pennington, J., Socher, R., Manning, C.D.: GloVe: global vectors for word representation (2013)
14. Zare, A., Ozdemir, A., Iwen, M.A., Aviyente, S.: Extension of PCA to higher order data structures: an introduction to tensors, tensor decompositions, and tensor PCA. Proc. IEEE **106**(8), 1341–1358 (2018)
15. Shlens, J.: A tutorial on principal component analysis, April 2014
16. Hinton, G., Roweis, S.: Stochastic neighbor embedding (2002)
17. Van Der Maaten, L., Hinton, G.: Visualizing data using t-SNE (2008)
18. Liu, S., Bremer, P.-T., Thiagarajan, J.J., Srikumar, V., Wang, B., Livnat, Y., Pascucci, V.: Visual exploration of semantic relationships in neural word embeddings. IEEE Trans. Vis. Comput. Graph. **24**(1), 553–562 (2018)
19. Pan, M., Jiang, J., Kong, Q., Shi, J., Sheng, Q., Zhou, T.: Radar HRRP target recognition based on t-SNE segmentation and discriminant deep belief network. IEEE Geosci. Remote Sens. Lett. **14**(9), 1609–1613 (2017)
20. Cheng, J., Liu, H., Wang, F., Li, H., Zhu, C.: Silhouette analysis for human action recognition based on supervised temporal t-SNE and incremental learning. IEEE Trans. Image Process. **24**(10), 3203–3217 (2015)
21. Ghag, K.V., Shah, K.: Comparative analysis of effect of stopwords removal on sentiment classification. In: 2015 International Conference on Computer, Communication and Control (IC4), pp. 1–6 (2015)
22. TFLearn: TFLearn (2018). http://tflearn.org/data_utils/. Accessed 07 Feb 2018
23. Yu, L.-C., Wang, J., Lai, K.R., Zhang, X.: Refining word embeddings using intensity scores for sentiment analysis. IEEE/ACM Trans. Audio Speech Lang. Process. **26**(3), 671–681 (2018)

Hybrid Instrumental Means of Predictive Analysis of the Dynamics of Natural and Economic Processes

Elena Popova[1](✉) ⓘ, Luís de Sousa Costa[2] ⓘ,
and Alfira Kumratova[1] ⓘ

[1] Kuban State Agrarian University,
13 Kalinina Str., 350044 Krasnodar, Russian Federation
{elena-popov, alfa05}@yandex.ru
[2] CIMO – Centro de Investigação de Montanha,
Departamento Ambiente e Recursos Naturais, Polytechnic Institute of Bragança,
Campus de Santa Apolónia, 5300-302 Bragança, Portugal
lcosta@ipb.pt

Abstract. The purpose of the presented research is the development and adaptation of mathematical and instrumental methods of analysis and risk management through the forecasting of both economic and natural time series with memory based on the application of new mathematical methods of investigation. The paper poses the problem of developing a constructive method for predictive analysis of time series in the framework of the currently emerging trend of using so-called "graphical tests" in the process of time series' modeling using nonlinear dynamics methods. The main purpose of using graphical tests is to identify both stable and unstable quasiperiodic cycles (quasi-cycles), the whole set of which includes a strange attractor (if one exists). New computer technologies that made it possible to study complex phenomena and processes "on a display screen" were used as instrumentation for the implementation of methods of non-linear dynamics. The proposed approach differs from classical methods of forecasting by new implementation of accounting trends (the evolution of centers and sizes of dimensional rectangles), and appears to be a new tool for identifying cyclic components of the time series in question. As a result, the person, that is making decision has more detailed information, which is impossible to obtain by the methods of classical statistics. The work was supported by Russian Foundation for Basic Research (Grants № 17-06-00354 A).

Keywords: Effluents volumes · Harvest · Prediction · Time series ·
Phase portrait · Bounding box · Quasicycles · Phase analysis

1 Introduction

The relevance of the presented study, that is based on the application of the methods of phase analysis to the problems of economic and mathematical modeling of the agro-industrial complex and natural risk factors, is beyond doubt. The authors used in the complex known methods of classical statistics [1–4] and methods of nonlinear dynamics.

© Springer Nature Switzerland AG 2020
A. M. Madureira et al. (Eds.): HIS 2018, AISC 923, pp. 31–39, 2020.
https://doi.org/10.1007/978-3-030-14347-3_4

The transition to a new economy calls for the development of new software tools for economic and mathematical modeling, including instruments for risk assessment (prediction and pre-forecast analysis), in particular, such as phase analysis, fractal analysis, linear cellular automaton and dynamic chaos methods.

Concerning the subject of the study - the time series "flow volumes of the Kuban mountain river ", we note the following: in conclusion on the investigation of the flood cause in June 2002 it is said that as a result of rainfall falling in the mountainous areas of the Kuban River basin (with a catchment area of 57,900 km^2 the total length of the rivers of the basin is 38 325 km, and the total number of rivers is equal to 13 569, the length of the Kuban is 870 km. [5]), there was a flood, which has no analog for almost a hundred-year period of observations, both in terms of maximum flow and rise equal, and for damage caused to the population and enterprises. As a result, 246 settlements were damaged in the South of Russia, more than 110 km of the gas pipeline, 269 bridges, 1,490 km of roads, 102 people were killed. The total number of victims in the Southern Federal District reached 340 thousand people, and material damage exceeded 15 billion rubles. Data on the flow volumes of the Kuban River are provided by the Karachaevo-Cherkess Center of Hydrometeorology and Environmental Monitoring - a branch of the Federal State Budget Institution "North Caucasus Department for Hydrometeorology and Environmental Monitoring" of Russia.

Concerning the relevance of the utility of forecasting the values of time series (TS) of crop yields, it can be noted that the importance of planning, achieving and maintaining the development of economic growth rates also in order to ensure a high standard of living for the population, is constantly increasing. It should be noted that planning and forecasting of the enterprise's activities is an objective necessity for any economic system.

2 Materials and Methods

The paper presents a comparison of the results of a pre-forecast analysis of time series of a different nature of cyclicity (wheat prices and runoff volumes of the mountain river Kuban) obtained on the basis of phase analysis. We'll demonstrate the method of phase analysis based on the TS of the monthly flow volumes of the mountain river Kuban [5] for the period from 1926 to 2003 (further TS "Kuban"), which has a clear annual cycle [6]. As a comparison, we'll consider the price of wheat per bushel in the American cents from January 1993 to December 2014 (further referred as TS "Wheat"). It should be noted, that a bushel is a unit of volume used in English system of measures. It is mainly used for measuring the volume of agricultural products. One bushel of wheat approximately equals to 27.2 kg.

Let's denote TS "Kuban" as follows:

$$z = \langle z_i \rangle, \ i = 1, 2, \ldots, N, \tag{1}$$

where N – is the number of TS levels.

In the investigation of this TS, it is sufficiently expedient and informative to construct phase portraits of TS (1) in the phase space $F(Z)$ [6, 7] of dimension 2:

$F(z) = \{(Z_i, Z_{i+1})\}$, $i = 1, 2, \ldots n - 1$. Figure 1 shows the phase trajectory of the TS "Kuban".

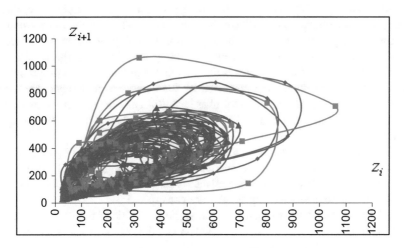

Fig. 1. Phase trajectory of the TS "Kuban"

Following Packard [7], Peters [8] for TS (1) for its phase space we use the formula:

$$\Psi_\rho(Z) = \{Z_i, Z_{i+1}, \ldots, Z_{n-\rho+1}\}, \ i = 1, 2, \ldots, n - \rho + 1. \tag{2}$$

The question of TS dimension ρ is a fundamentally important in the construction of the phase space (2) for a particular TS. This dimension must be no less than the dimension of the attractor of the observed series. As the dimension of the attractor, it can be used the fractal dimension C of this series. The value of this dimension, as noted in [8], is determined by the following formula:

$$D = 2 - H \tag{3}$$

Since the value of the Hurst exponent of the TS's is in the vicinity of (0; 1), we can obtain the estimate $D < 2$ [9–11], respectively. From this we can conclude that it is sufficient to use a phase space of dimension for this investigation $\rho = 2$.

The definition of the concept of "quasicycle" is described in detail in [7, 11].

3   Results

3.1   Approbation of the "Phase Analysis" Program on Real Time Series

An important and noteworthy feature of TS "Kuban" prediction is that the phase portrait consists of a continuous sequence of disjoint quasicycles whose dimension is equal to a year (exactly 12 months). In general, the trajectory of the phase portrait for

TS "Kuban" consists of disjoint quasicycles, C_r, $r = 1, 2, \ldots, 16$. Quasicycles are built from February to January, thus forming 12-monthly cycles.

Most of classical research methods have become available due to statistical application packages. The computer began to perform all the laborious, routine and volumetric work on the calculation of various statistical indicators, construction of diagrams and graphs. The researcher mostly performs creative work: setting of the task, determining the method for its solving, obtaining and analyzing the results of the research. Automation of this features is the goal that is pursued by the program "Phase Analysis" (C++), developed and presented by the authors in this paper; the program can be used for automatic realization of phase analysis calculations (in accordance with Fig. 4). The program has a convenient interface and enables the economist-expert to implement the analytical process.

In terms of the phase analysis' tools, a separate annual cycle belonging to TS Z (1) is presented as a typical quasicycle inherent to TS "Kuban" (Fig. 2(b)).

Along with TS "Kuban" (in accordance with Fig. 3(a)) phase portraits for TS "Wheat" (in accordance with Fig. 3(b)) were constructed.

The latter is due to the existence of the lag in the work of the sequential analysis algorithm. In this case, the size of lag is 3. The mentioned above lag is represented by three points 13, 14 and 15 (in accordance with Fig. 2(a)).

The definition of the term "quasicycle" is in a sense close to the definition of the generally accepted concept of a "cycle". The difference between these two concepts is that the initial and final points of the quasicycle do not necessarily have to coincide. The end point of a quasicycle is determined by its occurrence in a neighborhood of the initial point. In this case, self-intersection of the initial and final links of the quasicycle is allowed, if this leads to the best approximation of its initial and final points. Figure 3 shows examples of quasicycles that are obtained after the decomposition of the phase portrait in Fig. 1 into quasicycles. Table 1 shows the dimensions L_k of all 15 quasicycles.

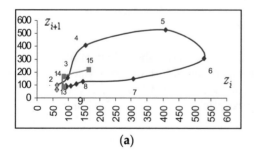 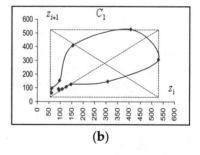

(a)   (b)

Fig. 2. (a) The first quasicycle of the phase portrait $\Psi_2(Z)$, including the lag – 13, 14, 15; (b) The first quasicycle TS Z, whose size is 12

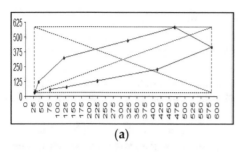 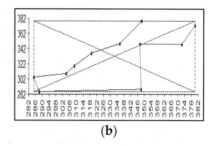

(a)   (b)

Fig. 3. The "Phase Portraits" tab of the "Phase Analysis" program with automatic construction of dimensional rectangles (represented by dotted lines): (a) TS "Kuban"; (b) TS "Wheat".

Table 1. The dimensions of all quasicycles

№ C_k	1	2	3	4	5	6	7	8	9	10	11	12	13	14	15
L_k^r		12	12	12	12	12	12	12	12	12	12	12	12	12	12

We denote Z_k by this segment BP Z, which is obtained by the removal procedure from all observation Z points related to quasicycles $C_1, C_2, \ldots, C_{k-1}$; according to this definition $Z_1 = Z$.

From Table 1 we can make a conclusion that the presence of long-term memory in the considered TS, along with other factors, is due to the cyclic component of this TS.

For each quasicycle C_k, a "overall rectangle of a quasicycle C_k" is constructed [10]. The intersection of the diagonals of the overall rectangle defines the so-called "center of rotation" of the quasicycle O_k, whose coordinates are denoted by $O_k(x_k, y_k)$. The construction of such a rectangle consists of the following operations. First, in the considered quasicycles K_r^1 two points are distinguished: the first with the minimum abscissa value, the second with the maximum abscissa value through these selected points we draw dashed lines which are parallel to the axis of ordinates. Further, in the quasicycle C_1 two points are distinguished: the first - with the minimum value of the ordinate, the second - with the maximum value of the ordinate; through these selected points we draw with dashed lines segments, parallel to the axis of abscissae. The intersection of the constructed two pairs of parallel lines forms the required dimensional rectangle for the quasicycle in question C_1; the center of this quasicycle is represented by the point of intersection of the diagonals of the overall rectangle (example is shown on Fig. 3).

Step-by-step algorithm for phase analysis

Based on the analysis carried out by the authors, the following approach for predicting the TS of the considered type is proposed; it consists of the following steps:

1. Conducting a fractal analysis of TS (1) in order to establish the presence of a long-term memory to evaluate its depth ρ. Wherein we get a fuzzy set of estimates $L = L(Z) = \{(l, \mu_l)\}$ of the depth of the memory of TS z.
2. The construction of a phase portrait for the indicated TS.

3. Decomposition of the phase portrait into quasicycles C_r.
4. Analysis of the movement of the centers of quasicycles $O_r(x_r, y_r)$, the movement of the dimensions of the overall rectangles' of quasicycles areas, and also the directions of rotation of the links of quasicycles.
5. Construction of the forecast by the principle of continuation (completion) of the corresponding quasicycle using the results of stage 4 for two cases, when the last quasicyclic is:
    (a) incomplete (we use the overall dimensions and the pattern of the rotation of quasicycles, considering the sector of the overall rectangle to which the predicted point belongs);
    (b) completed (we use the overall dimensions and the pattern of the rotation of quasicycles, considering the evolution of the centers and the transitions from the final point of one cycle to the starting point of a new cycle).

4 Discussion

Let's list the revealed features of the phase trajectories of the TS "Kuban".

1. The phase portrait of the TS "Kuban" consists of quasicycles with a dimension that equals 12. This fact does not contradict the results of a fractal analysis on the evaluation of the depth of the memory of TS [6].
2. It was considered that each link of all quasicycles rotates clockwise. The overall rectangle can be divided into 4 sectors by straight lines.
3. The centers of quasicycles $O_k(x_k, y_k)$, in the order of their numbering $k = \overline{1,72}$, evolve along a definite trajectory, whose points are located in a sufficiently small vicinity of the bisector of the positive orthant of Cartesian coordinates.
   The coordinates of the centers of all quasicycles define the points of the bisector of the positive orthant $x_k x_{k+1}$ of the Cartesian coordinates (in accordance with Fig. 4). Note that the trajectory of the motion of these coordinates is characterized by an wide range $K \approx 550 - 200 = 350$, which exceeds the minimum point by more than 1.5 times.

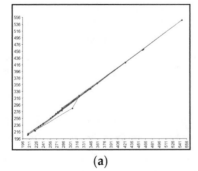 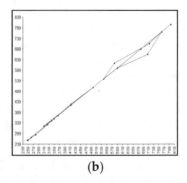

(a)   (b)

Fig. 4. Motion of the centers of quasicycles (the values of the coordinate axes for TS "Kuban" are represented by the values of river flow volumes (thousand cubic meters), for "Wheat" – by the price for bushel in US cents): (a) TS "Kuban"; (b) TS "Wheat".

In connection with the above mentioned, it is practical interesting to determine the long-term trends that govern the motion of the centers of the dimensional rectangles. For this purpose, the division for the periods was performed (Fig. 5): from 1926 to 1940 (in accordance with Fig. 5a); from 1946 to 1987 (in accordance with Fig. 5b); from 1988 to 2003 (in accordance with Fig. 5c). Visualization of Fig. 4 reveals the following trend: at approximately the same value min $\approx$ 200, in the process of time the value of the range increases in the following ratio:

$R_1 \approx 350 - 200 = 150$ (in accordance with Fig. 5a)
$R_2 \approx 450 - 200 = 200$, (in accordance with Fig. 5b)
$R_3 \approx 550 - 200 = 350$ (in accordance with Fig. 5c). A well-known climatologists' statement says about the existence of a general trend of climate warming in the northern hemisphere confirms the above-mentioned range of magnitude values as filling of mountain rivers is determined by the intensity of the glaciers' melting, especially in summer months.

4. The movement of the dimensions (area) of the overall rectangles of quasicycles is cyclical, which is confirmed by the visualization of Fig. 6.

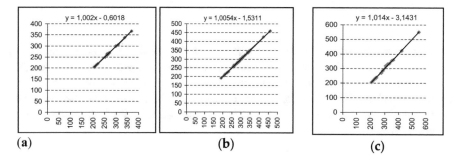

(a)   **(b)**   **(c)**

Fig. 5. Evolution of the area of the overall rectangles, considering time parameter

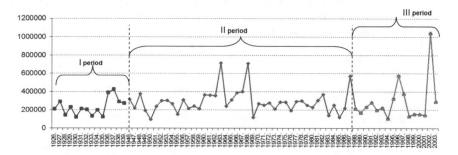

Fig. 6. Movement of areas of dimensional rectangles (taking into account time parameter)

It is important to mention that authors carried out graphical testing of time series, both natural, economic, and financial. The fact of confirming a 12-month cycle in natural and economic series with a pronounced annual seasonal cyclicity based on phase analysis allowed to use it for other time series. For example, financial monthly TS of pairs euro/ruble, dollar/ruble, gold, **silver**, palladium, platinum have quasi-cycles of various lengths from 3 to 8 months (Table 2). From the obtained results of phase analysis, it is reasonable to expect a sufficiently high degree of reliability in predicting general trends for TS elements of the financial market using models based on the use of long-term memory, in particular using a cellular automaton predictive model [1].

Table 2. Results of phase analysis (PhA) for the elements of the financial market time series

Financial market element	Mean length of quasicycles (PhA)	Quasi-cycles lengths' depth of memory (PhA)
Euro/ruble	5.83	5–6
Dollar/ruble	7.2	7–8
Gold	6.22	7–8
Silver	6.81	**5–6**
Palladium	6	6–7
Platinum	6.58	6–7

In this way, the proposed approach differs from the classical methods of forecasting by the new trend's registration implementation (evolution of the centers and sizes of the overall rectangles), represents a new tool (phase portraits) for identifying the cyclic component of the TS, which in its turn makes it possible to identify the prognostic properties of the series under study [2] and thus identify ways to reduce socio-economic risks.

References

1. Adamowski, J., Chan, H.F., Prasher, S.O., Sharda, V.N.: Comparison of multivariate adaptive regression splines with coupled wavelet transform artificial neural networks for runoff forecasting in Himalayan micro-watersheds with limited data. J. Hydroinf. **14**, 731–744 (2012). https://doi.org/10.2166/hydro.2011.044
2. Yaseen, Z.M., Kisi, O., Demir, V.: Water Resour. Manag. **30**, 4125 (2016). https://doi.org/10.1007/s11269-016-1408-5
3. Ahani, A., Shourian, M., Rahimi Rad, P.: Water Resour. Manag. **32**, 383 (2018). https://doi.org/10.1007/s11269-017-1792-5
4. Parmar, K.S., Bhardwaj, R.: Appl. Water Sci. **4**, 425 (2014). https://doi.org/10.1007/s13201-014-0159-9
5. Popular science encyclopedia "Water of Russia". http://water-rf.ru. Accessed 11 July 2018
6. Vintizenko, I.G.: Deterministic forecasting in economic systems. New technologies in management, business and law. Nevinnomyssk, pp. 163–167 (2003)
7. Packard, N., Crutchfield, J., Farmer, D., Shaw, R.: Geometry from a time series. Phys. Rev. Lett. **45**, 712 (1980)

8. Peters, E.: Chaos and Order in the Capital Markets. A New Analytical View of Cycles, Prices and Market Volatility, p. 333. Wiley, Hoboken (2000)

9. Kumratova, A.M.: The study of trend-seasonal processes using classical statistics. Polythematic Netw. Electron. Sci. J. KubSAU, Krasnodar, Russia **103**(09), 312–323 (2014)

10. Kumratova, A.M., Popova, E.V., Kurnosova, N.S.: Reduction of economic risk based on pre-forecast analysis. In: Modern Economy: Problems and Solutions, Voronezh, Russia, pp. 18–27 (2015)

11. Kumratova, A.M., Popova, E.V.: Evaluation and risk management: analysis of time series by methods of nonlinear dynamics, 212 p. KubSAU, Krasnodar, Russia (2014)

Optimizing Routes for Medicine Distribution Using Team Ant Colony System

Renan Costa Alencar⬤, Clodomir J. Santana Jr.⬤,
and Carmelo J. A. Bastos-Filho$^{(\boxtimes)}$⬤

Polytechnic University of Pernambuco, University of Pernambuco, Recife, Brazil
{rca2,cjsj,carmelofilho}@ecomp.poli.br

Abstract. Distributing medicine using multiple deliverymen in big hospitals is considered complex and can be viewed as a Multiple Traveling Salesman Problem (MTSP). MTSP problems aim to minimize the total displacement of the salesmen, in which all intermediate nodes should be visited only once. The Team Ant Colony Optimization (TACO) can be used to solve this sort of problem. The goal is to find multiple routes with similar lengths to make the delivery process more efficient. Thus, we can map this objective in two different fitness functions: minimizing the longest route, aiming to be fair in the allocation of the workload to all deliverymen; and decreasing the total cost of routes, seeking to reduce the overall workload of the deliverymen. However, these objectives are conflicting. This work proposes the use of swarm optimizers to improve the performance of the TACO concerning these two objectives. The results using global optimizers for the parameters far outperformed the original TACO for the case study.

Keywords: Multiple Traveling Salesman Problem ·
Swarm intelligence · Hospital logistics

1 Introduction

The Travelling Salesman Problem (TSP) and the Knapsack Problem (KP) are two of the most studied combinatorial optimization problems so far. TSP and KP belong to the set of NP-complete problems. In spite of the fact that they are tough to be solved even separately, there are some real-world problems that can be mapped as a combination of them, resulting in a complex task. There is a TSP variation, which multiple salesmen engage in building a solution, the so-called Multiple Salesman Problem (MTSP). According to Bektas [3], there is a variety of real-life problems, which are considered as MTSP, e.g., Routing Vehicle Problems (PRVs) with solution constraints. As stated in the MTSP definition, there are $m > 1$ salesmen located initially in the same city which defines the depot. The other cities of the instance are defined as intermediate nodes. The MTSP aims to minimize the sum of route lengths with a constraint, which every

© Springer Nature Switzerland AG 2020
A. M. Madureira et al. (Eds.): HIS 2018, AISC 923, pp. 40–49, 2020.
https://doi.org/10.1007/978-3-030-14347-3_5

route should begin and end at the depot node, and all intermediate nodes should visit only once. Besides, the MTSP problem has another constraint which there should be at least an intermediate node but the depot for each salesman route. The KP is a combinatorial problem, which allocates space in a knapsack in advance according to an object selection. Hence, the total value of all chosen objects is maximized in the knapsack. Martello and Toth [12] state that KP is a very often problem which appears in business, e.g., economic planning, and industry as cargo loading, cutting stock, and bin packing problems. They define the KP problem as an n-object vector of binary variables $x_i(i = 1, \ldots, n)$, in which the object i has a weight w_i and the knapsack has a capacity M. If a fraction x_i; $0 \le x_i \le 1$, is placed in the knapsack, then a profit, $g_i x_i$, is earned. The KP aims to find a combination of objects that maximizes the total profit from all chosen objects in the knapsack. While the knapsack's capacity is M, the total weight of all chosen objects to be at most M.

Although having those two problems associated is not unusual, they can appear in complex scenarios. For instance, the medicine distribution at big hospital centers can be viewed as an MTSP-KP instance. Both separation and distribution processes are seen as KP and MTSP instances, respectively. They have a high number of combinations and, even separated, those tasks are hard to be optimized, requiring sophisticated tools to tackle them. The hospital logistics can ensure patient's safety; however, it is one of the most challenging problems faced by hospital managers, especially in Brazil, due to meeting the organizational needs in a fast, accurate and efficient way. Furthermore, the financial resources addressed to hospital logistics should be implemented efficiently due to the low budget of Brazilian public hospitals. Some approaches, based on mathematical methods and evolutionary algorithms, are used to optimize MTSP and KP instances. We can cite the following examples Genetic Algorithm - GA [8], Ant Colony Optimization - ACO [6] and Artificial Bee Colony - ABC [9]. These approaches aim to solve problems of steel production [14], cigarette distribution [11], service orders [1], sensor network routing [16], among others.

For the best of our knowledge, there are no reports in the literature regarding the use of global optimization processes to improve the performance of meta-heuristics deployed to solve the MTSP or MTSP-KP problem. Thus, there is still room for optimizing MTSP parameters with meta-heuristic algorithms aiming to achieve specific goals, especially when one needs to balance the length of the routes and the number of deliveries per agent simultaneously. Then, this work proposes a methodology to optimize MTSP parameters through global population-based optimizers, based on swarm intelligence based algorithms. We use the medicine distribution process with multiple routes as a case study.

The remainder of this paper is structured as follows. Section 2 introduces the basic concepts of some swarm intelligence-based algorithms to solve combinatorial problems. Section 3 describes the related works to the MTSP. Sections 4 and 5 depict the proposed model and the problem instance to be optimized. Section 6 outlines the scenarios and their settings to minimize the MTSP instance. Sections 7 and 8 present the obtained results and the conclusions.

2 Background

2.1 Team Ant Colony Optimization

The Ant Colony Optimization (TACO), proposed by Vallivaara [15], is based on the Ant Colony System (ACS) to solve MTSP instances. This basic generalization is made by replacing N ACS ants, which build solutions for TSP, with N teams of m members. An ant team represents a salesman in building the MTSP solution, and each team has its taboo list.

All ants of every team are placed at the depot at the beginning of the route construction. To distribute the workload, an ant k with the shortest partial route chooses its next city j, at any moment of the building process, according to the Transition State Rule (TSR) equation, as shown in (1).

$$j = \begin{cases} \text{argmax}_{l \in J_k} \{\tau_{il}[\eta_{il}]^\beta\}, & \text{if } q \le q_0 \\ J, & \text{otherwise;} \end{cases} \tag{1}$$

After choosing the next city, it is checked if another ant l could add the chosen city to its route and end up with better total route length. If so, the ant k can make its move first not choosing j. This checkpoint avoids the algorithm to force non-optimal solutions.

TACO has several parameters which are responsible for its behavior while building solutions. The initial probability q_0 determines whether the ants' initialization has only deterministic or random choices ($0 \le q_0 \le 1$). The pheromone parameters α and β define the weight of the pheromone trail and the visibility, respectively, in the choice of the next node by the ant. The parameter ξ controls the pheromone persistence when the Pheromone Update Rule (PUR) takes place locally, just after an ant moves from one city to another, that is, it includes one more edge on its route. Likewise, ρ regulates pheromone persistence for global PUR, i.e., at the end of each cycle of the algorithm.

2.2 Particle Swarm Optimization

Kennedy and Eberhart [10] proposed the Particle Swarm Optimization (PSO) method based on bird flocking. PSO is suitable for the optimization of continuous variables in a high-dimensional search space and presents high precision. It performs searching via a swarm of particles through an iteration process. Each particle moves towards its previous best (P_{best}) position and the global best (G_{best}) position in the swarm to achieve the optimal solution.

The solution represents the particle position in the search space, a vector x_i. For each step, the particles have their positions according to their velocity vector v_i. The velocity clamping, an upper bound for the velocity parameter avoids particles flying out the search space. Likewise, the "constriction coefficient" strategy, proposed by Clerc and Kennedy [5], constricts the velocities through the dynamic swarm analysis.

2.3 Fish School Search

Bastos-Filho et al. [2] developed a population-based search algorithm inspired by fish swimming behavior, which expands and contracts while looking for food. The Fish School Search (FSS) algorithm considers the individual and collective fish movements. This optimization algorithm does not present the same exploitation capability of the PSO, but it has the capability to find good solutions in a search space with many local minima. Each fish, in n-dimensional location, represents a feasible solution for the problem. Its success is measured by its weight, a cumulative account of how successful the search for each fish in the school has been. The fishes not only store information about their weight but also position in the search space. FSS consists of moving and feeding operators. On the individual movement, each fish randomly moves towards a position in its neighborhood looking for promising regions. After moving to new positions, all fishes have their weights updated. The weight update is determined by the individual movement success, which is computed through the fitness of current and new positions. After feeding all fishes, the collective-instinctive movement takes place. All fishes move towards an influence vector. Those fishes that improved their fitness in the current iteration generate this vector. At the end of the current iteration, the school contracts or expands according to the volitive-collective movement operator. The school's contraction results in an exploitation search whereas its expansion make the school explore the search area avoiding local minima. Thus, the volitive operator computes the school's barycenter. This last operator gives to the FSS the capability to self-adjust the granularity of the search along the optimization process.

3   Related Works

We did not find any approach related to medicine distribution until the present moment, but there is a set of computational problems, which are transversal to the tackled problem. Those problems have been studied to improve commercial production and distribution under one or more objective constraints. Somhom, Modares and Enkawa [13] use a Competition-Based Neural Network (CBNN) to minimize the longest route in an MTSP (based on the TSPLIB) instance with a single depot closed routes. Tang et al. [14] use a Modified Genetic Algorithm (MGA) to improve production scheduling of hot rolling from an iron and steel industry in China. Carter and Ragsdale [4] use a Genetic Algorithm (GA) with a new chromosome and MTSP operators. Wang et al. [16] use the Ant Colony System (ACS) to group and route sensor nodes from a wireless multimedia network with a limited time interval as a constraint. Vallivaara [15] proposes the Team Ant Colony Optimization, based on ACS rules to manage routes with multiple robots in a hospital environment. The problem constraint is to minimize the longest route and the sum of routes. Liu et al. [11] use the ACS and the Max-Min Ant System to solve the distribution of cigarettes in a Chinese company. Barbosa and Kashiwabara [1] use the Single Team Ant Colony System (STACS), based on TACO, to route the service orders in an energy distribution company.

4   Proposed Model

The proposed model consists of using a base algorithm to compute the best solutions for an MTSP instance while having a global optimizer algorithm to seek the best parameters sets for the base algorithm. In this work, we deploy the Team Ant Colony Optimization (TACO) as the base algorithm. Moreover, the Particle Swarm Optimization (PSO) and the Fish School Search take part in the optimization of the values for TACO parameters.

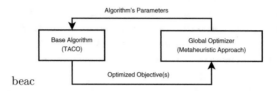

beac

Fig. 1. Proposed model for optimizing the TACO

As shown in Fig. 1, the Global Optimizer (GO) starts to generate a set of values for the TACO's parameters. In this case, the optimized parameters from TACO are α (pheromone relevance), β (relevance of pheromone's visibility), ξ (pheromone persistence, local update) and ρ (pheromone persistence, global update). Then, TACO builds a set of best solutions based on the optimized parameters. Next, TACO returns to GO the set of best solutions, which are evaluated to the meta-heuristic algorithm. As the iterations happen, the GO keeps a record of the best set of parameters, which has been found so far. The execution finishes when the GO reaches the limit of iterations.

5   Problem Instance

A Medicine Distribution Center (MDC) at a public hospital in Brazil usually executes the orders manually, both packaging the orders and building the routes. It does not matter how long it would take for the deliverymen to do that. Due to the considerable time variation for answering the orders, it is difficult to determine the delivery capacity for each agent to minimize individual costs of the built routes.

That real-world problem highlights a set of objectives, which can be optimized. One of those objectives is to reduce the total sum of the routes without worrying about the work balance between the deliverymen. Another objective is to have balanced routes at the end of order execution, which also results in more orders being executed at the same time interval, i.e., in this case, the target is to avoid a partially inoperative deliveryman because it has a route significantly shorter than the others.

Figure 2 highlights 16 pharmacies and an MDC, represented as $V1$, of a real hospital environment from Brazil represented as a graph with costs associated with distances for the deliveries. The MDC is the depot where the deliverymen start and finish their routes. A matrix, called cost matrix, contains the data of each pharmacy like identification (ID), latitude, longitude and its distance. The matrix helps to calculate the route cost. It is important to highlight that this graph shows the topology to reach the pharmacies, but do not explicit represent the geographical position.

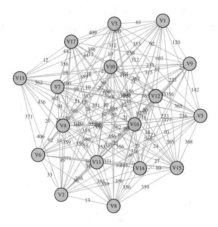

Fig. 2. Graph model

Likewise, another matrix, called the data matrix, contains the delivery orders for each node in the built model. Each order includes IDs of deliveryman and pharmacy, the initial and final distances when the deliveryman left the MDC and the duration of the order. The data matrix helps to simulate a day of ordering in the depot. Also, there are four deliverymen to deliver the requests to the pharmacies.

As stated in Sect. 4, a global optimizer, based on swarm intelligence, is also used to improve the TACO results by optimizing its parameters. FSS and PSO are taken as a global optimizer of the TACO parameters. We assessed the PSO because of its capacity to refine the values for the parameters (good exploitation capability) and the FSS because of its capacity to escape from local optima during the optimization process. The standard values for the set of used parameters are: number of deliverymen $M = 4$, the number of teams $N = 10$, initial probability $q_0 = 0.5$, pheromone relevance $\alpha = 0.5$, visibility relevance $\beta = 1.0$, pheromone persistence for local update $\xi = 0.1$, pheromone persistence for global update $\rho = 0.1$. A parameter test obtained those default values. The stop criterion is 1000 iterations for each independent run. The global optimizers optimize the subset of parameters $P = \{\alpha, \beta, \xi, \rho\}$ in a range $0 \leq P \leq 2$.

6   Experimental Setup

The experiments presented in this section aims to show the effectiveness of the developed methodology applied to the real problem. In the first scenario, the algorithms are configured to minimize the total cost of the solutions, without considering the distribution of the workload between the teams, as in the general description of the MTSP. In the second scenario, the algorithms minimize the cost of the largest individual route of the solutions, aiming the construction of solutions formed by routes with equal costs among the deliverymen, as in the MTSP with workload balance.

All experiments were executed on a MacBook Pro (13-inch, Late 2011) with 2.4 GHz Intel Core i5 CPU, 16 GB of RAM (1333 MHz DDR3) and macOS High Sierra (version 10.13.4) operating system. We used the database presented in Sect. 4. Then, we performed 30 independent runs of the algorithms for each experiment. TACO algorithm was coded in Java based on the proposed algorithm by Vallivaara [15]. The FSS is based on Bastos-Filho et al. [2] version. The single objective PSO were taken from the jMetal framework [7].

TACO takes part in the experiment as a base algorithm. Before starting the experiments of the proposed model, a base test has to be taken to prove the effectiveness and robustness of the chosen algorithm. The default values for both scenarios are stated in Sect. 5. The FSS as a GO is executed with a stop criterion of 1000 iterations per run, and it has 30 independent runs. The values of FSS parameters were taken from [2]. Similarly, the PSO has the same values of stop criterion and independent runs. The values of PSO parameter remained the same as in [7].

7   Results

The first set of experiments was carried out to minimize the Total Cost of Routes (TCR), i.e., the sum of each team's route. By comparing the results, the solution, which has the lowest total cost, is considered the best one. This scenario can be applied to real life when we aim to reduce the total amount of the deliverymen routes instead of prioritizing the work balance.

The second set of experiments aimed to minimize the Longest Route (LR) of the solutions keeping the same values of the parameters in the first scenario. That case is suitable for real situations when we prioritize the balance among the individual routes (work balance) rather than the total sum of routes.

7.1   Experiments Without Global Optimizers

Table 1 shows the average execution time of the algorithm for each case. We obtained this value from the average of the time spent to execute the 1000 iterations with 30 independent runs of the algorithm, for one working day.

TACO presented satisfactory results for the two MTSP variations: minimizing the total cost of the solution and minimizing the cost of the largest individual

Table 1. Comparison of the results with average of 30 independent runs

Approach	Minimizing TCR				Minimizing LR			
	TCR	S.D	LR	S.D	TCR	S.D	LR	S.D
TACO	1.8	0.25	0.721	0.003	1.223	0.050	0.756	0.012
FSS	1.0797	0.002	0.747	0	1.944	0.335	0.719	0.001
PSO	1.115	0.032	0.750	0.003	2.048	0.333	0.720	0.005

solution route (Table 1). The small values of the standard deviations in the two tables confirm the robustness of the algorithm when generating solutions with costs close to the average of the 30 executions. Although LR is being minimized, TCR varies throughout the iterations. The same situation happens when TACO minimizes TCR. Bearing this in mind, there is still room to make improvements in the results of both objectives.

7.2 Experiments with Global Optimizers

The experiments carried out with global optimizer showed better results for both scenarios, as displayed in Table 1. The approach having FSS as a GO (FSS-TACO) presented a better improvement when minimizing the longest route in comparison to the base algorithm. This result is due to the capability of the FSS-TACO to explore the search space. FSS-TACO has a lower standard deviation which corroborates to its robustness.

The approach having PSO as a GO (PSO-TACO) also had better results comparing to the TACO approach without global optimizers (see Table 1). Its standard deviation also shows the robustness and effectiveness when optimizing TCR and LR.

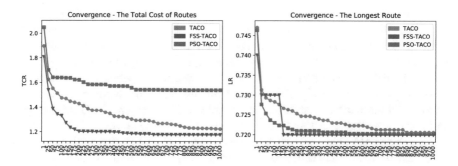

Fig. 3. Comparison among the three approaches when minimizing TCR and LR

7.3   Comparison

All the previous results are compiled in Table 1 for a better understanding of them. Comparing the three approaches, FSS-TACO got better results when minimizing both TCR and LR. It also had the lowest result with a standard deviation close to zero. As seen in Fig. 3, in both scenarios, FSS-TACO converged earlier than PSO-TACO and the base-algorithm. Results of PSO-TACO were also better than base-algorithm results and converged earlier than TACO as expected.

Figure 4 shows the boxplots for both TCR and LR. When minimizing the TCR, we can notice that the results for the FSS-TACO far outperformed the others, and PSO-TACO varies more than TACO. When minimizing the LR, both FSS-TACO and PSO-TACO approaches outperformed TACO.

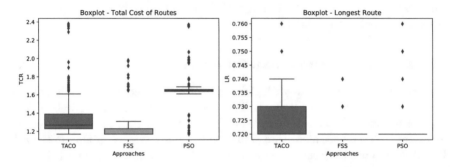

Fig. 4. Boxplots for TCR and LR considering TACO, FSS-TACO and PSO-TACO.

8   Conclusions

The proposed model described was efficient in the distribution of work orders among the deliverymen and in the creation of optimized routes to carry out the services. However, some questions need to be analyzed, such as multi-objective optimization of the total cost of the routes and the longest route and optimization of the deliverymen's knapsacks. For the creation of solutions optimized for MTSP in this work, an algorithm based on the ACO metaheuristic was implemented and associated with a global optimizer. The FSS and PSO were responsible for optimizing the TACO parameters to improve the results. As shown in Sect. 6, FSS-TAO achieved the best results for both scenarios with the lowest minimization results for the Total Cost of Routes (TCR) and the Longest Route (LR) with standard deviations around zero. In those two cases, FSS-TACO as a global optimizer converged earlier than the other two approaches.

Another possible approach to the MTSP problem is to optimize both objectives simultaneously, TCR and LR. To achieve this goal, we aim to use a multi-objective optimization algorithm called Multi-Objective Fish School Search (MOFSS) as a global optimizer for TACO's parameters.

References

1. Barbosa, D.F., Silla Jr, C.N., Kashiwabara, A.Y.: Aplicaçao da otimizaçao por colônia de formigas ao problema de múltiplos caixeiros viajantes no atendimento de ordens de serviço nas empresas de distribuiçao de energia elétrica. Anais do XI Simpósio Brasileiro de Sistemas de Informaç ao, pp. 23–30 (2015)
2. Bastos Filho, C.J., de Lima Neto, F.B., Lins, A.J., Nascimento, A.I., Lima, M.P.: A novel search algorithm based on fish school behavior. In: 2008 IEEE International Conference on Systems, Man and Cybernetics, SMC 2008, pp. 2646–2651. IEEE (2008)
3. Bektas, T.: The multiple traveling salesman problem: an overview of formulations and solution procedures. Omega **34**(3), 209–219 (2006)
4. Carter, A.E., Ragsdale, C.T.: A new approach to solving the multiple traveling salesperson problem using genetic algorithms. Eur. J. Oper. Res. **175**(1), 246–257 (2006)
5. Clerc, M., Kennedy, J.: The particle swarm-explosion, stability, and convergence in a multidimensional complex space. IEEE Trans. Evol. Comput. **6**(1), 58–73 (2002)
6. Dorigo, M., de Oca, M.A.M., Engelbrecht, A.: Particle swarm optimization. Scholarpedia **3**(11), 1486 (2008)
7. Durillo, J.J., Nebro, A.J.: jMetal: a java framework for multi-objective optimization. Adv. Eng. Softw. **42**(10), 760–771 (2011)
8. Holland, J.H.: Genetic algorithms. Sci. Am. **267**(1), 66–73 (1992)
9. Karaboga, D.: An idea based on honey bee swarm for numerical optimization. Technical report-tr06, Erciyes University, Engineering Faculty, Computer Engineering Department (2005)
10. Kennedy, J., Eberhart, R.C.: The particle swarm: social adaptation in information-processing systems. In: New Ideas in Optimization, pp. 379–388. McGraw-Hill Ltd., London (1999)
11. Liu, W., Li, S., Zhao, F., Zheng, A.: An ant colony optimization algorithm for the multiple traveling salesmen problem. In: 4th IEEE Conference on Industrial Electronics and Applications, ICIEA 2009, pp. 1533–1537. IEEE (2009)
12. Martello, S., Toth, P.: Knapsack Problems: Algorithms and Computer Implementations. Wiley, New York (1990)
13. Somhom, S., Modares, A., Enkawa, T.: Competition-based neural network for the multiple travelling salesmen problem with minmax objective. Comput. Oper. Res. **26**(4), 395–407 (1999)
14. Tang, L., Liu, J., Rong, A., Yang, Z.: A multiple traveling salesman problem model for hot rolling scheduling in Shanghai Baoshan iron & steel complex. Eur. J. Oper. Res. **124**(2), 267–282 (2000)
15. Vallivaara, I.: A team ant colony optimization algorithm for the multiple travelling salesmen problem with minmax objective. In: Proceedings of the 27th IASTED International Conference on Modelling, Identification and Control, pp. 387–392. ACTA Press (2008)
16. Wang, X., Wang, S., Bi, D., Ding, L.: Hierarchical wireless multimedia sensor networks for collaborative hybrid semi-supervised classifier learning. Sensors **7**(11), 2693–2722 (2007)

Extending Flow Graphs for Handling Continuous−Valued Attributes

Emilio Carlos Rodrigues[1,2(✉)] and Maria do Carmo Nicoletti[1,3]

[1] Centro Universitário C. Limpo Paulista (UNIFACCAMP),
Campo Limpo Paulista, SP, Brazil
emiliorcl986@gmail.com, carmo@cc.faccamp.br
[2] Instituto Federal de São Paulo (IFSP), Bragança Paulista, SP, Brazil
[3] Universidade Federal de S. Carlos (UFSCar), São Carlos, SP, Brazil

Abstract. This paper describes a research work that focuses on a suitable data structure, capable of summarizing a given supervised training data set into a weighted and labeled digraph. The data structure, named flow graph (FG), has been proposed not only for representing a set of supervised training data aiming at its analysis but, also, for supporting the extraction of decision rules, aiming at a classifier. The work described in this paper extends the original FG, suitable for discrete data, for dealing with continuous data. The extension is implemented as a discretization process, carried out as a pre-processing step previously to learning, in a two-step hybrid approach named HFG (Hybrid FG). The results of the conducted experiments were analyzed with focus on both, the induced graph-based structure and the performance of the associated set of rules extracted from the structure. Results obtained using the J48 are also presented, for comparison purposes.

Keywords: Flow graphs · Supervised machine learning ·
Discretization data process · Data structure · Hybrid systems

1 Introduction and Motivations

The concept of Flow Graph (FG) was first proposed in 2003 by Pawlak [11], as part of a mathematical formalism for representing and exploring the characteristics associated with a set $X = \{x_1, x_2, ..., x_N\}$ of N data instances. Such a set, according to Pawlak's conception, is typically characterized as a training set in supervised machine learning (see [10]), where each instance $x_i \in X$ ($1 \leq i \leq N$) is described by values associated with a fixed set of M attributes $\{a_1, a_2, ..., a_M\}$ and, also, by a class, which is one out of K available classes $\{c_1, c_2, ..., c_K\}$.

The formalism and the procedure for representing training sets as FGs have, as their main goal, to ease and promote the analysis of flow distribution of attribute values that describe the training data instances, taking into account their classes. The FG data structure also supports the extraction of decision rules, which can be used to classify new data instances that have no associated class. As stated in several works, some options for representing data (*e.g.*, those based on the Bayes rule) have a probabilistic interpretation, since their results are strongly linked to probability. The results of

© Springer Nature Switzerland AG 2020
A. M. Madureira et al. (Eds.): HIS 2018, AISC 923, pp. 50–60, 2020.
https://doi.org/10.1007/978-3-030-14347-3_6

analyzes performed taking into account an induced FG structure, in turn, have a deterministic interpretation, since they effectively reflect the data flow and not just a probability related to them. Like any inductive approach to data/knowledge representation, the training set has a deep impact on the suitability of FGs, when used as a framework for supporting the extraction of decision rules.

Based on the available literature, it is clear that the formalism supporting the induction of FGs and, subsequently, the extraction of decision rules from them, is not as popular as several other propositional supervised formalisms, such as Decision Trees [13, 14], Neural Networks [2, 10], etc. Perhaps one of the main reasons why FGs are not popular is due to the fact that several works found in the literature, including the theoretical ones that formally introduce the FG approach, restrain themselves to the use of discrete-valued training sets, which confines the use of FGs mostly to artificially generated data. The research work reported in this article proposes an extension of FGs, to represent and handle continuous data in a more efficient way, allowing FG structures to be used as a framework for learning classifiers in continuous domains. This is done by using a hybrid approach, the HFG, which pre-processes the training data attribute values using a discretization algorithm [6, 8].

The reminder of the paper is organized as follows. Section 2 introduces the main concepts related to FGs as well as a description of how to construct such graph (actually, a digraph) to summarize a given training data. Section 3 presents a small example of the process involved in creating a FG. Section 4 briefly presents the discretization algorithm used as a pre-processing step to the HFG construction. In Sect. 5 a set of experiments is reported, using the sets of rules extracted from conventional FGs, HFGs, as well as the decision trees induced by the J48 [7], as classifiers; the section ends by presenting a few conclusions based on the work done and pointing out a possible line of research for refining HFGs, considering their results versus the results obtained using the J48. In the text, depending on the context, FG refers to the algorithm or the structure that represents a conventional flow graph and HFG refers to the algorithm or the structure that represents the hybrid flow graph.

2   Defining and Constructing FGs

The definition of flow graph (FG) involves the definition of directed graphs (or digraphs); it is important, therefore, to present both concepts, graphs and digraphs, before the formal definition of FGs. As defined in [4], a graph $G = (V, E)$ consists of two finite sets in which (1) $V \neq \emptyset$ is the set of vertices (or nodes) of the graph and (2) E is the set of edges of the graph. Each edge $a \in E$ represents an unordered pair of nodes of V, noted by (u, v), where $u, v \in V$. The unordered pair representing the edge a, $a = (u, v)$, indicates that, through a, the node v can be reached from node u and vice-versa. It is important to note that if the set $E = \emptyset$, the graph is called the *null graph*. If, instead of unordered pairs (u, v), edges are defined by ordered pairs of nodes, noted by <u, v>, they are called *directed edges* (or *arcs*) and the structure is a particular graph named *digraph*. The arc $a = <u, v>$ indicates that a has its origin at node u and destination at node v, but not vice versa.

When modeling data using graphs, it is often necessary to assign values to both, nodes and edges (arcs). Particularly, when values are assigned to edges (arcs), such graphs (digraphs) are also referenced as weighted graphs (digraphs).

As defined in [11, 12], a *flow graph* G is denoted by $G = (N, \beta, \varphi)$, where: (1) $N \neq \varnothing$ is a finite set of nodes, (2) $\beta \subseteq N \times N$ is a finite set of arcs that connect nodes $x \in N$ and (3) $\varphi : \beta \to R^+$ is a function that associates with every arc of G a positive real number. In a flow graph, an arc is represented by the ordered pair <x, y>, where x is the *source node* and y is the *target node* of the arc and x, y $\in$ N. In the terminology associated with FGs, it is also said that given an arc <x, y>, the node x is an *input* of node y and that node y is an *output* of node x.

Given a set of data instances X (training set), $|X| = N$, each instance described by M attributes, A_1, A_2, ..., A_M, and an associated class (from a set with K classes), the construction of a FG that represents X can be carried out by the following steps:

(1) for each attribute A_i ($1 \leq i \leq M$) that describes instances, a set of distinct nodes are created, each representing a possible value of A_i in X. A training set described by M attributes and an associated class will be represented by a FG containing M + 1 layers, where the M first layers, L_1, L_2, ..., L_M, are composed by nodes representing the possible values of attributes A_1, A_2, ..., A_M in X, respectively. The class attribute is represented by layer L_{M+1}, which has as many nodes as there are class values (*i.e.*, K). In short, each one of the M layers has as many nodes as there are different values of the attribute associated with the layer. Layers 1 and M + 1 are known as *input* layer and *output* layer, respectively.

(2) consider that each attribute A_i ($1 \leq i \leq M$) has as possible values the elements in set $\{A_{i_1}, A_{i_2}, \ldots, A_{i_{ji}}\}$. Therefore, the layer L_i that represents attribute A_i ($1 \leq i \leq M$), will have $\left|\{A_{i_1}, A_{i_2}, \ldots, A_{i_{ji}}\}\right|$ nodes.

(3) nodes in a given layer L_i ($1 \leq i \leq M$) can relate to one or more nodes in layer L_{i+1}, depending on the values that attributes A_i and A_{i+1}, that define both layers, respectively, have in X. The relationship between nodes $x \in L_i$ and $y \in L_{i+1}$ is represented in the FG under construction by the arc <x, y>.

(4) for each arc <x, y> constructed in (3), the number of occurrences of the relationship between the two values of attributes (*i.e.*, x and y) that define the arc, in the training set, is counted. Such number is denoted by $\varphi(<x,y>)$ and is associated with the arc as a weighting label.

(5) two numbers, *inflow* ($\sigma_+(x)$) and *outflow* ($\sigma_-(x)$), are associated with each node x constructed in (2). The inflow of x represents the sum of the weights of all arcs that reach x and the outflow, the sum of the weights of all arcs that leave x. The inflow of nodes in layer L_1 is always zero and the outflow of nodes in layer L_{M+1} is always 0. All nodes x belonging to internal layers are such that $\sigma_+(x) = \sigma_-(x) = \sigma(x)$, where $\sigma(x)$ is named the *throughflow* of x. These number are used as a label associated with each node.

Once the set of nodes and set of arcs have been constructed and the associated labels and weights have been assigned to nodes and arcs, the construction of the basic version of the FG is finished. Finally, the flow of the FG, noted by $\varphi(FG)$, is defined as the number of instances (N) summarized by the FG. A normalized version of FG can be obtained by dividing each label/weight by the value given by $\varphi(FG)$.

3 Exemplifying the Induction of a FG

The example presented and discussed in this section considers a training set having 14 data instances, $X = \{x_1, ..., x_{14}\}$, each of them described by 4 attributes, A_1, A_2, A_3, and A_4, where the value of A_4 is the associated class of the instance. Table 1 describes the 14 instances in X and Fig. 1 shows the flow graph FG obtained from the training set shown in Table 1.

Table 1. Training set with 14 instances described by 3 attributes and a class $(A_4) \in \{1, 2, 3\}$.

	x_1	x_2	x_3	x_4	x_5	x_6	x_7	x_8	x_9	x_{10}	x_{11}	x_{12}	x_{13}	x_{14}
A_1	0	0	1	0	0	0	1	0	0	1	0	0	0	1
A_2	1	1	0	0	1	1	0	0	1	0	0	1	1	0
A_3	C	S	V	D	D	C	F	C	S	D	S	V	S	S
A_4	3	2	1	2	2	2	1	2	2	1	2	2	3	1

Considering that each attribute of X defines a layer in the FG, the FG to be constructed should have four layers. Based on the description of X shown in Table 1 and considering that the values assigned to attribute A_1 are $\{0, 1\}$, to A_2 are $\{0, 1\}$, to A_3 are $\{C, S, V, D, F\}$ and to A_4 are $\{1, 2, 3\}$, the four layers to be constructed will have, respectively, 2, 2, 5 and 3 nodes.

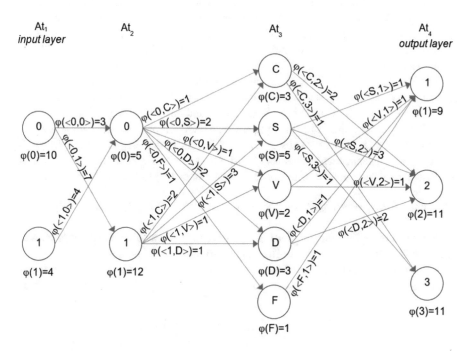

Fig. 1. The FG structure that represents the set of data instances in Table 1.

It can be observed in Table 1 that there are three instances having $A_1 = 0$ and $A_2 = 0$. The relation between $A_1 = 0$ and $A_2 = 0$ defines an arc <0, 0> in the FG, having an associated weight of 3 ($\varphi(<0, 0>) = 3$). Similarly, there are four instances in X where $A_1 = 1$ and $A_2 = 0$, so an arc <1, 0> is established between the two nodes, having an associated weight of 4 ($\varphi(<1, 0>) = 4$). The procedure is repeated for all pairs of nodes belonging to adjacent layers (the first node of the pair in layer J and the second in layer J + 1), which are represented in the training set. At the end of the process, the FG is constructed and it can be observed that its structure, in a way, condenses the information contained in the training set and thus, can be used for data analysis, as well as for the extraction of decision rules, among others.

4   The Pre-processing Phase – Discretizing Continuous-Valued Attributes

In the experiments described in Sect. 5, the *Minimum Description Length Principle Cut Criterion (MDLPC)*, proposed and described in [6], has been used as the first-step of the HFG proposal, for dealing with continuous-valued attributes. The MDLPC is one of the most commonly used discretization algorithms in supervised machine learning. It is based on two concepts, the entropy and the information gain measure, to determine the best cut point for splitting a continuous interval. The algorithm is recursive and has time complexity of $O(N \log N)$, where N is the number of instances in the data set. For finding a suitable partition of the range of values associated to a particular continuous attribute, the information gain measure is used to evaluate splitting point candidates in that range. The splitting point candidate with larger information gain is evaluated according to the MDLP criterion. If the splitting point candidate passes the MDLP criterion it becomes a cut point and the discretization method is recursively executed for both subsets of values induced by the new cut point. Otherwise, the algorithm ends.

Let $X = \{x_1, \ldots, x_N\}$ be a given data set with N instances, each described by M attributes, $A_1, A_2, \ldots, A_M$ and an associate class, from a set of K possible classes, $\{C_1, C_2, \ldots, C_K\}$. Consider A one of the M attributes and consider its range of values in X. Let CP be a cut point that splits the range of A, which will provoke a splitting in X. The gain in splitting X into two subsets, X_1 and X_2, such that $|X_1| = N_1$ and $|X_2| = N_2$ where X_1 has instances whose values of attribute A are less than or equal to CP and X_2 has instances whose values of attribute A are greater than CP is given by Eq. (1).

$$\text{Gain}(A, CP, X) = \text{entropy}(X) - (N_1/N) \times \text{entropy}(X_1) - (N_2/N) \times \text{entropy}(X_2) \quad (1)$$

For calculating the entropy of set X, it is necessary to calculate the prior probability of each one of the K classes in X. The probability of class C_i, noted by $p(C_i)$, $1 \leq i \leq k$, is given by dividing the number of instances of class C_i, by the total number of instances in X *i.e.*, N. Equation (2) shows how the entropy of a set X is calculated.

$$\text{entropy}(X) = \sum_{i=1}^{K} p(C_i) \log_2 p(C_i) \quad (2)$$

Taking into account the attribute A, the MDLPC criterion establishes that a cut point candidate CP is a cut point if and only if Eq. (3) holds.

$$\text{Gain}(A, CP, X) > \frac{\log(N - 1)}{N} + \frac{\Delta(A, CP, X)}{N} \tag{3}$$

Expression $\Delta(A, CP, X)$ is defined by Eq. (4) where Y_1 and Y_2 represent the number of classes in X_1 and X_2, respectively.

$$\Delta(A, CP, X) = \log(3^K - 2)$$
$$- [K \times \text{entropy}(X) - Y_1 \times \text{entropy}(X_1) - Y_2 \times \text{entropy}(X_2)] \tag{4}$$

The MDLPC criterion only splits an interval into two subintervals at a time. To split an interval into more than two subintervals, the procedure should be recursively called, until a cut point candidate no longer satisfies Eq. (3).

5   Experiments, Results and Discussion

To compare performances of both algorithms, the FG and the HFG, five synthetic and two real data sets were used in the experiments; their main characteristics are listed in Table 2. They are well-known data sets used in machine learning for evaluating algorithms. The five synthetic data sets were: Ruspini [15], Mouse-Like, Spherical_6_2 [1], 3MC [16] and Long Square [9]; the real data set Iris was downloaded from the UCI Repository [5] and the Fossil data set was obtained from [3]. The five synthetic bi-dimensional data sets are typically used in experiments related to clustering and their instances have no associated class. For the experiments conducted in this article, however, the instances of each of the five synthetic data sets have been assigned a class, corresponding to the group they belong to in a clustering identified by a visual analysis of the corresponding plotting of each data set.

Table 2. Data sets characteristics where DS: data set identification, #NI: no. of data instances, #NA: no. of attributes, #NG: no. of groups and G_Id = #NI: no. of instances per group where G_id represents the group identification.

DS	#NI	#NA	#NG	G_id = #NI
Ruspini (Ru)	75	2	4	1 = 20, 2 = 23, 3 = 17, 4 = 15
Mouse-Like (ML)	1,000	2	3	1 = 200, 2 = 200, 3 = 600
Spherical_6_2 (Sp)	300	2	6	1 = 50, 2 = 50, 3 = 50, 4 = 50, 5 = 50, 6 = 50
3MC (MC)	400	2	3	1 = 120, 2 = 170, 3 = 110
Long Square (LS)	900	2	6	1 = 147, 2 = 155, 3 = 150, 4 = 148, 5 = 150, 6 = 150
Iris (Ir)	150	4	3	1 = 50, 2 = 50, 3 = 50
Fossil (Fo)	87	6	3	1 = 40, 2 = 34, 3 = 13

For each data set $X \in$ {Ruspini, Mouse-Like, Spherical_6_2, 3MC, Long Square, Iris, Fossil}, in Table 2, the methodology for comparing the performances of FG and HFG implemented a 5-fold cross-validation process, by sequentially going through the following steps.

(1) X was partitioned into five parts *i.e.*, X_1, X_2, X_3, X_4, X_5.
(2) Using the FG algorithm and using the subsets of X, {X_1, X_2, X_3, X_4, X_5}, a 5-fold cross-validation process was conducted.

    (2.1) In each one of the 5 steps of the 5-fold cross-validation, four subsets were used for inducing the FG and, then, the set of decision rules embedded in the FG was extracted and evaluated using the fifty subset that was not used for training. The structure of the induced FG was then 'measured' in relation to its number of: nodes, arcs, existing paths from the initial nodes to the ending nodes and of extracted decision rules.

    (2.2) At the end of the 5-fold cross-validation, the average and corresponding standard-deviation of classification rates of the set of rules extracted from the corresponding induced FGs, taking into account the values obtained in each one of the five steps of the process, as well as the average and standard deviations associated with the collected metrics used to describe the induced FGs, are calculated.

(3) For each attribute and for each set in {X_1, X_2, X_3, X_4, X_5}, attributes were discretized, producing the discretized set versions {$DX_1, DX_2, DX_3, DX_4, DX_5$} and a 5-fold cross-validation process, as above, was conducted and final results obtained in a similar way as in (2).

As stated in the methodology, after the induction of a flow graph (FG or HFG), a process for extracting a classifier, represented as a set of decision rules, from the induced flow graph was conducted; each complete path within the FG structure gives rise to a rule. There are two numbers associated with the number of extracted rules namely, the total number of extracted rules (#TDR in Table 4) and the number of consistent rules (*i.e.*, rules have only one associated class), represented by #DR in Table 4. The presentation of the experiment results that follows has been split into two tables: Table 3 presents the results related to the structure of the induced flow graphs, taking into account a few measurable characteristics, and Table 4 presents the performance of the classifiers extracted from the induced flow graphs.

Table 3 shows the values of four characteristics used for evaluating the flow graph structures induced by processes that implement the conventional Flow Graph (FG) and the Hybrid Flow Graph (HFG), for each data set in Table 2. The characteristics were: its size (given by the number of nodes plus the number of arcs) and the number of complete paths (*i.e.*, paths starting in the first layer (L_1) and ending in the M + 1 layer (L_{M+1})), accordingly to their formal descriptions given before. The values in the table are the average and the corresponding SD of measurements related to the four characteristics, based on a 5-fold cross-validation process, as described at the end of Sect. 4. The original sequence of attributes (*i.e.*, attributes order) that describes the instances in each original data set was maintained for the experiments.

As shows Table 3, the structures induced by the conventional FG algorithm are far more bulky as far as the number of nodes in each of their M first layers is concerned, than those induced by the HFG algorithm. The number of arcs follows the same tendency, considering that such number is related to the number of nodes in the graph. The number of nodes in HFGs were, at least, 7.3 times lower than the number of nodes in FGs (as in data set Fo). The highest reduction in the number of nodes provided by HFGs was by 145.7 times, in the Mouse-Like data set. The highest reduction in number of arcs reached by HFGs was in the Long Square data set; on average, the number of arcs in HFGs in such data was 95.93 times lower than those in FGs. The lowest reduction occurred in the Fo data set; the number of arcs in HFGs was, on average, 6.66 times lower than those found in FGs.

Values related to the size characteristic, given by the number of nodes plus the number of arcs, reflect what was described before, in relation to both characteristics *i.e.*, number of nodes and number of arcs. The highest reduction happened in the Mouse-Like data set where, on average, HFGs had sizes 106.7 times smaller than those induced by FGs. The lowest reduction occurred in the Fo dataset, where, on average, HFGs were 6.87 times smaller than FGs.

Table 3. FG structure characteristics obtained in the 5-fold experiments using conventional flow graph (FG) and hybrid flow graph (HFG), in each one of the 7 data sets, where DS stands for the identification of the data set and Alg. for the method used.

DS	Alg.	# nodes	# arcs	Size	# complete paths
Ru	FG	99 (4.47)	109 (3.16)	208	61 (2.24)
	HFG	12 (0.00)	12 (2.24)	24	10 (3.61)
ML	FG	1,603 (0.00)	1,600 (0.00)	3,203	800 (0.00)
	HFG	11 (0.00)	19 (2.24)	30	30 (4.47)
Sp	FG	453 (6.56)	471 (4.69)	924	255 (3.74)
	HFG	14 (0.00)	12 (0.00)	26	10 (0.00)
MC	FG	609 (6.32)	613 (4.36)	1,222	323 (4.24)
	HFG	12 (0.00)	15 (0.00)	27	11 (0.00)
LS	FG	1,445 (1.41)	1,439 (1.41)	2,884	720 (0.00)
	HFG	13 (0.00)	15 (1.73)	28	15 (2.24)
Ir	FG	120 (3.16)	310 (5.57)	430	2649 (433.74)
	HFG	15 (0.00)	26 (1.73)	41	60 (8.49)
Fo	FG	190 (8.12)	360 (15.36)	550	16,114 (5669.16)
	HFG	26 (2.00)	54 (6.24)	80	399 (103.80)

The lowest reduction in the number of complete paths occurred in Ruspini data set where, on average, the number of paths in HFGs was 6.10 times smaller than the number of paths FGs. The highest reduction was in the Long Square data set where, on average, the number of complete paths extracted from HFGs was 48 times smaller than the number of paths extracted from FGs.

Table 4 presents the performance of the set of rules extracted from FGs and HFGs in each of the 7 data sets. As could be inferred based on the number of nodes and arcs between them, discussed in the previous paragraphs, the numbers of decision rules extracted from induced HFGs are much lower than those extracted from induced FGs. Although this is quite common, it cannot be considered as rule since there are situations in which the number of rules extracted from HFGs may, eventually, be greater than those extracted from their FGs counterparts.

The software that implements both algorithms for creating FGs and HFGs, when extracting the set of rules from either structure, only considers consistent rules *i.e.*, rules that have a unique class associated. However, from the set of inconclusive rules *i.e.*, rules that share the same antecedent but differ from each other in relation to the associated class, only the rule that has the highest performance is kept. For instance, in the Sp data set and considering the FG, the total number of extracted rules was 255 and, after removing the inconclusive ones, the number went down to 240, implying that among all extracted rules, 15 were inconclusive; now considering the HFG, among the 10 rules extracted, 4 were inconclusive. As can be confirmed in Table 4, in 6 out of 7 data sets, the set of rules extracted from FGs was not able to correctly classify any testing instance; only in one data set (Ir) the set of rules managed to have a performance around 8% which is highly inefficient. The FG based results clearly show that the FG proposal is totally unsuitable as a framework for extracting rules that serve as classifiers when continuous values are involved. As expected, the use of FGs in continuous data is totally inadequate, as shows the last column of Table 4. The only data set having an associated value of %acc different from zero is the Iris; this is due to the presence of repeated data instances. The behavior of a FG in continuous data is similar to the behavior of the NN algorithm, except that the new instance to be classified needs to exactly match a stored instance, part of the set of instances representing the concept.

Table 4. Classification performance of sets of rules extracted from FGs and HFGs. DS stands for data set, Alg. for the employed algorithm (FG or HFG), #TDR for the average number of total decision rules, #DR for the average number of consistent decision rules, #cc/#total for the average of number of correct classifications/average of number of instances (standard deviation) and %acc for the average percentage of accuracy values.

DS	Alg.	#TDR	#DR	#cc/#total (sd)	%acc
Ru	FG	61	60	0/15 (0.00)	0.00%
	HFG	10	7	12/15 (5.66)	80.00%
ML	FG	800	800	0/200 (0.00)	0.00%
	HFG	30	12	105/200 (116.11)	52.90%
Sp	FG	255	240	0/60 (0.00)	0.00%
	HFG	10	6	41/60 (6.99)	69.67%
MC	FG	323	320	0/80 (0.00)	0.00%
	HFG	11	9	69/80 (5.18)	87.25%
LS	FG	720	720	0/0 (0.00)	0.00%
	HFG	15	9	118/180 (8.37)	65.56%
Ir	FG	2,649	2034	2/30 (2.68)	8.00%
	HFG	60	35	21/30 (14.93)	70.67%
Fo	FG	16,114	16,114	0/17 (0.00)	0.00%
	HFG	399	399	14/17 (4.77)	87.06%

Although the performance of the HFG was quite acceptable for most data sets, in the ML data set the results were far from being satisfactory; in this case a further investigation about the reasons for that outcome need to be conducted. Aiming at a further investigation of the performance of the HFG, its results in the seven data sets were compared to those obtained by the J48 algorithm, which is considered the Weka 3.8 version of the well-known C4.5 algorithm [9, 14], are shown in Table 5.

Table 5. Average of accuracy rate results obtained by HFG and J48 algorithms for data sets presented in Table 2.

ID	HFG accuracy	J48 accuracy
Ruspini	80.00%	98.67%
Mouse-Like	52.90%	97.90%
Spherical_6_2	69.67%	100.00%
3MC	87.25%	100.00%
Long Square	65.56%	99.89%
Iris	70.67%	94.00%
Fossil	87.06%	100.00%

Although the J48 has comparatively achieved better results in all the seven data sets, the HFG has obtained promising results in four out of the seven data sets. The remaining three results, comparatively to those obtained by the J48, are very poor and, definitely, the HFG approach needs improvements. It is a fact, though, that the HFG construction can be highly influenced by the order in which the attributes (that define the layers of the digraph) are considered. As far as the continuation of the work described in this article is concerned, we intend next to investigate the influence of the order used for creating the layers of the FG/HFG. A few experiments have already shown that the order in which layers are created has an impact on the final structure of the graph and, consequently, on the set of rules extracted from it. The software implementing FGs and HFGs has been developed in Java, on a Windows platform.

Acknowledgments. The authors thank UNIFACCAMP and CNPq for their support and the anonymous reviewers for their suggestions. This study was financed in part by the Coordenação de Aperfeiçoamento de Pessoal de Nível Superior – Brasil (CAPES) – Finance Code 001.

References

1. Bandyopadhyay, S., Maulik, U.: Genetic clustering for automatic evolution of clusters and application to image classification. Pattern Recogn. **35**, 1197–1208 (2002)
2. Bishop, C.M.: Neural Networks for Pattern Recognition. Oxford University Press, Oxford (2005)
3. Chernoff, H.: Metric considerations in cluster analysis. In: Proceedings of the Sixth Berkley Symposium on Mathematical Statistics and Probability, pp. 621–629. UCLA Press, Berkley (1972)

4. Clark, J., Holton, D.A.: A First Look at Graph Theory, 2nd edn. World Scientific, Singapore (1998)
5. Dua, D., Karra Taniskidou, E.: UCI Machine Learning Repository. University of California, School of Information and Computer Science, CA, Irvine (2017). https://archive.ics.uci.edu/ml
6. Fayyad, U.M., Irani, K.B.: Multi-interval discretization of continuous-valued attributes for classification learning. In: Proceedings of the International Joint Conference on Artificial Intelligence, pp. 1022–1029 (1993)
7. Frank, E., Hall, M.A., Witten, I.A.: The WEKA Workbench. Online Appendix for "Data Mining: Practical Machine Learning Tools and Techniques". Morgan Kaufmann, Burlington (2016)
8. García, S., Luengo, J., Sáez, J.A., López, V., Herrera, F.: A survey of discretization techniques: taxonomy and empirical analysis in supervised learning. IEEE Trans. Knowl. Data Eng. 25(4), 734–750 (2013)
9. Handl, J., Knowles, J.: Multiobjective clustering with automatic determination of the number of clusters. Tech. report. UMIST, Manchester, TR-COMPSYSBIO-2004-02 (2004)
10. Mitchell, T.M.: Machine Learning. McGraw-Hill, New York (1997)
11. Pawlak, Z.: Flow graphs and decision algorithms. In: Wang, G., et al. (eds.) LNAI, pp. 1–10. Springer, Berlin (2003)
12. Pawlak, Z.: Flow graphs - a new paradigm for data mining and knowledge discovery. In: Proceedings of the KSS2004, JAIST Forum 2004 - Technology Creation Based on Knowledge Science: Theory and Practice, Jointly with The 5th International Symposium on Knowledge and Systems Science, pp. 147–153 (2004)
13. Quinlan, J.R.: Induction of decision trees. Mach. Learn. 11, 305–318 (1986)
14. Quinlan, J.R.: C4.5 Programs for Machine Learning. Morgan Kaufmann Publishers, Burlington (1993)
15. Ruspini, E.H.: Numerical methods for fuzzy clustering. Inf. Sci. 2, 319–350 (1970)
16. Su, M.C., Chou, C.H., Hsieh, C.C.: Fuzzy C-means algorithm with a point symmetry distance. Int. J. Fuzzy Syst. 7(4), 175–181 (2005)

Modelling and Predicting Individual Salaries in United Kingdom with Graph Convolutional Network

Long Chen[✉], Yeran Sun, and Piyushimita Thakuriah

UBDC, 7 Lilybank Gardens, Glasgow, UK
{long.chen,yeran.sun,piyushimita.thakuriah}@glasgow.ac.uk
https://www.ubdc.ac.uk/

Abstract. Job Posting Sites, such as Indeed and Monster, are specifically designed to help users obtain information from the market. However, at the moment, only approximately half of the UK job postings have a salary publicly displayed. Therefore, the aim of this research is to model and predict the salary of a new job, so as to improve the performance of job search and help a vast amount of job seekers better understand the market worth of their desirable positions. In order to effectively estimate the salary of a given job, we construct a graph database based on job profiles of each posting and build a predictive model through machine learning based on both metadata features and relational features. Our results reveal that these two types of features are conditionally independent and each of them is sufficient for prediction. Therefore they can be exploited as two views in graph convolutional network (GCN), a semi-supervised learning framework, to make use of a large amount of unlabelled data, in addition to the set of labelled ones, for enhanced salary classification. The preliminary experimental results show that GCN outperforms the existing ones that simply pool these two types of features together.

1 Introduction

With the advent of Web 2.0, job posting sites such as Indeed[1] and Monster,[2] have become one of the central paradigms for users to search their ideal jobs. However, since the information provided by the employers is often biased or incomplete, there is often a expectancy gap between the job seekers and the employers.

To this end, this paper introduces a new deep learning method for predicting a new job's salary, with empirical results based on real world datasets[3]. We first learn graph representations of each job by modeling both metadata features and relational features, where the former encode job types and skills and the

---

[1] www.indeed.co.uk/.
[2] www.monster.co.uk/.
[3] https://www.kaggle.com/c/job-salary-prediction/data.

© Springer Nature Switzerland AG 2020
A. M. Madureira et al. (Eds.): HIS 2018, AISC 923, pp. 61–74, 2020.
https://doi.org/10.1007/978-3-030-14347-3_7

later encode the relations with other jobs. We build a predictive model through machine learning based on both of these two types of features. Our investigation reveals that these two types of features are conditionally independent, and each of them is sufficient to make prediction, therefore they can be exploited as two views in graph convolutional network (GCN) – a semi-supervised learning framework – to make use of large amount of unlabelled data, in addition to the small set of manually labelled jobs, for enhanced classification.

Our proposed model is a general one, so in addition to the salary prediction problem, it can be applied into neural language models to facilitate a series of downstream tasks, such as web query classification and document retrieval. To the best of our knowledge, this is the first work that formally formulates the job salary prediction problem as a machine learning task and develops new features and semi-supervised deep learning techniques to address this problem in a principled manner.

The rest of this paper is organized as follows. We firstly formally defines the problem of salary prediction with deep learning in Sect. 2. Then, we introduce the related work in Sect. 3. Section 4 analyses both metadata features and relational ones used in our experiment. Section 5 systematically presents the proposed framework for graph classification. The experimental results of job classification are reported in Sect. 6. Finally, we present our conclusion and future work in Sect. 7.

2   Preliminary

In this section, we formally introduce several concepts and notations. Then Graph Fourier Transform is shortly recapitulated.

Definition 1 (Job Database): A **Job Database** D is an ordered pair $P = (V, E)$ comprising a set V of vertices and a set E of edges. To avoid ambiguity, we also use $V(P)$ and $E(P)$ to denote the vertex and edge set. Notice that there are two types of databases, namely training database and test database. There is a stark difference between these two databases: the former explicitly labeled all the job salaries while the latter are hidden for evaluation purpose.

Definition 2 (Job): A **job** J consists of V' and E', the former is a set of skills and the latter is a set of edges representing the relations between the jobs. For instance, an edge $< u, v >$ is a binary directed relation from node u to node v, where we use $w(u, v)$ to denote the weight of $< u, v >$. Two jobs are connected in D if they share a common attribute, such as skill, location or company name. The weight is indicated by the sum of the shared attributes.

Salary Categories: A **job** will fall into our predefined six salary **categories**, namely, 0–20,000, 20,000–30,000, 30,000–40,000, 40,000–50,000, and 50,000 above. Notice that in the original Kaggle task, this is defined as a regression task. However, since salaries are not evenly distributed[4], we argue that it

---

[4] https://www.kaggle.com/c/job-salary-prediction/discussion/4208.

is better to define it as a classification. Although there are some quantization errors, classification can generate local properties to the model, which eventually leads to more robust prediction. Now we can formulate our task of salary prediction as follows. Given a large connected graph from database D of size n, and a collection of new jobs from Q of size n_Q with $n > n_Q$, we aim to find the appropriate **categories** for each of the **job** in Q.

Our work is on the basis of the semi-supervised framework GCN [5], which introduced spectral networks for filtering the convolutional layer. A spectral filter generalizes a convolutional network through the Graph Fourier Transform, which is a generalization of the graph Laplacian function.

Definition 3 (Graph Fourier Transform): Let W be a $N \times N$ similarity matrix representing an undirected graph G, and $L = D^{-1/2}/WD^{-1/2}$ is the graph Laplacian with $D = W$ eigenvectors $U = (u_1, ..., u_N)$. Then a graph convolution of input jobs x with filters g on G is defined by $x*Gg = U^T(U_x \odot U_g)$, where $\odot$ indicates a point-wise product.

3   Related Work

As we transformed each job as a graph representation (c.f., Sect. 2), the performance of job classification is heavily related to the graph feature learned from the datasets. The most famous unlabeled strategy is arguably Fast Frequent Subgraph Mining (FFSM) [7], which is the first algorithm that employs depth-first search strategy to identify all connected graphs that appear in a large proportion of graph collection. Another important one is gSpan [24], which first creates a lexicographic graph set, and then converts each graph to a unique minimum DFS code as its canonical label. Based on this order, gSpan also uses the depth-first search algorithm to identify frequent connected subraphs. There have been many other approaches for frequent subgraph feature extraction as well, e.g. AGM [8], FSG [11], MoFa [3], and Gaston [16]. Moreover, the supervised subgraph feature extraction problem has also been well studied, such as LEAP [23] and CORK [22], which search informative subgraph patterns for graph classifications. In addition to subgraph frequencies, some other important filtering criteria include coherence, closeness, maximum size, and density.

Another popular way of feature extraction is via the dimensionality reduction of vector spaces. The state-of-the arts machine learning algorithms, such as SVM or Decision Trees, can not resolve this problem very well since they fail to consider the relational features. On the other hand, the current relational classifiers, e.g., ICA [19], can fit well into a local classifier, but the graph embedding is indirectly learned in a incremental fashion, which results in a suboptimal prediction. A good way is to adopt pairwise constraints as weak supervision to reduce vector dimensionality, for example, the must-link [2] constraints (pairs of instances comes from the same class) and cannot-link constraints [21] (pairs of instances comes from different classes). Feature selection methods in vector spaces have also been developed [16], which select informative features within the candidate feature collection. One of the drawback of these methods is that they heavily relied on the assumption of a set of candidate features before the feature selection operation. Therefore, conventional semi-supervised feature selection approaches cannot be directly applied to graph data, since it is usually infeasible to generate all the subgraph features of a graph dataset before the feature selection. In addition, the preprocessing phase of the conventional approach simply squashed the graph information to a simpler representation, e.g., a vector of reals and then applied the list-based data techniques. As a result, some important information, e.g., the topology and density of information may be lost during the preprocessing stage, which leads to a suboptimal result.

Fortunately, the recent advances in deep learning with semi-supervised setting has provided new opportunities to resolve this issue. In the seminal work of GNN [18], namely Graph Neural Network Model, the authors directly map graphs into vectors by employing recurrent neural network, from which the last layer is combined with the normalized neighborhood graphs, which is then used as the output representation. GGNN [12] later on is proposed, which modifies GNN to use gated recurrent units and Adam Gradient [9] optimization techniques and then extend to output sequences. More recently, PATCHY-SAN [15], is proposed by employing a convolutional neural network, which utilizes normalized neighborhood information as the receptive field for each candidate node. On another line, instead of directly resolving this problem, there are methods aim to learn a good graph representation first, which in turn is used as graph-level features feed to a softmax function for classification. For example, Planoid [25] is proposed by integrating graph matrix into a semi-supervised graph embedding framework. Deepwalk [17] employs the embedding of a node to predict its context in the graph, where the context is defined by a probabilistic model, random walk. Line [20] was later on proposed to consider high-order relationship of nodes and apply it to a large graph database for training.

4   Features Analysis

4.1   Metadata Features

We are interested in learning informative features from the datasets so that the classification accuracy can be maximized. In particular, it is well known that new attribute values derived from the graph structure of the network in the data, such as the betweenness centrality, are beneficial to the learning algorithms. We extracted 7 metadata features overall for the classification task. The metadata features used in our model includes Location, Sectors, Contract Type, Contract Time, Number of Distinct Skills, and Job Description (Fig. 1).

Locations. The average salary in England is £31,554.441. The cities with the highest averaged salaries are "Huntingdon" and "Bedfordshire", and "Southwest London", which are all well above £42,000. However, the lowest salary is in "Whitby", "Bramhope", and "Chorley", which are just above £15,000. The

Fig. 1. The jobs with the highest and lowest salary in UK.

highest salary area is Southern England and the lowest salary area is Northern England. This is contradictory to the long-term conception that London has the highest income in UK. We conjecture that this is because very low and very high salary jobs are not posted on the sites. In addition, the salary of many jobs are usually a hidden status and is negotiated with employer after the interview. It is also interesting to find that the average salary in "Channel Islands" is about £71,000, which is £40,000 above average; however, the rest of the cities are all £10,000 above or below the average.

Sectors. Table 1 shows the job salary distribution over varying sectors. As expected, one can easily observe that the highest salary sector is "Energy, Oil & Gas Jobs", and the lowest one is "Part Time Jobs". Another major finding was that the maximum standard deviation among the salaries based on Job titles was found in 'Reconciliations Accountant' with the deviation of £85,920. The maximum salary for this job is £199,860 and minimum is £27,840.

Table 1. The salary distribution over varying sectors

Sector	Salary (£)	Sector	Salary (£)
Accounting & finance jobs	38622.456172	**IT jobs**	44081.780990
Admin jobs	20916.362130	Legal jobs	42350.550210
Charity & voluntary jobs	28200.204936	Maintenance jobs	17533.320433
Consultancy jobs	36374.208544	Manufacturing jobs	25653.276129
Creative & design jobs	32585.487409	Other/general jobs	34361.716029
Customer services jobs	19795.438648	**Part time jobs**	10514.285714
Energy, oil & gas jobs	45384.110103	Travel jobs	23913.173143
Engineering jobs	35608.058499	Property jobs	32167.056647
Graduate jobs	28677.438703	Maintenance jobs	17533.320433
HR & recruitment jobs	32386.298070	Social work jobs	32324.809030
Healthcare & nursing jobs	32203.188169	Teaching jobs	27240.637182

Contract vs. Permanent. Another important indicator is whether people work on contract or are permanent. The initial instinct is that the people who work on contract would be paid more, since they do not have any job security and are supposed to get paid higher than permanent workers. Surprisingly, there is not much difference in the income of the people on contracts and permanent employee in UK. People on contract earn about £36,099.67 whereas permanent employee get around £35,327.47, the difference is almost negligible.

Job Description. The text features of a job are extracted from the text field of the job description after standard pre-processing steps (tokenization, lower-casing, stopwordremoval, and stemming) [13]. Finally each job is represented as a vector of terms weighted by TF×IDF [13]. To have a rough idea about each category of jobs, we sort all word/phrase features in terms of information gain for job classification, and show the most discriminative ones in Fig. 2. It is clear that Python, Scientist, and Machine learning are the best predictors of high salary. This may be due to the fact that these keywords are better correlated with data science roles. It is also clear that positions such as Manager or Director are associated with higher salaries. Higher degrees are associated with higher salaries. Interestingly, work experience of 5 years is even more valuable than that of 7 years, which we find to be quite difficult to explain. Perhaps after a certain threshold, the longevity of work experience doesn't matter anymore.

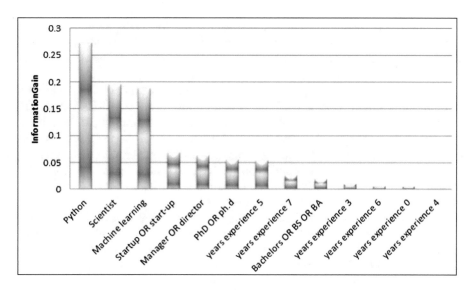

Fig. 2. The highest and lowest salary distribution in UK.

4.2   Relationship Features

Measuring job similarity and diversity plays a key role in various steps of the graph classification. The core idea is that jobs with similar graph structure may share some common properties as well, such as degree distribution, diameter, eigenvalues [?]. The relationship features are extracted and filtered through a pipeline of operations.

1. Given the graph representation of each job(see Sect. 2), we adopt the method in [6] to convert graph into matrix representation, which is extensively used in applications such as virtual screening and chemical database clustering. Notice that we take the average score of the entire matrix to represent the graphical similarity between two jobs.
2. The relational features are then extracted from graph matrix. Graph features we considered include: Number of Nodes, Number of Edges, AvgDegreePerN-ode, Variance, which are indirectly integrated into the model as a adjacency matrix (see Sect. 5).
3. Feature selection is conducted by using[5], which uses frequent subgraphs as relational features.
4. We then feed the relational features into DEEPWALK [17] as an input, which produces a latent representation as an node embedding.

5   Graph Convolutional Classification

It is time-consuming and expensive to manually estimate the job salary via crowd-sourcing. On the other hand, a small set of labelled jobs would seri-

---

[5] https://github.com/matt-gardner/pra.

ously limit the success of supervised learning for job classification. However, usually, obtaining unlabelled jobs are quite easy. So it is promising to apply semi-supervised learning with the aim to make use of a large amount of unlabelled data in addition to the small set of labelled data.

There are many semi-supervised learning techniques available. For this problem of job classification according to salary level, we think the *GCN* [10] approach is particularly suitable. Basically, *GCN* is a semi-supervised learning framework that requires two types of data: each job is described by both metadata features and relational features that provide different, complementary information about it. Our method is on the basis of spectral graph convolutional neural networks, which is introduced in Bruna et al. [4] and later extended by Defferrard et al. [5] with fast localized convolutions.

Let L and U be the number of labeled and unlabeled jobs. Let $x_{1:L}$ and $x_{L+1:L+U}$ denote the feature vectors of labeled and unlabeled jobs respectively. In addition to labeled and unlabeled jobs, graph relationship features, denoted as a $(L+U) \cup (L+U)$ matrix A, is also feed to the graph-based semi-supervised model. In this work, we are particularly interested in the scenario that a graph is explicitly given with additional information not present in the feature vectors. The loss function of *GCN* learning in the binary case can be given as:

$$\sum_{i=1}^{L} \ell(y_i, f(x_i)) + \sum_{i,j} a_{ij} \|f(x_i) - f(x_j)\| \tag{1}$$

$$= \sum_{i=1}^{L} \ell(y_i, f(x_i)) + \lambda f^T \Delta f \tag{2}$$

where ℓ is the squared loss and the second term is the graph Laplacian operator, which will cause a large penalty when similar nodes with a large wij are predicted to have different labels $f(x_i)! = f(x_j)$. The graph Laplacian matrix Δ is defined as $\Delta = A - D$, where D is a diagonal matrix that defined as $dii = \sum_j a_{ij}$. λ is a constant weighting factor.

Another important function is the spectral filter. Let's recall the definition here and its properties. A spectral network produces a convolutional network through the Graph Fourier Transform, which is relied on the Laplacian operator on the grid to the graph Laplacian. An input vector x_i is considered as a signal defined on a graph G. Let's have W as a lattice, from the definition of L (See Definition 3), we would like to recover the discrete Laplacian operator Δ. Learning filters on a graph thus amounts to learning spectral multipliers $w_g = (w_1, \ldots, w_N)$, $x * Gg = U^T(diag(w_g)U_x)$. It follows that the eigenvectors of Δ are given by the Discrete Fourier Transform (DFT) matrix [5].

The *GCN* algorithm which implements the graph convolution is described in Algorithm 1. So the algorithm can be divided into two distinct pass namely, the forward pass and the backward pass. In the forward pass, a graph batch x is learned and passed through the whole network. On our first training stage, since all of the weights or filter values were randomly assigned, the output will be an equal weighted vector that doesn't give preference to any node in particular.

The network, with its current weights, isn't able to look for those low level features, and thus isn't able to make any reasonable conclusion about what the classification should go. This goes to the loss function part (see Eq. 1) of back propagation. Backward pass through the network, which determines which weights contributed most to the loss and finding ways to adjust them so that the loss decreases. Once we compute this derivative, we then update the weight, where all the weights of the filters are integrated and updated so that they are in the opposite direction of the gradient.

Algorithm 1. The Process of Training GCN Model

---
1 1: Random initialize the parameters Θ
2 2: $t \leftarrow 0$;
3 **while** $t < MaxIteration$ **do**
4   **for** *Forward Pass* **do**
5   Input batch x and gradients w.r.t outputs ∇y.
6   Compute interpolated weights
7   Compute Output
8   **end**
9   **for** *Backward Pass* **do**
10   Compute gradient w.r.t input
11   Compute gradient w.r.t interpolated weights and update them with Equation 1
12   Compute gradient w.r.t weights with Spectral Filters in Definition 3
13   **end**
14       $t \leftarrow t + 1$
15 **end**

---

6   Experiment Setup

6.1   Datasets

We have chosen datasets which (a) have been made publicly available, so as to enable possible direct comparisons with other studies, and (b) have key characteristics covering a large part of the design space (e.g., regarding graph size and density). Table 2 summarizes the characteristics of the dataset we used, namely Adzuna for salary predication. Adzuna is a job search engine which collected job information from all major job posting sites, such as Indeed and Monster. In addition, the statistics are generated with GraphGen [2], allowing various parameters of interest to be specified. We follow a similar experimental setup in [25]. A more detailed description of how GraphGen constructs the dataset can be found in [9].

Table 2. Statistic of job dataset

Job statistic	0–10,000	10,000–20,000	20,000–30,000	30,000–40,000	40,000–50,000	50,000 above
#nodes	6,340	53,210	65,286	58,927	53,739	10,671
#edges	13,293	210,071	251,231	243,274	173,217	35,279

Approach. We train a two-layer GCN as in [25] and evaluate prediction accuracy on a test set of 1,000 labeled examples. We choose the same dataset splits as in [25] with an seperate validation set of 300 labeled examples for hyperparameter optimization. We do not use the validation set labels for training. We train all models for a maximum of 200 epochs (training iterations) using Adam (Kingma and Ba 2015) with a learning rate of 0.01 and early stopping with a window size of 10, i.e. we stop training if the validation loss does not decrease for 10 consecutive epochs. We initialize weights as random values between 0 and 1.

In our graph classification experiments, we compare the following four approaches:

1. the baseline approach graph convolutional network (GCN), which is a semi-supervised learning algorithm that is based on an efficient variant of convolutional neural networks;
2. the baseline approach Deep Walk [17], which exploit the embedding of a node to predict the context, where the context is generated by random walk in the graph;
3. the baseline approach graph convolutional networks with fast localized spectral filter (GCN-FLSF [5]).
4. the baseline approach iterative classification algorithm (ICA)[14] with a logistic regression wrapper [19].
5. the baseline approach, Predicting Labels And Neighbors with Embeddings from Data(Planetoid) [25], where we always choose their best performing model variant (transductive vs. inductive) as a baseline.

The baseline approaches include graph convolutional network (GCN) and (DeepWalk), iterative classification algorithm (ICA), and Planetoid.

We first train the local classifier using all labeled training set nodes and use it to bootstrap class labels of unlabeled nodes for relational classifier training. We run iterative classification (relational classifier) with a random node ordering for 10 fold cross validation. L2 regularization parameter and aggregation operator (count vs. prop, see [19]) are chosen based on validation set performance for each dataset separately.

6.2   Parameter Settings

For supervised learning experiments, we compared a series of off-the-shelf models, which are reported in Table 3. CNN used in our work is a simple neural network with one hidden layer with the same number of neurons as there are inputs (43 features, see Sect. 4. A rectifier activation function is used for the neurons in the hidden layer. A softmax activation function is used on the output layer to turn the outputs into probability-like values and allow one class of the 10 to be selected as the output prediction. Logarithmic loss is used as the loss function and the efficient ADAM gradient descent algorithm is used to learn the weights. The Linear SVM parameters are set to their default values except that the class weights are optimised for each job category by 5-fold cross-validation.

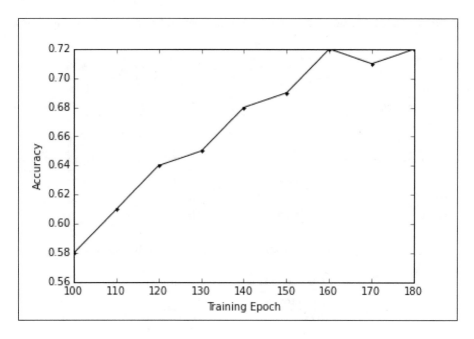

Fig. 3. The performance of GCN over iterations.

Figure 3 shows the performance of *GCN* over iterations with the optimal incremental size. One can observe that the optimal performance is achieved at the 165th iteration, after which point the performance get converged. Hence, we empirically set the number of iteration t as 200 in the following experiment.

Table 3 shows the performance accuracy of salary classification through supervised learning with different sets of features. It is clear that ensemble approaches, such as Random Forest, and Decision Boosting Machine, outperform the other types of learning models since they use multiple learning algorithms to obtain better predictive performance than the single classifiers. Furthermore, one can see that using both metadata features and relational features works better than using metadata features or relational features alone, for all learning models. Last but not least, one can also observe that deep learning techniques, such as CNN and RNN, fail to yield good results. The reason is probably that the dataset is too small for deep learning techniques to unlock its power. One possible way to remedy this is to carry out hot encoding on all the features so that the feature space can be expanded for learning. We left this for future work.

Instead of directly applying supervised deep learning, in this work, we focus on semi-supervised learning approach with deep learning to exploit the power of unlabelled data. Table 4 shows the performance of GCN versus other semi-supervised learning algorithms. As expected ICA achieved a better performance than Planetoid with additional relational features. GCN displays a very competitive performance as to ICA. More importantly, it can be seen that GCN with Spectral Filter consistently can achieve even better result, which shows the power of the spectral filtering process.

Table 3. The performance of supervised learning with different sets of features (accuracy)

Learning models	Metadata features	Relational features	All features
SVM	0.75	0.36	0.74
Decision tree	0.64	0.38	0.74
Logistic regression	0.67	0.37	0.72
Random forest	0.76	0.34	0.71
Gradient boosting machine	0.73	0.37	0.70
CNN	0.62	0.29	0.68
LSTM	0.61	0.30	0.65
DNN	0.63	0.32	0.65

Table 4. The comparison of varying semi-supervised learning models (accuracy)

Learning models	Accuracy
Planetoid	0.65
ICA	0.73
GCN	0.76
GCN + Spectral Filter	0.77

7   Conclusion

The major contribution of this paper is threefold. First, we formally defined the problem of salary prediction as a machine learning task. Second, we identified several relational features which can be used together with standard metadata features by learning algorithms to classify jobs in terms of their context. Third, we demonstrate that it is better to exploit both metadata features and relational features through the semi-supervise learning framework, *GCN*, rather than simply pooling them in supervised learning, since the former can make use of the hidden information within a large amount of unlabelled data.

There are several interesting and promising directions in which this work could be extended. First, in this work we only focused on graphical convolutional network, it will be interesting to learn the performance of some advanced graph classificaiton algorithms, such as BB-Graph [1]. Second, with regard to the feature selection, we only consider job content and job-job relationship features, in the future, we would like to explore other types of features as well, such as, one's social network and searching history. Lastly, GCN in the current form relies on one of the simplest convolutional neural network, which makes sense as a first step towards integrating job's context into learning model, but of course we could consider using more sophisticated neural network like RCNN.

References

1. Asiler, M., Yazıcı, A.: BB-graph: a new subgraph isomorphism algorithm for efficiently querying big graph databases. arXiv preprint arXiv:1706.06654 (2017)
2. Bar-Hillel, A., Hertz, T., Shental, N., Weinshall, D.: Learning a mahalanobis metric from equivalence constraints. J. Mach. Learn. Res. **6**(Jun), 937–965 (2005)
3. Borgelt, C., Berthold,M.: Finding relevant substructures of molecules: mining molecular fragments. In: Proceedings of the 2002 IEEE International Conference on Data Mining (ICDM 2002), pp. 51–58 (2002)
4. Bruna, J., Zaremba, W., Szlam, A., LeCun,Y.: Spectral networks and locally connected networks on graphs. arXiv preprint arXiv:1312.6203 (2013)
5. Defferrard, M., Bresson, X., Vandergheynst, P.: Convolutional neural networks on graphs with fast localized spectral filtering. In: Advances in Neural Information Processing Systems, pp. 3844–3852 (2016)
6. Foggia, P., Sansone, C.,, Vento, M.: A database of graphs for isomorphism and sub-graph isomorphism benchmarking. In: Proceedings of the 3rd IAPR TC-15 International Workshop on Graph-based Representations, pp. 176–187 (2001)
7. Huan, J., Wang, W., Prins, J.: Efficient mining of frequent subgraphs in the presence of isomorphism. In: 2003 Third IEEE International Conference on Data Mining, ICDM 2003, pp. 549–552. IEEE (2003)
8. Inokuchi, A., Washio, T., Motoda, H.: An apriori-based algorithm for mining frequent substructures from graph data. In: Principles of Data Mining and Knowledge Discovery, pp. 13–23 (2000)
9. Kingma, D., Ba, J.: Adam: a method for stochastic optimization. arXiv preprint arXiv:1412.6980 (2014)
10. Kipf, T.N., Welling,M.: Semi-supervised classification with graph convolutional networks. arXiv preprint arXiv:1609.02907 (2016)
11. Kuramochi, M., Karypis, G.: Frequent subgraph discovery. In: 2001 Proceedings IEEE International Conference on Data Mining, ICDM 2001, pp. 313–320. IEEE (2001)
12. Li, Y., Tarlow, D., Brockschmidt, M., Zemel,R.: Gated graph sequence neural networks. arXiv preprint arXiv:1511.05493 (2015)
13. Manning, C.D., Raghavan, P., Schütze, H., et al.: Introduction to Information Retrieval, vol. 1. Cambridge University Press, Cambridge (2008)
14. Neville, J., Jensen, D.: Iterative classification in relational data. In: Proceedings of AAAI-2000 Workshop on Learning Statistical Models from Relational Data, pp. 13–20 (2000)
15. Niepert, M., Ahmed, M., Kutzkov, K.: Learning convolutional neural networks for graphs. In: International Conference on Machine Learning, pp. 2014–2023 (2016)
16. Nijssen, S., Kok, J.N.: A quickstart in frequent structure mining can make a difference. In: Proceedings of the Tenth ACM SIGKDD International Conference on Knowledge Discovery and Data Mining, pp. 647–652. ACM (2004)
17. Perozzi, B., Al-Rfou, R., Skiena, S.: Deepwalk: online learning of social representations. In: Proceedings of the 20th ACM SIGKDD International Conference on Knowledge Discovery and Data Mining, pp. 701–710. ACM (2014)
18. Scarselli, F., Gori, M., Tsoi, A.C., Hagenbuchner, M., Monfardini, G.: The graph neural network model. IEEE Trans. Neural Netw. **20**(1), 61–80 (2009)
19. Sen, P., Namata, G., Bilgic, M., Getoor, L., Galligher, B., Eliassi-Rad, T.: Collective classification in network data. AI Mag. **29**(3), 93 (2008)

20. Tang, J., Qu, M., Wang, M., Zhang, M., Yan, J., Mei, Q.: Line: large-scale infor-
    mation network embedding. In: Proceedings of the 24th International Conference
    on World Wide Web, pp. 1067–1077. International World Wide Web Conferences
    Steering Committee (2015)
21. Tang, W., Zhong, S.: Pairwise constraints-guided dimensionality reduction. In:
    Computational Methods of Feature Selection, pp. 295–312. Chapman and Hall,
    CRC (2007)
22. Thoma, M., Cheng, H., Gretton, A., Han, J., Kriegel, H.-P., Smola, A., Song, L.,
    Yu, P.S., Yan, X., Borgwardt, K.: Near-optimal supervised feature selection among
    frequent subgraphs. In: Proceedings of the 2009 SIAM International Conference on
    Data Mining, pp. 1076–1087. SIAM (2009)
23. Yan, X., Cheng, H., Han, J., Yu, P.S.: Mining significant graph patterns by leap
    search. In: Proceedings of the 2008 ACM SIGMOD International Conference on
    Management of Data, pp. 433–444. ACM (2008)
24. Yan, X., Han, J.: gSpan: graph-based substructure pattern mining. In: 2002 IEEE
    International Conference on Data Mining, ICDM 2003, pp. 721–724. IEEE (2002)
25. Yang, Z., Cohen, W.W., Salakhutdinov, R.: Revisiting semi-supervised learning
    with graph embeddings. In: International Conference of Machine Learning (2016)

Predicting the Degree of Collaboration of Researchers on Co-authorship Social Networks

Doaa Hassan$^{(\boxtimes)}$

Department of Computers and Systems, National Telecommunication Institute,
5 Mahmoud El Miligy Street, 6th District-Nasr City, Cairo, Egypt
doaa@nti.sci.eg

Abstract. Academic social networks play a role in establishing the research collaboration by connecting researchers with each other in order to allow them sharing their professional knowledge and evaluating the performance of each individual researcher. In this paper, we mine the academic co-authorship social network in order to predict the potential degree of collaboration for each individual researcher on that network. Where predicting the degree of collaboration for each individual researcher is considered as one of the indicators for evaluating his/her potential performance. The main approach of this paper relies on using the sum of link weights specified by the total number of papers that each researcher has published with other co-authors in order to identify his/her initial degree of collaboration on the co-authorship network. This is achieved by collecting the total of link weights that each researcher has with others on the co-authorship network. Next, the obtained values of collaboration degree are used with other features extracted from the co-authorship dataset to train a supervised-learning regression model in the training phase. Consequently, the learning regression model will be able to predict potential degree of collaboration for every individual researchers on the co-authorship network in the test phase. Empirical experiments are conducted on a publicly available weighted academic co-authorship networks. The evaluation results demonstrate the effectiveness of our approach. It achieves a high performance in predicting the potential degree of collaboration of every researcher on the co-authorship network with 0.46 root mean square error (RMSE) using the linear regression model.

Keywords: Social networks · Co-authorship · Collaboration degree

1 Introduction

Research collaboration aims to establish links between various researchers in order to develop new projects or publishing papers together. The scientific co-authorship network is considered as one type of the social networks in which

© Springer Nature Switzerland AG 2020
A. M. Madureira et al. (Eds.): HIS 2018, AISC 923, pp. 75–84, 2020.
https://doi.org/10.1007/978-3-030-14347-3_8

the vertices/nodes of this network represent the researchers or authors and the edges/links represent the established scientific collaboration among them. Two researchers on this network are connected by an edge if they have collaborated by being co-authors in one or more paper. Therefore, the academic co-authorship network is build by extracting information about either authors or publications [6,7]. Thus, the scientific co-authorship/collaboration network is a network of researchers, and their collaborations represents a direct reflection of the publications that they have published with each other [2].

Research collaboration has been found to be very correlated to the scientific productivity of each researcher on the co-authorship network [8,9]. In addition, the collaboration is considered as a great indicator of the scientist success and hence his/her performance [13]. Therefore, there was a great demand for estimating the potential degree of collaboration of every individual researcher on the co-authorship network. This can be very important factor when recommending researchers to others on that network [3]. In addition, degree of collaboration can be used with other factors for estimating the productivity of researcher on the scientific collaboration network [16]. To this end, in this paper, we develop an approach for mining the academic collaboration networks in order to predict the potential degree of collaboration of every researcher on this type of social networks. Our approach relies on extracting the weights of edges between researches on the co-authorship network, where the weight of each edge/link represents the number of publications that each pair of researchers on the network have published together. For instance, if the weight of the link between two researcher is two, then this means that those two researchers have collaborated on two papers. Our approach collects the weights of links that each researcher has with other co-authors on the co-authorship network, then uses this result to express his/her initial degree of collaboration with others (represented by numerical value). Next, the calculated initial degree of collaboration is used with other features extracted from the co-authorship dataset including author information and network structure measures features for training a supervised learning regression model in the training phase. Consequently, this model will be able to predict the potential degree of collaboration of every researcher on the co-authorship network in the testing phase. We have conduced the experiment of the paper approach on the author and co-author datasets which are two subsets of AMminer dataset collection [1] that are publicly available for doing research on academic social networks [4]. The evaluation results demonstrate the effectiveness of our approach. It achieves a high performance in predicting the potential degree of collaboration of every researcher on the co-authorship network with 0.46 root-mean-square error (RMSE) using the linear regression model.

The structure of this paper is organized as follows: in Sect. 2, we discuss the related work. In Sect. 3, we present the methodology of our approach for predicting the potential degree of collaboration of individual researchers on the academic collaboration network. In Sect. 4, we conduct the basic experiment of the paper approach and evaluate the performance of our approach as well as the obtained results. Finally, in Sect. 5 we conclude the paper with some directions for future work.

2 Related Work

There have been many research studies that investigated the collaboration in the context of co-authorship social network, using some common centrality metrics [10] in the literature such as degree [11], betweenness [14] and eigenvector [15]. However, none of these approaches studied how to measure the degree of collaboration for each researcher on the collaboration network. Perhaps, the most relevant approaches to ours are those that were presented in [8,9]. Both approaches investigated how to measure the collaboration on co-authorship network in the form of networking intensity of a researcher on that network. Following we summarize both approaches.

Ioannidis [8] presented R metric for measuring the effective co-authorship networking size for adjusting the citation impact of the groups or institutions. He claimed that this metric can be also used for adjusting the citation impact of a scientist on the co-authorship network. This metric is determined by taking the power exponent of the relationship between two values: The first is I-index which is defined as the number of authors who appear in at least I papers of a certain scientist or the number of authors who appear in at least I_n papers that have the affiliation of a group of scientists. The second is the number of publications of a certain scientist or an affiliated group of scientists/institutions. Based on the value of R metric, the scientist can be categorized into one of five categories: solitary, nuclear, networked, extensively networked, and collaborators.

Later Parish et el. [9] extended Ioannidis's approach by presenting an approach for assessing the factors that are relevant to the collaborative behavior of the scientist on co-authorship network. They also used R to express the degree of collaboration for every researcher on the collaboration network, where higher values of R indicates lower collaborative behaviour. They also, investigated how this value varies across scientific fields and change over time. They also studied the correlation between R value and the citation impact of the scientist or the chance of being a principle investigator.

Both of the two previously mentioned approaches consider collaboration as continuous working that occurs between a scientist and a repeated group of scientists on the collaboration network (i.e, redundant co-authorship relation). However and different from both approaches, our approach defines the collaboration as the accumulations of the number of times that each researcher has collaborated with other co-authors on the co-authorship network even if he only collaborates once with anyone of them. Moreover, none of both approaches investigated how to predict the potential degree of collaboration (i.e., R in both approaches) for every researcher on the collaboration network.

Our approach is inspired by the work presented in [12] which presented an approach that uses the multivariate regression model (MVRM) for testing the effects of co-authorship network measures (e.g., centrality degree, centrality betweenness) on the author's performance measures (specified by his/her citation-based g-index.). One of the co-authorship network measures that was used in this approach was the average weights of co-authorship links. This measure is determined by dividing the sum of the link weights for each scholar (i.e., the number of collaborations of the scholar) by the total number of links established between that scholar and different co-authors (i.e., degree of author node). Different from this approach, our approach uses the sum of weights of co-authorship links for each scholar as an initial collaboration degree for that scholar. This initial collaboration degree is used as feature with other features extracted from the co-authorship network for training a regression model in order to make it able to predict the potential degree of collaboration of every scholar on the co-authorship network in the testing phase. The choice of sum weights of co-authorship links as a feature is because it precisely evaluates the net result of scholar's collaboration regardless if this was obtained by establishing many collaborations with different co-authors or by doing a repeated co-authorship to a group of co-authors.

3   Basic Approach

Our proposed approach relies on assigning an initial value for the degree of collaboration for every individual researcher on the co-authorship network. Such a value is equal to the sum of weights of links that each individual researcher establishes with other co-authors on the network. Each of those link weights refers to the number of publications that this researcher has published in collaboration with another co-author. Therefore, the first step in our approach is to extract the weights of edges/links that connects each researcher with other co-authors on the co-authorship network. By extracting those link weights, we can determine the initial value of the collaboration degree metric for every researcher on the co-author network. For example, if a researcher A has collaborated with other 5 co-authors and the weight for the edges that connects him/her with those co-authors were 4, 1, 3, 6, and 2, then the initial value of the collaboration degree for that researcher is simply calculated as follows:

Initial value of collaboration degree of researcher $A = 4 + 1 + 3 + 6 + 2 = 16$.

The initial values of collaboration degree metric for individual researchers on the co-authorship network are used to manually label the author nodes samples in the co-authorship network/dataset. Each of those author nodes has a vector of features extracted from the author information in addition to other two features extracted from the coauthor network graph structure: namely the node degree and the average link strength as it will be described in details in the next section. Next the obtained initial values of collaboration degrees of individual researchers are used in combination with the feature vectors of author nodes that represents those researchers to train a supervised learning regression model in the training phase. Using this combination for training the regression model, it will be able to predict the potential collaboration degree for each researcher on co-authorship network in the testing phase. Figure 1 shows the basic architecture of our proposed approach.

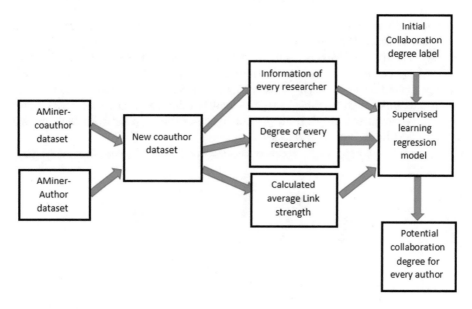

Fig. 1. The proposed approach for predicting the collaboration degree.

4   Basic Experiment

4.1   Dataset

The basic experiment of this paper is conducted on the AMiner-author and AMiner-Coauthor datasets that are two subsets of the AMinernetwork dataset collection [4]. This collection has been used for link prediction using supervised learning such as in [5], but to our knowledge it has not been used for predicting the degree of collaboration of individual researchers on co-authorship networks using the same technique. The collection consists of three datasets: namely AMiner-Paper, AMiner-Author, and AMiner-Coauthor. The dataset collection is publicly available at [1].

As for the Aminer-author dataset, it saves all authors information including the ID, affiliation, count of published papers, total number of citation, H-index, pi-index (i.e., P-index with equal A-index of this author), and upi-index (i.e, P-index with unequal A-index of this author) as well as the research interests for this author[1].

As for AMiner-Coauthor dataset, it saves all the established collaboration links among the authors in AMiner-Author dataset as well as their weights. The dataset has 4,258,615 collaboration links, established between 1,712,433 authors. Each sample in the dataset has three entries: the index of the first

---

[1] The A-index is used to compute the weighted total of peer-reviewed publications and journal impact factors C- and P-indexes that refers to the collaboration and productivity, respectively. More information about A-index, is available at [16].

author, the index of the second author, and the number of collaboration that occurred between both authors.

By utilizing the AMiner-Coauthor dataset, we can visualize the co-authorship network by generating a weighted undirected graph and plotting it. The generated graph connects researchers according to the number of collaboration among them (i.e., link weights). Since AMiner-Coauthor dataset is massive, Fig. 2 shows only the constructed graph for a part of the co-authorship network that utilizes a part of AMiner-Coauthor dataset.

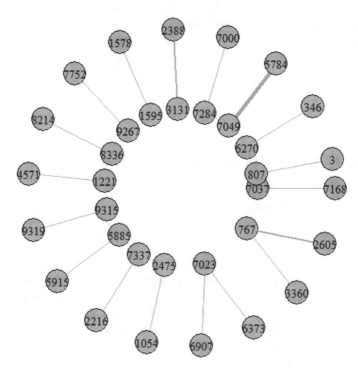

Fig. 2. A constructed graph for a part of the co-authorship network generated by AMiner-Coauthor dataset.

When observing the generated graph, we notice that the widthes of the established links among authors on the co-authorship network are different. This is because they have different weights, where the thick links indicates a large number of collaboration occurred between the two authors, while the thin links indicates a smaller number of collaboration between two authors on the network. Also, we notice that many nodes in the graph are disconnected which means there was no collaboration between the authors that represent these nodes.

Both of the Aminer-author and Aminer-coauthor datasets have been merged for generating the new co-author dataset that represents the regression dataset as shown previously in Fig. 1. Due to the memory limitations we have taken the first 50000 authors from the AMiner-author dataset and examined the number of links that they have with each other in AMiner-coauthor dataset. Therefore, the new generated co-author dataset has 1468 collaboration links established between 2618 coauthors (i.e., author nodes) in the AMiner-coauthor dataset. Each author node in the new generated co-author dataset has a vector of features. Some of those features are extracted from the author information in the AMiner-author dataset for each author in the AMiner-coauthor network dataset. This includes the count of published papers, total number of citation, H-index, pi-index, and upi-index for each author. The remaining features are extracted from co-authorship network structure including the degree of the author node (i.e., the number of co-authors that are directly connected to that node) and the average link strength between each author and his/her coauthors (i.e., average link weight [12]). The choice of the average tie/link strength and degree as features is because both of them has a positive correlation to the scholar performance which is also affected with his/her collaboration performance [12].

This dataset is input to the supervised learning regression model that will predict the potential collaboration degree for each author in the co-authorship network in the testing phase.

The new generated dataset has been used for training and testing the experimental regression learning model in 10-fold cross validation process, where 10% of samples of the dataset is randomly selected for testing, while the remaining 90% of dataset samples are kept for training the learning model.

4.2 Experimental Settings

R [17] has been used for generating and plotting the graph that represents the experimental co-authorship network and calculating the degree of each author node in the graph. Also R has been used for generating the adjacency matrix from the generated graph that is needed for calculating the sum of link weights for each node in the co-authorship network graph.

Moreover, python [18] has been used for preprocessing the AMiner-author dataset for feature extraction by extracting the information of each author as features. Python has been also used for merging the author and co-author datasets in order to generate the experimental dataset that has been input to the learning model.

Finally, Weka [19], a free open source software machine learning tool has been used for generating different learning regression models from different categories, and training them on the experimental dataset.

The proposed approach for predicting the collaboration degree has been run on laptop machine with 2.6 GHZ processor Intel core (TM)i5, 8 G Memory Rams with Windows 7 operating system.

4.3 Potential Collaboration Degree Prediction

The prediction of the potential degree of collaboration occurs in two steps. In first step, we build the training model by feeding the generated regression co-author dataset to a supervised learning regression model. Initially all data points in the dataset are labeled with initial collaboration degree values assigned to each author to be equal to the sum of weights of links between this author and his/her coauthors (i.e., collaborators). The combination of the extracted features and the initial value of collaboration degree for all data points are used to train the learning model. Consequently, in second step the learning model will be able to predict the potential collaboration degree of every researcher in the testing phase.

Five learning regression models have been generated using Weka for predicting the potential degree of collaboration for every author in the coauthor network. This includes linear regression (LR), k-Nearest Neighbors (KNN), Decision Tree (DT), Support Vector regression (SVR), and the neural network (NN) [20]. Namely, LinearRegression, IBK, REPTree, SMOreg, and MultilayerPerceptron respectively in Weka.

4.4   Performance Evaluation Results

We have used the root mean square error (RMSE) [21] as a metric for evaluating the performance of the generated regression models in predicting the potential degree of collaboration for every researcher on the co-authorship network. The choice of RMSE as a performance evaluation metric is because it represents the root of the average squared difference between the predicted values and what is actually predicted. As for regression model, RMSE is simply calculated as follows:

$$RMSE = \sqrt{\frac{\sum_{i=1}^{m}(y_i' - y_i)^2}{m}}$$

where y_i' is a vector of m predictions generated from a sample of m data points on all features/variables, and y_i is the vector of observed values of the feature/variable being predicted.

Table 1 shows the obtained RMSE for each of the generated learning regression models in the basic experiment of this paper. Clearly, the results in Table 1 show that the performance of LR outperforms other learning regression models as it achieves the lowest RSME of 0.46.

Table 1. The root mean square error of the experimental learning regression models.

Regression model	RMSE
LR	0.46
KNN	5.09
DT	19.70
SVR	0.53
NN	2.05

Also by analyzing the results of running regression algorithms on the experimental coauthor dataset, we found that there is a high positive correlation of 0.99 between the average link strength and the collaboration degree of researcher using the linear regression algorithm. Lower positive correlation of 0.064 is found between other features (including the node degree and author information features) and the collaboration degree using the same algorithm. This means that net result of scholar's collaboration with other coauthors (represented by the collaboration degree) is affected too much by establishing strong links (i.e, with large weights) between this scholar and other coauthors. The same observation is obtained with other regression algorithms used in the basic experiment of this paper, but this is omitted due to space.

5   Conclusions and Future Work

In this paper, we have developed a linear regression model to predict the degree of collaboration of researchers on the co-authorship networks. We have conducted our experiment on AMiner public dataset collection. Our approach relies on assigning an initial value of collaboration degree for every author in collaboration network specified by the sum of weights of links that each author has with other co-authors on that network. Next the initial values of collaboration degrees are used to manually label the coauthor dataset. Each data point (author node) in this dataset has a vector of features extracted from the author info as well as the coauthor network graph structure (i.e., the average link strength and degree). The combination of those extracted features with the initial values of collaboration degrees are used to train our model to make it able to predict the potential collaboration degree in the testing phase. Our developed model has achieved a high performance in predicting the potential degree of collaboration of every researcher on the co-authorship network with RMSE of 0.46.

Our future work has various directions. First, we are looking forward to extending our approach for predicting the degree of collaboration for individual researchers of various scientific fields. Second, our approach can be extended by investigating the change of the degree of collaboration that each researcher has according to the network size and over time [9]. Also investigating the change in the collaboration degree according to the coauthor country looks interesting [14]. Finally, we are also looking forward to predicting the degree of collaboration using the similarity between research interests of scholars as a feature.

References

1. Extraction and Mining of Academic Social Networks. https://aminer.org/aminernetwork
2. Newman, M.: Networks: An Introduction. Oxford University Press Inc., New York (2010)

3. Konstas, I., Stathopoulos, V., Jose, J.M.: On social networks and collaborative recommendation. In: Proceeding of the 32nd International ACM SIGIR Conference on Research and Development in Information Retrieval (SIGIR 2009), pp. 195–202 (2009)

4. Tang, J., Zhang, J., Yao, L., Li, J., Zhang, L., Su, Z.: ArnetMiner: extraction and mining of academic social networks. In: Proceedings of the Fourteenth ACM SIGKDD International Conference on Knowledge Discovery and Data Mining, pp. 990–998 (2008)

5. de Sa, H.R., Prudencio, R.B.C.: Supervised link prediction in weighted networks. In: Proceedings of the International Joint Conference on Neural Networks (2011)

6. Newman, M.E.J.: Coauthorship networks and patterns of scientific collaboration. PNAS **101**(Suppl. 1), 5200–5205 (2004)

7. Sameer Kumar Co-authorship networks: A review of the literature. Aslib J. Inf. Manag. **67**(1), 55–73 (2015)

8. Ioannidis, J.P.A.: Measuring co-authorship and networking-adjusted scientific impact. PLOS ONE **3**(7), e2778 (2008)

9. Parish, A.J., Boyack, K.W., Ioannidis, J.P.A.: Dynamics of co-authorship and productivity across different fields of scientific research. PLOS ONE **13**(1), e0189742 (2018)

10. Hagan, R., Feng, Y., Bush, J.: Centrality Metrics. University of Tennessee, Knoxville (2015)

11. Bordons, M., Aparicio, J., Gonzáalez-Albo, B., Diaz-Faes, A.A.: The relationship between the research performance of scientists and their position in co-authorship networks in three fields. J. Inf. **9**(1), 135–144 (2015)

12. Abbasi, A., Altmann, J., Hossain, L.: Identifying the effects of co-authorship networks on the performance of scholars: a correlation and regression analysis of performance measures and social network analysis measures. J. Inf. **5**(4), 594–607 (2011)

13. Abbasi, A., Altmann, J., Hwang, J.: Evaluating scholars based on their academic collaboration activities: two indices, the RC-index and the CC-index, for quantifying collaboration activities of researchers and scientific communities. Scientometrics **83**(1), 1–13 (2010)

14. Lee, S., Bozeman, B.: The impact of research collaboration on scientific productivity. Soc. Stud. Sci. J. **35**(5), 673–702 (2005)

15. Ponomariova, B.L., Boardman, C.: Influencing scientists' collaboration and productivity patterns through new institutions: university research centers and scientific and technical human capital. Res. Policy J. **39**(5), 613–624 (2010)

16. Stallings, J., et al.: Determining scientific impact using a collaboration index. Proc. Nat. Acad. Sci. U.S.A. **110**(24), 9680–9685 (2013)

17. The R Project for Statistical Computing. https://www.r-project.org/

18. Software Foundation: Python Language Reference, version 2.7. http://www.python.org

19. WEKA Manual for Version 3-6-8. University of Waikato, Hamilton, New Zealand (2012)

20. Mohri, M., Rostamizadeh, A., Talwalkar, A.: Foundations of Machine Learning. MIT Press (2012)

21. http://statweb.stanford.edu/~susan/courses/s60/split/node60.html

Deterministic Parameter Selection of Artificial Bee Colony Based on Diagonalization

Marco Antonio Florenzano Mollinetti[1(✉)], Mario Tasso Ribeiro Serra Neto[2], and Takahito Kuno[1]

[1] Systems Optimization Laboratory, University of Tsukuba, Tsukuba, Japan
`marco.mollinetti@gmail.com`, `takahito@cs.tsukuba.ac.jp`
[2] Science Faculty, University of Porto, Porto, Portugal
`mariotrsn@gmail.com`

Abstract. Artificial Bee Colony (ABC) is a bee inspired swarm intelligence (SI) algorithm well-known for its versatility and simplicity. In crucial steps of the algorithm, employed and scout bees phase, parameters (decision variables) are chosen in a random fashion. Although this randomness may apparently not influence the overall performance of the algorithm, it may contribute to premature convergence towards bad local optima or lack of exploration in multimodal problems featuring rugged surfaces. In this study, a deterministic selection method for decision variables based on Cantor's proof of uncountability of rational numbers is proposed to be used in the aforementioned steps. The approach seeks to eliminate stochasticity, enhance the exploratory capabilities of the algorithm by verifying all possible variables, and provide a better mechanism to displace solutions out of local optima, introducing more novelty to solutions. In order to analyze potential benefits brought by the proposed approach to the overall performance of the ABC, three variants featuring modifications discussed in this work were designed to be compared in terms of efficiency and stability against the original ABC on 15 instances of unconstrained optimization problems.

Keywords: Artificial Bee Colony · Parameter selection · Diagonalization

1 Introduction

In the last decade, there has been an increase in the popularity of Swarm Intelligence algorithms (SI), nature inspired heuristics based on the collective behavior of groups of organisms found in nature. The goal of SI is to provide better solutions to nonlinear, non-smooth problems that derivative-based methods does not fare well. ABC is a widely studied SI algorithm inspired by the foraging behaviour of swarms of honey bees to solve numerical optimization problems [8]. Popularity of this technique is due to its effectiveness and simplicity in its design

© Springer Nature Switzerland AG 2020
A. M. Madureira et al. (Eds.): HIS 2018, AISC 923, pp. 85–95, 2020.
https://doi.org/10.1007/978-3-030-14347-3_9

and implementation, making room for easy parallelization, improvements in its mechanisms or even hybridization with other optimization techniques.

In the first update step of ABC, the employed bees phase, decision variables of candidate solutions are randomly chosen to be updated. While at first, introducing randomness in this stage may contribute to the diversity of solutions. Potential downsides can be observed such as a lack of participation of the candidate solution in the improvement of the objective function and lack of exploration in high dimensional problems. Taking these issues in account, we propose a non-random selection procedure of decision variables and solution generation to be taken part in the employed and scout bees phase of the algorithm, respectively. Said approach is based on Cantor's diagonalization argument used to prove the uncountability of real numbers. This approach premise is to enhance the exploratory capabilities of the algorithm, and above all, introduce an increased novelty factor to solutions. To observe whether the proposed method influences the behavior of the ABC, three variants of the ABC employing the proposed ideas are developed and tested on several unconstrained continuous benchmark landscape functions. The original ABC is used as a baseline of comparison against the variants.

The rest of this work is organized as follows. Section 2 presents a brief explanation and some remarks regarding the ABC algorithm. Section 3 firstly provides theoretical foundation of Cantor's diagonal, then illustrates the idea integrated to ABC. Section 4 delineates the design of the experiment and discusses the obtained results. Finally, conclusions are made in the last section.

2   Artificial Bee Colony

Artificial Bee Colony (ABC) is regarded as a metaheuristic capable of producing acceptable solutions to nonlinear, non-smooth or ill-conceived problems that classical gradient-based optimization methods usually fail to converge to good solutions. ABC is classified as a bee-inspired Swarm Intelligence (SI) algorithm based on the mathematical model of foraging behavior of honey bees established by Tereshko and Loengarov [11], firstly introduced in a technical report written by Karaboga [8] in 2005. ABC stands as the most popular Bee-inspired algorithm, and a prominent SI technique due to its efficiency despite its simple design [2]. Although initially devised to solve continuous unconstrained problems, the implementation of the algorithm allows researchers to easily adapt it to tackle different families of problems.

Solutions are sampled and improved by local and global search procedures, employed, onlooker and scout bees phase, cyclically performed until a stopping criteria is met, as described in Algorithm 1. Three parameters regulate the algorithm: number of candidate solutions SN; maximum number of iterations MCN; and solution stagnation threshold Lit. For the sake of clarity, we denote $X = \{x_1, x_2, \ldots, x_{sn}\}$ as the set of solutions sampled by the algorithm.

When no partial information is provided, solution initialization is performed by drawing values from a uniform distribution (denoted by $U(\alpha, \beta)$) ranging

---

Algorithm 1. Artificial Bee Colony

Input: Objective function $f(x)$, SN, MCN, Lit
Initialization : for all $x^i \in X, i = 1, \ldots SN$, $x_j^i = U(x_{\min_j}^i, x_{\max_j}^i)$ $j = 1, \ldots, D$
while *stopping condition not satisfied* **do**
$\quad$ | $EmployedPhase(X)$
$\quad$ | $OnlookerPhase(X)$
$\quad$ | $ScoutPhase(X, Lit)$
$\quad$ | **if** X *does not converge* **then**
$\quad$ | $\quad$ | **return** X
end
return X

---

between the feasible bounds of each decision variable. Each step of the algorithm is explained as follows.

Employed Bees Cycle. A local search step is carried for each candidate solution of set X, a decision variable is shifted a random step size towards the neighborhood of another randomly chosen solution x^k. For each solution x^i, $i = \{1, 2, \ldots, N\}$, a decision variable j is updated by (1):

$$x_j^{i^{(t+1)'}} = x_j^{i^{(t)}} + \phi(x_j^{i^{(t)}} - x_j^{k^{(t)}}), \tag{1}$$

Where t is the time step; $k \neq i$ a randomly chosen index; ϕ is a uniformly distributed real random number between $[-1, 1]$. After a solution is updated by (1), a greedy selection takes place:

$$x_j^{(t+1)} = \begin{cases} x_j^{i^{(t+1)'}} & \text{if } f(x^{i^{(t+1)'}}) \leq f(x^{i^{(t)}}) \\ x_j^{i^{(t)}} & \text{otherwise} \end{cases}. \tag{2}$$

if the objective value $f(x^{i^{(t+1)'}})$ obtained by modifying decision variable x_j^i with (1) is worse than the original value $f(x^{i^{(t)}})$, the update step is invalidated and the former decision variable is kept. If this happens, it means that no improvement was observed in that solution, so the number of unsuccessful iterations associated to this solution is increased by 1.

Onlooker Bees Phase. Each solution is assigned a probability value p^i and selected to undergo the same procedure carried in the employed bees phase (updated by (1) and (2)) by means of a weighted roulette (similar to the fitness proportionate roulette selection mechanism found in Genetic Algorithms). Repetition of solutions is allowed. According to Akay and Karaboga in one of their most recent works [3], the onlooker bee phase intensifies local searches in the vicinity of more promising solutions, resulting in a positive feedback feature. Probability p^i is calculated for each candidate by (3):

$$p^i = \frac{F(x^i)}{\sum_{i=1}^{sn} F(x^i)}, \tag{3}$$

where $F(x^i)$ is the objective function value of candidate solution x^i, computed by (4):

$$F(x^i) = \begin{cases} \frac{1}{1+f(x^i)} & \text{if } f(x^i) \geq 0 \\ |1 + f(x^i)| & \text{otherwise.} \end{cases} \tag{4}$$

Scout Bees Phase. In the scout bees phase, an attempt is made to displace solutions which stopped converging during the prior steps in order to prevent premature convergence to bad local optima. The authors refer this procedure as a "negative feedback" produced by solutions [3]. Solution stagnation is determined if no improvement in the objective function value of a candidate solution x^i ($f(x^{(t+1)}) \geq f(x^{(t)})$) is observed for more than Lit iterations consecutively. If a solution is deemed stagnated, a new candidate solution is sampled by (5).

$$x^{new} = x^i_{\min_j} + \text{U}(0,1)\left(x^i_{\max_j} - x^i_{\min_j}\right) \quad j = 1, \ldots, D. \tag{5}$$

Lit is usually set to $(N * D)$, where D is the dimension of the problem. If multiple solutions surpassed Lit number of iterations without improving at the same iteration, only the worst is sampled by (5) [8,9]. Previous versions stated that x^{new} is subject to greedy selection (2) [8]. However, recent works consolidate the ABC with only one solution at the scout step, to reduce the burden on the complexity of the algorithm.

Focusing on the local and global search process conducted in the employed and onlooker bees phase, a single decision variable x_j of multiple solutions is randomly selected to be updated by (1) and (2). The randomness in the choice of parameters (decision variables) may apparently have no significance in the overall performance of the ABC. However, selecting a random j incur in several potential downsides. First, the chosen decision variable might have little to no effect in improving the objective function value of the solution, contributing to the increase in the number of failed iterations, should it be chosen repeatedly. Consequently, decision variables may never be selected for update, hindering the global search capability of the algorithm, especially in high dimensional problem instances. Lastly, moving the wrong decision variable would force solutions out of a potentially good neighborhood, leading towards bad accumulation points, such is the case of when dealing with multimodal functions with deceptive local optima. The same issues could be said to the candidate solutions that are subject to (5) during the scout bees phase. Generating new solutions by means of a random distribution may indeed increase solution variability, but not without the risks mentioned before.

Taking into account the negative aspects of the randomization of the selection of decision variables, we propose a simplistic method that address the stochasticity in the choice of decision variables in the employed and scout bees phase. The selection method takes inspiration from a well-known argument proposed by German mathematician, Georg Cantor, to attempt to prove the uncountability of rational sets, the Diagonalization argument.

3   Proposed Approach

Firstly, Cantor's Diagonal argument is introduced in order to support the proposed changes to the algorithm. Then, the deterministic parameter selection step of the employed and scout bees phase is explained step-by-step.

3.1   Cantor's Diagonal Argument

In 1874, Cantor proposed an ingenious argument to demonstrate the uncountability of the set of real numbers. Provided that no infinite list of decimals can contain representations of all real numbers and diagonals can be extracted from any list, the decimal constructed from the diagonal of the list does not pertain to it, and therefore, represents a real number that does not constitute any other member of the list [4]. Weisstein [12] formalizes the theorem as follows: Given a countably infinite list of elements from a set S, each of which is an infinite set (in case of real numbers, the decimal expansion of each real number). A new member S' of S is then created by arranging its n-th term to differ from the n-th term of the n-th number of S, showing that S is uncountable, since any attempt of a one-to-one correspondence with the integers will fail to include some elements of S.

Extending this notion, solutions in the ABC for continuous optimization problems are encoded as vectors in $\mathbb{R}^n$. Therefore, it is possible to interpret the concatenation of all solutions in a matrix A as a list of reals which is a subset of S, where each solution is represented as a column of A. For ease of reading, entries of A are described as a_{ij}; an entire row of A as a_i and an entire column as a^j. New solutions are generated by the following operator:

$$A = diag \begin{pmatrix} a_{11} & a_{12} & \dots & a_{1n} \\ a_{21} & a_{22} & \dots & a_{2n} \\ \vdots & \vdots & \ddots & \vdots \\ a_{d1} & a_{d2} & \dots & a_{dn} \end{pmatrix}$$

where $diag(\cdot)$ stands for the operator that extracts the main diagonal of a matrix. Although A is a square matrix in this case, the selection method is extended to rectangular matrices with more columns than rows. In case of the parameters of the ABC, $SN \geq D$.

3.2   Integrating the Diagonal Argument to the ABC

In this section, we propose modifications based on the diagonal argument to the selection of decision variables to be updated in the employed bees phase, and solution generation in the scout bees phase. For convenience, we refer to any variant of the ABC that employs the ideas proposed hereafter as diagonal ABC (dABC). The dABC, much like the original ABC, is designed to solve continuous problems featuring relatively low dimension ($D \leq 100$) due to the solution space of continuous problems being $\mathbb{R}^n$, and because of the computing power overhead brought by enforcing $SN \geq D$.

Cantor's diagonal is proposed to be used in the following steps of the ABC: choice of which decision variable of each candidate solution to be updated is now entirely based upon the diagonal of the list of solutions to be carried the employed bees phase; displacement of solutions trapped in local optima i.e., exceeded the threshold of failed iterations in the scout bees phase, is now done by extracting the main diagonal of the matrix of candidate solutions. Possible advantages that this approach offer are twofold. First, the assurance that every decision variable will be chosen at least once along the iterations in an attempt to reduce the number of failed iterations due to randomness. Second, in case of solutions updated in the scout phase, an improvement to the exploratory aspect of the algorithm by introducing more novelty to solutions when trying to displace them from bad local optima. It is worth mentioning that changes are restricted to these steps in order to closely examine potential consequences of the approach to the algorithm. Because the employed bees phase emphasizes global search, any alteration to the onlooker bees stage would influence the results of this study.

Foremost, integrating the aforementioned concepts require a significant change to the arrangement of solutions. Solutions must be represented in a $n \times m$ matrix A, where $n = D$ and $m = SN$. Each row i, $i = 1, \ldots n$ describe a decision variable while each column j, $j = 1, \ldots m$ is a solution itself. Matrix A must be square or wide rectangular ($n \leq m$), that is, the number of candidate solutions is greater than or equal the dimension of the problem. In the employed bees phase, chosen entries of A that undergo the update step are now part of a $m \times 1$ column vector X. If A is square, the entries of X are simply the diagonal of the matrix, $X = \{a_{11}, a_{22}, \ldots, a_{nm}\}$. However, when $n < m$, likely to be the most frequent case, X will be comprised of the main diagonal together with the components of A superdiagonals offset n units to the right. For instance, suppose A is a 2×6 matrix, X will be:

$$A = \begin{bmatrix} a_{11} \ a_{12} \ a_{13} \ a_{14} \ a_{15} \ a_{16} \\ a_{21} \ a_{22} \ a_{23} \ a_{24} \ a_{25} \ a_{26} \end{bmatrix} \quad X = \{a_{11}, a_{22}, a_{13}, a_{24}, a_{15}, a_{26}\}$$

This way, X has as its entries a single decision variable from each candidate solution. Having a column vector of solutions X makes possible for the update step (1) to be carried simultaneously for all columns of A by employing simple vector operations as follows:

$$X' = X + \Phi \odot (X - K). \tag{6}$$

Where $\odot$ stands for element-wise product, Φ is an uniform column vector ($m \times 1$) with values sampled between $[-1, 1]$, and K is a column vector where each element k_{i1} is a decision variable j of a random solution $k_{i1} \neq x_{i1}$ associated to each variable x_{i1} of X. Clearly, X' is also a column vector whose entries x'_{i1} replace entries x_{i1} of X in A if $f(a^i \cup x'_{i1}/a_{ij}) \leq f(a^i)$, in accordance to greedy selection (2). To exemplify this, suppose that in matrix A of the previous example, entries

a_{11} and a_{22} had their objective function values worse that the entries in X', so vector X' is now $\{x'_{1i}, x'_{2i}, a_{13}, a_{24}, a_{15}, a_{26}\}$. To guarantee that every decision variable of each candidate solution is updated at least once along the iterations, at the end of the employed bees phase, a safeguard step is performed. The last column a^m of A replaces the first column of the matrix while the remaining columns are shifted one position to the right, in the following way:

$$A = \begin{bmatrix} x'_{i1} & a_{12} & a_{13} & a_{14} & a_{15} & a_{16} \\ a_{21} & x'_{i2} & a_{23} & a_{24} & a_{25} & a_{26} \end{bmatrix} \longrightarrow \begin{bmatrix} a_{16} & x'_{i1} & a_{12} & a_{13} & a_{14} & a_{15} \\ a_{26} & a_{21} & x'_{i2} & a_{23} & a_{24} & a_{25} \end{bmatrix}$$

This step ensures that every decision variable is updated by (6) every m number of iterations. Main steps of the modified selection of the employed bees phase is depicted step-by-step in Algorithm 2.

Algorithm 2. Deterministic selection in the employed bees phase

---

1 **if** $n == m$ **then**
2 $\quad\mid\quad X \leftarrow \mathrm{diag}(A)$
3 **if** $n > m$ **then**
$\quad\mid\quad$ // main and superdiagonals offset by D
$\quad\mid\quad X \leftarrow \mathrm{superdiag}(a^1, a^{D-1}, a^{2D-1}, \ldots, a^{\frac{SN}{2}D-1})$
4 $X' \leftarrow$ UpdateSolutions$(X \cup A)$ // update with (6)
$\quad$ **for** $a^j \in A \quad j = 1, \ldots, m$ **do**
$\quad\mid\quad$ Evaluate$(x'_{j1} \cup a^j / a_{ij})$
$\quad\mid\quad a_j \leftarrow$ GreedySelection(a_j) // apply (2)
$\quad$ **end**
$\quad A \leftarrow$ UpdateMatrix$(X' \cup A)$ // include successful entries x'_{j1}
$\quad$ // Safeguard step
$\quad a^1 \leftarrow a^m$ // Shift the last column of A to the first position
$\quad$ **for** $j = 1, \ldots, m - 1$ **do**
$\quad\mid\quad a^{j+1} \leftarrow a^j$ // Shift the columns of A one column to the right
$\quad$ **end**

Addressing the changes to scout bees cycle, a candidate solution that has exceeded the limit of Lit iterations failing the greedy selection process (2), have new decision variables values drawn in an alternative way. Instead of random uniform values using (5), a new solution is now generated as the main diagonal of matrix A, without loss of generality.

$$x^{new} = \mathrm{diag}(A) \tag{7}$$

Choosing the diagonal of A to be x^{new} implies in introducing novelty to the solution space by generating an element that was not in A. The reasoning behind this is simple, taking into consideration Cantor's diagonal argument, letting A be a subset of an uncountable set S containing all possible candidate solutions in $\mathbb{R}^n$, then the diagonal of A is a solution that is not present in A that still belongs to S. Adopting this strategy early on also contributes to the avoidance of the unlikely occurrence of all SN solution to converge to a single point in the space, since it is guaranteed that the new solution is not present in the A, as long as the $SN - D$ solutions are not the same.

As a final remark, concerning the complexity of the dABC, the original authors of the ABC [1] state that the total number of function evaluations of the algorithm is approximately $2SN + 1$ per iteration. Surely, it can be understood that this amount is the same for the dABC.

4   Experiments and Results

To analyze whether the proposed approach exert any influence to the performance of the original ABC, a three-factor experiment is conducted. Partial and full modifications are included in three distinct variants of the dABC:

- **dABC1:** deterministic selection in the employed bees phase.
- **dABC2:** modified scout bees phase.
- **dABC3:** deterministic selection in the employed bees phase and modified scout bees phase.

Performance of the three variants is assessed by 15 instances of well-known landscape functions designed to validate the capability of metaheuristics to handle various challenging aspects when optimizing a continuous function, such as ruggedness, multimodality and deceptive global optima. Functions chosen for this experiment are grouped by their characteristics, following the guidelines of Molga and Smutnicki [10] and Jamil et al. [7]: Differentiable or Non-differentiable (D or Nd); Separable or Non-separable (S or Ns); Unimodal or Multimodal (Um or Mm). Dimension, range, characteristic and global optimum of each function is displayed in an orderly fashion in Table 1. Multiple instances of the Rastrigin and Rosenbrock function were chosen due to the difficulty of finding the global optimum be inversely proportional to the dimension.

Results obtained by the variants of the dABC are solely compared against the canonical implementation of the ABC. This choice is due to the fact that the state-of-the-art versions of ABC highly deviate from the original idea, providing an inadequate baseline of comparison. Furthermore, this experiment holds the premise that if the inclusion of the diagonalization influences positively in the overall performance of the ABC, integration of the proposed idea to the state-of-the-art would be seamless, since the majority of them incorporate changes to the update step (1) of the employed and onlooker bees phase.

Regarding the parameters of the experiment, hyperparameters of all four algorithms are: number of solutions SN set to 40 and Lit defined as $SN * D$ for all instances. Stopping criteria was defined as maximum number of function evaluations $(FE's)$ in accordance to the works of Akay and Karaboga [1,3] and set to 400.000. Sample size was 30 times for each problem $(n = 30)$ with distinct seeds. The following statistics are measured for the results: best run, worst run, mean and standard deviation of the runs. Non-normality of the data was asserted

Table 1. Landscape functions according to the guidelines of [10] and [7]

Name	No.	Dim	Range	Definition	Opt
De Jong	F1	30	[−100, 100]	DSUm	0
Sphere	F2	30	[−100, 100]	DSMm	0
Ackley	F3	2	[−32, 32]	DNsUm	−200
Easom	F4	2	[−100, 100]	DNsMm	−1
GoldsteinPrice	F5	2	[−2, 2]	DNsMm	3
Camelback	F6	2	[−5, 5]	DNsMm	−1.0316
Rastrigin	F7/F8/F9	10/20/30	[−5.12, 5.12]	DNsMm	0
Rosenbrock	F10/F11/F12	10/20/30	[−5, 10]	DNsMm	0
Colville	F13	4	[−10, 10]	NdSMm	0
Wood	F14	4	[−10, 10]	NdSMm	0
Schwefel	F15	30	[−500, 500]	NdNsMm	0

Table 2. Results of experiments. For each function, the following is cataloged: (a) best solution (b) worst solution (c) mean (d) standard deviation

		ABC	dABC1	dABC2	dABC3
F1	Best	**0.0000000**	**0.0000000**	**0.0000000**	**0.0000000**
	Worst	0.3512997	0.8749444	**0.0000000**	**0.0000000**
	Mean	0.0348416	0.0472532	**0.0000000**	**0.0000000**
	Std. Dev	0.0962468	0.1598244	**0.0000000**	**0.0000000**
F2	Best	0.0000000	0.0000000	0.0000000	0.0000000
	Worst	0.0000000	0.0000000	0.0000000	0.0000000
	Mean	0.0000000	0.0000000	0.0000000	0.0000000
	Std. Dev	0.0000000	0.0000000	0.0000000	0.0000000
F3	Best	**-200.0000**	**-200.0000**	**-200.0000**	**-200.0000**
	Worst	-199.98641	-199.91891	**-200.0000**	**-200.0000**
	Mean	-199.99940	-199.99669	**-200.0000**	**-200.0000**
	Std. Dev	0.0025307	0.0145834	**0.000000**	**0.0000000**
F4	Best	**-1.000000**	**-1.000000**	**-1.000000**	**-1.000000**
	Worst	**0.0000000**	**0.0000000**	**0.0000000**	**0.0000000**
	Mean	-0.277839	-0.318555	**-0.200016**	-0.266720
	Std. Dev	0.4190777	0.4363626	**0.3999918**	0.4421840
F5	Best	**3.0000000**	**3.0000000**	**3.0000000**	**3.0000000**
	Worst	3.0169933	**3.0001466**	84.000000	30.000000
	Mean	3.0005664	**3.0000055**	18.300000	13.800000
	Std. Dev	0.0030504	**0.0000264**	21.693547	13.227244
F6	Best	1.0316284	1.0316284	-1.0316284	-1.0316284
	Worst	1.0316284	1.0316284	-1.0316284	-1.0316284
	Mean	1.0316284	1.0316284	-1.0316284	-1.0316284
	Std. Dev	0.0000000	0.0000000	0.0000000	0.0000000
F7	Best	**0.0000000**	**0.0000000**	**0.0000000**	**0.0000000**
	Worst	0.0073776	0.9953192	**0.0000000**	**0.0000000**
	Mean	0.0002846	0.0663556	**0.0000000**	**0.0000000**
	Std. Dev	0.0013265	0.2482284	**0.0000000**	**0.0000000**
F8	Best	**0.0000000**	**0.0000000**	**0.0000000**	**0.0000000**
	Worst	0.0091530	0.2776341	**0.0000000**	**0.0000000**
	Mean	0.0005767	0.0135547	**0.0000000**	**0.0000000**
	Std. Dev	0.0018983	0.0530280	**0.0000000**	**0.0000000**

		ABC	dABC1	dABC2	dABC3
F9	Best	**0.0000000**	**0.0000000**	**0.0000000**	**0.0000000**
	Worst	1.024343	0.0005054	**0.0000000**	**0.0000000**
	Mean	0.035320	0.0000193	**0.0000000**	**0.0000000**
	Std. Dev	0.183765	0.0000912	**0.0000000**	**0.0000000**
F10	Best	0.0007185	0.0000826	**0.0000129**	0.0000912
	Worst	**0.1775829**	0.1847511	5.1899202	2.7817019
	Mean	0.0244622	**0.0175540**	1.3708137	0.2754877
	Std. Dev	0.0354981	0.0342402	1.9309282	**0.5896565**
F11	Best	**0.0001144**	0.0002162	0.0005270	0.0004651
	Worst	0.1457762	**0.0461462**	0.2099147	0.2171571
	Mean	0.0235054	**0.0109893**	0.0246267	0.0324192
	Std. Dev	0.0336631	**0.0113240**	0.0424297	0.0546658
F12	Best	0.0012400	0.00060129	**0.0002671**	0.0005631
	Worst	0.2096818	0.12118576	0.2548033	**0.1101160**
	Mean	0.0272272	0.02143225	0.0359651	**0.0148429**
	Std. Dev	0.0439847	0.02431783	0.0619632	**0.0217649**
F13	Best	0.0087550	0.0107647	**0.0012005**	0.0282847
	Worst	0.3492618	0.2768628	3.3588345	**0.1138077**
	Mean	0.1156770	0.1164427	0.1773360	**0.0704852**
	Std. Dev	0.0799586	0.0717128	0.5915495	**0.0217889**
F14	Best	0.1847750	0.2974708	0.0001535	**0.0000007**
	Worst	1.2071960	**1.2054814**	392.06496	5.2899286
	Mean	0.9051700	0.9029995	44.143876	**0.5769940**
	Std. Dev	0.2900020	**0.2855825**	78.218139	0.9586741
F15	Best	0.0000000	0.0000000	0.0000000	0.0000000
	Worst	0.0000000	0.0000000	0.0000000	0.0000000
	Mean	0.0000000	0.0000000	0.0000000	0.0000000
	Std. Dev	0.0000000	0.0000000	0.0000000	0.0000000

by a Shapiro-Wilk test, therefore ANOVA statistical test is used to verify statistical significance the results. All three variants are tested against the ABC as the control population. Significance (α) is set to 95%, while power (β) is set to 0.85. All implementations were written in Python programming language and run on a 4.2 GHz Intel processor and 16 GB of RAM. Results are summarized in Table 2.

Results indicate that the proposed approach influenced the performance of the ABC in several instances. Excluding cases where all algorithms reached the same global optimum (F2, F6 and F15), ANOVA ($p < 0.05$), has pointed statistical significance, for better or for worse. The variants were able to reach values closer to the optimum than the original algorithm, albeit unreliably in some cases, such as in F5 and F13. The unreliability in these instances can be explained by the nature of the functions, low dimensional and multimodal problems with many deceptive local optima in which dABC2 and dABC3 had issues to displace solutions out of local optima due to premature convergence. The same could be said about F10 and F11, which underperformed in comparison to the ABC. On the other hand, in high dimension cases, such as F1, F9 and F12, the performance of the variants dominated over the canonical algorithm, what could be attributed to the novelty factor in the scout cycle.

5   Conclusion

In this work, a decision variable selection method based on Cantor's diagonalization argument was proposed to be integrated in the employed and scout bees steps of the Artificial Bee Colony Algorithm. The new mechanism was tailored to deterministically choose decision variables to be updated in the first phase and introduce novelty to stagnated solutions that undergo the scout bees phase. Potential positive outcomes to the ABC were investigated by testing the variants against the canonical algorithm on several instances of continuous unconstrained problems. Results indicate general better performance in multimodal problems with high number of dimensions, where the proposed idea implemented in the variants of the ABC were able to contribute to the robustness and solution quality.

Future research includes the integration of the concepts of the diagonalizations to popular adaptations of the ABC to be tested against the state-of-the-art. One such is the gbest ABC of Zhu and Kwong [13] analyzed by Akay and Karaboga in one of their recent works [3]. Additionally, analysis and experimentation of the parameter selection for continuous constrained and combinatorial optimization problems is considered due to the possibility of promising results brought by the novelty factor to the candidate solutions by extracting the diagonal. Another research avenue is to employ the proposed method for shallow networks [5,6] optimization. Such networks may benefit from our optimization strategy, since it tackles small sample size problems.

References

1. Akay, B., Karaboga, D.: Artificial bee colony algorithm for large-scale problems and engineering design optimization. J. Intell. Manuf. **23**(4), 1001–1014 (2012)
2. Akay, B., Karaboga, D.: A survey on the applications of artificial bee colony in signal, image, and video processing. English. Sig. Image Video Process. **9**(4), 967–990 (2015)
3. Akay, B.B., Karaboga, D.: Artificial bee colony algorithm variants on constrained optimization. Int. J. Optim. Control: Theor. Appl. (IJOCTA) **7**(1), 98–111 (2017)
4. Dauben, J.W.: Georg Cantor: His Mathematics and Philosophy of the Infinite. Princeton University Press, Princeton (1990)
5. Gatto, B.B., dos Santos, E.M.: Discriminative canonical correlation analysis network for image classification. In: 2017 IEEE International Conference on Image Processing (ICIP), pp. 4487–4491. IEEE (2017)
6. Gatto, B.B., de Souza, L.S., dos Santos, E.M.: A deep network model based on subspaces: a novel approach for image classification. In: IAPR International Conference on Machine Vision Applications (MVA). IEEE (2017)
7. Jamil, M., Yang, X.S.: A literature survey of benchmark functions for global optimization problems. Int. J. Math. Model. Numer. Optim. **4**(2), 150–194 (2013)
8. Karaboga, D.: An idea based on honey bee swarm for numerical optimization. Tech. report. Erciyes University (2005)
9. Karaboga, D., Basturk, D., Ozturk, C.: Artificial bee colony (ABC) optimization algorithm for training feed-forward neural networks. In: Modeling Decisions for Artificial Intelligence, vol. 4617, pp. 318–319 (2009)
10. Molga, M., Smutnicki, C.: Test functions for optimization needs, p. 101 (2005)
11. Tereshko, V., Loengarov, A.: Collective decision-making in honey bee foraging dynamics. Comput. Inf. Syst. **9**, 1–7 (2005)
12. Weisstein, E.W.: CRC Concise Encyclopedia of Mathematics. Chapman and Hall/CRC, London (2002)
13. Zhu, G., Kwong, S.: Gbest-guided artificial bee colony algorithm for numerical function optimization. Appl. Math. Comput. **217**(7), 3166–3173 (2010)

An Ensemble of Deep Auto-Encoders for Healthcare Monitoring

Ons Aouedi[1]([✉]), Mohamed Anis Bach Tobji[1,2], and Ajith Abraham[3]

[1] ESEN-University of Manouba, Manouba, Tunisia
{ons.aouedi,anis.bach}@esen.tn
[2] LARODEC Laboratory, ISG-University of Tunis, Tunis, Tunisia
[3] Machine Intelligence Research Labs (MIR Labs),
Scientific Network for Innovation and Research Excellence, Auburn, WA 98071, USA
ajith.abraham@ieee.org

Abstract. Ambient Intelligence (AmI) is a new paradigm that redefines interaction between humans, sensors and flow of data and information. In AmI, environment is more sensitive and more responsive to the user who acts spontaneously in the foreground, while sensors, machines and intelligent methods act in background. Behind the AmI interfaces, a huge volume of data is collected and analysed to make decision in real time. In the field of health care, AmI solutions prevent the patients from emergency situations by using data mining techniques. This paper assess the performance of deep learning against traditional dimensionality reduction and classification algorithms for healthcare monitoring in a hospital environment. An ensemble method is proposed using three classifiers on a reduced version of the dataset, where the dimensionality reduction technique is based on deep learning. The evaluation based on different performance metrics like accuracy, precision, recall and f-measure illustrated reliable performance of the proposed ensemble method.

1 Introduction

According to the World Health Organization, coronary heart disease and stroke are the major cause of death in the world and reached more than 26% in the last 15 years [1]. This figure shows that traditional health care services are inefficient to handle classical diseases we can prevent in advance. Thus, Ambient intelligence (AmI) for healthcare may be an effective solution to support people with chronic illness, identify abnormal activities for elderly persons and detect emergency situations that require intervention. AmI introduces a new way of interaction between people and technology where a human behaves spontaneously while machines and artificial intelligence act in the background to improve his comfort and to reduce the risks he may face.

AmI is based on Internet of Things where the collection and analysis of data make it pro-active and effective to predict abnormal situations. Indeed, AmI for health care generates a massive data from heterogeneous sources such as body, environment and other discrete sensors. Collected data need to be extracted and

A. M. Madureira et al. (Eds.): HIS 2018, AISC 923, pp. 96–105, 2020.
https://doi.org/10.1007/978-3-030-14347-3_10

analysed upstream, to allow then prediction of the nature of new situations to provide accurate decisions. Although Data mining is plentiful of classification techniques, new methods are emerging to improve quality of prediction, either for dimensionality reduction, especially for data that include a huge number of attributes, or in the classification techniques themselves. In this context, deep learning techniques showed last years very interesting results for these two data mining tasks.

This paper analysed a hospital health care monitoring dataset based in the Baraha Medical City in Khartoum North in Sudan [19]. The aim is to assess the performance of Deep Autoencoders against traditional dimensionality reduction algorithms such as PCA and univariate feature selection. These methods are applied on the aforementioned dataset. The comparison considered the effect of the selected/extracted features on the prediction accuracy of several classifiers. We constructed a new ensemble method by combining the best classifiers based on Deep Autoencoders. The classification quality of the proposed method outperformed the existing ones as shown in the experimental results where we used several common classification measures such as accuracy, precision, recall, F-measure, sensitivity, specificity, loss, and ROC Area.

The rest of the paper is organized as follows: Sect. 2 presents the main dimensionality reduction and classification methods used throughout this paper. Section 3 presents the experimental results performed on our healthcare dataset for classification, while Sect. 4 discusses the different results obtained in the experiments. Finally, Sect. 5 summarizes the paper.

2 Preliminaries

2.1 Dimensionality Reduction Techniques

In data sets with a very high number of attributes, dimensionality reduction techniques are a must to produce an accurate model of classification in a reasonable time.

2.1.1 Deep Learning and Autoencoder

Deep learning (DL) is a machine learning method based on the Artificial Neural Network (ANN). DL has been applied to several fields such as natural language processing, speech recognition, image recognition and produced very interesting results that approached human expert classification. Based on the traditional ANNs, deep neural networks contain a great number of intermediate layers, where in each level, the DL method learns to transform input data into a composite representation. The main difference with the ANNs is the fact it eliminates the task of feature extraction for the handled dataset. Recently, DL has become an efficient solution for machine health monitoring system. It is used for early diagnosis of Alzheimer's disease [17]. In [20] the authors apply deep learning algorithms to identify autism spectrum disorders (ASD) patients from the large brain imaging dataset, based solely on the patient's brain activation patterns.

DL models have different variants, like Deep Autoencoder, Deep Belief Network, Deep Boltzmann Machines, Convolutional Neural Networks and Recurrent Neural Network. In this research, we used Deep Autoencoder to extract features.

An autoencoder is one of the several architectures of deep learning algorithms [21]. It is an unsupervised learning method [18]. It is a feedforward network trained to reproduce its input at the output layer which compresses the original data to produce the code through the encoder, then the decoder decompress the code to reconstruct the output [7]. Thereby, it consists of two core segments placed back to back:

1. The encoder: Takes input data and maps it to be hidden representation, when the neurons of this hidden layer are less dimension than the input data, the encoder reduces the dimensionality.
2. The decoder: Uses the hidden representation to reconstruct the input.

2.1.2   Principal Component Analysis

Principal Component Analysis (PCA) is a feature extraction method that was developed by Pearson in 1901 [13] and Hotelling in 1993 [15]. This method acts on the numerical dataset. It allows to reconstruct K uncorrelated variables from n correlated variables, while $K < n$. These K variables are the principal components.

2.1.3   Univariate Feature Selection

Univariate feature selection is fast and easy to calculate and to run on data [6]. It allows to select the best variables based on statistical tests. In other words, it chooses the variables that have the strongest relationship with the target variables.

2.2   Classification Methods

Classification is a data mining task. It takes some kind of input data and map it to a discrete label. When the classifier is trained accurately, it has the ability to generate correct output given a new input [16]. In this section, we briefly introduce the classification methods considered in our experiments.

2.2.1   Decision Tree

A decision tree is tree-like structure where each node represents a test and each branch represents the outcome of the test. It is constructed by learning from a set of data, and is used to classify new instances. In this paper, we used the CART binary tree. It can be used for regression and classification problem. It was introduced in 1984 [3]. The impurity measure used in the CART algorithm is *Gini index*.

2.2.2   Random Forest

Random Forest (RF) is a supervised learning algorithm used for regression and classification task [5]. It is a combination of uncorrelated decision tree without pruning and it gets more accurate and stable prediction. Every decision tree classifies the new instance and the RF assigns the most frequent class among these classifications.

2.2.3   k-Nearest Neighbors

k-Nearest Neighbors (KNN) is a non-parametric, lazy learning algorithm and widespread technique used in classification [9]. It predicts the class of new input by calculating the distance between the new input and each training instance. There are several distances metric used with KNN but the most common distance utilized is the Euclidean distance [2].

2.2.4   Support Vector Machine

Support Vector Machine (SVM) is nonlinear and supervised machine learning algorithm [8]. The goal of SVM is to design hyperplane that classifies all training vectors in n-class (n is the number of classes). The best hyperplane is the one that leaves the maximum margin with the classes.

2.2.5   Logistic Regression

Logistic Regression (LR) is a linear algorithm used for binary classification [10]. LR is a linear method, but the prediction is converted by logistic function. It is named for this function used at the core of the method.

2.2.6   Extra-Tree

Extra-Tree is a supervised learning algorithm used for both regression or classification [14]. It combines ensemble of decision tree, but the major differences with other tree based ensemble methods are that it splits nodes by choosing cut-points fully at random and uses all learning sample (rather than a bootstrap replica) to grow the trees.

2.2.7   AdaBoost

AdaBoost is a simple algorithm used to boost the weak classifiers [12]. It trains a series of model using a weighted trainer. The final classification will be decided by a weighted vote of the classification.

2.3   Ensemble Methodology

Ensemble methods are machine learning techniques that generate multiple learners and then combine output's learners in order to get better prediction. To obtain a good ensemble, the best classifiers are selected. No method is perfect for many problems, but if we have multiple classifiers, we can run the classification on all of them and then choose the most frequent results. Bagging [4] and Boosting [11] are ensemble machine learning algorithms that are mostly used. To combine ensemble's output prediction used the majority voting method.

Voting method is generally used for classification problems. It can be weighted or not. In non-weighted voting, all the classifiers have equal weight and the predicted class is the one that has the majority votes. In weighted case, we give different weights to the classifiers. There are several ways in which we can weight the classifiers such as assigning a weight proportional to the accuracy of each classifier.

3   Experimentation and Results

In these experiments we use scikit-learn libraries and python as a programming language. Scikit-learn libraries cover all data mining steps, from loading, pre-processing to mining data. It also includes a vast collection of machine learning algorithms and most of them with minimal code adjustments. 80% of the dataset was used for training and 20% for testing.

The purpose of these experiments is the following: The first phase is to assess the performance of Deep Autoencoder with traditional dimensionality reduction techniques. The second phase is to compare the performance of classifiers with the best reduction methods obtained in the first phase. Finally, we combine the best three classifiers using the voting method to get a better classification [19].

3.1   Dataset Description

The dataset used in this work consists of 745 instances, 300 attributes and a binary attribute class (Normal/Abnormal) [19]. It was constructed by collecting 300 vital signals of patients wearing sensors in the Baraha medical city in Shambat, Khartoum North, Sudan. The hospital receives patient with chronic illness. The dataset included 30 patients in the sample. They focused on patients who suffer from chronic illness. The idea is to model the performance of the hospital as a binary classification problem (performing well or not).

3.2   Attribute Selection

We compare the efficiency of dimensionality reduction techniques on our dataset. The methods used are PCA, univariate feature selection and Deep Autoencoder. These methods were used for the same number of output features. Also, the comparison is based on classifier accuracy which use the features generated by the dimensionality reduction techniques.

For Deep Autoencoder we used keras with the tensorflow framework as a backend. Its architecture consists input layer, 3 hidden dense layers and output layer. The parameters of the Deep Autoencoder layers were as follows: the input layer has the original features of our dataset, the second layers with 100 neurons, the third one with 70 neurons, the last hidden layers with 40 neurons and the output layers with 20 neurons. The activation function is *ReLU*, and the kernel initializer is *uniform*.

The results illustrated in Table 1, show the accuracy obtained for different classifiers with the various dimension reduction methods. The best performance was demonstrated by the Random Forest using Deep Autoencoder. In addition, the classifiers using Deep Autoencoder have better results than the PCA and the univariate feature selection method in 5 out of 7 cases. Figure 1 confirms this interpretation. Through the use of different selection/extraction methods we manage to reduce this dataset to 20 input features.

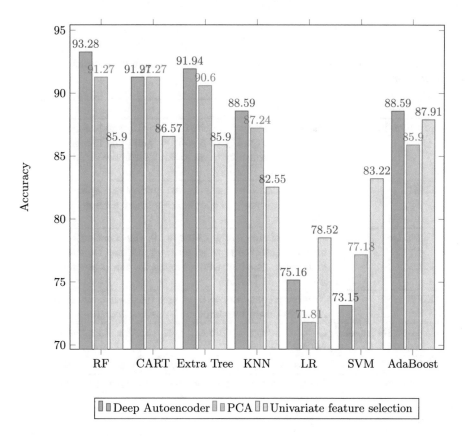

Fig. 1. Performance of classification methods with feature selection/extraction methods

3.3   Classification

We used the features generated through Deep Autoencoder to compare the performance of different classifiers. The objective is to create an ensemble method,

Table 1. Classification accuracy with Deep Autoencoder, PCA and Univariate feature selection method

Classifiers	Accuracy		
	Autoencoder	PCA	Univariate
Random Forest	93.28%	91.27%	85.90%
CART	91.27%	91.27%	86.57%
Extra tree	91.94%	90.60%	85.90%
KNN	88.59%	87.24%	82.55%
Logistic Regression	75.16%	71.81%	78.52%
SVM	73.15%	77.18%	83.22%
AdaBoost	88.59%	85.90%	87.91%

Table 2. The list of classifiers used and their parameters (default parameters available in scikit-learn).

Classifiers	Parameters
Random Forest	Number of the tree in the forest = 10; the function that measure the quality of a split = 'gini'; the number of features to consider when looking for the best split = $\sqrt{number\ of\ feature}$
CART	The function that measure the quality of a split = gini; the strategy used to choose the split at each node is "best"
Extra tree	Number of tree in the forest = 10; the function that measure the quality of a split = 'gini'; the number features to consider when looking for the best split = $\sqrt{number\ of\ feature}$
KNN	Number of neighbor = 5; weight function = uniform; algorithm used to compute the nearset neighbor = 'auto'
Logistic Regression	C = 1.0; solver = liblinear; tolerance for stopping criteria = 0.0001
SVM	The kernel type used in the algorithm is 'rbf'; C = 1.0
AdaBoost	Base_estimator = decision tree classifier; the number of estimators = 50

which combines the three most accurate classifiers. Table 2 summarizes the different parameters of each classifier used in this paper.

Tables 3 and 4 illustrate the performance of the classifiers by several measures: Accuracy, precision, recall, f-measure, specificity, sensitivity, loss, ROC Area and the execution time for each classifier.

As evident in Table 3, Random Forest, CART and Extra Tree have the highest precision and recall in comparison with the rest classifiers. The execution time of Logistic Regression is lower for classification than other classifiers. Random Forest has lower loss in comparison with the rest classifiers, while the CART algorithm has the highest loss value.

Table 3. Classifiers performance in terms of precision, recall, f-measure, loss and time required to build the models (in second)

Classifiers	Precision	Recall/Sensitivity	F-measure	Loss	Time
Random Forest	0.9323	0.9323	0.9323	0.3943	0.0631
CART	0.9117	0.9126	0.9121	3.0134	0.0286
Extra tree	0.9188	0.9188	0.9188	1.4641	0.0473
KNN	0.8938	0.8797	0.8833	1.5428	0.0314
Logistic Regression	0.7608	0.7408	0.7425	0.5210	0.0214
SVM	0.7352	0.7224	0.7237	0.5683	0.0993
AdaBoost	0.8866	0.8932	0.8945	0.6203	0.1795

Table 4 indicates that Random Forest, CART and Extra Tree have the highest specificities and accuracies than the rest of classifiers. In addition, Random Forest, Extra Tree and CART have the highest ROC Area. Therefore, these methods can distinguish between the situation of the patient more than SVM, KNN, AdaBoost and Logistic Regression.

According to these results, we created an ensemble using Random Forest, Decision Tree and Extra Tree, to obtain better classification results using the majority vote to predict the class labels.

Table 4. Classifiers performance in terms of accuracy, specificity and ROC Area

Classifiers	Accuracy	Specificity	ROC Area
Random forest	0.9328	0.9382	0.932
Extra tree	0.9194	0.9259	0.9313
CART	0.9127	0.925	0.919
KNN	0.8859	0.8651	0.880
AdaBoost	0.8859	0.8845	0.8809
Logistic regression	0.7516	0.7291	0.7291
SVM	0.7315	0.720	0.720

Table 5 depicts the performance of the new ensemble in term of accuracy, precision, f-measure, loss and execution time. In this table the results indicate that our new method has better accuracy, precision and recall than other classifiers used in this research. But, it has higher execution time than other classifiers used in this paper except SVM and AdaBoost. Also, it has high loss value.

Table 5. Performance of the new ensemble in term of accuracy, precision, recall, f-measure, loss and time required to build the model

Accuracy	Precision	Recall	F-measure	Loss	Time
94.63%	0.95	0.95	0.95	1.8544	0.06618

Table 6 depicts the performance of the new ensemble in term of specificity and ROC Area. It is inferred from this table that the new ensemble has higher sensitivity and specificity values than individual classifiers used in this paper. Also, we can notice that the new ensemble has the highest ROC Area.

Table 6. Performance of the new ensemble in term of specificity and ROC Area

Specificity	ROC Area
0.9397	0.94

The results presented in Tables 5 and 6 indicate that the new intelligent ensemble is able to offer a better decision that is more accurate than the other machine learning classifiers.

4   Discussion

The experimental result presented in the previous section suggests that the features extracted through the Deep Autoencoder are more informative than those obtained by PCA and the univariate feature selection method. However, Deep Autoencoder spends more time to generate new features than traditional feature reduction methods. In the second phase of this work, we have noticed that the worst classifiers with autoencoder are SVM, KNN and Logistic regression. So, we can infer that the modern classifiers are better with Deep Autoencoder than the traditional methods. As has been stated, the results of the new ensemble improve those obtained by individual classifier in term of the all metrics except loss and the time required to build the model. Thus, our new intelligent ensemble has the ability to offer better classification and so improves the decision support healthcare systems.

In conclusion, the new intelligent ensemble with Deep Autoencoder is an appropriate method to classify the situation of patient hospital based on vital signs from wearable sensors.

5   Conclusions

This paper attempted to improve the classification quality of performance monitoring in a hospital environment [19]. We constructed an ensemble method that combines the best three classifiers. These classifiers use the features generated by the Deep Autoencoder and we were able to reduce to 20 features (original dataset has 300 features). Performance measures using accuracy, precision, recall, f-measure, sensitivity, specificity, ROC Area and execution time illustrate that the ensemble performs better than individual classifiers.

References

1. https://www.worldlifeexpectancy.com/country-health-profile/tunisia
2. Agrawal, R., Ram, B.: A modified k-nearest neighbor algorithm to handle uncertain data. In: 2015 5th International Conference on IT Convergence and Security (ICITCS), pp. 1–4 (2015)
3. Breiman, L., Friedman, J., Stone, C., Olshen, R.: Classification and Regression Trees. Taylor & Francis, Milton Park (1984)
4. Breiman, L.: Bagging predictors. Mach. Learn. **24**(2), 123–140 (1996)
5. Breiman, L.: Random forests. Mach. Learn. **45**(1), 5–32 (2001)
6. Brewer, J.K., Hills, J.R.: Univariate selection: the effects of size of correlation, degree of skew, and degree of restriction. Psychometrika **34**(3), 347–361 (1969)
7. Charte, D., Charteb, F., Garciaa, S., del Maria, J., Jesusb, F.H.: A practical tutorial on autoencoders for nonlinear feature fusion: taxonomy, models, software and guidelines. Inf. Fusion **44**, 78–96 (2018)
8. Cortes, C., Vapnik, V.: Support-vector networks. Mach. Learn. **20**(3), 273–297 (1995)
9. Cover, T., Hart, P.: Nearest neighbor pattern classification. IEEE Trans. Inf. Theory **13**(1), 21–27 (1967)
10. Cox, D.R.: The regression analysis of binary sequences. J. Roy. Stat. Soc. Ser. B **20**(2), 215–242 (1958)
11. Freund, Y.: Boosting a weak learning algorithm by majority. Inf. Comput. **121**(2), 256–285 (1995)
12. Freund, Y., Schapire, R.E.: A decision-theoretic generalization of on-line learning and an application to boosting. J. Comput. Syst. Sci. **55**(1), 119–139 (1997)
13. Pearson, K.: LIII. on lines and planes of closest fit to systems of points in space. Lond. Edinb. Dublin Philos. Mag. J. Sci. **2**(11), 559–572 (1901)
14. Geurts, P., Ernst, D., Wehenkel, L.: Extremely randomized trees. Mach. Learn. **63**(1), 3–42 (2006)
15. Hotelling, H.: Analysis of a complex of statistical variables into principal components. **24** (1933)
16. Kotsiantis, S.: Supervised machine learning: a review of classification techniques. **31**, 249–268 (2007)
17. Liu, S., Liu, S., Cai, W., Pujol, S., Kikinis, R., Feng, D.: Early diagnosis of Alzheimer's disease with deep learning. In: 2014 IEEE 11th International Symposium on Biomedical Imaging (ISBI), pp. 1015–1018 (2014)
18. Nweke, H.F., Teh, Y.W., Al-garadi, M.A., Alo, U.R.: Deep learning algorithms for human activity recognition using mobile and wearable sensor networks: state of the art and research challenges. Expert Syst. Appl. **105**, 233–261 (2018)
19. Salih, A.S.M., Abraham, A.: Novel ensemble decision support and health care monitoring system. J. Netw. Innov. Comput. **2**, 041–051 (2014)
20. Salon, A., Franco, A., Craddock, C., Buchweitz, A., Meneguzzi, F.: Identification of autism spectrum disorder using deep learning and the ABIDE dataset. NeuroImage: Clin. **17**, 16–23 (2017)
21. Vincent, P., Larochelle, H., Bengio, Y., Manzagol, P.A.: Extracting and composing robust features with denoising autoencoders (2008)

A Real-Time Monitoring System for Boiler Tube Leakage Detection

Min-Gi Choi[1], Jae-Young Kim[1], In-kyu Jeong[1], Yu-Hyun Kim[1],
and Jong-Myon Kim[2(✉)]

[1] Department of Electrical and Computer Engineering, University of Ulsan,
Ulsan, South Korea
127cmk@naver.com, kjy7097@gmail.com,
jeonginkeyu@gmail.com, dbgus115@ulsan.ac.kr
[2] School of IT Convergence, University of Ulsan,
Ulsan, South Korea
jmkim07@ulsan.ac.kr

Abstract. In this paper, we propose a network-based boiler tube monitoring system that acquires raw data from multiple acoustic emission (AE) sensors. It can diagnose and monitor the status of boiler tubes in real time. Such a system can help prevent sudden breakdown of boiler tubes installed in thermal power plants. These measures, using the proposed network-based boiler tube monitoring system, can increase the productivity and security of power plants.

Keywords: Acoustic emission · Boiler tube · Condition monitoring ·
Deep neural network · Fault diagnosis

1 Introduction

In recent years, power generation efficiencies have increased due to the larger size of thermal power generation boiler tubes. However, due to aging, old boiler tubes undergo wear and tear that may lead to failure. The failure of boiler tubes halts the power generation process and results in huge economic losses [1]. Most large thermal power plants currently use boiler tube leakage detection systems, which rely on pressure and flow meters. Additionally, some power plants use boiler tube leakage detection systems that incorporate acoustic emission sensors. However, boiler tube leakage detection is hectic, requiring complex processes and keen knowledge to successfully accomplish the task. So far, only threshold-based techniques have been implemented in boiler tube leakage detection systems to provide an alarm when a certain threshold value is crossed. However, such techniques cannot be adapted to detect leaks in boiler tubes if certain environmental changes are encountered [2–7].

In this paper, a boiler tube leakage diagnosis and real-time monitoring system is proposed. Figure 1 shows the configuration of the proposed system. It stores the original data (i.e., the sound emission signal) collected through a data-collecting device in a

© Springer Nature Switzerland AG 2020
A. M. Madureira et al. (Eds.): HIS 2018, AISC 923, pp. 106–114, 2020.
https://doi.org/10.1007/978-3-030-14347-3_11

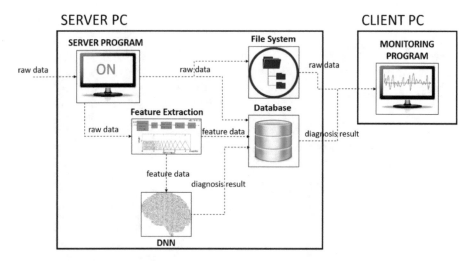

Fig. 1. Configuration of the proposed boiler tube leakage monitoring system.

database (DB) and analyzes the stored signals in the DB. The system consists of a server that extracts features from the signals and runs a deep neural network to detect boiler tube leaks. It also contains client stations that can monitor the diagnosis results in real time.

2 Boiler Tube Monitoring Server

The boiler tube monitoring server contains a server program, file system, database, feature-extraction algorithm, and deep artificial neural network. The acoustic emission (AE) signals received from the data-collection device are stored as a file, and the file path is stored in the database along with additional information, such as the time at which the signal was collected. In practice, observing only a raw AE signal is not sufficient to detect and diagnose a leak in a boiler tube. Therefore, a feature-extraction step is required to characterize signals from different states of the boiler tube. The features are stored in the database after completing the feature-extraction process. Later, the extracted features are fetched from the database and provided to a deep artificial neural network algorithm to classify the instances into respective classes. The characteristics of the original signal collected in real time are input into the learned neural network model to derive the diagnosis result. The result of the diagnosis is again stored in the database.

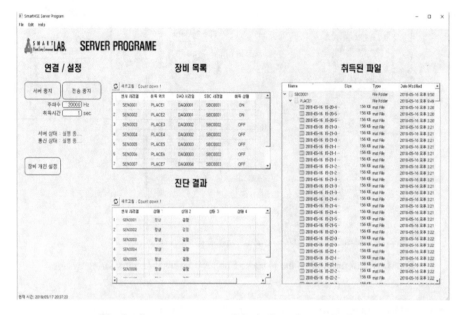

Fig. 2. Server program used for boiler tube monitoring.

2.1   Server Program Used for Boiler Tube Monitoring

Figure 2 shows a snapshot of the server program used for boiler tube monitoring, and performs several functions, such as interfacing with the data-collection device, managing the database, storing information from the data-collection device and the diagnosis results, and helping to confirm the collected data.

2.2   Database Used for Boiler Tube Monitoring

To implement the boiler tube leakage diagnosis system effectively, information management and real-time monitoring are required. These tasks cannot be achieved without storing the data reliably in a database. Figure 3 shows the architecture of the database used to store information regarding different entities related to boiler tube diagnosis. Information about the boiler tube, sensors, acquisition equipment, raw data, features, and results are stored in the database. For example, the device table stores information regarding the equipment through which the original data is acquired, and the data table stores the path associated with the location of the original data. The model information table is used to identify a model derived from the diagnosis results.

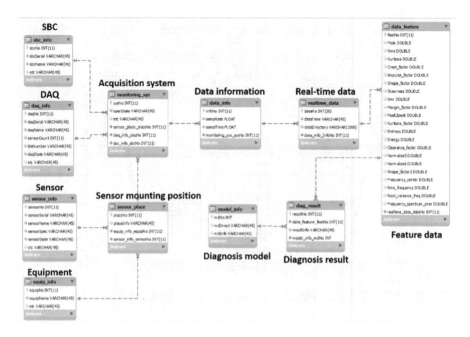

Fig. 3. Database structure used for boiler tube monitoring.

2.3   Feature Extraction

Figure 4 shows a flowchart of the feature-extraction process. The feature-extraction method is used to recognize the pattern according to the changing state of the raw data acquired from the boiler tube. Features are essential to obtain representative values that can be used to identify a defect from the signal. Moreover, the feature space is a compressed representation of the original signals that both boosts the classification performance and reduces the learning time. In the proposed method, features are extracted from time and frequency domains. In the time domain, features such as the margin factor, peak-to-peak value, and impulse factor are extracted. Alternatively, the AE signals are converted into the frequency domain with the help of the fast Fourier transform (FFT). In the frequency domain, features such as the frequency center and root variance frequency are extracted. Table 1 shows the definition and expressions of features extracted from the time and frequency domains.

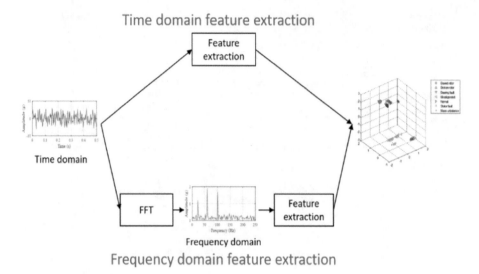

Fig. 4. Feature space representation using tine- and frequency-domain features.

Table 1. Features considered in this study

Name	Definition	Expression
Root mean square (RMS)	Meaning of the amount or the amount of the sound and the amount of the continuous waveform	$RMS = \sqrt{\frac{1}{N}\sum_{n=1}^{N}x(n)^2}$
Skewness	A measure of the asymmetry of the probability distribution	$Skewness = \frac{1}{N}\sum_{n=1}^{N}\left(\frac{x(n)-\bar{x}}{\sigma}\right)^3$
Impulse factor	An indicator of the magnitude of the largest impulse in the signal	$IF = \dfrac{\max(|x(n)|)}{\frac{1}{N}\sum_{n=1}^{N}|x(n)|}$
Kurtosis	An indicator of the degree of sharpness of the probability distribution	$Kurtosis = \frac{1}{N}\sum_{n=1}^{N}\left(\frac{x(n)-\bar{x}}{\sigma}\right)$
Kurtosis factor	A modified value of kurtosis that compensates for the fact that kurtosis is sensitive to the size of the whole signal	$KF = \dfrac{Kurtosis}{\left(\frac{1}{N}\sum_{n=1}^{N}x(n)^2\right)^2}$
Square mean root (SMR)	An indicator related to the amplitude or magnitude of the difference between the sound and the amount of the continuous waveform; similar to RMS, although SMR is more sensitive to the signal size than RMS	$SMR = \left(\frac{1}{N}\sum_{n=1}^{N}\sqrt{|x(n)|}\right)^2$
Peak-to-peak value	An indicator of the overall width of the signal, i.e., the difference between the smallest and largest in the signal	$PP = \max(x(n)) - \min(x(n))$

(continued)

Table 1. (*continued*)

Name	Definition	Expression
Margin factor	Significant difference between the minimum/maximum values compared to the average signal size	$MF = \dfrac{\max(\lvert x(n)\rvert)}{\left(\frac{1}{N}\sum_{i=1}^{N}\sqrt{\lvert x_i\rvert}\right)^2}$
Crest factor	Equivalent to the margin factor, this is the difference between the minimum and maximum values relative to the average size of the signal; it uses RMS instead of SMR for the average size	$CF = \dfrac{\max(\lvert x(n)\rvert)}{RMS}$
Shape factor	Related to the ratio of the magnitude of the continuous waveform and the average of the signal	$SF = \dfrac{RMS}{\frac{1}{N}\sum_{n=1}^{N}\lvert x(n)\rvert}$
Frequency center	The mean frequency domain	$FC = mean(S)$
RMS of frequency	The RMS value in the frequency domain	$RMSF = \sqrt{\frac{1}{N}\sum_{n=1}^{N}S(n)}$
Root variance frequency	Related to the dispersion of frequency values in the frequency domain	$RVF = \sqrt{\frac{1}{N_f}\sum_{n=1}^{N_f}(S(n)-mean(S))^2}$

2.4 Deep Neural Network

Figure 5 shows a flowchart of the deep neural network (DNN) used in this study to diagnose boiler tube leaks. In the model creation and learning process, features from the raw data are extracted and the DNN model is trained on the extracted features. In the real-time diagnosis process, characteristic features are extracted from the raw data, which is acquired in real-time. The DNN is a variant of a standard artificial neural network, in which multiple layers are stacked on each other to create a deep architecture. The DNN deals more efficiently with complex data, as compared to classical shallow machine learning algorithms. It consists of only fully connected layers, in which node-to-node connectivity exists between the layers. The numbers of nodes in the input layer and output layer are set to 21 and 2, respectively. There are three hidden layers in the developed DNN, each of which consists of 30 nodes. The dataset consists of instances taken from the boiler tube in normal and leakage states.

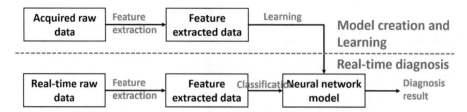

Fig. 5. Work flow of boiler tube leak detection using the DNN.

3   Client Program Used for Boiler Tube Monitoring

The client station receives the original signal, extracted features, and diagnosis results from the server and outputs them to the screen. Figure 6 shows a snapshot of the client program. It is interfaced with the server and is used to visualize the diagnosis results, the channel being used for data acquisition, the original files collected through different channels, and the features extracted from the original data. Moreover, when a specific channel is selected, information (e.g., the data file acquired through the channel, the original signal being acquired, and the frequency spectrum) is displayed on the screen. When a specific collected data file is selected, the feature value of the data can be visualized (Fig. 7).

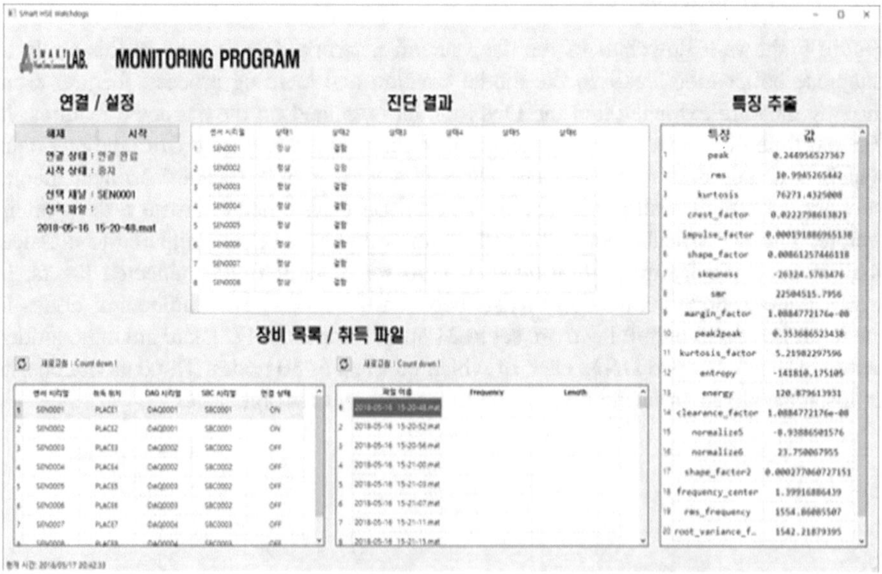

Fig. 6. Client program used for boiler tube monitoring.

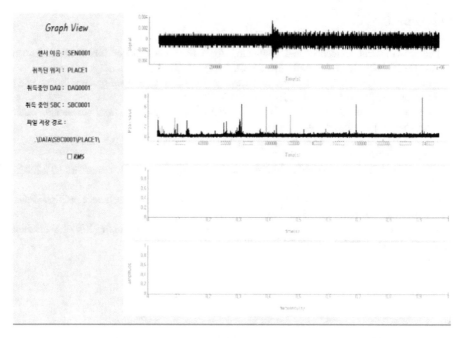

Fig. 7. Visualization results in the client program.

4   Conclusion

In this paper, a network-based boiler tube leakage diagnosis monitoring system was proposed. The proposed system consists of two parts: a server and client stations. AE signals associated with the normal and leakage states of a boiler tube were collected with help of AE sensors. The server station is used to run a server program, file system, database, feature extraction, and deep neural network for the diagnosis of boiler tube leaks. Additionally, the client station is used to visualize information related to data acquisition channels, acquired data, and extracted features. The proposed boiler tube monitoring system is effective for boiler tube leak detection and diagnosis and can be used to prevent sudden failure of boiler tubes installed in thermal power plants.

Acknowledgement. This work was supported by the Korea Institute of Energy Technology Evaluation and Planning (KETEP) and the Ministry of Trade, Industry & Energy (MOTIE) of the Republic of Korea (No. 20161120100350).

References

1. Lee, S.-G.: Leak detection and evaluation for power plant boiler tubes using acoustic emission. J. Korean Soc. Nondestruct. Test. **24**(1), 45–51 (2004)
2. Kim, J.-M., Kim, J.-Y.: Methods and devices for diagnosing facility conditions. Patent Registration No. 10-1818394 (2018)
3. Kim, J.-M., Kim, J.-Y.: Methods and devices for diagnosing machine faults. Patent Registration No. 10-1797402 (2018)
4. Kim, J.-M., Kim, J.-Y.: Probability density based fault finding methods and devices. Patent Registration No. 10-1808390 (2017)
5. Kim, J.-M., Kim, J.-Y.: Machine fault diagnosis method. Patent Registration No. 10-1808390 (2017)
6. Kim, J.-M., Kim, J.-Y.: Apparatus and method for monitoring machine condition. Patent Registration No. 10-1745805 (2017)
7. Kim, J.-M., Kim, J.-Y.: Method and apparatus for predicting remaining life of a machine. Patent Registration No. 10-1808461 (2017)

An Intelligent Data Acquisition and Control System for Simulating the Detection of Pipeline Leakage

Jaeho Jeong[1], In-Kyu Jeong[1], Duck-Chan Jeon[2],
and Jong-Myon Kim[3(✉)]

[1] Department of Electrical and Computer Engineering,
University of Ulsan, Ulsan, South Korea
jaeho9929@gmail.com, jeonginkeyu@gmail.com
[2] ICT Convergence Safety Research Center,
University of Ulsan, Ulsan, South Korea
dcjeon@ulsan.ac.kr
[3] School of IT Convergence, University of Ulsan, Ulsan, South Korea
jmkim07@ulsan.ac.kr

Abstract. The leakage of gas or oil pipelines causes tremendous economic losses and damages the environment. In this paper, we propose a method to control a testbed to simulate the pinhole condition of pipelines and a data acquisition system that can acquire data from the testbed. In the experiment, we artificially generate two different sizes of pinholes (2 mm and 0.3 mm) and collect four different types of data (pressure, flow, temperature, and acoustic emission) with and without pinholes to detect leaks in a pipeline. Experimental results show that all the sensors except the temperature sensor can detect a pinhole of 2 mm, but not a pinhole of 0.3 mm. Overall, we observe that the proposed method is affordable to simulate leaks in a pipeline.

Keywords: Fault diagnosis · Leakage detection · Pipeline testbed · Acoustic emission

1 Introduction

Due to the rapid increase of energy demand worldwide, various types of energy development, transportation, and utilization projects are underway [1, 2]. According to ExxonMobil, energy demand is expected to increase by 25% by 2040, especially in countries such as China and India, which are not part of the Organization for Economic Cooperation and Development [3]. Energy resources that are transported from production areas to consumption areas are in the gas or liquid state. For the safe transportation of these fluids, construction of energy transportation pipelines is essential. For example, in Europe, about 30% of total consumption is supplied through pipelines [4]. Therefore, the reliability of piping, which is a means of energy transport, is significant.

A pinhole is a fine hole on the pipeline surface that can occur during piping operation or manufacturing and that can lead to internal corrosion, fluid leakage, or explosion due to the flammable nature of the fluids inside. To detect leakage in a pipeline, data

© Springer Nature Switzerland AG 2020
A. M. Madureira et al. (Eds.): HIS 2018, AISC 923, pp. 115–124, 2020.
https://doi.org/10.1007/978-3-030-14347-3_12

collected from the pipeline with the pinholes is required. However, it is difficult to artificially simulate pinholes or to collect such data from the pipelines used in industry because the pipelines are not easily accessible. Therefore, it is necessary to design a testbed that can simulate leakage of a pipeline. That said, it is difficult to obtain data similar to actual leakage data without a proper control system. In this paper, we propose a testbed of control and data collection to simulate and evaluate leaks in a pipeline.

2 Related Work

2.1 Acoustic Emission Sensor

Ultrasonic waves resemble normal sound waves because of their physical properties with the exception that humans cannot hear them. An acoustic emission (AE) signal is a type of ultrasonic wave and can be thought of as an elastic energy wave generated by a sudden release of strain energy formed inside a medium. These emissions can be measured through an AE sensor that quantifies the elastic waves in the form of signals coming from openings caused by improper welding or structural cracks. In addition, it is used as a mean of structural safety monitoring to check the spalls and cracks of bearings in rotary machines [5].

Recently, a study has shown that AE signals can be effectively used for pinhole detection in a pipeline [6]. In the study, a fast Fourier transform of the acoustic emission signals was first calculated to characterize the natural frequency of the pipes used in the experiment. Then the processed signals were provided to the support vector machine to detect pinholes in the pipe. In addition, several studies explored the detection of corrosion as well as pinholes in a pipeline with the help of AE signals [7].

3 Proposed System

3.1 System Configuration

3.1.1 Sensor Unit

To simulate an artificial pinhole in a pipeline, four bolts with 0.3, 0.5, 1, and 2 mm pinholes was fabricated. The fabricated bolts with pinholes can be seen in Fig. 1. In addition, a solenoid water valve was installed at the center of the pipeline, and the pinhole was remotely simulated by opening and closing the solenoid water valve, as shown in Fig. 2.

Fig. 1. The fabricated bolts with four different size of pinholes.

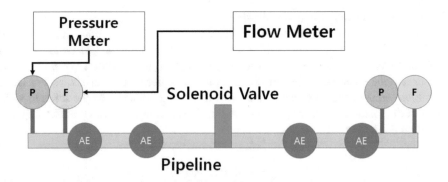

Fig. 2. Sensor unit configuration.

3.1.2 Control Unit

To control the 220-V-rated solenoid valve, we used an Arduino UNO as a controller and relay module, which received a 5 V DC signal as an input to match the level of general-purpose input/output signal.

3.1.3 Data Acquisition Unit

Figure 3 shows an overview of the data acquisition system. To acquire the AE sensor data, we used a PCI-2 AE, which is an AE-signal-collecting device manufactured by MISTRAS. The flow/pressure gauge data were obtained by converting 4–20 mA data into digital data by using a RADIONODE UA20. An external trigger mode was used to synchronize the solenoid valve operation time and data acquisition time. After triggering the system, the pressure/flowmeter and AE data were acquired. After a certain period, the solenoid valve was remotely controlled to generate the effect of a pinhole in the pipe. The collected data were used to detect a pinhole in the pipe.

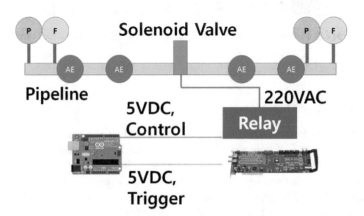

Fig. 3. Overview of the solenoid valve control and the data acquisition system.

3.1.4   Result of a Testbed Application

For compatibility with the PCI-2 AE collector, a WDI-AST 100–900 kHz Wideband Differential AE sensor from MISTRAS and a solenoid valve (MPW-2120) from HYOSIN Corporation were used in the experiment. In the proposed system, the AE sensor and the solenoid valve were attached to the center section, as shown in Fig. 4. The thermometer, pressure gauge, and flowmeter were attached to the left and right sides of the solenoid valve, as presented in Fig. 5, and compared with the data acquired from the AE sensor.

Fig. 4. Development of the testbed.

Fig. 5. Pipeline testbed implementation: left sensor unit (top), right sensor unit (bottom).

4   Results

4.1   Temperature Data

Temperature data were tested as potential indicators of 0.3 mm and 2 mm pinhole conditions. In total, 957 data samples over a 5 min interval were collected for each case. Steady state was simulated for 3 min and the pinhole state was simulated for the remaining 2 min by using the testbed control system. Figure 6 shows the results. The upper part of the graph shows the data obtained when a pinhole having a diameter of 2 mm was used, and the lower part shows data obtained when a pinhole having a diameter of 0.3 mm was used.

As time elapsed, the temperature rose regardless of the point at which the pinhole was generated. This was due to the heat generated by the pump motor, which produced the fluid pressure in the pipe. We observed that data obtained from the left sensor part was directly affected by the heat generation, but the right sensor part was less affected by the heat generation.

We could not detect pinholes with a diameter of 2 mm or less. The reason is that the characteristics corresponding to the time at which the pinholes were generated could not be visually confirmed.

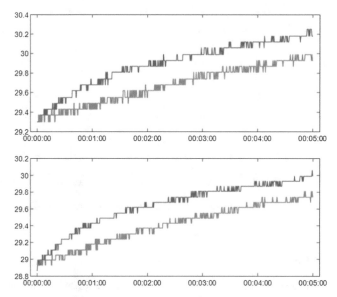

Fig. 6. The temperature data, in the case of the 2 mm pinhole (top), and the 0.3 mm pinhole (bottom).

4.2   Flowmeter Data

A flowmeter shows how much fluid flows in a pipe per unit time through a reference point. The pinhole was assumed to be 0.3 mm and 2 mm in diameter. In total, 957 data samples were recorded for 5 min, in which the steady state samples were recoded for 3 min and the pinhole state for 2 min. Figure 7 show the results. The magnitude shown in the lower graph decreased if a pinhole having a diameter of 2 mm were generated, while the magnitude in the upper graph increased at 3 min, which was the time when the pinhole was generated. Moreover, a small curve can also be observed in the upper graph. Thus, we assume that the data acquired through the flowmeter sensor cannot be used as a criterion for the detection of a pinhole having a diameter of 2 mm because it exhibited a relatively smaller change than the numerical value of the noise on the graph.

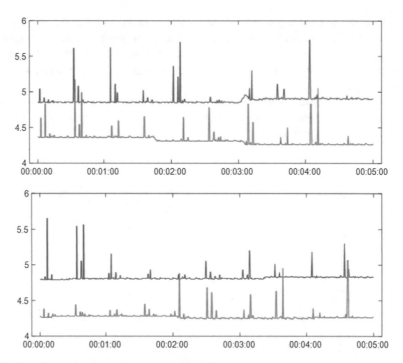

Fig. 7. The flowmeter data, in the case of the 2 mm pinhole (top) and the 0.3 mm pinhole (bottom).

4.3 Pressure Data

Pressure data was also recorded for pinholes with a diameter of 0.3 mm and 2 mm. In total, 957 data samples were recorded over 5 min in which steady-state data were recorded for 3 min and the pinhole state for 2 min.

The steady state and the pinhole state were simulated using a testbed control system. The results are shown in Fig. 8. For the experiment, pipes were filled with water having a pressure of 2.5 bar. In the case of a 2-mm-diameter pinhole, the pressure was increased to 2.4 bar after 3 min while it was 2.0 bar magnitude after 2.4 min of the data recording. There was a significant variation in the magnitude of pressure between the steady state and the pinhole state of the pipe filled with a fluid. Therefore, pressure data for the fluid flowing in the pipe can be used as reference data for determining the presence of a pinhole having a diameter of 2 mm. However, the pinhole having a diameter of 0.3 mm in the lower graph did not show any significant variation in the pressure values after 3 min of the data recording when the pinhole occurred in the pipe, as shown in Fig. 8. Hence, it is proved that the pressure data cannot be used to detect fine pinholes such as pinholes having 0.3 mm diameter or less.

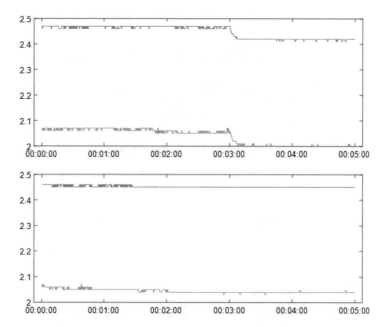

Fig. 8. The pressure data, in the case of the 2 mm pinhole (top and the 0.3 mm pinhole (bottom).

4.4   Acoustic Emission Data

The channel arrangement of the AE sensor is shown in Fig. 9. The sampling frequency for data acquisition was set to 1 MHz. To remove the influence of the solenoid control valve, the graph represents 0.1 s of steady-state data from 1 s before the valve was operated, and the data of the pinhole state is recorded for 0.1 s after the solenoid valve was operated. The results for the 2 mm diameter pinhole simulation are shown in Figs. 10 and 11.

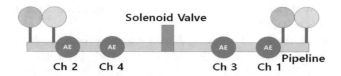

Fig. 9. Channel placement of AE sensors.

As can be seen from the data acquisition results, the data obtained through the AE sensor had a difference of at most a factor of ten depending on whether there was a pinhole or not. Therefore, it can be concluded that an AE sensor can be applied for the detection of a pinhole having a diameter of 2 mm.

Figures 12 and 13 show the data obtained from the testbed having a pinhole size of 0.3 mm. To reduce the effect of the solenoid valve operation, the data represents 0.1 s

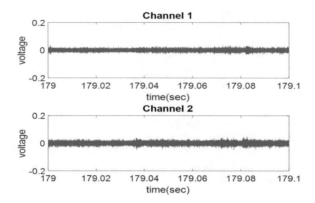

Fig. 10. State-state data acquisition results: AE sensor channel 1, AE sensor channel 2 (179.0–179.1 s).

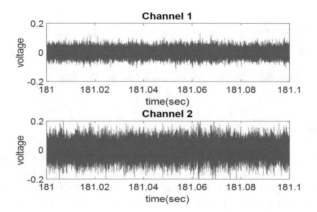

Fig. 11. 2 mm pinhole data acquisition results: AE sensor channel 1, AE sensor channel 2 (181.0–181.1 s).

of steady-state data from 1 s before the valve was operated, and for 0.1 s after the solenoid valve was operated.

It is clear that there was not a significant difference between the sensor values before and after the simulation of the pinhole state. This means that in the case of a pinhole having a diameter of 0.3 mm or less, visualizing signal values along the time axis is not sufficient, or analysis cannot be performed only by the AE sensor itself. Therefore, data from the other type of sensors combined with AE data are required to yield satisfactory results.

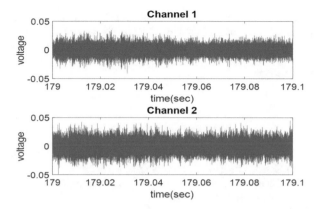

Fig. 12. Steady-state data acquisition results: AE sensor channel 1, AE sensor channel 2 (179.0–179.1 s).

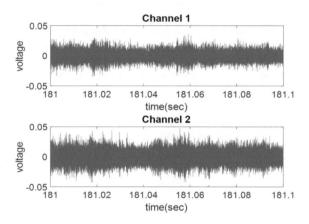

Fig. 13. 2 mm pinhole state data acquisition results: AE sensor channel 1, AE sensor channel 2 (181.0–181.1 s).

5 Conclusion

In this paper, a testbed control and data acquisition system for pipeline leakage simulation was implemented. It consisted of a sensor unit, a control unit, and a data acquisition unit. The data obtained through sensors with the help of the implemented system was analyzed, and the correlation between different sizes of pinholes and the sensor data was monitored on the time axis. Experimental results showed that all the sensors except the temperature sensor can detect a pinhole of 2 mm, but not a pinhole of 0.3 mm.

Acknowledgement. This work was supported by the Korea Institute of Energy Technology Evaluation and Planning (KETEP) and the Ministry of Trade, Industry, & Energy (MOTIE) of the Republic of Korea (No. 20172510102130). It was also funded in part by the Ministry of

SMEs and Startups (MSS), Korea, under the "Regional Specialized Industry Development Program (R&D or non-R&D, P0003086)" supervised by the Korea Institute for Advancement of Technology (KIAT), and in part by the Basic Science Research Program through the National Research Foundation of Korea (NRF) funded by the Ministry of Education (2016R1D1A3B03931927).

References

1. Cho, S., Kim, J.: Feasibility and impact analysis of a renewable energy source (RES)-based energy system in Korea. Energy **85**, 317–328 (2015)
2. Zou, C., Zhao, Q., Zhang, G., Xiong, B.: Energy revolution: from a fossil energy era to a new energy era. Nat. Gas Ind. B **3**, 1–11 (2016)
3. Ji, G.-S., Kim, W.-S.: Strain-based design method of energy pipeline. Mech. J. **54**(1), 32–37 (2014)
4. Outlook for Energy: A View to 2040, Exxon Mobil, 3 (2017)
5. Utrasonic and Acoustic Emission. http://astint.co.kr/archives/36443
6. Kim, J.Y., Jeong, I.-K., Kim, J.-M.: Acoustic emission based early fault detection and diagnosis method for pipeline. Asia-Pac. J. Multimed. Serv. Converg. Art Humanit. Sociol. **8** (3), 571–578 (2018)
7. Kim, J.Y.: Acoustic emission-based corrosion detection of spherical tank using deep learning technique. In: Annual Conference of the Korean Gas Society, vol. 5, p. 173 (2018). (n.d.)

Assessing Ant Colony Optimization Using Adapted Networks Science Metrics

Sergio F. Ribeiro[ID] and Carmelo J. A. Bastos-Filho$^{(\boxtimes)}$[ID]

Polytechnic University of Pernambuco, University of Pernambuco, Recife, Brazil
{sfr,carmelofilho}@ecomp.poli.br

Abstract. This paper presents a method to assess the state of convergence of Ant Colony Optimization algorithms (ACO) using network science metrics. ACO are inspired by the behavior of ants in nature, and it is commonly used to solve combinatorial optimization problems. Network Science allows studying the structure and the dynamics of networks. This area of study provides metrics used to extract global information from networks in a particular moment. This paper aims to show that two network science metrics, the Clustering coefficient and the Assortativity, can be adapted and used to assess the pheromone graph and extract information to identify the convergence state of the ACO. We analyze the convergence of the four variations of the ACO in the Traveling Salesman Problem (TSP). Based on the obtained results, we demonstrate that it is possible to evaluate the convergence of the ACO for the TSP based on the proposed metrics, especially the adapted clustering coefficient.

Keywords: Swarm intelligence · Ant Colony Optimization ·
Networks science · Network science metrics

1 Introduction

The convergence analysis of population-based algorithms from the computational intelligence field is a challenge. The detection of the convergence state is not trivial since in many cases the algorithm still has room for finding an even better solution in future iterations. However, in other cases, there is no diversity within the population to escape from a local minimum, and further iterations are unnecessary. Besides the ability to detect the convergence and avoid additional computations, the real-time knowledge regarding the algorithm convergence state can allow the use of adaptive mechanisms for the algorithms, thus creating the opportunity to develop more advanced algorithms.

Proposed by Dorigo *et al.* [3,5], the ACO is a family of meta-heuristics inspired by the behavior of ants in nature searching for food sources. For the sake of simplicity, we refer to the simple reactive agents, called Artificial Ants, as Ants in the rest of the paper. The ants search for the best paths in a graph in the ACO. They perform this search in the environment by hopping between the nodes. Along the movements, the ants can create trails by depositing pheromone

© Springer Nature Switzerland AG 2020
A. M. Madureira et al. (Eds.): HIS 2018, AISC 923, pp. 125–134, 2020.
https://doi.org/10.1007/978-3-030-14347-3_13

in the passing edges, and these trails can be used as a communication mechanism, representing an exchange of information among the ants. The pheromone graph is formed based on the pheromones dropped by the ants, and the encoded information of this pheromone graph can represent possible solutions.

The number of ants passing through different links can be diverse, and this generates a weighted pheromone graph. Besides, the amount of pheromone dropped by an ant can vary depending on the deployed variations of the ACO. At the end of each iteration, the pheromone weight in each link indicates the intensity of communication between the ants through the interaction with the environment, and useful information can be extracted from this [4].

The use of some network science tools allows the study of the structural and dynamic behavior of graphs using metrics that enable one to evaluate the distribution and evolution of the network weights along the iterations [2]. Metrics, such as the number of subgraphs and the degree of connectivity [10,12,14], can also directly provide global information about the current state of the network. Both the set of traveled paths or the set of pheromone trails can be viewed as networks. From this perspective, one can apply network science metrics to extract information. While the path graph represents the path traversed by each ant, allowing the knowledge of the best path, the pheromone graph describes how the communication between the ants is occurring and enables to analyze the behavior of the algorithm as a whole. The pheromone graph varies during the execution according to the indirect communication between ants, thus reflecting its current state. The application of network science metrics becomes interesting since the evolution of the pheromone graph demonstrates how the interaction evolves between the ants within the algorithm, turning possible the determine the convergence state by analyzing the pheromone graph.

This paper proposes a method to evaluate the state of convergence of the Ant Colony Optimization (ACO) algorithm [3] using adapted network science metrics [7]. The application of network science metrics is performed directly to the ACO pheromone graph that is generated by the ants during the search for a solution and represents the communication between the ants. The ACO used in this work is the Ant Colony System (ACS) [4], but one can be apply the proposal to any variation of the ACO.

The remainder of the paper is organized as follows: Sects. 2 and 3 present the theoretical background and related works; Sect. 4 introduces our proposal; Sect. 5 describes our simulation setup; Sect. 6 depicts the obtained results, and Sect. 7 presents the conclusions and future works.

2   Theoretical Background

Network Science is the study of the theoretical foundations of the network structure, its dynamic behavior and the application of networks to various subfields. Some networks have specific characteristics, both in their structure and in their behavior. Thus, one can use complex network metrics to analyze these characteristics [6,12]. A network is, in its purest form, a collection of nodes (also called vertices) joined in pairs by edges [7]. Many aspects of these systems are interesting to study, some people study the components individually, while others study connections and interactions between components. But there is a third aspect of the interaction of these systems, often neglected, but almost always crucial to the behavior of the system, which is the pattern of links between the components [1]. These patterns can be identified using network science metrics. Many metrics from network sciences can be used, such as Diameter, Average Path Length (APL), Density, Betweenness, Algebraic Connectivity, Number of Zero Eigenvalues of the Laplacian Matrix, among others. However, we have observed that two particular metrics, called Clustering Coefficient and Assortativity, can bring relevant information about the convergence state of the algorithm. The expected behavior of such metrics is detailed in Sect. 4.

Clustering Coefficient: In non-directed graphs, the Clustering Coefficient of a node is defined by $cc_i = \frac{2 \cdot c}{d_i \cdot (d_i - 1)}$, where d_i is the number of neighbors of node i and c is the number of pairs of nodes connected between all neighbors of node i. The Clustering coefficient of the network is given by the average values for the Clustering Cluster of all nodes in the network. Nodes with less than two neighbors have the Clustering Coefficient equal to zero [9]. This can be interesting for detecting convergence in ACO algorithms in TSP problems since a quasi-2-regular topology is expected at the end of the process.

Assortativity: The concept of assortativity was introduced by Newman [8] in 2002 and has been extensively studied since its proposal. Assortativity is a metrics that represents the extent to which nodes in a given network are associated with other nodes, being in an organization where nodes are connected to nodes with similar degree or not [11]. The assortativity of a network is determined based on the degree (number of direct neighbors of the node) of the nodes of the network [16]. The concept of assortativity can be applied to another characteristic of the node, such as the weight of the node, the betweenness, etc. Assortativity is represented by a scalar value r, where $-1 \leq r \leq 1$. A network is said to be assortative when nodes are connected to other nodes with a similar degree, and in this case the value of r is close to 1. The value of r near 0 represents a non-assortative network, i.e. random connections, and r near to -1 indicates a disassortative network (when nodes with high degrees are connected to nodes with low degrees) [18]. The equation that defines the correlation between the nodes was defined by Newman in [8] and it is presented in Eq. (1).

$$r = \frac{M^{-1} \sum_i j_i k_i - [M^{-1} \sum_i \frac{1}{2}(j_i + k_i)]^2}{M^{-1} \sum_i \frac{1}{2}(j_i^2 + k_i^2) - [M^{-1} \sum_i \frac{1}{2}(j_i + k_i)]^2}, \tag{1}$$

where j_i and k_i are the degrees of the vertices that are connected by the edge i and M is the number of edges.

3   Related Works

To the best of our knowledge, Oliveira et al. [13] were the first to investigate the flow of information between simple reactive agents in a swarm intelligence algorithm. In their approach, a graph is formed depending on the actual information regarding the transference of positional records from one particle to another. The authors analyzed the number of null eigenvalues and the R-value to assess the performance of the algorithms. In [14], the same authors presented a more comprehensive study including the analysis of a density spectrum and cumulative measures using a windowed approach.

In [15], the authors analyzed the communication diversity in particle swarm optimization (PSO), concerning positional information inside the swarm to assess the diversity of the swarm, thus trying the detect stagnation processes.

In [17], the authors presented a network-theoretic approach combined with spectral graph theory tools to assess swarm dynamics.

In summary, there are just some few papers in the literature regarding this topic and most of them are focused on the PSO approach, that is used for the optimization of continuous variables. This means that none of them threats the indirect communication established by pheromone trails in ACO and there is room for this kind of analysis.

4   Our Proposal

This work proposes the evaluation of the convergence state of the ant colony optimization algorithm through the use of network science metrics. To use the metrics, we had to adapt them for tackling weighted graphs. The pheromone graph evolves to a quasi-2-regular network, where from each node leaves two edges with a high amount of pheromone and all the rest with a low amount of pheromone. The metrics used to assess the convergence state of the ACO were variations of the Clustering Coefficient and Assortativity. The original version of the metrics was developed for application in unweighted networks, so it was necessary to adapt in this work such metrics for use in the ACO.

The clustering coefficient analyzes the neighboring nodes of a given node and whether those neighbors are also neighbors to determine a coefficient. For the proposal of this work this approach would not work, since in the TSP all the nodes are neighbors of all, the graph is totally connected. Thus it was necessary to develop a variation of this metric so that it adapted the application in weighted graphs. The coefficient of clustering used in this work follows Eq. (2), that results in a variation between 0 and 1 as well as the original coefficient. However, according to the weights of the edges between the main node and its neighbors, and the weight of the edge between the neighbors, we define the value of the new coefficient.

$$cc_x = \frac{2 \cdot c_x}{d_x \cdot (d_x - 1)}, \tag{2}$$

Where d_x is the number of neighbors of node x and cc_x is the clustering coefficient of node x. The adaptation takes place on c_x. In the original clustering coefficient, c represents the number of neighbors connected to each other. In the new version, we adapted the metrics, and c_x is represented by Eq. (3) since all neighbors are always connected.

$$c_x = ((w_{xy} + w_{xz})/2) \cdot w_{yz}, \tag{3}$$

where w_{xy} is the weight of the edge between the nodes x and y.

Assortativity analyzes the co-relation between nodes of a network, the assortativity coefficient r has a value between -1 and 1 and represents how assortative or disassortative the network is. When a network has hubs attached to nodes with low degrees the value of r tends to -1, when nodes with similar degrees are connected, the value of r tends to 1. The pheromone graph used in this proposal is a weighted graph, so it was necessary to adapt the assortativity. For this, it was necessary to use the weight of each edge in the equation. The use of weight w_{jk} in the formula generalizes the equation, so Eq. (4) can be used for both weighted and unweighted networks.

$$r = \frac{M^{-1} \sum_i j_i k_i w_{jk} - [M^{-1} \sum_i \frac{1}{2}(j_i + k_i) w_{jk}]^2}{M^{-1} \sum_i \frac{1}{2}(j_i^2 + k_i^2) w_{jk} - [M^{-1} \sum_i \frac{1}{2}(j_i + k_i) w_{jk}]^2}. \tag{4}$$

In the adapted assortativity, the degree of a node is the sum of all the weights of the edges connected to it. It is as if in the original proposal each edge had weight 1 when there is a connection or 0 when the connection is non-existent. In our proposal, since all edges always exist, the weight w was placed in the formula to balance the degree of the nodes.

For a better understanding, consider two nodes with high degree, but the connection between them is low, almost nonexistent. In the original proposal, considering that this low connection would not exist, these nodes would not influence each other in the calculation of assortativity. However, the nodes are connected and would influence multiplying the degrees of these nodes by the weight of the edge that connects them reducing the influence on the assortativity.

According to the behavior of the pheromone graph, it is possible to infer how the metrics should evolve along with the pheromone graph. The clustering coefficient, which evolves according to the connection between the neighbors of a node, initially has its maximum value equal to one. This occurs since all nodes are connected by edges with similar weights, so the neighbors of a node are connected by an edge with a similar weight to the weight of the edge that connects the neighbor to the node under analysis. During the execution of the algorithm, the clustering coefficient tends to decrease, since the ants tend to deposit more pheromone in the edges that compose the best path while the pheromone in the other edges tends to evaporate. The behavior of the assortativity tends to be similar to the clustering coefficient in the beginning. This occurs since all the nodes are connected to nodes with a similar degree. During the execution of the algorithm, the assortativity behaves differently, since it tends to decrease and then increase. This occurs because only two edges that are connected to each node receives a higher amount of pheromone. Then, the graph tends to be assortative next to the stagnation of the algorithm since it tends to a quasi-2-regular network.

5   Simulation Setup

We used a set of computational experiments to evaluate our proposal. For the implementation of the ACO, we used two instances of the traveling salesman problem. We chose the oliver30 problem, which was used by Dorigo. Oliver30 is a TSP that has 30 cities and has as the best-known result the value 423.741.

This work does not aim to make a comparison with other proposals, the analysis regards on the evolution of the values of the metrics and the evolution of the ACO, showing that the metrics stagnate at the same time that the algorithm also stagnates. All tests were performed 30 times, and we used as the stopping criterion 2,500 iterations. The traveling salesman problem (TSP) is a well-known combinatorial optimization problem. To define the TSP, consider a fully connected graph with n nodes. Each edge has a weight, representing the distance. A candidate solution is a path where nodes are visited in the graph exactly once and returns to the initial node [19].

The values for the parameters used in all trials are $\alpha = 1$ (Influence of pheromone), $\beta = 2$ (Influence of the distance), $c = 0.01$ (Pheromone initialization constant), $\delta = 0.1$ (Pheromone decay), 10 ants, $q_0 = 0.9$, $\rho = 0.1$.

6   Results and Discussion

The following images illustrate the evolution of the pheromone graph during a trial for the **oliver30** problem. Figure 1 introduces the pheromone graph in the first iteration of the algorithm execution. After 50 iterations, the pheromone graph assumes the form shown in Fig. 2. In iteration 100, the pheromone graph is as in Fig. 3. In all figures, the pheromone intensity is grayscale, so the darker an edge, more pheromone can be found at this edge.

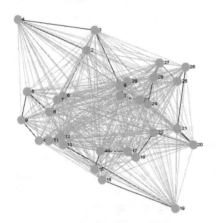

Fig. 1. Pheromone graph in oliver30 problem after the first iteration.

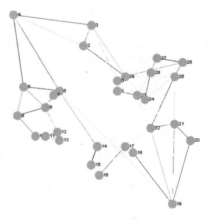

Fig. 2. Pheromone graph in oliver30 problem after 50 iterations.

Figure 4 shows the evolution of the Assortativity during the execution of the ACS in the oliver30 problem. Figure 5 shows the evolution of the Clustering

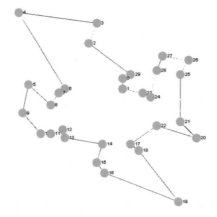

Fig. 3. Pheromone graph in oliver30 problem after 100 iterations.

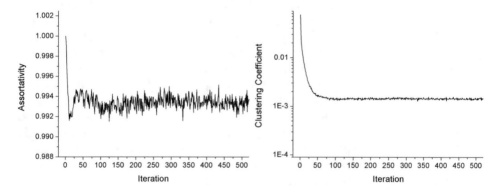

Fig. 4. Evolution of the assortativity in the oliver30 problem running the ACS.

Fig. 5. Evolution of the clustering coefficient in the oliver30 problem running the ACS.

Coefficient during the execution of the ACS in the oliver30 problem. To validate that the metrics stagnated at the moment of the algorithm's stagnation, Fig. 6 shows the evolution of the best solution found in the oliver30 problem during the execution of the ACS.

Note the behavior of the clustering coefficient relative to the lowest path found. One can observe that the clustering coefficient indicates stagnation at the same iteration that the graph representing the solutions found by the ants also enters. The assortativity goes through a small variation and then continues to suffer minor changes, but always close to the value one.

It is possible to observe in all the presented results that the clustering coefficient evolves according to the best solution found. When the best-found solution stops to improve, it means that the algorithm is stagnated, and it is in a state of convergence. The clustering coefficient stagnated at the same time as the best-found solution, showing that the metrics value can indicate the convergence

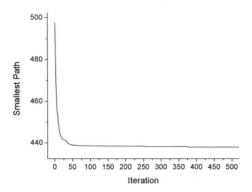

Fig. 6. Evolution of the smallest path in the oliver30 problem running the ACS.

state of the algorithm. After the point where the stagnation is determined, the clustering coefficient still goes through small variations. This occurs since the pheromone graph is always being updated and since it is not mandatory that all the ants follow the same path, even when there are edges with a lot of pheromone, there exists the possibility for the ant to choose another edge to follow. It occurs in edges that are not part of the best path to receive pheromone of some ant, and this generates small variations in the clustering coefficient.

The assortativity presented a great difference in the experiments that used the variation AS maintaining its result constant in one. This is because in AS only the global update is performed, so the deposition and evaporation of pheromone do not cause the nodes to cease to have a similar degree, keeping the assortativity constant even with the algorithm evolving as usual. In the other experiments, the assortativity showed a behavior similar to that expected but presented small variations during the algorithm, which do not guarantee that this metrics alone can define the convergence state of the ACO. Assortativity is a metrics that can have a value of -1 up to 1 and in the results presented the variation of the metrics is less than 0.1 in most of the presented results. Thus, assortativity does not define the state of convergence but can serve as a complementary metrics to be used in conjunction with a metrics that presented a better behavior as the clustering coefficient. Although the assortativity did not vary as expected, it can be used as a validation metrics, since it is impossible for the algorithm to be in a state of convergence with this metric presenting values closer to 0 and -1.

7   Conclusions

The extensive use of the ACO makes this algorithm of swarm intelligence more known in the research area, which makes a more obtainable information on the state of algorithm convergence feasible and quite attractive. This paper contributes with the swarm intelligence algorithm ACO in the line of detecting convergence state, and this extraction of information regarding the convergence state of the algorithm from the use of network science metrics. The application of network science metrics in swarm intelligence algorithms is simple for ACO, but we had the challenge to adapt the metrics in weighted networks. The adaptation of the metrics used for weighted networks is also a significant gain for this area of research, considering that most of the real networks have weights at their edges. Two metrics were adapted and used in this work, the clustering coefficient, and the assortativity. With the use of these adapted metrics, we verified that it is possible to define the state of convergence of the algorithm only based on the information given by the metrics. This occurs especially for the clustering coefficient, which according to the results was more promising. Concerning the use of these two metrics, we can conclude that the clustering coefficient presented satisfactory results regarding the definition of the convergence state of the algorithm, while the assortativity, even presenting near-expected behavior, did not show considerable variation and could be used as a complement to another metrics that behaved better. Also explained previously, the application of science metrics of the networks in the ACO is feasible due to the existence of the existing pheromone network in the algorithm, which leads to the conclusion of the gain that has with this application. The application of science metrics of the networks in the ACO is feasible due to the existence of the pheromone network in the algorithm.

For future works, we intend to apply a similar methodology to show the metrics can be useful to identify the convergence state for other classical problems solved by the ACO.

References

1. Albert, R., Barabasi, A.L.: Statistical mechanics of complex networks. Rev. Mod. Phys. **74**, 47–97 (2002)
2. Barabási, A.L., Albert, R.: Emergence of scaling in random networks. Science **286**(5439), 509–512 (1999)
3. Dorigo, M.: Optimization, learning and natural algorithms. Ph.D. thesis. Dipartimento di Elettronica, Politecnico di Milano, Milão, Italy (1992). (in Italian)
4. Dorigo, M., Gambardella, L.M.: Ant colony system: a cooperative learning approach to the traveling salesman problem. IEEE Trans. Evol. Comput. **1**(1), 53–66 (1997)
5. Dorigo, M., Maniezzo, V., Colorni, A.: Ant system: optimization by a colony of cooperating agents. IEEE Trans. Syst. Man Cybern. Part B: Cybern. **26**(1), 29–41 (1996)
6. Lewis, T.G., Pickl, S., Peek, B., Xu, G.: Network science [guest editorial]. IEEE Netw. **24**(6), 4–5 (2010). https://doi.org/10.1109/MNET.2010.5634435
7. Newman, M.: Networks: An Introduction. OUP, Oxford (2010)
8. Newman, M.E.: Assortative mixing in networks. Phys. Rev. Lett. **89**(20), 208701 (2002)
9. Newman, M.E.: Random graphs as models of networks. arXiv preprint cond-mat/0202208 (2002)
10. Newman, M.E.: The structure and function of complex networks. SIAM Rev. **45**(2), 167–256 (2003)
11. Noldus, R., Van Mieghem, P.: Assortativity in complex networks. J. Complex Netw. **3**(4), 507–542 (2015)

12. Oliveira, M., Bastos-Filho, C.J., Menezes, R.: Assessing particle swarm optimizers using network science metrics. In: Complex Networks IV, pp. 173–184. Springer, Heidelberg (2013)
13. Oliveira, M., Bastos-Filho, C.J., Menezes, R.: Towards a network-based approach to analyze particle swarm optimizers. In: 2014 IEEE Symposium on Swarm Intelligence, SIS, pp. 1–8. IEEE (2014)
14. Oliveira, M., Bastos-Filho, C.J., Menezes, R.: Using network science to assess particle swarm optimizers. Soc. Netw. Anal. Min. **5**(1), 1–13 (2015)
15. Oliveira, M., Pinheiro, D., Andrade, B., Bastos-Filho, C., Menezes, R.: Communication diversity in particle swarm optimizers. In: International Conference on Swarm Intelligence, pp. 77–88. Springer (2016)
16. Piraveenan, M.: Topological analysis of complex networks using assortativity. School of Information technologies, The University of Sydney, Doutorado (2010)
17. Sekunda, A., Komareji, M., Bouffanais, R.: Interplay between signaling network design and swarm dynamics. Netw. Sci. **4**(2), 244–265 (2016)
18. Thedchanamoorthy, G., Piraveenan, M., Kasthuriratna, D., Senanayake, U.: Node assortativity in complex networks: an alternative approach. Proc. Comput. Sci. **29**, 2449–2461 (2014). https://doi.org/10.1016/j.procs.2014.05.229. 2014 International Conference on Computational Science
19. Weise, T., Chiong, R., Lassig, J., Tang, K., Tsutsui, S., Chen, W., Michalewicz, Z., Yao, X.: Benchmarking optimization algorithms: an open source framework for the traveling salesman problem. IEEE Comput. Intell. Mag. **9**(3), 40–52 (2014)

User Modeling on Twitter with Exploiting Explicit Relationships for Personalized Recommendations

Abdullah Alshammari[✉], Stelios Kapetanakis, Roger Evans,
Nikolaos Polatidis, and Gharbi Alshammari

School of Computing, Engineering and Mathematics,
University of Brighton, Brighton, UK
{A. Alshammari1, S. Kapetanakis, R. P. Evans, N. Polatidis,
G. Alshammari}@brighton. ac. uk

Abstract. The use of social networks sites has led to a challenging overload of information that helped new social networking sites such as Twitter to become popular. It is believed that Twitter provides a rich environment for shared information that can help with recommender systems research. In this paper, we study Twitter user modeling by utilizing explicit relationships among users. This work aims to build personal profiles through a alternative methods using information gained from Twitter to provide more accurate recommendations. Our method exploits the explicit relationships of a Twitter user to extract information that is important in building the user's personal profile. The usefulness of this proposed method is validated by implementing a tweet recommendation service and by performing offline evaluation. We compare our proposed user profiles against other profiles such as a baseline using cosine similarity measures to check the effectiveness of the proposed method. The performance is measured on an adequate number of users.

Keywords: Recommender systems · User modeling · User profiling · Explicit relationships · Twitter · Influence score

1 Introduction

The real-time web is growing as a new technology that allows users to communicate and share information in multi-dimensional contexts such as Twitter. Twitter is a very well-known social media micro-blogging platform used by hundreds of millions of users. It is a social network that allows users to post and exchange short messages, called tweets, of up to 280 characters [20]. It has become a very important source of shared information and breaking news, and it works quickly and effectively. Also, it can be considered an example of a distinctive type of social networking platforms that presents relationships based on the following strategy, which makes it different from other classic social networking platforms that based on reciprocal network structure such as Facebook. The relationships among users of Twitter can be social or informational or both because users follow other users not only to maintain social links but also to receive interesting information generated by others. Twitter features lead to the

© Springer Nature Switzerland AG 2020
A. M. Madureira et al. (Eds.): HIS 2018, AISC 923, pp. 135–145, 2020.
https://doi.org/10.1007/978-3-030-14347-3_14

possibility of using it as a primary source for modeling a specific user who participates in a network characterized by relationships and interactions [1, 20].

According to [1], some studies have shown that Twitter is seen as a valuable resource for many powerful approaches such as recommender systems. Recommender systems have proven to be powerful parts of many web and mobile applications. The main goal of such systems is to provide real-time, context-aware, personalized information to help increase sales and user satisfaction. Many studies have used Twitter to model users, build user profiles, and deliver accurate recommendations.

In this paper, we investigate modeling Twitter users by exploiting relationships to improve recommendations in recommender systems based on short-text-based profiles within short-term (recent tweets such as tweets within the last 2 weeks). User profiles in our approach are built from tweets of the user's following list (i.e., friends). Our model redefines the rule of influence and use it to identify influential friends and use their tweets to build the user profile by examining all incoming tweets within short-term time. As a result, our approach can deliver more accurate recommendations to users in comparison with profiles from the literature.

2   Related Work

The authors of [12] proposed a method to improve the accuracy of recommended news articles based on Twitter tweets. The user profile was built by extracting the nouns in the user's tweets and re-tweets. Tweet refers to the action of writing a post whereas re-tweet refers to the action of sharing a tweet of another user. Their results showed that Twitter-activity-based recommendations were more accurate than the random recommendations. The TRUPI system was proposed in [9], combining the history of the user's tweets and the social features. It also catches the dynamic level of user's interests in several topics to measure changes in the user's interests over time.

In [1], the authors analyzed temporal dynamics in Twitter profiles for personalized recommendations in the social web. They build two different types of profiles, based on hashtags and entities (e.g., places and celebrities), taking into consideration time-sensitivity, enrichments (using external resources such as Wikipedia), and the activity of the user. The results showed that the entity-based profile, which was built with short-term time (i.e., last two weeks) and enrichment, performed better than other profile types in the news recommender system based on Twitter activities. Furthermore, the problem is that for many users there is not enough data in their recent activities to create a reliable user profile. Authors in [15] showed that using a decay function in long-term profiles that gives higher weight to recent topics of interest than older topics of interest showed better performance in delivering recommendations than the long-term profiles without the decay function. Moreover, authors in [2] proved that a short-term profile (last week, for example) is better than the complete profile.

One of the solutions is to enrich user profile with other data. The authors in [2] modeled user profiles in Twitter with different dimensions and compared the quality of each one and enrichments was one of the dimensions. Results showed that using external resources, such as news articles, is better than relying solely on Twitter.

Enriching user profile with data have been done in different ways: exploiting URLs in tweets or using textual external resources (such as articles or Wikipedia). In exploiting URLs in tweets, a CatStream system was proposed in [10] that uses traditional classification techniques to profile Twitter users based on URLs in their tweets. However, the system only focuses on URLs in tweets and it is not suitable for users who do not have enough tweets that include URLs. In [3], the authors categorized a set of tweets as interesting or uninteresting by using crowdsourcing. The method showed that the existence of a URL link is a successful feature in selecting interesting tweets with high accuracy. However, the shortcoming of this factor is that it might incorrectly categorize an uninteresting tweet that links to useless content [11].

In using external resources, authors in [1, 10] used external resources such as Wikipedia and news articles. The enriched user profiles with external resources performed better than profiles that built with only Twitter activities. These methods can be useful to supply the user profile with more information and thus improve the accuracy of the recommender system. However, some data gained from external resources might have no relevance to the user's interests and it might affect the performance of the recommender system. Also, many users do not provide enough URL links in their tweets.

A field that remains to be investigated is exploiting the network of relationships between users in Twitter to characterize a specific user and improve the performance of recommender systems based on short-text activities. It is obvious that a single user who generates short-text data (i.e., tweets and retweets) can be characterized based on his history and behavior by collecting history data (i.e., timeline) that the user himself has generated. However, to acquire sufficient information for profiling, this method may need to go a long way back into the past and the collected information might not form a very coherent set and up to date. Furthermore, the problem is that for many users there is not enough data and URLs in their recent activities to create a reliable profile. To address this lack of data, we propose using explicit links between users (e.g., following links) to expand the set of relevant recent activities. The advantage of this method is that there is more recent data to build profiles from, which will allow us to improve the performance of short-text-based recommender systems.

Exploiting following links can be achieved by finding influential users in the friends list. Most of the researches focuses on how influential a user is based on his popularity, as indicated by the number of followers and followees (friends) and his interactions with others [4, 5, 8, 21]. Authors in [18] collected and classified different Twitter influence measures in the literature. However, from the perspective of a normal user who follows other identified influential users, it is believed that the influence rule can vary from one user to another in the list of followees (friends). For example, some accounts followed by a user, such as close and active friends, that are not identified as influential by the influence rule, but they might be more influential than the identified influential users such as celebrities. In [18] authors stated that naming a person in a social network as an influential is a conceptual problem and there is no agreement what the influential user should be. Therefore, there is a need to create a method that can generate an influence score from the user's perspective relating to the user's behavior and interactions.

After identifying the influence score, there is a need to classify incoming tweets into categories such as relevant and not relevant. Researchers proposed techniques to predict the likelihood of a tweet being retweeted based content-based features [14], collaborative tweet ranking (CTR) [7] and a coordinate ascent (CA) algorithm [19].

3   Proposed Method

In general, the recommender system includes two stages: (1) user profiling and (2) item ranking. In this work, user profiles are built by extracting information from tweets in the user's timeline and tweets by his following list (friends). Then, recommendation items are ranked based on the user profile. Figure 1 shows the general process of our proposed work. The user's information is collected using the Twitter API. This collected information is processed to identify important keywords posted by the user's friends. The following subsections explain in detail the steps of the two stages.

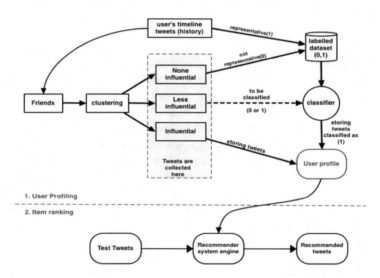

Fig. 1. The general steps of the proposed method.

3.1   User Profiling Stage

In this stage, we develop user profiles that contains important information about the user. These profiles are built from tweets by other users with explicit relationships to the user in Twitter (e.g., user's friends list). All profiles are built as key-words profiles (bag of words); pre-processing must be carried out before the recommendation process based on the steps suggested by Micarelli and Sciarrone [13]. The aim is to filter tweets and then extract important content. The tweets generated by the user himself and re-tweeted tweets explicitly represent the user's interests, whereas the received and collected tweets from explicit links need to be examined and classified.

Before building the profiles, and for each user, data about the user is collected from the user's Twitter timeline, including friends list, timeline tweets (posted and re-tweeted tweets), and favorited tweets. Then, we calculate an influence score between the user and his/her friends to help us rank his/her friends based on their importance and then choose and collect appropriate content. The influence score, which considers actions such as following, re-tweeting, replying, and favoriting, might be a good attribute to find friends who are influential to the user. The influential users, who are followed by the examined user, are found by using the influence score from the user's perspective. The following simple Eq. (1) calculates the influence score, where the original user, denoted user1 (u1) follows user2 (u2):

$$\text{Influence Score } (u1, u2) = \left(\frac{\sum \text{RTs } (u2)}{\sum \text{RTs p(u1)}} + \frac{\sum \text{RTs } (u2)}{\sum \text{Ts p(u2)}} + \frac{\sum \text{MT(u2)}}{\sum \text{MT p(u1)}} + \frac{\sum \text{FV(u2)}}{\sum \text{FV p(u1)}} \right) \times \frac{1}{4}$$

(1)

$\sum$RTs (u2) represents the total number of tweets posted by user2 and re-tweeted by user1. $\sum$RTs p(u1) represents the total number of re-tweets in the user1 profile. $\sum$Ts p (u2) represents the number of tweets in the user2 profile. $\sum$MT(u2) represents the number of replies (mentions) that user1 posted to user2. $\sum$MT p(u1) represents the total number of mentions in user1's profile. $\sum$FV(u2) represents the total number of tweets from user2 that user1 has favorited. $\sum$FV p(u1) represents the total number of favorited tweets in user1's profile. Finally, 1/4 is used to normalize the score between 0 and 1.

The friends list is divided into three groups based on influence score: influential users, less influential users, and non-influential users, by K-means clustering algorithm.

All tweets from the influential group are added to the user profile whereas tweets from the non-influential group are not added. The tweets from less influential users are classified into representative (re-tweetable) or not representative (not re-tweetable). Different classifiers are used in this process: Naïve Bayes, Random Forest, Support Vector Machine, Decision Tree, K-Nearest Neighbor and Neural Networks. All classifiers trained using the same labeled dataset from the user's timeline (the history of the user's tweets and re-tweets) and the tweets of the non-influential users. The tweets from the user timeline are labeled as representative (re-tweetable) and the tweets from non-influential users are labeled as not representative (not re-tweetable). The dataset is divided into training and testing sets. The latter is used to calculate the accuracy of each classifier. The classifier with the highest accuracy is chosen automatically to classify less influential user tweets. This step ensures that tweets are classified with the most accurate classifier. After classifying less influential users' tweets, the tweets classified as representative (re-tweetable) are stored in the user's profile with influencers' tweets.

3.2   Items Ranking Stage

At this stage, the recommendation items are a set of tweets that the user might show some interest in by re-tweeting. Vector space model representation is used, which considers user profile and recommendation items as vectors and then calculates the angle between them. The closer the item is to the user profile, the more relevant it is. Cosine similarity is used here to calculate the angles.

The user profiles built as described in the previous subsection are used as the base for ranking the set of tweets in the recommendation items. Every tweet in the recommendation items is evaluated based on its similarity to the user profile. This process is applied following the steps of [13] and additionally, tweets are excluded if the remaining text contains fewer than three words. Finally, the similarity score between the tweet profile and the user profile is calculated by the cosine similarity equation. All the recommendation items are ranked, and top-k tweets are recommended to the user.

4   Experiment and Results

To validate the advantages of our proposed method, a tweets recommender system was implemented, and an offline evaluation was performed with an adequate group of users. Using the Twitter API provided in the development section of Twitter's website, the timelines of 29 randomly chosen users were collected and examined. For the recommendation items, test tweets (recommendation items) in which the user had shown some interest by re-tweeting were collected from the user's timelines. The next subsection will explain how these test tweets were collected.

4.1   Experiment Setup

After collecting the timeline of the examined user and before calculating the influence score and clustering users into the three mentioned groups, the dataset is divided into three time periods as it is seen in Fig. 2.

The tweets in the first period (the previous week) are used as test items (test tweets); it is already known that the user showed interest in them by re-tweeting them. This is like going back on a time machine into the past to predict a future that is already known, which can help in the evaluation process. The tweets from the second period (between 1 week ago and 3 weeks ago) are used to build user profiles from different sources. The third period (more than 3 weeks ago) is used along with the second period to calculate the influence score from the timeline of the examined user. Also, the user' timeline in this period is used in the machine learning classification.

Profiles: For each user of the examined users, and after calculating the influence score between the examined user and his friends and classifying them as influential, less influential, and non-influential, tweets are collected from influential and less influential friends to build different profiles that are examined and compared in the experiment. These profiles are:

- Baseline: includes all tweets from the user's timeline. It contains both the tweets that the user posted and the tweets the user re-tweeted.
- BLCinf: includes all tweets from the user's timeline, short-term tweets by influential friends (second time period) and short-term tweets by less influential friends that are classified as representative (second time period).

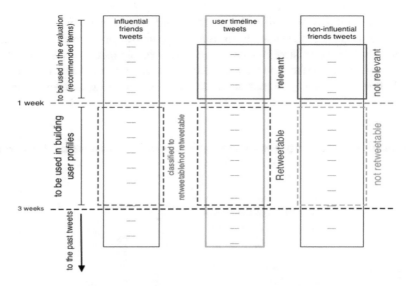

Fig. 2. Dividing the user timeline into three evaluation time periods.

- STBLCinf: includes tweets only from the second time period (short-term), and includes tweets from the user's timeline, tweets by influential friends, and tweets by less influential friends that are classified as representative.
- STBLinf: includes tweets only from the second time period (short-term) and includes tweets from the user's timeline and all tweets by influential and less influential friends without consideration of classification.
- BLinf: includes all tweets from the user's timeline and all short-term tweets by influential and less influential friends. Classification of the tweets of less influential friends is not considered in this profile.
- Followers count: this profile is built in the same way of STBLCinf. In contrast, the clustering is applied based on followers count instead of influence score. Some literature researches used it in their experiments such as [5, 6, 16, 18, 21].

Test Tweets: Test tweets are used to evaluate the accuracy of the recommender system. They are a collection of tweets from the first period (1 week ago) as explained previously, and they are used as recommendation items in the recommender system. This collection contains both relevant and non-relevant tweets. From the timeline of the user, it is already known which items the user showed some interest in by re-tweeting, and these are considered relevant items. Tweets from non-influential users in the same period are used as non-relevant items. Thus, the set of the recommended items contains a mixture of relevant and non-relevant items, which can help the recommender system to measure the accuracy when it runs on different user profiles. It also allows a comparison between the built profiles and the baseline profiles.

Evaluation Metrics: In this study, offline evaluation is used to measure the accuracy of the recommender system with different user profiles [19]. As in our methodology, various user profiles are used in the recommender system and then compared. The metrics are used in this research to measure the accuracy of the performance of the system are average precision @ k (AP) and mean average precision (MAP). Average precision is used to measure how good the system is in retrieving top-k relevant items and MAP is used to measure how good the system in retrieving all relevant items.

4.2   Results

In the metric of the Average Precision @top-k recommendations, the tested values of k are: 1, 3, 5, 10, 15 and 20. (See Fig. 3a).

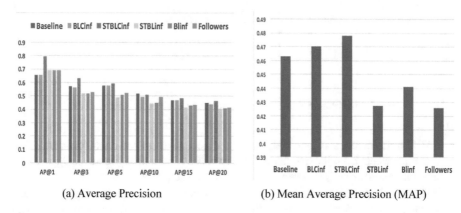

(a) Average Precision   (b) Mean Average Precision (MAP)

Fig. 3. The average precision @1, 3, 5, 10, 15 and 20, and the Mean Average Precision (MAP) of profiles.

In the top-k = 1, 3 and 5, results showed that the profile STBLCinf outperformed all other strategies of building user profiles. On the other hand, the baseline and BLCinf profiles has the lowest average precision on top-1.

In the top-k = 10, the results showed that the baseline outperformed all profiles. Whereas the STBLCinf came in the second place. On the other hand, STBLinf achieved the lowest Average Precision.

When the top-k is set to 15 and 20, profile STBLCinf outperformed all other profiles. However, again profile STBLinf achieved the lowest average precision, which might mean that non-relevant tweets affected the accuracy of the recommender system. STBLCinf and STBLinf were built similarly but the difference is that the latter included all tweets from influential and less influential users. Therefore, this small difference can affect the performance and make it the worst profile.

In the Mean Average Precision (MAP), Fig. 3b shows the results that the profile STBLCinf achieved the highest performance against all profiles. Also, profile BLCinf achieved better performance than the baseline and that might mean enhancing the baseline profiles with some related data can improve the performance of the recommender system. Also, profile BLinf was built similarly to BLCinf achieved less MAP than the latter and the baseline profile. This may clarify that enriching profile with none related data (unclassified tweets in this case) can reduce the quality of recommendations even worse than the baseline. Also building profiles based only on timeline (baseline) cannot deliver more relevant recommendations. On the other hand, profiles STBLinf and followers achieved the lowest MAP. Table 1 shows the average number of tweets in each profile. Followers count profile has the lowest tweets whereas BLinf has the highest tweets. In contrast, enriching the profile with good chosen tweets can improve its quality such as STBLCinf.

Table 1. Average number of tweets in each profile.

	Baseline	BLCinf	STBLCinf	STBLinf	BLinf	Followers
Average	1991	4195	2491	4335	6039	1098

To validate how strong our proposed influence score that the profile STBLCinf built based on, we included the followers count profile and we built it in the same way of the former profile. Results shows that our influence score achieved better performance than the classic method of considering users as influential based on their followers.

5   Conclusion

To conclude, in this work we presented a new way to build users' profiles by exploiting a Twitter-explicit network structure to improve the performance of short-text-based recommender systems. This new user profile exploits the following links between users and their friends to collect a set of short-term tweets and then uses these tweets to build profiles. Furthermore, the influence rule in Twitter is redefined, which helped us to cluster the following list into three groups: influential, less influential, and non-influential. As a result, influential users' tweets are stored directly to the user profile whereas non-influential users' tweets are excluded. Classifiers classify less influential users' tweets to store the representative tweets into the users' profiles.

The advantage of this method has been validated by an offline evaluation experiment over a prototype tweets-recommender system. The discriminative power of our method has been presented by testing and comparing our method against baseline profiles and followers count profiles. Therefore, the proposed method outperformed other profiles. In the future, regarding explicit relationships, we may explore additional types of relationships between a user and friends of his/her friends. Also, we may consider the similarity between the user and his followers to expand the set of tweets that represent the user's interests.

References

1. Abel, F., Gao, Q., Houben, G.J., Tao, K.: Analyzing temporal dynamics in Twitter profiles for personalized recommendations in the social web. In: Proceedings of the 3rd International Web Science Conference, p. 2. ACM (2011)
2. Abel, F., Gao, Q., Houben, G.J., Tao, K.: Twitter-based user modeling for news recommendations. IJCAI **13**, 2962–2966 (2013)
3. Alonso, O., Carson, C., Gerster, D., Ji, X., Nabar, S.U.: Detecting uninteresting content in text streams. In: SIGIR Crowdsourcing for Search Evaluation Workshop (2010)
4. Anger, I., Kittl, C.: Measuring influence on Twitter. In: Proceedings of the 11th International Conference on Knowledge Management and Knowledge Technologies, p. 3. ACM (2011)
5. Bakshy, E., Hofman, J.M., Mason, W.A., Watts, D.J.: Everyone's an influencer: quantifying influence on Twitter. In: Proceedings of the Fourth ACM International Conference on Web Search and Data Mining, pp. 65–74. ACM (2011)
6. Cha, M., Haddadi, H., Benevenuto, F., Gummadi, P.K.: Measuring user influence in Twitter: the million follower fallacy. Icwsm **10**(10–17), 30 (2010)
7. Chen, K., Chen, T., Zheng, G., Jin, O., Yao, E., Yu, Y.: Collaborative personalized tweet recommendation. In: Proceedings of the 35th International ACM SIGIR Conference on Research and Development in Information Retrieval, pp. 661–670. ACM (2012)
8. Chen, C., Gao, D., Li, W., Hou, Y.: Inferring topic-dependent influence roles of Twitter users. In: Proceedings of the 37th International ACM SIGIR Conference on Research and Development in Information Retrieval, pp. 1203–1206. ACM (2014)
9. Elmongui, H.G., Mansour, R., Morsy, H., Khater, S., El-Sharkasy, A., Ibrahim, R.: TRUPI: twitter recommendation based on users' personal interests. In: International Conference on Intelligent Text Processing and Computational Linguistics, pp. 272–284. Springer (2015)
10. Garcia Esparza, S., O'Mahony, M.P., Smyth, B.: Catstream: categorising tweets for user profiling and stream filtering. In: Proceedings of the 2013 International Conference on Intelligent User Interfaces, pp. 25–36. ACM (2013)
11. Karidi, D.P., Stavrakas, Y., Vassiliou, Y.: A personalized tweet recommendation approach based on concept graphs. In: 2016 International IEEE Conferences Ubiquitous Intelligence & Computing, Advanced and Trusted Computing, Scalable Computing and Communications, Cloud and Big Data Computing, Internet of People, and Smart World Congress (UIC/ATC/ScalCom/CBDCom/IoP/SmartWorld), pp. 253–260. IEEE (2016)
12. Lee, W.J., Oh, K.J., Lim, C.G., Choi, H.J.: User profile extraction from twitter for personalized news recommendation. In: Proceedings of 16th Advanced Communication Technology, pp. 779–783 (2014)
13. Micarelli, A., Sciarrone, F.: Anatomy and empirical evaluation of an adaptive web-based information filtering system. User Model. User-Adapted Interact. **14**(2–3), 159–200 (2004)
14. Naveed, N., Gottron, T., Kunegis, J., Alhadi, A.C.: Bad news travel fast: a content-based analysis of interestingness on Twitter. In: Proceedings of the 3rd International Web Science Conference, p. 8. ACM (2011)
15. Piao, G., Breslin, J.G.: Exploring dynamics and semantics of user interests for user modeling on Twitter for link recommendations. In: Proceedings of the 12th International Conference on Semantic Systems, September 2016, pp. 81–88. ACM (2016)
16. Razis, G., Anagnostopoulos, I.: InfluenceTracker: rating the impact of a twitter account. In: IFIP International Conference on Artificial Intelligence Applications and Innovations, pp. 184–195. Springer (2014)
17. Ricci, F., Rokach, L., Shapira, B.: Recommender systems: introduction and challenges. In: Recommender Systems Handbook. Springer (2015)

18. Riquelme, F., González-Cantergiani, P.: Measuring user influence on Twitter: a survey. Inf. Process. Manag. **52**(5), 949–975 (2016)
19. Uysal, I., Croft, W.B.: User oriented tweet ranking: a filtering approach to microblogs. In: Proceedings of the 20th ACM International Conference on Information and Knowledge Management, pp. 2261–2264. ACM (2011)
20. Vosoughi, S.: Automatic detection and verification of rumors on Twitter. Doctoral dissertation. Massachusetts Institute of Technology (2015)
21. Weng, J., Lim, E.P., Jiang, J., He, Q.: Twitterrank: finding topic-sensitive influential Twitterers. In: Proceedings of the Third ACM International Conference on Web Search and Data Mining, pp. 261–270. ACM (2010)

Hybrid Approaches for Time Series Prediction

Xavier Fontes[1,2(✉)] and Daniel Castro Silva[1,2]

[1] Faculty of Engineering of the University of Porto,
Rua Dr. Roberto Frias s/n, 4200-465 Porto, Portugal
{xavier.fontes,dcs}@fe.up.pt
[2] Artificial Intelligence and Computer Science Laboratory,
Rua Dr. Roberto Frias s/n, 4200-465 Porto, Portugal

Abstract. The focus of this work is the development of various hybrid prediction models, capable of predicting a given variable in the context of a time series with sporadic external stimuli. As a case study, we use data from a university student parking lot, together with other events and information. Working on top of previous research, we used Gradient Boosting models, Random Forests and Decision Trees in three proposed hybrid approaches: a voting-based combination of models, an approach based on pairs of models working together and a third novel approach, based on social dynamics and trust in human beings, called Evolutionary Directed Graph Ensemble (EDGE). Results show some promise from these methods, in particular from the EDGE approach.

Keywords: Hybrid models · Gradient Boosting · Random Forest · Decision Tree · Ensemble methods

1   Introduction

In the past years, prediction has become of major importance in several fields. One particular case is prediction over a time series, made even more complex when external stimuli come into play. The work presented herein is part of a continuum of work with the goal of achieving hybrid models capable of performing better than their base models in prediction over a time series with sporadic external stimuli. In previous work we established the groundwork for the hybrid approach, by using a university parking lot as a case study [5]. We tested five different methods – Convolutional Neural Networks (CNNs), Long Short-Term Memory networks (LSTMs), Decision Trees (DTs), Random Forests (RFs) and Gradient Boosting models (GBs) – and after some parameter tuning we arrived at the best performing configurations for each method.

CNNs date back to at least 1988 [7] and started gaining traction in image classification problems. They have also achieved good results in time series problems [2]. LSTMs are a type of Recurrent Neural Networks [12] and have also been

The first author was supported by the Calouste Gulbenkian Foundation, under a New Talents in Artificial Intelligence Program Grant.

A. M. Madureira et al. (Eds.): HIS 2018, AISC 923, pp. 146–155, 2020.
https://doi.org/10.1007/978-3-030-14347-3_15

used in this type of problem [6]. Since at least 1979 [18], DTs have shown good results in time series problems [10]. RFs rely on the idea of averaging out possible mistakes [9], and have been used for time series regression such as in [19]. GB models, like RFs, are an ensemble-type models, and have been used in predicting waste patterns [11] or travel times [21], among others.

Using the three best performing methods (DT, RF and GB), this paper explores three approaches at combining them, using voting schemes, pairs of models and an ensemble-based trust-inspired method.

The rest of this paper is organized as follows. In Sect. 2 we discuss some works that use hybrid approaches. Section 3 introduces the work already done in hybrid prediction models that serves as the basis for this work. Section 4 presents our approaches to creating hybrid models and the corresponding experimental results are discussed in Sect. 5. In Sect. 6 we present some conclusions regarding our work and some lines for future work.

2   State of the Art

Hybrid approaches are typically capable of building on the advantages of the models they are based on, at the same time mitigating their disadvantages.

A combination of ARIMA and ANNs is employed in [20] where the ARIMA model is used to forecast the linear component of the data while the ANN is used to model the residuals, the nonlinear component of the data. This approach is tested on three known datasets and results show that the hybrid achieves smaller mean squared error (MSE) and mean absolute deviation (MAD) values, even when taking into account forecasting data points further in the future.

Tackling stock price forecasting, a combination of ARIMA and Support Vector Machine (SVM) models is introduced in [17]. ARIMA is a strong linear model for time series forecasting but has difficulty against nonlinear patterns, thus being complemented with a SVM model, which is successfully used in nonlinear regression problems. The ARIMA model is used for the linear part and the SVM for the residuals, similar to what was done in [20]. The hybrid is shown to reduce the errors in forecasting when compared against each of its components alone.

A combination of a Radial Basis Function (RBF) Neural Network and an Auto Regressive (AR) model is applied in [22] to forecasting in time series, decomposing the problem in several parts. The datasets are first smoothed with a binomial smoothing technique [15] and then modeled by a RBFNN. The residuals are modeled by the AR model and the overall prediction comes from aggregating the results from the two models. Tested on the Canadian Lynx dataset, the results show the hybrid approach can significantly reduce forecasting errors.

Working with data from porosity and permeability of wells, a combination of Functional Networks (FNs) and SVMs was used in [1] to better predict properties of petroleum reservoirs. One interesting property of this model was its execution time being lower than a standalone SVM model, due to part of the hybrid reducing the dimensionality of the dataset.

In [13] a hybrid neural network approach is used in photovoltaic (PV) panels. Instead of computing five parameters in the problem domain, which is computationally costly, two of them are predicted using the neural network while the other three are computed by reduced form. This allows gains in terms of computation power required while maintaining a good level of accuracy. Their validation was made on PV panels from the California Energy Commission.

Telescope [16] is a framework developed in R for univariate time series that makes use of explicit time series decomposition and several different models as basis such as ANN, K-Mean Clustering, ARIMA and XGBoost. The result is a multi-step-ahead forecasting while maintaining a short runtime. Telescope is tested on a trace of transactions from an IBM z196 Mainframe and a trace of monthly international airline passengers from 1949 to 1960. The results show this framework is competitive in terms of both results and runtime.

Multi-dimensional Bayesian network classifier trees (MBCTree) [8] is another hybrid approach worth mentioning. A MBCTree is a classification tree with multi-dimensional Bayesian classifiers (MBCs) in the leaves. An internal node of a MBCTree corresponds to a feature variable and a leaf is a MBC over all the class variables and the features not present in the path from the root to the leaf.

3   Previous Work

In previous work [5], we created an artificial university student parking lot dataset, based on the real data (which had severe problems that prevented it from being used directly), and generated based on the actual student roster, their respective schedules and exam dates, weather information from a weather station about 100 m from the parking lot and national and regional holidays. The parking lot occupancy rate was discretized into 10 classes. Along the time axis we see different patterns of park occupancy rate. For example at night the park is usually empty (occupancy class 0–10%) and during the day it peaks at the class 90–100% due to the park being completely full.

We studied the use of different kinds of models in similar problems, eventually deciding on an experimental setup consisting of five models, based on their performance as found in the literature: CNN, LSTM, DT, RF and GB. For each model we tuned the parameters for maximizing performance. LSTM performed worse than expected: out of 1728 parameter combinations, the best one had only an accuracy of 54.6%. We think this happened due to the way the inputs were fed to the network and its structure. The best CNN combination achieved 70.9% accuracy, a better result than LSTM but still not what was expected, which might point to similar modeling problems. With better results, the best DT model achieved an accuracy of 86.4%, a significant improvement over the CNN and LSTM models. The best performing RF model achieved an impressive 89.4% of accuracy, the best results among the five tested methods. Also with good results, the GB achieved an accuracy of 86.8%, establishing itself as one of the top 3 models from the ones tested.

Detailed results from testing these five methods are shown in [5], and constitute the basis for the work described in this paper.

4   Approaches to Hybridization

Our hybrid experiments start with a basic approach of combining the predictions of the three methods in a voting manner (Sect. 4.1) and then introducing weights in the predictions of each model. Later on, we introduce pairs of models, where the output of the first model is connected as an additional input to the second model (Sect. 4.2). Lastly, we present a concept for an ensemble model that is relatively more complex than the initial approaches (Sect. 4.3).

4.1   Voting

As described in Sect. 3, we have previously tested different methods for prediction. The three models with the best results, GB, RF and DT, are now used for hybridization, using the parameters that maximized their performance, with the exception of the training percentage, which we experimented with 70, 75, 80, 85 and 90%. A diagram of the model is presented in Fig. 1(a).

In the first approach, called **fair voting**, for each data point in the test portion of the dataset we collect each models' predicted class and in case of a majority, that class is chosen. Otherwise, we perform a weighted average of all predicted classes, where each model's weight is proportional to that model's average accuracy in the baseline results.

The next approach, named **bully voting**, is very similar to fair voting, with the exception that when no majority is found, the RF prediction is chosen as the final prediction, due to its higher average accuracy.

Weighted Voting. Similarly to the approach above, we take each model's configurations and trained instances with 70, 75, 80, 85 and 90% of the dataset. We then split the testing data, using 40% for weight tuning and 60% for testing the approach. By weight tuning we mean we took 40% of the data that the models had never seen and made them each give answers to which class every data point belonged to, then for each class in each model we computed its recall and precision score. The recall and precision vector scores served as the weights for two variants of the Weighted Voting model, one using recall and the other using precision. These weight vectors, for each model, are multiplied element-wise with the one-hot encoded answer of the corresponding model. Summing up all the results from each model, the index of the max valued entry corresponds to the class to be taken as final answer of the voting mechanism.

4.2   Pairs Hybrids

The idea for these hybrids comes from reflecting on the inner workings of Boosting techniques, where at each stage one model is used to compensate where another model fails to correctly classify the input data.

We start with two models, one **top model** and one **bottom model**. The idea is to see if we can train the bottom model to understand when the top model fails and improve the classification results of the test portion of the dataset.

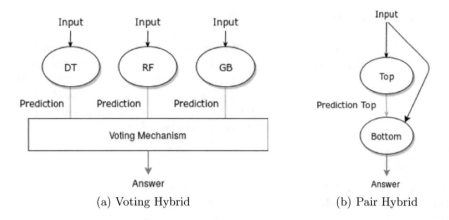

(a) Voting Hybrid   (b) Pair Hybrid

Fig. 1. Voting and pair hybrids

The top model is trained on part of the dataset and then is made to predict the classes of unseen data. With this information we train the bottom model and then we test the pair by using data neither of the models has seen before. The train portion of the dataset was divided three different ways in an effort to see if using more data to train the top model or to train the bottom model was better than using the same percentage of data to train each model.

Besides the combinations involving the division of the dataset we also tested three configurations for the pairs' top and bottom models and also experimented combinations of the models' parameters. The combinations for the Pair Hybrids are showcased in Table 1 and a diagram of the model is presented in Fig. 1(b).

4.3   Evolutionary Directed Graph Ensemble

We believe Evolutionary Directed Graph Ensemble (EDGE) is the cornerstone of what we are trying to achieve and a good place to further explore hybridization.

We can think of a Directed Acyclic Graph (DAG) as a network of nodes and edges representing a network of influence and trust about a certain opinion. For example, a given node can be thought of as being an individual that for a certain fact (a data point X) gives an answer he thinks is correct (his opinion Y) and shares it with another individual directly connected to him in the form of a directed edge. The receiving individual has a certain trust in the opinion of the first node (a weight W). These nodes can be full-fledged, pre-trained models. We start with only two nodes connected directly to one another, the first one giving its answer to a given input while the other weights its own answer with the previous node's answer and gives the final prediction to the data point.

We are currently only setting the base configuration for an EDGE model, this is, having only two directly connected nodes but the next step would be growing the network, which means adding nodes and connections in such a way that the ensemble itself evolves and improves on its performance by changing the topology of the network and the weights (trust) of each node in its predecessors.

Table 1. Pair hybrids combinations

Parameter	GB-RF	GB-DT	RF-DT
Training examples (%)		70, 80, 90	
Top-Bottom Train Division		3-5, 4-4, 5-3	
Bottom's Input		Input + Top's Prediction	
GB - Loss		Deviance	
GB - Learning Rate	0.001, 0.01, 0.1	0.001, 0.01	
GB - Estimators		10, 100, 200	
GB - Max Depth		5, 15	
GB - Criterion		MSE, Friedman Improved MSE	
GB - Max Features		All Features, Log_2(Features)	
RF - Estimators	10, 100		10, 100
RF - Criterion	Gini Impurity, Information Gain		Gini Impurity, Information Gain
RF - Max Depth	10, 50, 200		10, 50, 200
RF - Max Features	All Features, Log_2(Features)		All Features, Log_2(Features)
DT - Criterion		Gini Impurity, Information Gain	
DT - Splitter		random, best	
DT - Max Depth		10, 100, 1000	
DT - Max Features		All Features, Log_2(Features)	

This approach requires several types of models to be created, trained and then inserted in a reservoir that will act as the source for new nodes. We are aware of the increased time and space complexity but are primarily focused on modeling a trust network that can be used to improve an ensemble's performance in data classification.

There is also the possibility of using this hybrid not only in classification problems but also in regression ones and another interesting possibility is having the network adjust its structure in a dynamically evolving problem.

This whole idea stems not only from Graph Theory but also from Social Sciences and Human Behaviour and Psychology, as this model tries to emulate what happens when a certain individual has to give an opinion on a certain topic taking into account his own beliefs as well as what the people around him believe, being that these people also have other people in which they deposit part of their trust. A representative diagram can be found in Figs. 2(a) and (b). The different colored edges represent the different weights in the connections.

5 Results

The results from the voting hybrids are presented in Table 2. Even though the voting mechanisms could not outperform the RF baseline model, they still achieved good results, and it is worth mentioning that the voting consisted only of 3 parties, which might have been too low a number to make a difference.

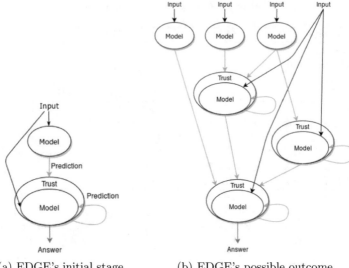

(a) EDGE's initial stage   (b) EDGE's possible outcome

Fig. 2. EDGE hybrid

Table 2. Voting results table (WV - Weighted Voting)

Metric	Bully voting	Fair voting	WV (recall)	WV (precision)
Average accuracy	0.89	0.88	0.88	0.88
Off by 1 avg. accuracy	0.96	0.96	0.95	0.95
Precision	0.88	0.88	0.88	0.87
Recall	0.89	0.88	0.88	0.88
F1-score	0.88	0.88	0.87	0.87
Log loss	3.94	4.04	4.03	4.17
Hamming loss	0.11	0.12	0.12	0.12
Training examples (%)	90	90	90	90

The results for the hybrid pairs, presented in Table 3, are divided among the three choices for the hybrid's components. Each hybrid is named as the abbreviations of the top and bottom model that act as components. The table also presents the results for the three individual models used to compose the hybrids.

Seeing the results, the Pair Hybrids scored worse results than the base components but this is somewhat expected when considering that the parameters for the models were chosen in the beginning as standalone models. A new search of parameters could reveal that the Pair Hybrid does indeed have some merit but as of now we don't think this is the best choice to follow.

Table 3. Pair hybrids results table

Metric	DT	GB	RF	RF-DT	GB-RF	GB-DT
Average accuracy	0.87	0.89	0.90	0.82	0.88	0.83
Off by 1 avg. accuracy	0.94	0.96	0.96	0.91	0.95	0.91
Precision	0.87	0.88	0.89	0.82	0.87	0.83
Recall	0.87	0.89	0.90	0.82	0.88	0.83
F1-score	0.87	0.89	0.89	0.82	0.87	0.83
Log loss	4.60	3.75	3.53	6.30	4.26	5.88
Hamming loss	0.13	0.11	0.10	0.18	0.12	0.17

The results for the EDGE hybrid are presented in Table 4. EDGE was instantiated using an RF and a GB model with the best configuration found when only looking at each model individually for the parking dataset In the case of the GB the number of estimators was changed to 5. EDGE was also tested on three other datasets, one with Anuran calls [4], the known MNIST handwritten digit dataset [14] and a dataset about appliances energy use in a low energy building [3]. The results shown are the average of 10 runs for each dataset.

The Anuran and MNIST datasets are not of time series but they were chosen on purpose to determine if this approach might have potential in areas outside of time series. The Appliances dataset was chosen because it represents a time series, just like the original parking lot dataset, and we also discretized its predicting variable, the used energy, the same way as for the parking lot, for comparison purposes.

The results show little or no improvement (even lowering the average accuracy score against one of its components in our dataset) but we think this is due to EDGE being in its early stage, without applying topology or weight evolution.

Table 4. EDGE results table (NA - Not Applicable)

Dataset	Park. lot	Appliances	Anuran	MNIST
Average accuracy (%)	87.33	86.09	97.08	95.90
Off by 1 avg. accuracy (%)	95.16	94.25	NA	NA
Precision (%)	86.33	83.32	97.10	95.90
Recall (%)	87.33	86.09	97.08	95.90
F1-score (%)	86.56	84.40	97.04	95.89
Log loss	4.38	4.80	1.01	1.42
Hamming loss	0.13	0.14	0.03	0.04
Avg acc. gained WRT best node (%)	−0.02	0.29	0.13	0.13
Avg acc. gained WRT worst node (%)	16.39	3.82	6.51	8.46

6   Conclusions and Future Work

In this work we present three hybrid approaches to be used in prediction over time series with sporadic external stimuli. We have previously tested five different models in order to find the best configurations when dealing with the parking lot dataset. In this work the three best models were used in three approaches: Voting Hybrid, Pair Hybrid, and the novel model, EDGE, which stems from the thought of using models as nodes in a evolving trust network of individuals, where each one has a certain opinion and is influenced by the opinion of others.

Out of the three presented approaches, Pair Hybrids were that ones with worse performance, never achieving better results than the best standalone models. The Voting Hybrid might be interesting to explore but with more voters and different models to have more testing variety. The EDGE approach demonstrated potential and we argue that it might lead to a path of fruitful hybridization even though its present configuration is only in its initial state. We can see that the improvements on the results were small and in one case even performing worse (although only by 0.02%, which is an order of magnitude smaller than the improvements over the best performing nodes in other cases). As it is, it might as well be called only Directed Graph Ensemble.

In the future we will search for other ways of hybridizing models and search for better performing combinations. One insight we would like to pursue would be to combine different models in a way that emulates the way the human nervous system works with its complex structure. We believe EDGE should also be a key focus of future work. Its development to a more mature stage might lead us to tackling a richer range of problems with good results. To first step in further development will be growing the base state of the model. Our thoughts to achieve this is by using an evolutionary approach at first and then migrate to user-defined heuristics or meta-heuristics that allow a better exploration of the space state.

References

1. Anifowose, F., Labadin, J., Abdulraheem, A.: A hybrid of functional networks and support vector machine models for the prediction of petroleum reservoir properties. In: Proceedings of the 11th International Conference on Hybrid Intelligent Systems, HIS 2011, Melacca, Malaysia, 5–8 December, pp. 85–90 (2011)
2. Borovykh, A., Bohte, S., Oosterlee, C.W.: Conditional time series forecasting with convolutional neural networks. In: Proceedings of the 26th International Conference on Artificial Neural Networks, ICANN 2017, Part I, Alghero, Italy, 11–14 September, pp. 729–730 (2017)
3. Candanedo, L.M., Feldheim, V., Deramaix, D.: Data driven prediction models of energy use of appliances in a low-energy house. Energy Build. **140**, 81–97 (2017)
4. Colonna, G., Nakamura, E., Cristo, M., Gordo, M.: Anuran Calls (MFCCs) Data Set (2017). https://archive.ics.uci.edu/ml. Accessed 10 Nov 2018
5. Fontes, X., Silva, D.C.: Towards hybrid prediction over time series with non-periodic external factors. In: Proceedings of the 5th International Conference on Time Series and Forecasting, ITISE 2018, Granada, Spain, 19–21 September, pp. 1431–1442 (2018)

6. Fu, R., Zhang, Z., Li, L.: Using LSTM and GRU neural network methods for traffic flow prediction. In: Proceedings of the 31st Youth Academic Annual Conference of Chinese Association of Automation, YAC 2016, Wuhan, China, 11–13 November 2016 (2016)

7. Fukushima, K.: Neocognitron: a hierarchical neural network capable of visual pattern recognition. Neural Netw. **1**(2), 119–130 (1988)

8. Gil-Begue, S., Larrañaga, P., Bielza, C.: Multi-dimensional Bayesian network classifier trees. In: Proceedings of the 2018 Intelligent Data Engineering and Automated Learning, IDEAL 2018, pp. 354–363 (2018)

9. Hastie, T., Tibshirani, R., Friedman, J.: The Elements of Statistical Learning - Data Mining, Inference and Prediction. Springer, New York (2009)

10. Hills, J., Lines, J., Baranauskas, E., Mapp, J., Bagnall, A.: Classification of time series by shapelet transformation. Data Min. Knowl. Discov. **28**(4), 851–881 (2014)

11. Johnson, N.E., Ianiuk, O., Cazap, D., Liu, L., Starobin, D., Dobler, G., Ghandehari, M.: Patterns of waste generation: a gradient boosting model for short-term waste prediction in New York City. Waste Manag. **62**, 3–11 (2017)

12. Karim, F., Majumdar, S., Darabi, H., Chen, S.: LSTM fully convolutional networks for time series classification. IEEE Access **6**, 1662–1669 (2017)

13. Laudani, A., Lozito, G.M., Riganti Fulginei, F., Salvini, A.: Hybrid neural network approach based tool for the modelling of photovoltaic panels. Int. J. Photoenergy **2015**, 10 (2015)

14. LeCun, Y., Bottou, L., Bengio, Y., Haffner, P.: Gradient-based learning applied to document recognition. Proc. IEEE **86**(11), 2278–2324 (1998)

15. Marchand, P., Marmet, L.: Binomial smoothing filter: a way to avoid some pitfalls of least-squares polynomial smoothing. Rev. Sci. Instrum. **54**(8), 1034–1041 (1983)

16. Züfle, M., Bauer, A., Herbst, N., Curtef, V., Kounev, S.: Telescope: a hybrid forecast method for univariate time series. In: Proceedings of the 2017 International Work-Conference on Time Series, ITISE 2017, Granada, Spain, 18–20 September, July 2017

17. Pai, P.F., Lin, C.S.: A hybrid ARIMA and support vector machines model in stock price forecasting. Omega **33**(6), 497–505 (2005)

18. Winston, P.: A heuristic program that constructs decision trees. Artificial Intelligence Memo 173. MIT (1969)

19. Wu, H., Cai, Y., Wu, Y., Zhong, R., Li, Q., Zheng, J., Lin, D., Li, Y.: Time series analysis of weekly influenza-like illness rate using a one-year period of factors in random forest regression. Biosci. Trends **11**(3), 292–296 (2017)

20. Zhang, G.P.: Time series forecasting using a hybrid ARIMA and neural network model. Neurocomputing **50**, 159–175 (2003)

21. Zhang, Y., Haghani, A.: A gradient boosting method to improve travel time prediction. Transp. Res. Part C: Emerg. Technol. **58**, 308–324 (2015)

22. Zheng, F., Zhong, S.: Time series forecasting using a hybrid RBF neural network and AR model based on binomial smoothing. World Acad. Sci. Eng. Technol. **75**, 1471–1475 (2011)

Clustering Support for an Aggregator in a Smart Grid Context

Cátia Silva, Pedro Faria$^{(\boxtimes)}$, and Zita Vale

GECAD – Research Group on Intelligent Engineering and Computing
for Advanced Innovation and Development, Rua Dr. António Bernardino
de Almeida, 4200-072 Porto, Portugal
{cvcds,pnf,zav}@isep.ipp.pt

Abstract. The future of the industry foresees the automation and allocation of more intelligence to processes. A revolution in relation to the present. With this, new challenges and consequently more complexity is added to the management of the sectors. In the electric sector is introduced the theme of the Smart grids and so all the concepts aggregated with it. The possibility of the existence of demand response programs and the expansion of the distributed generation units for small players are key concepts and with enormous influence in the management of the markets belonging to this sector. Thus, a method is proposed that would help manage these resources through their aggregation, opening a new port for business models based on this idea. The benefit will be to take advantage of a more effective and efficient way the energy potential present in each group that is formed. Thus, in this paper will be explored the potential of clustering methods for the aggregation of resources.

Keywords: Clustering · Smart grid · Aggregation · Business model

1 Introduction

One of main drivers of the revolution that the world industry is going through is environmental concern. With the increase in the campaign to reduce the emissions of harmful gases, the solution will have to go through resources and processes that are more environmentally friendly, [1]. With regard to the electric sector, the expansion and promotion of the idea of endogenous renewable energies makes the concept of generation distributed one of the variables that will certainly have to be included in the new market models, [2]. In addition, the concept of Smart grids opens the possibility for consumers to have more access to information about their energy. It can be pointed out that they have access to the real-time price or even the possibility of selling to the energy market produced through the micro-production that they possess.

However, all these new hypotheses are variables of complication for the business models in the electricity markets, [3]. From now on, demand response programs will have to be considered and, given that the wholesale market is defined by competition between several companies, as efficiently as the solution found that includes these variables, the better will be the results. However, the uncertainty that these variables introduce, for example the microgeneration, may hinder this task, [4].

© Springer Nature Switzerland AG 2020
A. M. Madureira et al. (Eds.): HIS 2018, AISC 923, pp. 156–165, 2020.
https://doi.org/10.1007/978-3-030-14347-3_16

Regarding the infrastructure of the existing distribution network, it is not yet prepared for the future that lies ahead despite efforts. It will still be necessary to go a long way to allow conventional networks to be able to feed their loads in a controlled and smarter way and, in addition, to enable the addition of the elements distributed, [5].

However, there is a prediction of multiple advantages in the promotion and implementation of these concepts. One can speak for example in reducing the losses in the transport of energy in the distribution network. With the possibility of the energy being consumed immediately by consumers closest to the production sites, the distance that it travels will be lower, thus reducing losses. In this way there is an enormous advantage in the promotion of distributed generation and an incentive for producers and consumers to increase their participation in demand response programs.

Regarding consumers, this new chance of realizing in real time the price of energy in the market could change their behavior. Through signs and changes, both reductions and increases, the end-user can review their consumption. Thus, virtual power players will need to understand the changes in the load diagram and work to achieve the goal of reducing the load peak. Efficient management will be crucial to achieving all the goals, [6].

In this way the methodology proposed in [7], used in this paper, allows the virtual power player to aggregate small resources, including consumers participating in demand response programs.

Different works, like the authors of this paper, try to fill this gap in the actual business models, including in their studies the concepts introduced by Smart Grids, like in [8–11] and many more. For example in [8] was developed a solution through a hierarchical multi-agent system with 3 levels to control renewable energy resources and demand response units, including advanced inverters for PVs. One of those layers makes the schedules for flexible energy sources, based on its optimization criteria, either maximizing the profit or minimizing energy differences between local production and consumption. The results from scheduling depends on the system rules and boundary conditions, which include technical, economic, market, and regulatory provisions.

The method proposed in this paper will be presented in more detail in the following section and it has been designed in such a way as to allocate the fairer remuneration for each group of aggregate resources, adding a different perspective to the actual business models. Different numbers of groups and aggregation methods are tested in this paper through a tool that was designed to understand what will be most beneficial in prior decision making of the optimum number that enables the minimization of operating costs by the aggregator. In this way, the results from optimization phase are used as input for this tool. This paper will present the tool and results from one chosen k cluster. The main objective is to show the viability of it and how can be useful for the aggregator.

2   Approach

Through the methodology proposed in [7] it will be possible to understand the role of the aggregator in the network infrastructure and, thus, a way of dealing with the challenges that the new paradigm is imposing. Figure 1 presents each of the phases that make up the methodology and its logical sequence of events.

Fig. 1. Proposed methodology

At an early stage, an optimal scheduling of all resources associated with the aggregator will be carried out, namely, distributed generation units (DG), consumers who belong to the demand response and supplier programs and suppliers. The last one may be used if DG units do not meet the needs of the consumer. Profiles used as input data for this optimization include, for example, the elasticity of the demand price, the levels of comfort sought by each of them, the possibilities of direct control of load and the existence of production of either heat or electricity, … Price and operation constraints were considered in this optimization, also operational constraints imposed by virtual power player to achieve its goals. The MATLAB software potentials were used through the toolbox, TOMLAB, for the optimization.

The second phase represents the aggregation of resources. Through a method of clustering are defined groups considering the results obtained in the first phase – schedule power for each resource. By grouping these small resources, VPP will have a considerable amount of energy for the negotiations in the market and allows the expansion of their participation in DR programs. It will be an innovative thing since so far, the participation of such resources would only be made in an indirect way. In [12] classification methods were studied using this methodology. In this paper and to test the potential of the clustering methods, four were compared for several k – number of clusters. The methods and the case study will be presented in the following sections more explicitly. This part of the study was carried out using software R and Shiny package.

The last phase, the remuneration stage, is used for continued collaboration of all resources with the aggregator in the network operation, it is used as compensation. Here, taking into account the tariffs that were designed specifically for each group, through the maximum of each one, all resource will be paid at the end of the schedule.

3   Case Study

The case of study presented to test the feasibility of the proposed method is composed of a real distribution network composed of about 548 distributed producers and 20310 end-user consumers. To manage the consumers from demand response programs, two main programs were considered. The first one is called Real Time Pricing (RTP), based on price, where consumers change their load by responding to changes in the price of electricity in real time. The second one is based on incentives, in which consumers are paid at a fixed price per kW of reduced load. Incentive based programs (Reduce, Cut) and RTP were applied in this study to distinct types of consumers.

Regarding the distributed generation, this case was studied Wind, Biomass, Small hydro, co-generation, Photovoltaic, Fuel cell and Waste-to-energy. Table 1 presents the detailed information of these units, showing the unit number by type, the unit operating price in m.u./kWh and the total available capacity.

Table 1. Distributed generation characterization

Designation	N° of units	Capacity (kWh)	Price (m.u./kWh)
Wind	254	5 866.09	0.071
Co-generation	16	6 910.10	0.00106
Waste-to-energy	7	53.10	0.056
Photovoltaic	208	7 061.28	0.150
Biomass	25	2 826.58	0.086
Fuel cell	13	2 457.60	0.098
Small hydro	25	214.05	0.042
Total DG	548	25 388.79 kWh	

Regarding the Table 2. shows the characterization for the types of consumers that participate in demand response programs: Domestic (DM), Small Commerce (SM), Medium Commerce (MC), Large Commerce (LC) and industrial (ID). It also shows the different possibilities of participation.

Table 2. Demand response consumers characterization

Designation	Reduce	Cut	RTP	Initial price (m.u./kWh)
Domestic (DM)	●			0.12 (0.20)
Small commerce (SM)	●			0.18 (0.16)
Medium commerce (MC)		●		0.2 (0.20)
Large commerce (LC)		●		0.19 (0.20)
Industrial (ID)			●	0.15 (0.53)
Total N° of DR	19 996	167	147	20 310
Total capacity (kWh)	8 676	1 106	11 571	21 354.36

The main goal is to aggregate the resources considering the results of the optimization performed in the first phase. In this paper, the authors choose to study the behave of the tool considering only consumers. So, for testing, in this paper will only be exposed part of the aggregated resources (DG and DR units) and will not be considered the ones that got a null result in the optimization phase. Consequently, in the chosen scenario, only two types of consumers who belong to programs of DR were tested. The incentive-based program were studied, resulting in consumers belonging only to Small Commerce and Medium Commerce.

4 Demonstration

Through Fig. 2 will be able to perceive the tool interface built for comparing clustering methods.

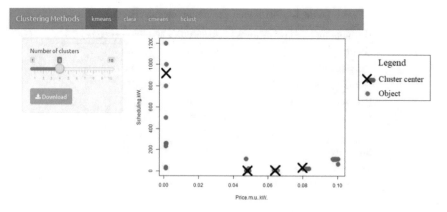

Fig. 2. Clustering methods: tool to compare different clustering methods

This interactive tool has been developed by giving use to the software R through the Shiny package. The main intention of this tool will be to facilitate and assist in the task of the Analyst in this type of business model. Thus, a clearer vision and analysis will be possible for the decision in the selection of the method to be used as the problem it faces.

One of the advantages goes through help in the decision of the optimal number of clusters, since it is usually a task of the analyst, through the visualization of the database graphically. As the number of K cluster changes, it also varies the group to which each point belongs and the center of the cluster. There is the possibility of variation between 1 and 10 clusters.

Thus, four clustering methods were selected to be used and compared in this tool: two methods of partitional clustering – k-means and CLARA, one method of fuzzy clustering – c-means and one method of hierarchical clustering.

After finding the appropriated k cluster, this tool also the possibility to download the results to an external Excel file to be used later.

Firstly, K-means is presented as it is one of the most common unsupervised machine learning methods. The variation of this method used in this tool is the one presented by Hartigan-Wong and its used by default in R software, [13]. The algorithm is defined as the total variation within a cluster is then taken as the sum of the squares of Euclidean distance between a point and the center of the cluster, and then assigns the point to the nearest cluster. In this paper this method was performed only one time to obtain results. Each cluster is represented by a new center, the centroid, that is represented by a cross in Fig. 2, which corresponds to the average of the points assigned to the k cluster in question. This method, however, presents a problem that can be considered serious for some cases since it cannot handle noise and outliers.

CLARA (Clustering Large Applications) is an extension of Partitioning Around Medoids (PAM), for a larger database. The algorithm is defined by the iterative search of objects that can represent a cluster, medoids, also represented by a cross. The exchange between a medoid and a non-medoid is validated only when there is an improvement in the criterion of the objective function - the minimization of the sum of the dissimilarities of all objects relative to the nearest medoid, [14]. As already mentioned, the advantage of this method is that it can deal with larger sets than PAM. However, the efficient performance of CLARA depends upon the size of the data base and a biased sample data may lead to poor results, [15].

Regarding C-means, this method is considered as a soft clustering that, unlike realizing whether or not an object belongs to a cluster, assigns a likeness of this point to the assigned cluster, [16]. To perform this clustering method in R software was chosen the algorithm *cmeans*. The parameter m is the degree of fuzzification and only support real values greater than 1. The bigger it is, the more fuzzy the membership values of the clustered data points are. In this paper, $m = 20$. Similar to k-means, this type of fuzzy clustering method can also represent the centers of each cluster, again presented with a cross.

Finally, hierarchical clustering (hclust), as the name implies, structures the database in a hierarchical way, according to a proximity matrix. This clustering method can be divided into hierarchical agglomeration algorithms and divisive clustering. In R, an agglomeration method was used and was chosen *complete* linkage for this paper. In this way, each object is assigned to cluster and then, iteratively, join the two most similar clusters, continuing until there is just a single cluster with all the objects. At each iteration, distances between clusters are recomputed by the Lance–Williams dissimilarity update formula. It will not be possible to decide a priori a number of the clusters. So, the analyst according to the division that finds most interesting for the situation, can split the resulting dendrogram, [17].

As previously said, the usefulness and feasibility of this tool was tested in the second phase of the methodology where it is proposed to compare several clustering methods for the aggregation of the various resources associated with the aggregator, specifically in this paper, small and medium commerce type of consumers. Initially, a database with the results obtained in MATLAB of the scheduling proposed in phase one, and with the tool, all the resources were grouped according to the number k of clusters.

5   Results

In this paper and through the following figures, the authors choose to expose the results of the aggregation for only two types of consumers associated with the programs of DR: Small Commerce and Medium commerce. These consumers represent the highest percentage in the data base so, a total of 9910 consumers were analyzed. The figures represent the total power according to each type of consumer for each of the groups formed.

Using the results obtained in the clustering tool, the four methods group these types of consumer according to the results obtained in the first phase of the proposed methodology. Thus, the inputs for this tool were the scheduled power of each of the resources. In this paper, the authors choose to only study one of the k clusters for the different methods chosen, to compare them in the same base. Since there are two types of consumers to be studied, the results for k = 2 clusters are presented through this section to understand how these methods behave.

Figure 3 refers to the K-means method. This method included all small commerce consumers in Group 1. Regarding the consumers of medium commerce, only one minority was included in the group 1 The remaining composes the group 2.

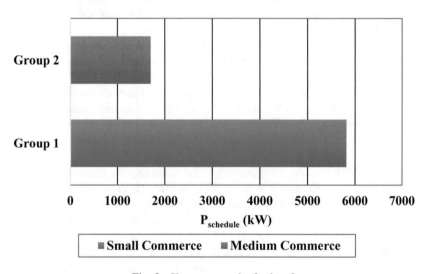

Fig. 3. K-means: results for k = 2

Figure 4 shows the results for the C-means method. Group 1 is composed, in majority, by small commerce elements, presenting only a minimum percentage, 4 elements, of medium commerce. The group 2 is constituted in its entirety by elements of medium commerce.

The results obtained through the hierarchical method are shown in Fig. 5. As it is possible to repair, group 1 is a major part of the small consumers. Group 2 is only made

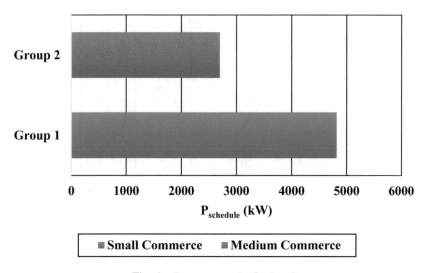

Fig. 4. C-means: results for k = 2

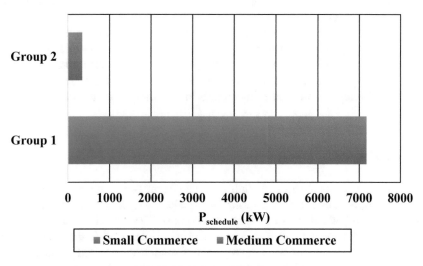

Fig. 5. Hclust: results for k = 2

up of 3 elements that add up to each other about 343.94 kW. In this paper, since the purpose was to compare k = 2 clusters for all the methods, the result dendrogram is not analyzed.

About the CLARA method, Fig. 6 shows its results to K = 2. Group 1 is again constituted mostly by Small Commerce. The medium commerce represents only 0.006% of group 1. In relation to the other group, the remaining 78 elements of medium commerce are what constitute it.

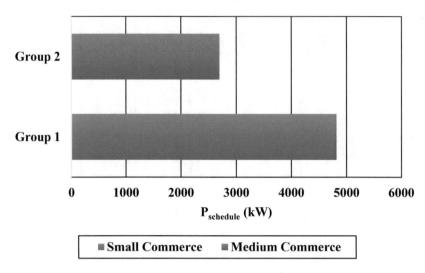

Fig. 6. CLARA: results for k = 2

Now, comparing and analyzing the four methods presented, we can conclude that CLARA and c-means almost divided the data base by type of consumer, attributing group 1 to small commerce and group 2 to medium commerce. Regarding k-means and hclust, both assign a lot of members of medium commerce to group 1. One way to provide better results on hclust may be changing the distance metric and the linkage criteria. Regarding k-means, the initial seeds have a strong impact in the final result, in this way, changing and add some randomness to start may result in different outcomes, [15].

6   Conclusions

A radical change approaches all industry. Intelligent and automatic processes are required for the future. The structuring of the electricity sector is now changing to achieve a full introduction of the smart grid context. The market is not an exception, and players of the market will have to create new business models that include all the variables. Thus, the suggestion of aggregation of small resources and the idea of using their potential is suggested as a viable solution in a methodology proposed. The goal for the tool presented is to help decision making made by the virtual power player, as aggregator. By aggregating similar resources, the resulting groups will perform a fairer remuneration. The tool presented in this paper can simulates several situations, e.g., the number of clusters in which the database can be divided and compare different methods in order to decide the optimal solution.

Acknowledgments. The present work was done and funded in the scope of the following projects: CONTEST Project (P2020 - 23575), and UID/EEA/00760/2013 funded by FEDER Funds through COMPETE program and by National Funds through FCT.

References

1. Alahakoon, D., Yu, X.: Smart electricity meter data intelligence for future energy systems: a survey. IEEE Trans. Ind. Inform. **12**(1), 425–436 (2016)
2. Faria, P., Vale, Z., Baptista, J.: Demand response programs design and use considering intensive penetration of distributed generation. Energies **8**(6), 6230–6246 (2015)
3. Silva, C., Faria, P., Vale, Z.: Assessment of distributed generation units remuneration using different clustering methods for aggregation. In: IEEE International Conference on Communications, Control, and Computing Technologies for Smart Grids, Aalborg, Denmark (2018)
4. Clastres, C.: Smart grids: another step towards competition, energy security and climate change objectives. Energy Policy **39**(9), 5399–5408 (2011)
5. Deng, R., Yang, Z., Chow, M.-Y., Chen, J.: A survey on demand response in smart grids: mathematical models and approaches. IEEE Trans. Ind. Inform. **11**(3), 1 (2015)
6. Lin, S., Li, F., Tian, E., Fu, Y., Li, D.: Clustering load profiles for demand response applications. IEEE Trans. Smart Grid **3053**(c), 1–9 (2017)
7. Faria, P., Spínola, J., Vale, Z.: Aggregation and remuneration of electricity consumers and producers for the definition of demand-response programs. IEEE Trans. Ind. Inform. **12**(3), 952–961 (2016)
8. Zupancic, J., et al.: Market-based business model for flexible energy aggregators in distribution networks. In: International Conference European Energy Market, EEM (2017)
9. Rezania, R., Prüggler, W.: Business models for the integration of electric vehicles into the Austrian energy system. In: 9th International Conference European Energy Market, EEM 2012, pp. 1–8 (2012)
10. Ilieva, I., Rajasekharan, J.: Energy storage as a trigger for business model innovation in the energy sector. In: 2018 IEEE International Energy Conference ENERGYCON 2018, pp. 1–6 (2018)
11. Zhang, Q., Zhang, J., Han, X., Jin, X., Fu, H., Zang, T.: Business model status of distributed power supply and its concerns in future development. In: 2017 IEEE Conference on Energy Internet and Energy System Integration, EI2 2017 - Proceedings, vol. 2018, pp. 1–5 (2018)
12. Silva, C., Faria, P., Vale, Z.: Classification Approaches to Foster the Use of Distributed Generation with Improved Remuneration (2013)
13. Morissette, L., Chartier, S.: The k-means clustering technique: general considerations and implementation in Mathematica. Tutor. Quant. Methods Psychol. **9**(1), 15–24 (2013)
14. Kassambara, A.: Practical Guide to Cluster Analysis in R, 1st ed
15. Saket, S.J., Pandya, S.: An overview of partitioning algorithms in clustering techniques. Int. J. Adv. Res. Comput. Eng. Technol. **5**(6) (2016)
16. Jyoti Bora, D., Kumar Gupta, A.: A comparative study between fuzzy clustering algorithm and hard clustering algorithm. Int. J. Comput. Trends Technol. **10**(2), 108–113 (2014)
17. Wirasanti, P., Ortjohann, E., Schmelter, A., Morton, D.: Clustering power systems strategy the future of distributed generation. In: International Symposium on Power Electronics, Electrical Drives, Automation and Motion, pp. 679–683 (2012)

Economic Impact of an Optimization-Based SCADA Model for an Office Building

Mahsa Khorram, Pedro Faria$^{(\boxtimes)}$, Omid Abrishambaf, and Zita Vale

GECAD – Knowledge Engineering and Decision Support Research Centre,
Polytechnic of Porto (IPP), Porto, Portugal
{makgh, pnf, ombaf, zav}@isep.ipp.pt

Abstract. The daily increment of electricity usage has led many efforts on the network operators to reduce the consumption in the demand side. The use of renewable energy resources in smart grid concepts became an irrefutable fact around the world. Therefore, real case studies should be developed to validate the business models performance before the massive production. This paper surveys the economic impact of an optimization-based Supervisory Control And Data Acquisition model for an office building by taking advantages of renewable resources for optimally managing the energy consumption. An optimization algorithm is developed for this model to minimize the electricity bill of the building considering day-ahead hourly market prices. In the case study, the proposed system is employed for demonstrating electricity cost reduction by using optimization capabilities based on user preferences and comfort level. The results proved by the performance of the system, which leads to having great economic benefits in the annual electricity cost.

Keywords: Energy optimization · Renewable resources · SCADA

1 Introduction

Nowadays, Demand Response (DR), and Renewable Energy Resources (RERs) are considered as major concepts in energy research topics [1]. The importance of DR programs and RERs is due to the advantages for all parts of the community including energy producers, consumers, and the environment by reducing the dependency on fossil fuels [2]. There is global concern about increasing of CO_2 emissions, melting the glaciers, and the collapse of nature cycle [3], and hence, currently the demand of RERs and DR programs are increased [4]. In the DR programs, consumers are emboldened to change their electricity consumption pattern based on the variation of electricity price, or technical commands from the network operators [5]. DR programs can classify into two main incentive-based and price-based [6]. Real-Time Pricing (RTP), Time-Of-Use (TOU), and Critical-Peak Pricing (CPP) are included in the price-based programs. It should be noted that in RTP, the used prices are the day-ahead or hour-ahead basis [6].

The present work was done and funded in the scope of the project UID/EEA/00760/2013 funded by FEDER Funds through COMPETE program and by National Funds through FCT.

According to diverse surveys in [7], a large amount of energy consumption is dedicated to all types of buildings. In U.S. 35.5% of total electricity consumption belongs to the commercial buildings [8]. For taking advantages of DR programs, the buildings should be intelligent and equipped to several automation infrastructures [9]. Supervisory Control And Data Acquisition (SCADA) system is considered a part of DR implementation since it offers various advantages to have automatic load control in different types of buildings [10]. After equipping the buildings with the required infrastructures, an optimization algorithm is required to optimize and reduce the power consumption based on the existing conditions. For instance, in commercial buildings, Air Conditioners (ACs) and lighting systems can be considered as flexible and controllable loads [11, 12] due to their massive portion in electricity consumption [13].

This paper presents an optimization-based SCADA model, which is implemented in an office building. The main objective of the optimization algorithm is to minimize the electricity bill by using RERs and decreasing the power consumption according to day-ahead hourly electricity prices. The power consumption of the ACs and lighting system have been selected to apply the consumption optimization. All the devices in this system are categorized based on the priorities defined by each user for each device in order to observe preferences and comfort level.

Several research works have been done in this context. In [14], the authors applied for DR programs in a residential building by using two types of retail pricing and control Heating, Ventilation, and Air Conditioning (HVAC) system. An optimal DR scheduling model for HVAC has been presented in [15], considering the thermal comfort of the users. [16] proposes an energy optimization controller algorithm which is used for industrial and commercial equipment such as HVAC, based on hour-ahead RTP programs. In [17], a SCADA system is implemented which is connected to MATLAB software to control and integrate different information of intelligent buildings such as temperature, ventilation, and illumination to manage and maintain the user satisfaction. In [18] represents a real implementation of an optimization model supported by a SCADA system, which employs several controlling and monitoring methods in order to manage the consumption and generation of the building. The main contribution of this paper is to survey the annual impact of the developed optimization algorithm in implemented SCADA system of an office building by considering day-ahead electricity prices.

After this introductory section, the architecture of the developed model is described in Sect. 2. Section 3 represents a case study considered for the system, and the final results are presented in Sect. 4. Finally, Sect. 5 details the main conclusions of the work.

2 System Architecture

This section describes the SCADA model which is implemented in an office building for managing energy consumption, and then, it focuses on an optimization algorithm employed in the SCADA for energy optimization purposes.

2.1   SCADA Model

The SCADA model presented in this paper has been implemented in an office building, which includes eight offices, one server room, and a corridor. Every three offices have been categorized into a zone. Therefore, there are three independent zones including three offices. Only the relevant information regarding this SCADA system are mentioned in this section and more details are available on [18], which have been developed by the authors in the scope of their previous works.

There are three distributed based Programmable Logic Controllers (PLCs) for three zones, which control every three offices. Moreover, there is a main PLC associated with this model in order to supervise the other distributed based PLCs. In this paper, ACs and the lighting system of the building have been targeted for implementing the optimization purposes. The lighting system consists of 13 fluorescent lamps connected to the related PLC by Digital Addressable Lighting Interface (DALI). There are DALI ballasts installed for each lamp that allow the related PLC to fully control the intensity of the light as well as switching them ON/OFF. Furthermore, the SCADA system controls 9 ACs in the building. A microcontroller (Arduino® – www.arduino.cc), which equipped with an Ethernet Shield and an Infrared Light-Emitting Diode (IR LED) have been programmed and installed near to each AC. In fact, this controlling method emulates the remote control of ACs, somehow the SCADA takes decision for each AC and transmits the desired command to each AC controller (Arduino®) via Ethernet interface (MODBUS protocol). Then, Arduino® controls the ACs based on SCADA decision.

In this SCADA model, there is a Photovoltaic (PV) system, which supplies a part of building consumption, and in high generation periods, the surplus of energy will inject into the utility grid. Also, six energy meters have been employed in the SCADA model to measure the consumption and generation of building. All energy meters follow serial communication with MODBUS-RTU (RS485) protocol to transmit the information to the related PLC.

Finally, a unit so-called Optimizer is connected to the model, which is responsible for solving optimization algorithm (explained in the next section) associated for the SCADA and provide the optimized information to the PLCs in order to be performed.

2.2   Optimization Problem

This section shows the details about the optimization algorithm, which is employed by developed SCADA system. This algorithm has been solved via "lpsolve" tools of RStudio optimization environment (www.rstudio.com) in order to minimize the electricity cost by managing the power consumption of the building. The objective of the optimization problem is to optimize the power consumption of the ACs a lighting system, based on the priority of each device, electricity cost variation, and stochastic PV generation. The power reduction of the ACs and lighting system depend profoundly on the electricity price, and the priority of each device. Figure 1 shows the procedure of the optimization algorithm.

This optimization algorithm is a dynamic algorithm which runs online and updates the input parameters. As it can be seen in Fig. 1, the first part of the algorithm is

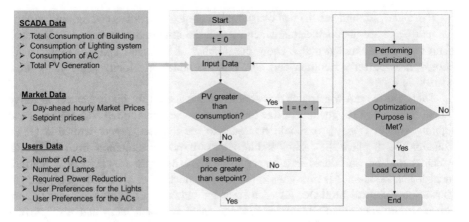

Fig. 1. Flowchart of the developed optimization algorithm.

dedicated to the definition of input data, such as total consumption of the building, total PV generation, consumption of the ACs, and consumption of the lighting system measured by SCADA system. In the meanwhile, the algorithm receives the information of the electricity market, since it is based on the day-ahead hourly electricity market. Then, the algorithm checks the amount of PV generation and compares that with the total consumption of ACs and lighting system. If the rate of consumption is greater than the PV generation, the algorithm continues to the next decision steps, otherwise, it will not enter into the optimization.

The algorithm also receives the hourly electricity prices one day in advance and it takes an average value from all prices for the entire day and chooses it as a setpoint for the next day in order to perform the optimization. If the real-time electricity price is greater than the setpoint price, the algorithm starts optimizing the consumption, and in periods with a lower price than the setpoint, the optimization is not required, and the algorithm will check the input data again and again till it achieves the desired condition. By this way, consumption would be automatically optimized in the expensive periods.

Furthermore, there are some required data which should be indicated by the users, such as user preferences for the devices, the number of ACs, number of lights, and the amount of required reduction. User preferences can be interpreted as the priority of each device for the user. The weight of these priorities can be converted to a range between 0 and 1. The priority numbers near 0 present the low priority devices from the user standpoint and the high priority devices have a number near to 1. The objective function of the optimization problem is demonstrated in Eq. (1):

Minimize

$$EC = \sum_{t=1}^{T}((\sum_{a=1}^{A} P_{AC(a,t)} \times I_{AC(a,t)} + \sum_{l=1}^{L} P_{Lamp(l,t)} \times I_{Lamp(l,t)}) - PV_{(t)}) \times COST_{(t)}$$

$$(1)$$

P_{AC} and P_{Lamp} indicate the real consumption of the AC devices and lighting system. I_{AC} and I_{Lamp}. are the coefficients which present the importance of each AC, and each lamp respectively and A and L show the number of ACs and lamps. PV stands for photovoltaic power generation and $COST$ is the hourly electricity price in the time horizon of t.

There are several constraints that should be considered for the presented objective function. Equation (2) guarantees that the sum of power reduction of ACs ($P_{Red.AC}$) and lighting system ($P_{Red.Lamp}$) should be equal to the required power reduction (P_{RR}). Equation (3) displays the technical limitations of power reduction from each AC, which should be considered as a binary value since the ACs are classified in curtailment loads. Equation (4) limits the total power reduction of all AC devices, and (5) shows the technical limitation of each light individually. Equations (4), and (5) are considered to have a minimum illumination level for each lamp and also prevent turning all AC devices OFF in order to respect to the user comfort level. In fact, (5) models the lighting system as flexible loads, which allows the algorithm to reduce their consumption.

$$\sum_{a=1}^{A} P_{Red.AC(a,t)} + \sum_{l=1}^{L} P_{Red.Lamp(l,t)} = P_{RR(t)} \tag{2}$$
$$\forall t \in \{1, \ldots, T\}$$

$$P_{Red.AC(a,t)} = P_{maxRed.AC(a,t)} \times X \tag{3}$$
$$\forall X \in \{0, 1\}$$
$$\forall t \in \{1, \ldots, T\}$$
$$\forall a \in \{1, \ldots, A\}$$

$$\sum_{a=1}^{A} P_{Red.AC(a,t)} \leq P_{maxTotalRed.AC(t)} \tag{4}$$
$$\forall t \in \{1, \ldots, T\}$$

$$0 \leq P_{Red.Lamp(l,t)} \leq P_{maxRed.Lamp(l,t)} \tag{5}$$
$$\forall t \in \{1, \ldots, T\}$$
$$\forall l \in \{1, \ldots, L\}$$

$P_{maxRed.AC}$ declares the maximum power reduction for each AC device, and $P_{maxTotalRed.AC}$ is the maximum available power reduction from all AC devices. Furthermore, $P_{maxRed.Lamp}$ is the maximum power that is allowed to be reduced from each light. Required power reduction (P_{RR}) is a flexible rate, which can be defined as any desired rate. However, in this paper, it is considered that the power reduction is the difference of the consumption of ACs and lighting system, and PV generation, in order to take advantages of PV generation in electricity bill. This definition is formulated by (6). In other words, (6) demonstrates that the PV generation supplies the consumption of ACs and lighting system as much as possible. Finally, (7) presents the limitation of importance coefficients.

$$P_{RR(t)} = \left(\sum_{a=1}^{A} P_{AC(a,t)} + \sum_{l=1}^{L} P_{Lamp(l,t)} \right) - PV_{(t)} \tag{6}$$
$$\forall t \in \{1, \ldots, T\}$$

$$0 \le I_{AC(a,t)} \le 1; \; 0 \le I_{Lamp(a,t)} \le 1 \tag{7}$$
$$\forall t \in \{1, \ldots, T\}$$
$$\forall a \in \{1, \ldots, A\}$$

As a summary, this section presented the technical specification of the SCADA model as well as the mathematical formulation of the proposed optimization algorithm. In the next section, this system is used for a case study in order to survey its annual economic impact.

3 Case Study

A case study is provided in this section to test and validate the functionalities of the developed system. For this purpose, the annual PV generation curve of the building and also annual consumption profiles of the ACs and lighting system are considered in order to perform the presented optimization problem. Figure 2 illustrates the consumption and generation profiles considered for this case study. These curves are real data measured by the SCADA system and stored in a database.

The average consumption of each of 13 AC is 1500 W and each of 9 lamps is 116 W. The consumption curve showed in Fig. 2, is the consumption of ACs and lighting system for an entire year, and non-controllable loads are not shown in this figure. Furthermore, energy consumption in cold weather is higher than in hot. It means ACs are mostly used in winter for heating purposes.

Also, the generation curve showed in Fig. 2, is the real PV generation of the building for a year. This produced power will be first used for building consumption, and then in the high generation and low consumption periods, the energy surplus will be injected into the main grid. The electricity that is purchased from the main grid for the ACs and lighting system, is shown in Fig. 3.

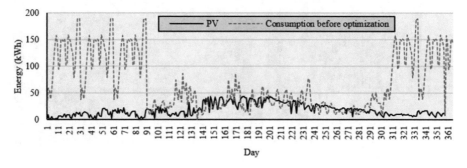

Fig. 2. Annual profiles considered for a case study for Consumption and PV generation.

In the periods that optimization is required, the algorithm attempts to reduce the consumption profile shown in Fig. 3 (Power purchased from the main grid). In other words, the consumption curve illustrated in Fig. 3 is the data that should be provided to the algorithm as required consumption reduction. Although, a maximum reduction level always is considered by the algorithm in order to not switching all the devices OFF and keep high priority devices switched ON for respect to user preferences and comfort.

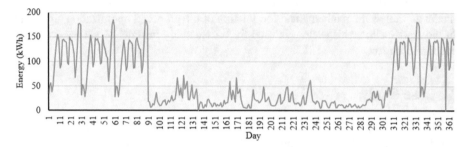

Fig. 3. Power purchased from the utility grid for ACs and lighting system electricity demand.

It should be noted that the prices used for defining setpoints, are for the entire year of 2016 and have been adapted from the Portuguese sector of Iberian Electricity Markets (MIBEL) [19].

4   Results

This section focuses on the results of the case study using the developed optimization-based SCADA model for the office building. Figure 4 illustrates the scheduled consumption profile after the optimization. In fact, the results that are shown in Fig. 4 is the output of the optimization algorithm regarding each AC and lamp.

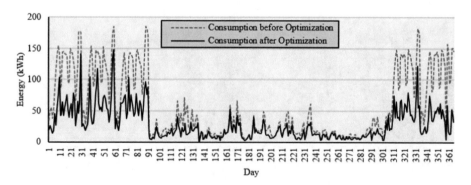

Fig. 4. Consumption optimization results for ACs and lighting system in the office building.

The consumption profile that is shown by red lines in Fig. 4 is the purchased energy from the utility grid for ACs and lighting system consumption. Therefore, the optimization algorithm minimized this purchased energy based on the hourly prices in order to minimize the total energy cost (as it can be seen in Fig. 4 with blue line).

Furthermore, in the periods that the price is greater than the setpoint price, and the optimization algorithm is applied in the building, the user's comfort level is not violated since the algorithm minimizes the consumption of ACs and lighting system based on the priorities that each user defines. Also, the maximum consumption reduction level considered in the algorithm keeps the high priority devices always switched ON.

The other noticeable point in Fig. 4 is that from the beginning of the April to the end of October, the optimization algorithm is not widely performed since there is a significant amount of PV generation, and the generation supported all the demand for some periods and a reasonable part of it in other periods.

In order to survey the economic impact, Fig. 5 illustrates the accumulated cost for an entire year under three different conditions.

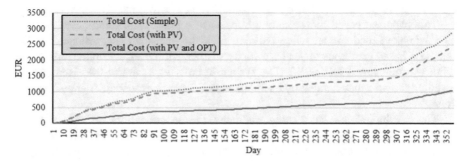

Fig. 5. Accumulated energy costs for one year considering three scenarios.

As it can be seen in Fig. 5, the red line indicates the purchased energy cost from the main grid for the ACs and lighting system while no PV and optimization capabilities are considered for the SCADA. In fact, this scenario is considered as a basis to be compared with the other scenarios.

The gap between the red and green lines in Fig. 5 indicates the cost that is gained due to the local PV generation. This gap is increased during the summer, where the PV generation is higher, and it supplies a great part of building consumption. Finally, by employing both PV system and optimization capabilities, the total electricity cost of the building is reduced significantly, which indicates the advantages of an optimization-based SCADA model equipped with a local energy resource (e.g. PV).

Table 1. Annual cost comparison for three scenarios.

Scenario	Annual cost (EUR)	Cost reduction (Compared to simple)
Simple (No PV, No OPT)	2890.92	–
With PV, No OPT	2455.49	15.06%
With PV and OPT	1048.99	63.71%

In order to clarify the annual cost comparison, Table 1 demonstrates the details of annual electricity costs for the presented scenarios.

As it is clear in Table 1, the amount of cost reduction with a PV system and without optimization capability is 15.06% compared to the simple scenario. While both the PV system and optimization are used by the SCADA, the cost is 63.71% is reduced compared to the simple scenario. This proves the functionalities and capabilities of the developed optimization-based SCADA system, which presents an acceptable performance in day-ahead and real-time electricity markets.

5   Conclusions

Electricity demand increment in the current state of energy system requires more survey and investigation regarding energy optimization and using renewable energy resources. Commercial buildings, especially office buildings are appropriate targets for optimizing the lighting systems and air conditioner devices. However, the need for real case studies is obvious for testing the business model before massive production.

This paper surveyed the economic impacts of a real optimization-based SCADA model implemented for an office building to make decisions and manage the consumption based on the day-ahead electricity market. Moreover, an optimization algorithm was developed for the SCADA system in order to minimize the electricity bill based on the priorities defined by each office user for each air conditioner and lighting system.

The results of the case study validated the functionalities of the presented optimization-based SCADA model in the day-ahead electricity market. Furthermore, a comparison of annual electricity costs of the building was demonstrated by considering different scenarios to show the economic benefits of applying the proposed optimization algorithm.

References

1. Arun, S., Selvan, M.: Intelligent residential energy management system for dynamic demand response in smart buildings. IEEE Syst. J. **12**(2), 1329–1340 (2018)
2. Khorram, M., Faria, P., Vale, Z.: Optimization-based home energy management system under different electricity pricing schemes. In: IEEE 16th International Conference of Industrial Informatics (IEEE INDIN) (2018)
3. Rotger-Griful, S., Welling, U., Jacobsen, R.: Implementation of a building energy management system for residential demand response. Microprocess. Microsyst. **55**, 100–110 (2017)
4. Rehmani, M., Reisslein, M., Rachedi, A., Erol-Kantarci, M., Radenkovic, M.: Integrating renewable energy resources into the smart grid: recent developments in information and communication technologies. IEEE Trans. Indus. Inform. **14**(7), 2814–2825 (2018)
5. Abrishambaf, O., Faria, P., Gomes, L., Spínola, J., Vale, Z., Corchado, J.: Implementation of a real-time microgrid simulation platform based on centralized and distributed management. Energies **10**(6), 806–820 (2017)

6. Faria, P., Vale, Z.: Demand response in electrical energy supply: an optimal real time pricing approach. Energy **36**(8), 5374–5384 (2011)
7. Yang, F., Guo, Q., Pan, Z., Sun, H.: Building energy management based on demand response strategy considering dynamic thermal characteristic. In: IEEE Power & Energy Society General Meeting (2017)
8. Yu, L., Xie, D., Jiang, T., Zou, Y., Wang, K.: Distributed real-time HVAC control for cost-efficient commercial buildings under smart grid environment. IEEE Internet Things J. **5**(1), 44–55 (2018)
9. Tsui, K., Chan, S.: Demand response optimization for smart home scheduling under real-time pricing. IEEE Trans. Smart Grid **3**(4), 1812–1821 (2012)
10. Khorram, M., Faria, P., Abrishambaf, O., Vale, Z.: Demand response implementation in an optimization based SCADA model under real-time pricing schemes. In: 15th International Conference on Distributed Computing and Artificial Intelligence (DCAI) (2018)
11. Faria, P., Pinto, A., Vale, Z., Khorram, M., de Lima Neto, F., Pinto, T.: Lighting consumption optimization using fish school search algorithm. In: 2017 IEEE Symposium Series on Computational Intelligence (SSCI) (2017)
12. Pan, X., Lee, B.: An approach of reinforcement learning based lighting control for demand response. In: PCIM Europe 2016, Nuremberg, Germany (2016)
13. Shi, J., Yu, N., Yao, W.: Energy efficient building HVAC control algorithm with real-time occupancy prediction. Energy Procedia **111**, 267–276 (2017)
14. Yoon, J., Bladick, R., Novoselac, A.: Demand response for residential buildings based on dynamic price of electricity. Energy Build. **80**, 531–541 (2016)
15. Kim, Y.: Optimal price-based demand response of HVAC systems in multi-zone office buildings considering thermal preferences of individual occupants. IEEE Trans. Indus. Inform. 1 (2018)
16. Abdulaal, A., Asfour, S.: A linear optimization based controller method for real-time load shifting in industrial and commercial buildings. Energy Build. **110**, 269–283 (2016)
17. Figueiredo, J., Costa, J.: A SCADA system for energy management in intelligent buildings. Energy Build. **49**, 85–98 (2012)
18. Khorram, M., Abrishambaf, O., Faria, P., Vale, Z.: Office building participation in demand response programs supported by intelligent lighting management. Energy Inform. **1**(9), 1–14 (2018)
19. MIBEL Electricity Market. http://www.omip.pt. Accessed 08 Aug 2018

Kernel Based Chaotic Firefly Algorithm for Diagnosing Parkinson's Disease

Sujata Dash[1(✉)], Ajith Abraham[2], and Atta-ur-Rahman[3]

[1] North Orissa University, Baripada, India
sujata238dash@gmail.com
[2] MIR Labs, Auburn, USA
abraham.ajith@gmail.com
[3] Imam Abdulrahman Bin Faisal University,
Dammam, Kingdom of Saudi Arabia
aaurrahamn@iau.edu.sa

Abstract. Parkinson's disease (PD) is prevalent all over the world and the amount of research carried out so far is not sufficient to precisely diagnose the disease. In this context, a novel kernel based chaos optimization algorithm is proposed by hybridizing a chaotic firefly algorithm (CFA) with kernel based Naïve Bayes (KNB) algorithm for diagnosing PD patients employing a voice measurement dataset collected from UCI repository. Six different chaotic maps are used to develop six CFA models and compared to find the best chaotic firefly model that can enhance the global searching capability and efficiency of CFA. To select the discriminative features from the search process, non-parametric kernel density estimation is used as a learning tool for the Bayesian classifier for evaluating the predictive potential of the features from four different perspectives such as size of the feature subset, classification accuracy, stability, and generalization of the feature set. The simulated results showed that logistic based CFA-KNB model outperformed other five chaotic map based CFA-KNB models. The generalization and stability of the predictive model is established by computing the model with four well-known representative classifiers viz., Naïve Bayes classifier, Radial Basis classifier, k-Nearest Neighbor, and Decision Tree in a stratified 10-fold cross-validation scheme. By finding appropriate chaotic maps, the proposed model could able to assist the clinicians as a diagnostic tool for PD.

Keywords: Chaotic mappings · Metaheuristic model ·
Kernel density estimation · Diagnostic tool · Firefly algorithm ·
Logistic model · ROC curve

1 Introduction

Parkinson's disease is a neurodegenerative disorder and a progressive disease caused due to the damage of a portion of brain cells that produce dopamine. The past research reveals that about 90% of PD patients are affected with speech/vocal impairments also known as dysphonia which is a motor related symptoms and have been used widely for diagnosing PD [1].

© Springer Nature Switzerland AG 2020
A. M. Madureira et al. (Eds.): HIS 2018, AISC 923, pp. 176–188, 2020.
https://doi.org/10.1007/978-3-030-14347-3_18

Many researchers have applied various machine-learning methods for identifying PD patients with dysphonic measurements. A support vector machine (SVM) with Gaussian radial basis kernel function [3] as learning algorithm used to predict PD dataset after identifying four different dysphonic features successfully. For early detection of PD, an enhanced instance based learning model is proposed [12] by integrating a chaotic bacterial foraging optimization (CBFO) with fuzzy k-nearest neighbor (FKNN). An enhanced chaos based firefly (ECFA) algorithm combined with RBF kernel based support vector machine (SVM) [13] builds an effective predictive model by detecting relevant speech patterns from PD dataset which in turn can help in developing telediagnosis and telemonitoring models. In paper [4], an information theoretic based Meta search method combined with chaos-based firefly algorithm has developed a model for identifying discriminative features from microarray dataset to build an effective classification model.

Metaheuristic search algorithms are computationally more efficient and can obtain optimal solution by making a trade-off between exploitation and exploration. Recently, a mathematical approach termed as 'chaos theory' is integrated into the domain of metaheuristics to enhance the performance. Chaotic maps are used to replace the random parameters [6] of the stochastic algorithms.

In firefly algorithm (FA), different chaotic maps tune the attractiveness parameter β [7] to achieve higher convergence rate and accuracy and also a chaos-enhanced firefly algorithm [8] was introduced to automate the tuning of parameters. A novel meta-heuristic algorithm namely, chaotic crow search algorithm (CCSA) developed by [9] to optimize the feature selection problems employing ten different types of chaotic maps to enhance the classification performance and identify a reduced feature set. In [10], a modified FA is developed by combining chaotic map to solve reliability and redundancy based optimization problem. In [11], four types of chaotic maps are used to initialize the population of fireflies and the constant value of absorption coefficient γ and the study indicates an increase in rate of convergence and accuracy. Some more applications of firefly algorithm in advanced problems such as handling financial option pricing has opted for parallel Firefly meta-heuristic algorithm [15] and as a recognizer for multi-lingual named entity recognition [16].

This paper introduces a novel hybrid approach that exploits the characteristics of kernel-based Naïve Bayes algorithm for evaluating the objective functions of six different chaotic map-embedded firefly algorithms and then test the performance of the chaotic search methodologies on the basis of most discriminative feature subsets obtained from the clinical dataset of PD. The last phase of work is to evaluate the effectiveness of the resulting feature subsets on the basis of generalization employing four different machine-learning algorithms. The rest of the paper is organized in the following way. A brief overview of working principles of kernel-based Naïve Bayes algorithm is presented in Sect. 2. A brief description of the firefly algorithm with six chaotic maps is provided in Sect. 3. The detailed description of the proposed hybrid algorithm CFA-KNB for feature selection is given in Sect. 4 and Sect. 5 introduces the dataset and experimental setting. The implications of the experimental results are presented in Sect. 6 and Sect. 7 concludes the paper with a future projection.

2   Kernel-Based Naïve Bayes Classifier

Naïve Bayes algorithm is considered as a simple probabilistic learning algorithm which is specifically designed for supervised learning problems. For a particular class value c, let $X = (x_1, x_2, x_3, ..., x_n)$ represents a vector of observed attribute values in the training dataset (X, C). The probability of each corresponding class can be computed using the following Eq. (1):

$$P(Y_j/X) = \frac{P(Y_j)P(X/Y)}{\sum_{i=1}^{c}(P(Y_j)P(X/Y_j))} \tag{1}$$

where $j = 1, 2,..., c$,

$P(Y_i)$ = prior probability of class Y_i, $p(Y_j/X)$ = class conditional probability density function. Here, the assumption of conditional independence assumes that each variable is conditionally independent of the other in the given dataset. So, the training dataset can be used to estimate the test dataset using Eq. (2).

$$p(X/Y_j) = \prod_{i=1}^{n} p(X_i/Y_j) \tag{2}$$

where $j = 1, ..., c$, X_i = ith attribute of X, n = No. of attributes and c_i = ith class. Then, the probability distribution over the set of features is calculated in the following Eq. (3).

$$p(X/Y_j) = \prod_{i=1}^{n} p(c_i p(X/c_i)) \tag{3}$$

In Naïve Bayesian approach, the probability of an observed value can be efficiently computed from the estimates. So, the continuous attribute can be represented using Eq. (4).

$$p(X = x|C = c) = g(x; \mu_c, \sigma_c) \tag{4}$$

where

$$g(x; \mu_c, \sigma_c) = \frac{1}{\sqrt{2\pi\sigma}} e^{-(\frac{(x-\mu)^2}{2\sigma^2})} \tag{5}$$

is the probability density function of a normal distribution. Kernel density estimation (KDE) function [18] is used for the present problem to estimate the density of each continuous features of PD dataset and then the estimated density will be averaged over a large set of kernels to identify patients with PD and healthy. It can be defined using the conditional probability $p_i(x_i|C = c)$ which can be estimated applying KDE [18] from a set of labelled training dataset (X, C). The mathematical equation for the same is given in Eq. (6).

$$p_i(x_i|C = c) = (nh)^{-1} \sum_j \frac{(x - \mu_i)}{h} \tag{6}$$

where $h = \sigma$ and $K = g\ (x, \mu, 1)$. Therefore, while computing $P(X = x \mid C = c)$ for continuous attribute to classify an unknown test instance, kernel density estimator-based NB computes n times for each observed value of X in class c $i.e.$, K is the number of possible unique feature values for input X.

3 Firefly Algorithm and Chaotic Maps

3.1 Firefly Algorithm

Firefly algorithm is a nature-inspired metaheuristic optimization algorithm that performs searching operation on a population of fireflies. The algorithm has been designed [17] by making the following assumptions on the behavior of real fireflies such as: (i) All fireflies assumed to be unisex. (ii) Attractiveness is relatively proportional to the brightness of fireflies. (iii) The brightness of the fireflies is evaluated by the landscape of the objective function.

It's understood from the assumptions that the light intensity $I(r)$ of fireflies is inversely proportional to the distance r. The coefficient of light absorption is denoted by γ. Therefore, the light intensity $I(r)$ of fireflies vary with distance r can be defined as shown in Eq. (7):

$$I(r) = I_0 e^{-\gamma r^2} \tag{7}$$

where I_0 is the initial intensity at the source. The attractiveness parameter β of each firefly can be defined as shown in Eq. (8):

$$\beta(r) = \beta_0 e^{-\gamma r^2} \tag{8}$$

where β_0 is the attractiveness at $r = 0$. Then firefly at position x_i gets attracted towards a brighter one at x_j and the movement between these two fireflies is computed using the following Eq. (9)

$$x_{i+1} = x_i + \beta_0 e^{-\gamma r^2} \left(x_j - x_i \right) + \alpha \varepsilon \tag{9}$$

where α denotes randomization and ϵ as a vector of random numbers derived from Gaussian distribution. The second term of the Eq. (9) represents attraction between fireflies. In this work, the initial population of FA is replaced with chaotic maps.

3.2 Chaotic Map

The recently used chaotic optimization theory [6] is very sensitive to the initial condition and it transforms the variables from chaotic space to solution space. The characteristics of randomness, ergodicity, and regularity of chaotic motion are used to achieve global optimal solution instead of trapping into local optima. Therefore, it enhances the exploration and exploitation capability of the integrated nature-inspired metaheuristics [7, 9, 10]. In this paper, six different chaotic maps are used to initialize the population of FA and update the constant value of absorption coefficient which is explained below:

Logistic Map. The chaotic sequence is generated using the following second order polynomial function:

$$x_{k+1} = rx_k(1 - x_k), k = 0, 1, 2, 3 \ldots \tag{10}$$

where r is the chaotic parameter and it is assumed that $0 \leq x_0 \leq 1$ and $0 \leq r \leq 4$.

Sine Map. It produces the following discrete time dynamical system:

$$x_{k+1} = \lambda \sin(\pi x_k), k = 0, 1, 2, 3 \ldots \tag{11}$$

where λ is the chaotic parameter and varies $0 \leq \lambda \leq 1$ and the ergodic area is [0, 1].

Chebyshev Map. The chaotic sequence is generated by the following iteration function

$$x_{k+1} = \cos(k \cos^{-1}(x_k), k = 0, 1, 2, 3 \ldots \tag{12}$$

where k is the iteration number and the ergodic area is [0, 1] for the map.

Circle Map. The chaotic sequence is generated by following the iteration function

$$x_{k+1} = x_k + b - (a - 2\pi) \sin(2\pi x_k) mod(1) k = 0, 1, 2, 3 \ldots \tag{13}$$

where a = 0.5 and b = 0.2. The ergodic area is bounded by [0, 1].

Gauss/Mouse Map. Gaussian function is used to define the following non-linear iterated function:

$$x_{k+1} = exp - \alpha x_k^2 + \beta \tag{14}$$

where α = 4.9 and β = −0.58. The deterministic chaotic sequence is produced in the interval, $x_k = [0, 1]$.

Piecewise Map. The chaotic sequence is generated using four linear pieces that are evaluated by the following iterated function:

$$x_{k+1} = \frac{x_k}{d}, \ 0 \leq x_k \leq k; \ x_k - \left(\frac{d}{0.5}\right) - d, \ d \leq x_k < \frac{1}{2}$$
$$1 - d - \left(\frac{x_k}{0.5} - d\right), \ \frac{1}{2} \leq x_k < 1 - d \tag{15}$$
$$1 - \frac{x_k}{d}, \ 1 - d \leq x_k < 1$$

where d is the endpoints of the four sub-intervals and can be set as d ϵ [0, 0, 5] and the ergodic area is bounded by [0, 1].

4 Proposed Kernel Based Chaos Optimization Algorithm (CFA-KNB)

The proposed stochastic optimization algorithm combines a chaotic search algorithm for optimizing the search process and a kernel based probabilistic learning algorithm as a fitness function to measure the performance of the combination of selected features. The chaotic-based firefly algorithm selects the initial population of fireflies using chaotic variables instead of randomly distributed variables. The chaotic maps develop chaotic sequences x_i to update the firefly position in the solution space and the value of absorption coefficient γ. Moreover, the chaotic mappings choose homogenously distributed important fireflies that enhance the convergence rate and precision of the coupling metaheuristic algorithms. Since the steps size of the random movement α affects the random vector term ϵ, so third and second term of Eq. (9) can be replaced by chaos time series as follows:

$$\epsilon_i = c_i^k \qquad \beta_i = \beta_0 c_i^k \tag{16}$$

where c_i^k is a chaotic parameter to be determined by a chaotic map (k). The attractiveness parameter β depends on the light absorption coefficient γ which in turn regulates the firefly social movement stated in Eq. (9). Generally, it assumes a fixed value in the entire optimization process but we have considered γ to be tuned for all six types of chaotic maps (k) instead of using a fixed value. Also from Eq. (9), two limiting cases can be observed when $\gamma \to 0$, β becomes β_0 that makes all fireflies to see each other and when $\gamma \to \infty$ all of them move randomly. To get rid of this, the social component term of Eq. (9) can be tuned with chaotic map (k) and can be represented as:

$$x_i^{(t+1)} = x_i^{(t)} - \beta exp^{-\gamma x^{(t)^2}} x_i^{(t)} \tag{17}$$

Therefore, in this paper, six different chaotic mappings are coupled with standard FA to identify the best chaotic maps which strongly influence the search process and develop a generalized predictive model. These maps are embedded into a wrapper based firefly algorithm to select an optimal subset of features that characterize the whole problem. The functional description of the model is given in Algorithm 1 which will be executed in a nested loop. The inner loop optimizes the process of feature selection via 10-fold cross validation (CV) and the outer loop executes the optimal subset of inner loop to perform the diagnostic classification of PD via stratified 10-fold CV method.

4.1   Initialization of Parameters

The initial population of fireflies are mapped to the chaotic sequences that have been created according to the Eqs. (10)–(15) and update the absorption coefficient γ in the search space. Then remaining parameters are initialized with the values following [5] and summarized in Table 1.

Table 1. Initial values of the experimental parameters of CFA

Parameters	Values	Parameters	Value
γ - light absorption coefficient	1.0	β_0 - attractiveness	0.8
r - chaotic parameter	0.1–4.0	Population size of CFA	50
K (for cross validation)	10	Search domain	[0.0, 1.0]
Ci(k) - chaotic map	Logistic, sine, gauss, Chebyshev, piecewise, circle		
Number of generation of CFA	20	Mutation probability	0.01
Problem cardinality	Total number of features in the problem		

Algorithm1. Kernel based Naïve Bayes-Chaotic Firefly Algorithm

Input: Population of fireflies $x = (x_1, x_2 \ldots \ldots x_n)$;
  Objective function $f(x_i), i = 1,2,3 \ldots$
Output: Best solution x_{best} and value of $f_{min} = minf\ (x_{best})$)
Set the initial values of β_0, γ, Max-Gen, t;
Initialize the position of fireflies x_i randomly
  $x^0 = (x_1^{(0)}, x_2^{(0)}, \ldots \ldots x_N^{(1)})$;
Compute the fitness value of each fireflies using fitness function $f_n(x_i^{(0)})$
and update the corresponding light intensity I_i, i=1,2,.....N;
Set t = 0;
  while (t < Max-Gen)
    Get value of chaotic map $C_i(k)$
    Tune absorption coefficient γ using C_i(k)
    Define attraction parameter β
    for i = 1:N (N= number of fireflies)
      for j= 1:N (N=number of fireflies)
        if ($I_i > I_j$)
          move firefly i towards j
        endif
        compute attractiveness parameter β which varies with distance r

$$r_{ij} = \left\| X_i - X_j \right\| = \sqrt{\sum_{k=1}^{D} X_{ik} - X_{jk}}$$

$$\beta = \beta_0 e^{-\gamma \times t^2}$$
        and position of fireflies
          $x_i^{t+1} = x_i^{(t)} - \beta exp^{-\gamma \times t^2} x_i^t$
        evaluate fitness function $f_n(x_i^{(t)})$ for new solutions
          and update the corresponding light intensity I
      end for j
    endfor i
      Rank fireflies based on fitness value and find the x_{best}
  t = t+ 1;
  end while

4.2   Fitness Function

The fitness function is evaluated to find the position of each candidate solution (firefly) at each iteration using 10-fold cross validation. 10-fold cross validation split the whole dataset randomly into two different sets called training and testing datasets and helps the model to achieve stability. The fitness function given in Eq. (18) is framed by combining the objectives of the proposal such as learning accuracy and size of feature subset. Each of them is assigned with a weight parameter considering their importance in the evaluation process. KNB is a probabilistic kernel density estimator based supervised algorithm which estimate the potential of the selected features in terms of classification accuracy and mean squared error. KNB computes n times for each observed value of X in class c $i.e.$, the kernel (K) is the number of possible unique feature values for input X.

$$F_n(x_i) = \delta p\left(\frac{Y}{X}\right) + (1 - \delta)\left(1 - \frac{SF}{TF}\right) \qquad (18)$$

where $P\left(\frac{Y}{X}\right)$ is the learning accuracy of KNB, SF and TF are the selected features and total features of the dataset respectively. In this work, the weight factor assigned to both the parameters is as follows: $\delta = 0.9$, $1 - \delta = 0.1$ to maximize the classification accuracy and minimize the size of the feature subset.

5   Experimental Results

5.1   Illustration of Parkinson Dataset

For this study, measurements of biomedical speech signals of 31 participants, 23 with PD and 8 healthy participants ranged from 46 to 86 years of age are considered. The dataset consists of total instances of 195 voice recordings with 22 types of voice measures represented in columns [F1–F22], collected from UCI repository and more is available in [2]. This dataset is used for discriminating PD patients from healthy one as per the status column that set 1 for PD and 0 for healthy.

Experiment Setup
The empirical experiment of the proposed feature selection and classification model CFA-KNB is implemented in Java using Weka API and executed in Windows 10, Intel (R), core-i7-7500U CPU @2.70 GHz and 12.0 GB RAM. Statistical Inter Quartile Range (IQR) method is applied as a pre-processing tool to remove outliers and extreme values from PD dataset. In this study, six different chaotic firefly search algorithms are developed by combining six chaotic maps such as logistic, sine, gauss/mouse, Chebyshev and piecewise with KNB to improve the diversity of the population. The algorithms are executed 20 times each as the generation is fixed to 20 with a population of size 50 along with other parameters given in Table 1. The results of the proposed methodology were evaluated by 10-fold CV [14] to guarantee robustness and reliability of the selected feature subset. The classification performance of the proposed chaos based hybrid model CFA-KNB and other comparative models are validated by a stratified 10-fold cross validation [14] which guarantees stable and generalized solutions by making all test sets independent from each other. In addition, as the proposed model is an approximation algorithm; the final solutions are obtained by averaging 10

independent iterations and selecting the configuration of chaos based CFA-KNB model which provides the best optimal solution.

6 Discussion

The performance of CFA-KNB model with six chaotic mappings are compared with each other in terms best fitness value, average fitness value, size of selected feature subset and p-value of Wilcoxon's rank sum test which is shown in Table 2 for finding the best chaotic feature selection model for small imbalanced dataset [13]. It can be seen that the model CFA-KNB with sine chaotic mapping has selected the least number of discriminative features viz.F1, F22 with best fitness value 0.893, average fitness value 0.882, light intensity 1.0 and p-value 0.187. Comparing the results of sine model with others, it is be observed that logistic has obtained the same best fitness value of 0.893 but the size of selected feature subset is four viz. F1, F8, F21, F22. The non-parametric Wilcoxon's rank sum test [19] calculates the p-values which are considered as statistically significant when $p < 0.05$ and highly significant when $p < 0.01$. The p-values of the models are presented in Table 2 and the comparison of p-values of logistic model i.e., 0.009 with the sine model i.e., 0.187 shows that logistic model is highly significant than sine model. Therefore, the above analysis indicates that even though the model CFA-KNB with logistic mapping selected more features than sine model but has achieved best significant optimal discriminative solution in the search space than sine model.

Additionally, Table 2 presents the frequency of selection of the features by all the six chaotic models in order to understand their significance in PD diagnosis process. The analysis shows, 10 features out of 22 are selected in different combination by the six chaotic models. However, only 6 features out of 10 have multiple occurrence in more than one selected subset such as: F1 – frequency of occurrence is 6 and appears in all subsets; F8 – occurs in logistic and circle mappings; F15 – occurs in circle and piecewise; F20 – occurs in Chebyshev, gauss and piecewise; F21 – occurs in logistic and gauss; F22 – occurs in logistic and sine. Most importantly, 4 features out of 6 i.e., F1, F8, F21, F22 selected by logistic mapping is repeatedly selected by others. Hence, these four features can be attributed to clinical markers for the diagnosis of PD. The experimental results signify the importance of the application of logistic chaotic map time series [6] which adds an additional diversity of population into the kernel based firefly algorithm and selects a good combination of discriminative features to design a PD diagnostic model.

In small clinical datasets, highest classification performances with potentially discriminating feature subsets are highly desirable for making quick decisions to detect disease. To establish the same, Table 3 depicted the classification performance KNB for all chaotic models on skewed PD dataset in terms of accuracy (ACC), sensitivity, F-measure, confusion matrix and Mathew correlation coefficient (MCC), area under ROC curve (AUC), model building time over 10 runs of stratified 10-fold CV. As the stratified 10-fold CV splits the data into folds of equal proportion of samples with a given class value, could able to handle 33.83% of skewness of the PD dataset. Each fold of stratified 10-fold CV is iterated 10 times and the average result of the experiment are reported as the performance measure in Table 3. It is observed that the discriminative ability of the logistic based CFA-KNB model is comparatively more significant than other models by achieving best performance for all metrics: ACC of 89.326, Sensitivity of 0.893, F-measure of 0.884, MCC of 0.707, AUC of 0.914 and least model-building time 7.14 s highlighted in bold letters in Table 3. The above results clearly established that logistic chaos based CFA-KNB model improves the classification performance of KNB by reducing the time and cost of the experiment and thereby the model could able to detect the PD patients very fast.

Further, the results of confusion matrix shown in Table 4 summarizes that the logistic map based model showed only one false negative/one PD patient diagnosed as healthy from total 133 PD patients and 18 patients diagnosed as false positive/diagnosed as PD from 45 total healthy patients. The false positive rate is very high in the remaining five chaotic map models as compared to logistic model. Hence, the results of the matrices proved the diagnostic efficiency of logistic based model in comparison to other five models. The visual presentation of receiver-operating characteristic curve (ROC) for healthy and PD patients for all six chaotic models are shown in Figs. 1 and 2 to evaluate the quality of the models. It is observed that the curve traced by CFA-KNB-logistic based model goes smoothly through the upper left corner and the corresponding area under ROC curve (AUC) is 0.914 reported in Table 3 which is highest among all the models.

Table 2. Features selected by CFA-KNB model using 10-fold CV and Wilcoxon's p-values

Chaotic maps features selected	Features selected	Best fitness	Average fitness	p-values
Logistic	F1, F8, F21, F22	0.893	0.888	0.009
Sine	F1, F22	0.893	0.882	0.187
Chebyshev	F1, F7, F18, F20	0.871	0.772	0.151
Circle	F1, F8, F15	0.855	0.766	0.159
Gauss	F1, F2, F16, F20, F21	0.854	0.769	0.276
Piecewise	F1, F15, F17, F20	0.862	0.767	0.353

So far, logistic based chaotic CFA-KNB model has proved to be the robust and reliable model for the problem undertaken in comparison to other chaotic models. So, to prove the generalizability of the logistic model, Table 5 summarizes the classification accuracies obtained by four different classifiers namely Naïve Bayes (NB), Radial Basis Function Classifier (RBFC), k-Nearest Neighbor (KNN) and Decision Tree (J48). Observing the results, it is quite apparent that logistic chaotic based CFA-KNB model is working efficiently for all the classifiers except RBFC. Logistic model has obtained highest accuracy of 89.888 for KNN, 89.326 for KNB and 83.708 for NB. The advantage of logistic model is it converges very fast and achieves accuracy significantly better than others achieve.

Table 3. Performance of KNB on CFA-KNB using stratified 10-fold CV

Chaotic maps	ACC	Sensitivity	F-Measure	MCC	AUC	Model building time (sec)
Logistic	**89.326**	**0.893**	**0.884**	**0.707**	**0.914**	**7.14**
Sine	85.955	0.860	0.851	0.587	0.866	58.19
Piecewise	84.832	0.848	0.839	0.569	0.899	55.38
Chebyshev	87.079	0.871	0.861	0.636	0.904	59.16
Circle	86.517	0.865	0.859	0.585	0.862	56.91

Table 4. Confusion matrix of six different CFA-KNB classification models

Chaotic models	Expected output	Predicted output	
Logistic	PD patients	132	1
	Healthy	18	27
Sine	PD patients	126	7
	Healthy	19	26
Chebyshev	PD patients	129	4
	Healthy	19	26
Circle	PD patients	127	6
	Healthy	20	25
Gauss	PD patients	124	9
	Healthy	21	24
Piecewise	PD patients	126	7
	Healthy	20	25

Table 5. Performance of NB, RBFC, KNN and J48 for logistic CFA-KNB model

	NB	RBFC	KNN	J48
ACC	83.708	85.955	89.888	82.584
Sensitivity	0.837	0.860	0.899	0.826
F-measure	0.834	0.851	0.900	0.822
MCC	0.580	0.688	0.628	0.530
AUC	0.874	0.914	0.890	0.788
Model build time (Sec)	7.02	7.08	7.04	7.03

Hence, the proposed feature selection algorithm CFA-KNB in combination with logistic chaotic map has achieved better generalization compared to others. Therefore, the efficiency of CFA-KNB-logistic based model can be attributed to the embedding of chaotic sequences into the searching iteration of CFA-KNB model which helps to achieve global optima faster by overcoming the limitation of local optima.

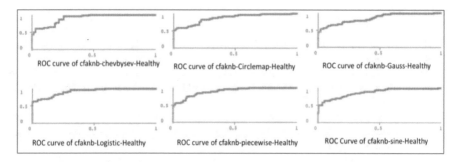

Fig. 1. ROC curve comparison of Healthy Patients for CFA-KNB model

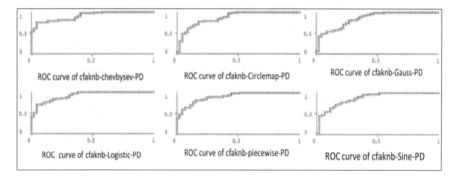

Fig. 2. ROC curve comparison of PD Patients for CFA-KNB model

7   Conclusion

The proposed work is a hybrid kernel based chaotic metaheuristic feature selection model. We designed six CFA-KNB models using six chaotic mappings and then compared them to find the best model. The chaotic maps are used to initialize the population of fireflies and the absorption coefficient and a non-parametric kernel based Naïve Bayes classifier acts as a fitness function. The simulation results show that logistic model outperformed other five chaotic models on the basis of achieving robust and reliable subset of features for PD dataset. The empirical analysis proves that the proposed model can be used as an alternative diagnostic tool for Parkinson's disease. Implementation of the proposed model in parallel computing platform for evaluating other medical decision making problems is considered for future work.

References

1. Ho, A.K., Iansek, R., Marigliani, C., Bradshaw, J.L., Gates, S.: Speech impairment in a large sample of patients with Parkinson's disease. Behav. Neurol. **11**(3), 131–137 (1998)
2. Little, M.A., McSharry, P.E., Hunter, E.J., Spielman, J., Ramig, L.O.: Suitability of dysphonia measurements for telemonitoring of Parkinson's disease. IEEE Trans. Biomed. Eng. **56**(4), 1015–1022 (2009)
3. Little, M.A., McSharry, P.E., Roberts, S.J., Costello, D.A.E., Moroz, I.M.: Exploiting nonlinear recurrence and fractal scaling properties for voice disorder detection. Biomed. Eng. Online **6**, 23 (2007)
4. Dash, S., Thulasiram, R., Thulasiram, P.: A modified firefly based meta-search algorithm for feature selection: a predictive model for medical data. Int. J. Swarm Intell. (2018, accepted)
5. Kennedy, J., Eberhart, R.: Particle swarm optimization. In: IEEE International Conference on Neural Networks, vol. 4, pp. 1942–1948 (1995)
6. Fister Jr., I., Perc, M., Kamal, S.M., Fister, I.: A review of chaos-based firefly algorithms: perspectives and research challenges. Appl. Math. Comput. **252**, 155–165 (2015)
7. Emary, E., Zawbaa, H., Hassanien, A.: Binary gray wolf optimization approaches for feature selection. Neurocomputing **172**, 371–381 (2016)
8. Taha, A.M., Mustapha, A., Chen, S.-D.: Naive Bayes-guided bat algorithm for feature selection. Sci. World J. **2013**, 9, Article ID 325973 (2013)
9. Sayed, G.I., Hassanien, A.E., Azar, A.T.: Feature selection via a novel chaotic crow search algorithm. Neural Comput. Appl. (2017). https://doi.org/10.1007/s00521-017-2988-6
10. Askarzadeh, A.: A novel metaheuristic method for solving constrained engineering optimization problems: crow search algorithm. Comput. Struct. **169**, 1–12 (2016)
11. Su, S., Su, Y., Xu, M.: Comparisons of firefly algorithm with chaotic maps. Comput. Model. New Technol. **18**(12C), 326–332 (2014)
12. Cai, Z., Gu, J., Wen, C., Zhao, D., Huang, C., Huang, H., Tong, C., Li, J., Chen, H.: An intelligent Parkinson's disease diagnostic system based on a chaotic bacterial foraging optimization enhanced fuzzy KNN approach. Comput. Math. Methods Med. **2018**, 24, Article ID 2396952 (2018)
13. Dash, S., Thulasiram, R., Thulasiram, P.: An enhanced chaos-based firefly model for Parkinson's disease diagnosis and classification. In: IEEE ICIT Conference, Bhubaneswar, Odisha, 21–23 December. IEEE Xplore (2017)
14. Kohavi, R.: A study of cross-validation and bootstrap for accuracy estimation and model selection. In: Proceedings of the 14th International Joint Conference on Artificial Intelligence (IJCAI 1995), Montreal, Canada, August, pp. 1137–1143 (1995)
15. Mather, K., Thulasiram, P., Thulasiram, R., Dash, S.: A parallel firefly meta-heuristics algorithm for financial option pricing. In: IEEE SSCI Conference, Hawaii, USA, 27th November–1st December 2017. IEEE Xplore (2017)
16. Biswas, S., Dash, S.: Firefly Algorithm Based Multilingual Named Entity Recognition For Indian Languages. Communications in Computer and Information Science. Springer (2018)
17. Yang, X.S.: Firefly algorithms for multimodal optimization. In: Watanabe, O., Zeugmann, T. (eds.) Stochastic Algorithms: Foundations and Applications, SAGA 2009. Lecture Notes in Computer Science, vol. 5792. Springer (2009)
18. Parzen, E.: On estimation of a probability density function and mode. Ann. Math. Stat. **33**, 1065–1076 (1962)
19. Wilcoxon, F.: Individual comparisons by ranking methods. Biom. Bull. **1**(6), 80–83 (1945)

Hybridization of Migrating Birds Optimization with Simulated Annealing

Ramazan Algin[1](✉), Ali Fuat Alkaya[1], and Vural Aksakalli[2]

[1] Computer Engineering Department, Marmara University, Istanbul, Turkey
algin.ramazan@gmail.com, falkaya@marmara.edu.tr
[2] School of Science, RMIT University, Melbourne, Australia
vural.aksakalli@rmit.edu.au

Abstract. Migrating Birds Optimization (MBO) algorithm is a promising metaheuristic algorithm recently introduced to the optimization community. Despite its superior performance, one drawback of MBO is its occasional aggressive movement to better solutions while searching the solution space. On the other hand, simulated annealing is a well-established metaheuristic optimization method with a search strategy that is particularly designed to avoid getting stuck at local optima. In this study, we present hybridization of the MBO algorithm with the SA algorithm by embedding the exploration strategy of SA into the MBO, which we call Hybrid MBO. In order to investigate impact of this hybridization, we test Hybrid MBO on 100 Quadratic Assignment Problem (QAP) instances taken from the QAPLIB. Our results show that Hybrid MBO algorithm outperforms MBO in about two-thirds of all the test instances, indicating a significant increase in performance.

Keywords: Migrating birds optimization ·
Computational optimization · Quadratic assignment problem ·
Simulated annealing · Hybrid algorithm

1 Introduction

Metaheuristics are widely used solution techniques for combinatorial optimization problems. They are more preferable while finding optimum or near-optimum solutions on larger instances when compared to the exact algorithms. However, for the problems with high complexity or large scale problem instances, the heuristics or metaheuristics may not be adequate to obtain satisfactory results. Therefore, especially for the last decade, researchers are trying to find new techniques to obtain better performance. One of the directions that the researchers take is to hybridize metaheuristics. Hybrid metaheuristics are obtained typically by combining the power of two or more metaheuristics or embedding a local search heuristic to a metaheuristic. In the literature, there are several studies that hybridize metaheuristics successfully. The most popular one is to hybridize

© Springer Nature Switzerland AG 2020
A. M. Madureira et al. (Eds.): HIS 2018, AISC 923, pp. 189–197, 2020.
https://doi.org/10.1007/978-3-030-14347-3_19

genetic algorithms with a local search procedure or with some other metaheuristics [5]. Other than the genetic algorithm oriented studies, we observe hybridization of other metaheuristics such as ant colony optimization algorithm with simulated annealing [3], simulated annealing with differential evolution [10], differential evolution with harmony search algorithm [6], etc.

In this study, we studied on embedding a different exploration strategy to the recently proposed migrating birds optimization (MBO) algorithm. The exploration strategy embedded in MBO nothing but the stochastic move tactic in the well known simulated annealing (SA) algorithm. In this way, MBO algorithm is expected to exploit a broader solution space and in one sense hybridized with the SA algorithm.

MBO algorithm is inspired from the V formation of the migrating birds in real life [8]. Solutions (corresponding to birds in the analogy) are initialized randomly in the solution space and they try to move to better positions by searching their neighborhood. Throughout the algorithm, the birds share their unused neighbors with the follower birds which are placed in V formation hypothetically. It is tested on quadratic assignment problem instances and it is compared with other metaheuristics. As a result, it was found to be performing at least the same as the simulated annealing and better than the guided evolutionary simulated annealing, tabu search, genetic algorithm, scatter search, particle swarm optimization and differential evolution algorithms. On the other hand, SA is said to be the oldest among the meta-heuristics and mimics the annealing process in metallurgy. The fundamental idea in SA is to apply an explicit strategy to escape from local minima [4].

Even though MBO is proven to be a good performing algorithm, it is designed in the way that it moves always to better solutions where it may get stuck in local minima. Having a swarm structure, it still has the chance to find the global minima. In order to use a better exploration strategy, thus avoiding to get stuck at local minima, hybridization of MBO with the SA is a novel and promising idea. Hence, we embedded a stochastic strategy in MBO where while moving to a new solution, there is a chance of moving to worse solutions for the sake of avoiding getting stuck at local minima. This new version of MBO is called Hybrid MBO throughout the manuscript.

The paper is organized as follows. Information about MBO algorithm and previous work are given in Sect. 2. Section 3 gives the details of the Hybrid MBO algorithm. Computational experiments and discussion are given in Sect. 4. Section 5 concludes the paper by summarizing the outputs of the paper and with some future work.

2   MBO and Previous Work

The MBO algorithm is one of the swarm intelligence techniques proposed recently inspired from the migrating birds and their V formation. There is a leader bird in the algorithm and its chosen randomly then other birds divided into two groups behind the leader bird as in V formation. Each solution generates a number of solution that determines the flock of the speed. Flock's search area depends on its speed. If the speed is higher this means flock can do search in wider area.

The algorithm works as follows. Initial solutions are generated by placing the birds to the search space randomly in a hypothetical V formation. Leader bird generates its neighbours and selects one of them that is better than the current solution. Then leader bird gives a parametric number of best unused solutions to the bird behind. This process repeats till leader bird becomes the last in the V formation. The stopping criteria in this algorithm is limited number of iterations. The pseudocode of the MBO algorithm is presented in Fig. 1. In this pseudocode notations are as follows:

n = the number of initial solutions (birds)
k = the number of neighbor solutions to be considered
x = the number of neighbor solutions to be shared with the next solution
m = number of tours.

---

1. Generate n initial solutions in a random manner and place them on an hypothetical V formation arbitrarily
2. **while** termination condition is not satisfied
3.   **for** m times
4.   Try to improve the leading solution by generating and evaluating k neighbors of it
5.   **for** each solution s_i in the flock (except leader)
6.   Try to improve s_i by evaluating $(k\text{-}x)$ neighbors of it and x unused best neighbors from the solution in the front
7.   **endfor**
8.   **endfor**
9.   Move the leader solution to the end and forward one of the solutions following it to the leader position
10. **endwhile**
11. return the best solution in the flock

Fig. 1. Pseudocode of the MBO

There are several studies about MBO in the literature. In [1], ant system (AS), genetic algorithm (GA), simulated annealing (SA) and migrating birds optimization (MBO) algorithms are developed for the Obstacle Neutralization Problem (ONP). Among these metaheuristic algorithms MBO and SA give competitive results and outperforms AS and GA. In another study, MBO algorithm is applied to 30 different functions on continuous domain. The contribution in this study is to develop an adaptive and effective neighbor generation function for MBO. The tests are conducted on 2, 10 and 30 dimensions [2].

In another study an improved MBO is proposed to minimize the total flow time for a hybrid flow shop scheduling problem, which has important practical applications in modern industry. In this study, several advanced and effective technologies are introduced to enhance the MBO algorithm, including a diversified initialisation method, a mixed neighbourhood structure, a leaping mechanism, a problem-specific heuristic, and a local search procedure [11].

In [7], MBO is used for solving credit card fraud detection problem. In this study standard MBO algorithm is improved by using some benefit mechanism. MBO, improved MBO and genetic algorithm with scatter search (GASS) are compared. Results shows that GASS is outperformed by MBO and improved MBO.

In the next section, we provide the details of the proposed Hybrid MBO algorithm.

3   Hybrid MBO

In this section we present the details of the hybridization of the MBO algorithm. In order to hybridize the MBO algorithm with the SA algorithm, we embedded a new exploration strategy and its related parameters into MBO.

Even though MBO is proven to be a good performing algorithm, it is designed in the way that it moves always to better solutions where it may get stuck in local minima. Nevertheless, having a swarm structure, it still has the chance to find the global minima.

In order to avoid getting stuck at local minima, hybridization of MBO with the SA is a novel and promising idea. Hence, we embedded a stochastic move strategy in MBO in which there is a chance of moving to worse solutions. That is, the best solution, s', in the neighbour set of a bird (s) is accepted as new bird depending on $f(s)$, $f(s')$ and T where f is the fitness evaluation function. s' replaces s if $f(s') < f(s)$ or, in case $f(s') \geq f(s)$, with a probability which is a function of T and $f(s') - f(s)$. The probability is computed following the Boltzmann distribution. Mathematically,

$$s \Leftarrow \begin{cases} s', \text{ if } f(s') < f(s) \\ s', \text{ else if } random(0,1) < exp(\dfrac{-|f(s') - f(s)|}{T}) \end{cases} \tag{1}$$

In this way, we expect to improve the exploration capability of MBO. Of course from the programmer point of view, the best solution found throughout the algorithm is traced and its fitness is reported as the output.

Hybrid MBO is presented in Fig. 2. It has three additional parameters:

- T: initial temperature
- a: decrease ratio of the temperature
- dp: place of the temperature decrementation operation

Values of T, a and other parameters with their best ones are explained in Sect. 4. On the other hand, we determined two possible locations for the decrementation operations (dp); one is line 9, just after the most inner for loop, and the other one is in line 13, just before moving the solution.

1. Generate n initial solutions in a random manner and place them on an hypothetical V formation arbitrarily

2. Initialize T

3. **while** termination condition is not satisfied

4.   **for** m times

5.   Try to improve the leading solution by generating and evaluating k neighbors of it using the new exploration strategy (Eq. 1)

6.   **for** each solution s_i in the flock (except leader)

7.   Try to improve s_i by evaluating $(k\text{-}x)$ neighbors of it and x unused best neighbors from the solution in the front using the new exploration strategy (Eq. 1)

8.   **endfor**

9.   **if** $dp==1$

10. $T = T/a$

11.   **endif**

12.   **endfor**

13.   **if** $dp==0$

14. $T = T/a$

15.   **endif**

16.   Move the leader solution to the end and forward one of the solutions following it to the leader position

17. **endwhile**

18. return the best solution in the flock

Fig. 2. Pseudocode of the hybrid MBO

A higher T value may keep the algorithm exploring continuously by moving to worse neighbour solutions, whereas a low T value may keep the algorithm exploiting around the given initial solution just like the original MBO. A high a value may decrease the temperature -the probability of moving to worse solutions- very swiftly, resulting in the original MBO. On the other hand, a low a value may decrease the temperature not as high as needed. When we consider the position of the temperature decrementation operation, it is performed m times more frequently when dp is 1 compared to when dp is 0, resulting in a quick decrease in T.

4   Computational Experiments and Results

In the first set of computational tests where we tried to determine the best values of the Hybrid MBO algorithm. The tests are conducted on the quadratic assignment problem (QAP) instances taken from QAPLIB [12].

QAP is a well-known combinatorial optimization problems and it is one of the most difficult problem to solve optimally. The QAP is stated for the first time by Koopmans and Beckmann [9]. The problem is defined as follows. There is a set of N possible locations and N facilities. The aim in this problem is assigning all of the facilities to the different locations with minimum total assignment cost. Let c_{ij} be the cost per unit distance between facilities i and j and d_{ij} be the distance between locations i and j. p is the permutation function. The total assignment cost (tac) is calculated according to the formula (2).

$$tac = \sum_{i=1}^{n} \sum_{j=1}^{n} c_{ij} d_{p(i)p(j)} \tag{2}$$

The experiments are run on an HP Z820 workstation with Intel Xeon E5 processor at 3.0 GHz with 128 GB RAM running Windows 7. The MBO and Hybrid MBO algorithms are implemented in Java language. The stopping criterion for the algorithms is a given number of neighbors generated. Specifically, the allowed number of neighbors generated is N^3 where N is the number of possible locations or facilities for QAP instances. In order to decide on the best performing values of the Hybrid MBO algorithm, we compared their performance with the best known solutions (BKS) given in the literature [12] by calculating their distance to BKS values in percentages.

In fine tune experiments, for n, k, m, x, T, a, and dp we used 5, 4, 1, 4, 19, 11 and 2 different values, respectively. Regarding the values of parameters T, a and dp, 3000, 1.06 and 0 are the best performing values, respectively. Remaining best performing values of the parameters are given in Table 1. For MBO algorithm same n, k, m, x values are tested and both Hybrid MBO and MBO algorithms have same best performing values for these parameters.

Table 1. Values of parameters used in the computational experiments. (Bold ones are the best performing values)

Parameters	Values
n	{**3**,5,11,25,51}
k	{**3**,5,7,9}
m	{**10**}
x	{**1**,2,3,4}
T	{100,200,300,...,900,1000,2000,**3000**,...,9000,10000}
a	{1.01,1.02,1.03,1.04,1.05,**1.06**,1.07,1.08,1.09,1.1,1.5}
dp	{**0**,1}

We conducted tests on 100 different QAP instances with 100 runs in each test. Results are given as average of 100 runs. According to the results, Hybrid MBO outperformed MBO on 62 instances, MBO outperformed Hybrid MBO on 29 instances and on 9 instances both Hybrid MBO and MBO had same performance. As a result, Hybrid MBO is better than original MBO where in 62% of the problem instances Hybrid MBO outperforms the original MBO. This can be explained with the improved exploration capability of Hybrid MBO. The details of the results are shown in Table 2. In this table % values are the distance to BKS values in percentages.

Table 2. Performance comparison of Hybrid MBO and MBO on the QAP instances

File name	Hybrid MBO	MBO	File name	Hybrid MBO	MBO
bur26a	**0.24%**	0.26%	nug28	**2.09%**	3.11%
bur26b	**0.31%**	0.37%	nug30	**2.06%**	2.71%
bur26c	**0.26%**	0.31%	rou12	**2.89%**	3.76%
bur26d	**0.28%**	0.39%	rou15	**4.44%**	4.98%
bur26e	**0.20%**	0.31%	rou20	**2.80%**	3.40%
bur26f	**0.33%**	0.34%	scr12	4.57%	**4.17%**
bur26g	**0.22%**	0.26%	scr15	**5.78%**	7.35%
bur26h	**0.23%**	0.34%	scr20	**3.95%**	6.10%
esc16a	3.55%	**0.15%**	sko100a	**1.29%**	1.50%
esc16b	0.00%	0.00%	sko100b	**1.34%**	1.51%
esc16c	2.64%	**0.00%**	sko100c	**1.53%**	1.69%
esc16e	4.76%	**0.28%**	sko100d	**1.39%**	1.51%
esc16f	0.00%	0.00%	sko100e	**1.49%**	1.74%
esc16g	4.55%	**0.00%**	sko100f	**1.42%**	1.51%
esc16h	0.02%	**0.00%**	sko42	**1.97%**	2.40%
esc32a	**9.03%**	10.43%	sko49	**1.62%**	2.18%
esc32c	0.00%	0.00%	sko56	**1.74%**	2.10%
esc32d	**1.55%**	1.96%	sko64	**1.64%**	1.91%
esc32f	0.00%	0.00%	sko72	**1.63%**	1.90%
esc32g	0.00%	0.00%	sko81	**1.52%**	1.59%
esc32h	**0.70%**	2.30%	sko90	**1.43%**	1.65%
had12	1.94%	**0.58%**	ste36a	**6.36%**	8.13%
had14	1.89%	**0.41%**	ste36b	**9.42%**	13.48%
had16	2.20%	**0.31%**	ste36c	6.89%	**6.30%**
had18	1.87%	**0.58%**	tai100a	3.08%	**3.07%**
had20	0.93%	**0.78%**	tai100b	**3.98%**	4.41%
kra30a	**4.42%**	5.47%	tai10a	**0.00%**	0.91%
kra30b	**2.83%**	3.42%	tai10b	**0.00%**	0.33%
kra32	**3.89%**	4.94%	tai12a	**4.69%**	5.84%
lipa20a	3.35%	**2.33%**	tai12b	8.18%	**7.83%**
lipa20b	**8.97%**	10.62%	tai15a	**2.80%**	3.39%

(*continued*)

Table 2. (*continued*)

File name	Hybrid MBO	MBO	File name	Hybrid MBO	MBO
lipa30a	**1.69%**	1.77%	tai15b	**0.40%**	0.42%
lipa40a	**1.33%**	1.35%	tai17a	**3.60%**	3.95%
lipa50a	1.17%	1.17%	tai20a	**3.85%**	4.28%
lipa60a	**1.00%**	1.01%	tai25a	**3.78%**	3.89%
lipa70a	**0.87%**	0.88%	tai30a	**3.74%**	3.86%
lipa80a	0.77%	0.77%	tai35a	3.81%	**3.78%**
lipa90a	0.71%	0.71%	tai35b	**6.59%**	6.97%
nug12	6.66%	**2.81%**	tai40a	**3.84%**	3.88%
nug14	7.48%	**3.15%**	tai40b	8.52%	**8.34%**
nug15	7.86%	**2.35%**	tai50a	**3.90%**	3.95%
nug16a	7.53%	**3.02%**	tai50b	**5.22%**	5.57%
nug17	7.62%	**2.10%**	tai60a	3.79%	**3.68%**
nug18	7.46%	**2.69%**	tai60b	6.05%	**5.28%**
nug20	4.36%	**2.89%**	tai64c	0.50%	**0.46%**
nug21	2.53%	**2.50%**	tai80a	3.38%	3.38%
nug22	**1.17%**	2.19%	tai80b	4.96%	**4.70%**
nug24	**1.82%**	2.64%	tho30	**2.25%**	3.34%
nug25	**1.26%**	1.99%	tho40	**3.03%**	3.48%
nug27	**1.85%**	2.85%	wil50	**0.75%**	1.21%

5 Conclusion

In this study, we studied on embedding a different exploration strategy to the recently proposed migrating birds optimization (MBO) algorithm. MBO is hybridized with simulated annealing algorithm (SA) by taking the stochastic part of SA to use it while moving to the next solutions. It is called Hybrid MBO algorithm and it is tested on 100 quadratic assignment problem instances taken from QAPLIB. Regarding to the experiment results, Hybrid MBO algorithm gives solutions better than MBO on about two thirds of the 100 QAP instances. As a future research, other metaheuristic algorithms or local research algorithms can be used to hybridized the MBO algorithm.

References

1. Alkaya, A.F., Algin, R.: Metaheuristic based solution approaches for the obstacle neutralization problem. Expert Syst. Appl. **42**(3), 1094–1105 (2015)
2. Alkaya, A.F., Algin, R., Sahin, Y., Agaoglu, M., Aksakalli, V.: Performance of migrating birds optimization algorithm on continuous functions. In: Advances in Swarm Intelligence, vol. 8795, pp. 452–459. Springer (2014)
3. Behnamian, J., Zandieh, M., Ghomi, S.F.: Parallel-machine scheduling problems with sequence-dependent setup times using an ACO, SA and VNS hybrid algorithm. Expert Syst. Appl. **36**(6), 9637–9644 (2009)
4. Blum, C., Roli, A.: Metaheuristics in combinatorial optimization: overview and conceptual comparison. ACM Comput. Surv. **35**(3), 268–308 (2003)
5. Drezner, Z.: Extensive experiments with hybrid genetic algorithms for the solution of the quadratic assignment problem. Comput. Oper. Res. **35**(3), 717–736 (2008)
6. Duan, Q., Liao, T., Yi, H.: A comparative study of different local search application strategies in hybrid metaheuristics. Appl. Soft Comput. **13**(3), 1464–1477 (2013)
7. Duman, E., Elikucuk, I.: Solving credit card fraud detection problem by the new metaheuristics migrating birds optimization. In: Advances in Computational Intelligence, vol. 8795, pp. 62–71. Springer (2013)
8. Duman, E., Uysal, M., Alkaya, A.F.: Migrating birds optimization: a new metaheuristic approach and its performance on quadratic assignment problem. Inf. Sci. **217**, 65–77 (2012)
9. Koopmans, T.C., Beckmann, M.: Assignment problems and the location of economic activities. Econometrica: J. Econometric Soc. **25**, 53–765 (1957)
10. Liao, T., Chang, P., Kuo, R., Liao, C.: A comparison of five hybrid metaheuristic algorithms for unrelated parallel-machine scheduling and inbound trucks sequencing in multi-door cross docking systems. Appl. Soft Comput. **21**, 180–193 (2014)
11. Pan, Q.K., Dong, Y.: An improved migrating birds optimisation for a hybrid flow-shop scheduling with total flowtime minimisation. Inf. Sci. **277**, 643–655 (2014)
12. QAPLIB: Quadratic assignment problem library. http://anjos.mgi.polymtl.ca/qaplib/

A Hybrid EDA/Nelder-Mead
for Concurrent Robot Optimization

S. Ivvan Valdez[1], Eusebio Hernandez[2(✉)], and Sajjad Keshtkar[3]

[1] Universidad de Guanajuato, DICIS, Salamanca, Mexico
ivvan@cimat.mx
[2] Instituto Politécnico Nacional, ESIME Ticoman, Mexico City, Mexico
euhernandezm@ipn.mx
[3] Tecnológico de Monterrey, Altamira, Mexico

Abstract. We introduce an optimization algorithm which combines an Estimation of Distribution Algorithm (EDA) and the Nelder-Mead method for global and local optimization, respectively. The proposal not only interleaves global and local search steps but takes advantage of the information collected by the global search to use it into the local search and backwards, providing of an efficient symbiosis. The algorithm is applied to the concurrent optimization of a rehabilitation robot design, that is to say, to the dimensional synthesis as well as the determination of control gains. Finally, we present an statistical analysis and evidence about the performance of this symbiotic algorithm.

Keywords: Estimation of Distribution Algorithm · Nelder-Mead ·
Concurrent optimization · Dimensional synthesis ·
Walking rehabilitation device

1 Introduction

Robotic systems for rehabilitation is an emerging field which provides a solution to automate rehabilitation therapy. A robotic rehabilitation device perform intensive repetitive motions for delivering therapy at a reasonable cost [1,2,7,9]. Traditional methods, usually based on treadmill training, are laborious, and often require more than two therapists to assist the legs and torso of a patient during the therapy. This fact imposes a big economic burden on any rehabilitation center, thus limits its clinical acceptance. Therefore, a solution that circumvents this issue should be affordable and available for a large number of patients, for a long period.

The conceptual design of a manipulator for rehabilitation is the process of determining the configuration which satisfies a set of task requirements. Commonly, designers should fulfill design criteria related to inertial properties, control, human anatomy, structure, and economics. In addition, a set of optimization criteria could be defined for the manipulator design, for instance, weight, dexterity, error to follow a trajectory, adequate dimensions, etc. In the context of

A. M. Madureira et al. (Eds.): HIS 2018, AISC 923, pp. 198–207, 2020.
https://doi.org/10.1007/978-3-030-14347-3_20

rehabilitation, various limits of the distribution of requirements for accuracy and rigidity between modules or their elements, as well as different walking habits, weights, and geometrical relationships, must be regarded. Based on such analysis, when using a conventional design methodology, it is necessary repeating the design cycle until finding adequate design parameters which fulfill all the involved criteria. Fortunately, current algorithmic and technological tools can assist engineers and decision makers on the design parameter selection. Regarding the complex architecture of a manipulator for rehabilitation and the difficulty for formalizing the whole structure requirements, one must first calculate some basic-but-essential dimensions of the manipulator elements, in such a way that the remaining dimensions are chosen constructively. Concurrent optimal design of a manipulator is defined as *finding optimal structural and control parameters for a given objective function during the same optimization process, which is dependent on the kinematic or dynamic model of the mechanism* [10]. Previous works [7,12,14] have contributed with an initial draft of the manipulator design, this design can be improved by optimizing some criteria, then developing a detailed production design.

Meta-heuristics has been largely applied to optimization of robots. In [10] the authors distinguish three kinds of problems according to the system of differential equations solved during the simulation of a robot, according to it, this work deals with a dynamic problem in which a second order system of differential equations must be solved to compute the manipulator performance, for this purpose a proportional derivative control is used. A concurrent optimization of control gains and dimensions has been tackled using Evolutionary Algorithms (EAs) in [8]. This reference is of the same class of optimization that we perform, nevertheless we deal with a more realistic case, which includes additional considerations such as dexterity issues. Furthermore the mechanism presented here is more complex and can be considered as a real-world case of a rehabilitation manipulator. There are other several surveys and reviews which report the most relevant proposals in this topic, [5,13].

The main contribution of this article is the introduction of a hybrid method for optimum design, which uses a global and local search. The first is carried out by an Estimation of Distribution Algorithm and the second by the Nelder-Mead simplex method [6]. The first, search for a solution using a Normal distribution, and provides to the second the step length and a starting point.

2   Walking Rehabilitation System and Kinematic Model

The walking rehabilitation system consists of a parallel robot with three degrees of freedom (DOF), which induces the required gait pattern by driving the foot of an individual, while the body weight is supported by a harness system. Each foot requires one parallel manipulator, consisting of three linear actuators aligned in one direction. The design concept of the robot is illustrated in Fig. 1. Each footplate is supported by three limbs, linked to both, the linear actuators and the bottom part of the footplate via revolute joints. While two of the linear actuators are aligned linearly, a third one is displaced backward to enlarge the workspace. The programmed movements of the linear actuators generate the required displacements and trajectories in footplate which follows a pre-calculated walking or climbing pattern.

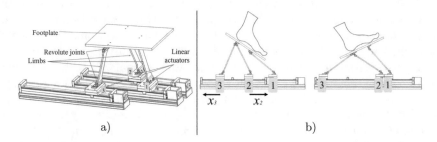

a)   b)

Fig. 1. (a) Design of the 3 DOF manipulator. (b) Rotation of the footplate by on actuation.

The training trajectory can be achieved by the combination of three types of movements: (1) translation along the actuators while maintaining the footplate's orientation and height, it is shown in Fig. 2(a). The translation is achieved by displacing all three limbs, connected to the linear actuators, in the same direction and with the same magnitude and velocity. (2) A pure variation in height of the footplate can be generated by displacing two parallel actuators in one direction, and the third in the opposite direction as shown in Fig. 2(b), with the same magnitude. (3) An orientation control can be obtained by displacing any of the actuators, while the other two are stopped or moved with a different velocity, as can be observed in Fig. 1(b).

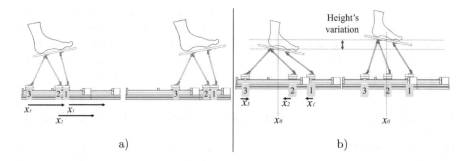

a)   b)

Fig. 2. (a) Displacement along the actuators direction. (b) Displacement along the vertical direction.

The schematic architecture of the parallel manipulator is shown in Fig. 3(b). The moving platform is supported by three limbs. Each limb connects the fixed

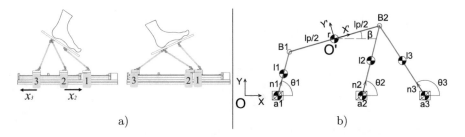

Fig. 3. (a) Rotation of the footplate by on actuation. (b) Schematic representation of the 3PRR manipulator.

base (a) to the moving platform by an actuated prismatic joint (P), a revolute joint (R), followed by another revolute joint (R). Hence, the total structure of the manipulator is 3PRR.

As it can be seen in Fig. 3(b), a_i, $(i = 1, 2, 3)$ denotes the passive revolute joints mounted on the prismatic joints and B_i the points of connection of the limbs to the moving platform. The limbs' with length l_i are inclined by the angle θ_i with respect to the base. The movement of the joints a_i installed on the prismatic joints by the values x_i will result in the change in the position and orientation of the moving platform. A fixed global reference system $O - XYZ$ is located at the edge of the base with the Y axis normal to the base plate and the X axis directed along the direction of the prismatic joints. Another reference frame, called top frame $O' - X'Y'Z'$ is located at the mass center of the moving platform. The Y' axis is perpendicular to the output platform and the X' axis is directed along the longitudinal axis of the platform.

The inverse kinematic analysis of the platform is based on the constraint in Eq. (1) of the platform.

$$r + b_i = x_{ai}e_1 + l_i n_i \qquad i = 1, 2 \tag{1}$$

where $r = [x, y]^\top$ is the position vector of point O' in the XY frame, x_{ai} is the X coordinate of the linear actuators, e_1 is the conventional unitary vector $e_1 = [1, 0]^T$, l_i is the length of the i-th link, b_i is the position vector of point B_i in the XY frame ($b_i = Rb'_i$, with b'_i as the position of B_i in the $X'Y'$ frame and R is the rotation matrix from the $X'Y'$ frame to the XY frame with angle β). The vector $n_i = \begin{bmatrix} \cos\theta_i & \sin\theta_i \end{bmatrix}^\top$ is the unitary vector of each link. By taking the product of (1) with e_1^T, is obtained the linear position of the actuators

$$x_{ai} = e_1^T r + e_1^T b_i - l_i e_1^T n_i \tag{2}$$

After deriving (1), (3) is obtained.

$$\dot{r} + \dot{\alpha} E b_i = \dot{x}_{ai} e_1 + l_i \dot{\theta}_i E n_i \tag{3}$$

where $E = \begin{bmatrix} 0 & -1 \\ 1 & 0 \end{bmatrix}$ is a skew symmetric matrix. By taking the product with n_i^T, and by defining $\delta x_A = \begin{bmatrix} \dot{x}_{a1} & \dot{x}_{a2} & \dot{x}_{a3} \end{bmatrix}^T$ and $\delta X = \begin{bmatrix} \dot{x} & \dot{y} & \dot{\beta} \end{bmatrix}^T$, is attained the

Jacobian expression

$$\delta x_A = \begin{bmatrix} 1 & \frac{n_1^T e_2}{n_1^T e_1} & \frac{n_1^T E b_1}{n_1^T e_1} \\ 1 & \frac{n_2^T e_2}{n_2^T e_1} & \frac{n_2^T E b_2}{n_2^T e_1} \\ 1 & \frac{n_3^T e_2}{n_3^T e_1} & \frac{n_3^T E b_2}{n_3^T e_1} \end{bmatrix} \delta X = J_A \delta X \tag{4}$$

Deriving (3) once more is also found the acceleration relation

$$\frac{d}{dt}\delta x_A = J_A \frac{d}{dt}\delta X + \begin{bmatrix} \frac{-\dot{\beta}^2 n_1^T b_1 + l_1 \dot{\theta}_1^2}{n_1^T e_1} \\ \frac{-\dot{\beta}^2 n_2^T b_2 + l_2 \dot{\theta}_2^2}{n_2^T e_1} \\ \frac{-\dot{\beta}^2 n_3^T b_2 + l_3 \dot{\theta}_3^2}{n_3^T e_1} \end{bmatrix} = J_A \frac{d}{dt}\delta X + \tilde{J}_A \tag{5}$$

The motion of the links can be studied in a similar way. Consider the constraint equations for the links

$$r_i = x_{ai} e_1 + \frac{l_i}{2} n_i, \quad i = 1, 2, 3 \tag{6}$$

where $r_i = [x_i, y_i]^T$ is the position vector of the mass center of the i-th link.

Deriving (6), is obtained

$$\dot{r}_i = (\dot{x} + \dot{y}\frac{n_i^T e_2}{n_i^T e_1} + \dot{\beta}\frac{n_i^T E b_i}{n_i^T e_1})e_1 + \frac{l_i}{2}\dot{\theta}_i E n_i \tag{7}$$

and by the product of (3) with e_2^T is obtained

$$\dot{\theta}_i = \dot{y}\frac{1}{l_i e_2^T E n_i} + \dot{\beta}\frac{e_2^T E b_i}{l_i e_2^T E n_i} \tag{8}$$

Defining $\delta x_i = \begin{bmatrix} \dot{x}_i & \dot{y}_i & \dot{\theta}_i \end{bmatrix}^T$, and by simple manipulations of (7) and (8), is obtained the expression

$$\delta x_i = \begin{bmatrix} 1 & \frac{n_i^T e_2}{2 n_i^T e_1} & \frac{2n_i^T E b_i - (n_i^T e_2)(e_2^T E b_i)}{2n_i^T e_1} \\ 0 & \frac{1}{2} & \frac{e_2^T E b_i}{2} \\ 0 & \frac{1}{l_i e_2^T E n_i} & \frac{e_2^T E b_i}{l_i e_2^T E n_i} \end{bmatrix} \delta X = J_i \delta X \tag{9}$$

By deriving (7) and (8) may be also found the acceleration relation

$$\frac{d}{dt}\delta x_i = J_i \frac{d}{dt}\delta X + \begin{bmatrix} \frac{-l_i \dot{\theta}_i^2 + \dot{\beta}^2((e_2^T b_i)(n_i^T e_2) - 2n_i^T b_i)}{2n_i^T e_1} \\ \frac{-\dot{\beta}^2 e_2^T b_i}{2} \\ \frac{-\dot{\beta}^2 e_2^T b_i + l_i \dot{\theta}_i^2 e_2^T n_i}{l_i e_2^T E n_i} \end{bmatrix} \tag{10}$$

$$= J_i \frac{d}{dt}\delta X + \tilde{J}_i$$

3   Objective Function and Numerical Simulation

The objective function is defined as in Eq. (11), where $x = [l_1, l_2, l_3, l_p, k_c]$, being l_i link lengths and l_p is the platform length, and k_c is the derivative control gain, and $k_c^2/4$ is the proportional control gain. The first part of the function returns 0 for all the configurations which dimensions do not produce a functional manipulator, with the purpose of discarding such configurations. The second part considers manipulators which can be simulated for a short time interval, before the simulation is stopped due to a singular position, the larger simulation interval the greater objective value. Finally, the third subfunction considers the feasible configurations, that is to say, those configurations which can be simulated during the whole time interval. A singular position is determined by the inverse of the condition number, K_c^{-1}, a large value (close to 1) correspond to a large dexterity. The last two subfunctions, depend on $tr(x)$ that is the number of time steps which fulfill the dexterity constraint divided by the total number of time steps. $Er(x)$ is the average of the absolute value of the sum of the position errors, and $\tau_e(x)$ is the average of the absolute value of the sum of the control signals of all the actuators, which is a measure of the energy consumed by the manipulator. The summations run over t for the time steps, and over i for three signals. The simulation time is of $10\,$s, with a time step of $1e-4\,$s.

$$\max_x f_{obj}(x) = \begin{cases} 0 & if \ tr(x) < 2 \cdot 10^{-4} \\ f^1_{obj}(x), & if \ 2 \cdot 10^{-4} < tr(x) < 1 \\ f^2_{obj}(x), & if \ tr(x) = 1 \end{cases} \quad (11)$$

Where $f^1_{obj} = \tau_C \left(9(10^3)/tr - (Er + \tau_e) \right), f^2_{obj} = \tau_C \left(10^4 - (Er + \tau_e) \right)$, and $tr(x) = \left(\sum_{t \in T} I(K_c^{-1}(J_A(x)) > 10^{-4} \& K_c^{-1}(M(x)) > 10^{-6}) \right)/10001$. For short we obviate the (x) notation, for example $tr = tr(x)$, $\tau_C = \tau_C(x)$, etc. Er is computed as follows: $Er(x) = \left(10^3/(10001 tr(x)) \right) \sum_{i=1..3} \sum_{t \in T} |error_i^{position}(t)|$, the sum of the absolute τ signals as $\tau_e(x) = (1/(10001 tr(x))) \sum_{i=1..3} \sum_{t \in T} |\tau_i(t)|$, and $\tau_C(x) = \left(1 - \sum_{i=1..3} \sum_{t \in T} I(\tau_i(t) > 500)/10001 \right)$.

4   Hybrid Design Methodology

This work combines the Nelder-Mead simplex method [6] with an Estimation of Distribution Algorithm (EDA) [3]. The EDA uses univariate normal distributions to sample candidate configurations which approximate the optimum, as it is shown in Algorithm 1.

Initialization. A random manipulator configuration from a uniform distribution is sampled, between the given search limits, if the configuration is evaluated with the subfunction 2 of (11), then, the configuration is stored in the current population, otherwise it is discarded, this step is repeated until n_{pop} solutions are integrated into the current population.

Algorithm 1. HEDA-NM, Hybrid Estimation of Distribution Algorithm with Nelder-Mead.

Require: n_{var}: number of optimization variables. n_{pop}: population size.

$\quad n_e$: elite set size . x_{inf}: array of inferior limits. x_{sup}: array of superior limits.

1: $t = 1$

2: X_1 =Initialize($x_{inf}, x_{sup}, n_{pop}, n_{var}$)

3: F_1 =evaluate(X_1)

4: $[\theta_{t+1}, X_t^{parent}, F_t^{parent}]$=truncate($X, F, t, \theta_t, n_e$)

5: $\{X_t^{elite}, F_t^{elite}\} = \{X_t^{parent}, F_t^{parent}\}[1:n_e]$

6: **while** stopCrit $\neq$ **true do**

7: $\quad [\mu_t, \Sigma_t]$ =MeansAndVar($X_t^{parent}, G_t^{parent}$)

8: $\quad$ **if** $unifRand(0,1) < exp(-1/(0.005t))$ **then**

9: $\quad\quad i_s = binaryTournament(F_t^{elite})$

10: $\quad\quad [X_t^{elite}[i_s], F_t^{elite}[i_s]] = NelderMead(X_t^{elite}[i_s], \delta = \sigma_t/n_{pop})$

11: $\quad$ **end if**

12: $\quad t = t + 1$

13: $\quad X_t = [X_{t-1}^e; sampling(\mu_{t-1}, \Sigma_{t-1})]$

14: $\quad F_t$ =evaluate(X_t)

15: $\quad [\theta_{t+1}, X_t^{parent}, F_t^{parent}]$=truncate($X, F, t, \theta_t, n_e$)

16: $\quad \{X_t^{elite}, F_t^{elite}\}=\{X_t^{parent}, F_t^{parent}\}[1:n_e]$

17: **end while**

18: **return** X_t^{elite}

Evaluation. It is done via the objective function in Eq. (11).

Truncation. At the first iteration the truncation threshold θ_t is equal to the worst objective value in the population. In the other cases, starting at position n_{pop} of the population sorted by its objective value, we find the first individual that is better than θ_t, then, the population is truncated at that position, and the threshold is updated to that objective value. If there is not a better objective value than θ_t, the best individual is returned.

Computation of Means and Variance and Stopping Criterion. In order to maintain the optimization variables between 0 and 1, we first normalize the individuals in the selected set, using the formula: $x = (x_{orig} - x_{inf})/(x_{sup} - x_{inf})$, x_{orig} is the original value of the parameter, then the means and variances are computed as follows:

$$\mu_i = \frac{\sum_{jinS} x_{j,i} G_j}{\sum_{jinS} G_j}, \tag{12}$$

$$\sigma_i^2 = \frac{\sum_{jinS}(x_{j,i} - \mu_i)^2 G_j}{\sum_{jinS} G_j}. \tag{13}$$

Where S is the selected set and i is a dimension. The weights G_j are $(n - j + 1)^{1.5}$, for sorted individuals according to its objective function, thus the best individual gets an $n^{1.5}$ weight, the second best $(n-1)^{1.5}$, and so on. n is the number of individuals in the selected, those which have not been truncated in the step above. Weighted estimators have been well defined and analyzed in [11]. We compute the norm of the standard deviation vector if it is lower than σ_{min}, then, the algorithm stops. In addition, if any standard deviation is less than $\sigma_{min}/(\sqrt{n_{var}} + \sigma_{min})$, then it is set to $\sigma_{min}/(\sqrt{n_{var}} + \sigma_{min})$.

Sampling. We sample from a normal univariate distribution with the parameters computed in the above subsection. Regarding that the normal distribution parameters are computed with normalized data, the sampled data must be translated and re-scaled as follows: $x = x_{sampled}(x_{sup} - x_{inf}) + x_{inf}$.

5   Concurrent Design of a Parallel Manipulator for Rehabilitation

The parameters inputed to the HEDA-NM are the following: $x_{inf} = [0.8, 0.8, 0.8, 1.0, 15]$, $x_{sup} = [1.11.11.11.3, 30]$, recall the variable order, $[l_1 l_2 l_3 l_{platform}, k_c]$. The population size is $n_{pop} = 50$, the number of elite individuals is 45, the stopping criterion is $\sigma_{min} = 10^{-4}$. For the sake of completeness, the best solution found in 15 executions of the algorithm is $[l_1 = 9.041661e - 01, l_2 = 8.716352e - 01, l_3 = 1.043077, l_p = 1.044829, k_c = 2.193619e + 01]$, with an objective function value of $9.849197e + 03$.

Typical Automated Design. Figure 5 shows a sequence of snapshots of the mechanism simulation, in order to demonstrate that it successfully perform its task. A more quantitative view is to analyze its positional error, it is shown in Fig. 4(a). According to these results, it is demonstrated that the methodology automatically delivers the essential parameter of well performed designs.

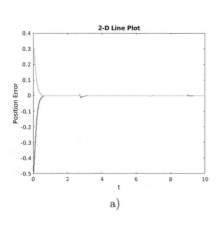

a)

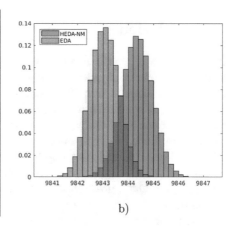

b)

Fig. 4. (a) Positional error for coordinates $(X(t), Y(t))$ of the center of the platform. (b) Histogram of 10^5 boostrap estimators of the mean of the best objective value found by HEDA-NM versus the standard EDA.

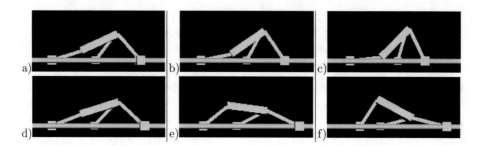

Fig. 5. The instance of a typical design in different instants of time.

Hypothesis Test. We perform a hypothesis test using the boostrap methodology [4], to test whether the HEDA-NM test is *significantly* better than the EDA without local search with exactly the same parameters and operators. According to this test using 2×10^6 boostrap samples HEDA-NM has a better performance with $p-value = 0.0626$. Figure 4(b) shows histograms of bootstrap estimators of the mean of the best objective value found by the HEDA-NM and the GA. That is to say, the HEDA-NM as well as the EDA deliver approximations to optimum designs, one approximation by execution, we measure the performance of these designs by using the objective function in Eq. (11), then we can compute estimators of the mean value of each set of performances using the boostrap method. We can compute as many estimators as desired, then we plot the histogram of the mean estimators of the HEDA-NM and the EDA to determine whether they are equal or not. As noticed, it is clear that the distribution of the HEDA-NM is in a region of higher performance than that of the EDA.

6   Conclusions

We introduce a hybrid algorithm for robotic optimization. According to our results the hybrid algorithm is significantly better than the very same algorithm without local search. The local search improves the algorithm performance, notice that the solution improved by local search is taken from the elite set of the EDA, and if the solution improved in the local search is used to update the EDA parameters, hence there is information interchange between both search methods. The local search is not applied every generation neither is applied on the best individual always, these approaches where tested but they are not reported because they where unsuccessful, applying the local search in all generations or on the best individual always requires a larger computational effort and reports a worst performance than the reported. We conclude that the improvement of hybrid searches must be validated statistically. Selecting different algorithms or search steps is not simple and the unique issue to solve, in addition adequate parameters, information interchange and timing of synchronization of the local search must be validated. Future work contemplate a complete statistical analysis of different proposals, for instance, our algorithm applies the local search with a higher probability at the final generations, other schedules could be investigated, such as varying the probability of application, limiting the number of evaluations for each search steps, etc. The results are promising but we need to explore and explain different hybridizations and elucidate which combinations are more prone to be successful and why.

Acknowledgments. Part of this work have been supported through grants AEM-Conacyt 262887, SIP-IPN 20181422.

References

1. Abedi, M., Moghaddam, M.M., Fallah, D.: A poincare map based analysis of stroke patients walking after a rehabilitation by a robot. Math. Biosci. **299**, 73–84 (2018)
2. Banala, S.K., Agrawal, S.K., Scholz, J.P.: Active leg exoskeleton (ALEX) for gait rehabilitation of motor-impaired patients. In: IEEE 10th International Conference on Rehabilitation Robotics, ICORR 2007, pp. 401–407. IEEE (2007)
3. Cheng, R., He, C., Jin, Y., Yao, X.: Model-based evolutionary algorithms: a short survey. Complex Intell. Syst. **4**, 283–292 (2018)
4. Efron, B., Tibshirani, R.: An Introduction to the Bootstrap. CRC Press, Boca Raton (1994)
5. Fong, S., Deb, S., Chaudhary, A.: A review of metaheuristics in robotics. Comput. Electr. Eng. **43**(Suppl. C), 278–291 (2015). http://www.sciencedirect.com/science/article/pii/S0045790615000154
6. Nelder, J.A., Mead, R.: A simplex method for function minimization. Comput. J. **7**(4), 308–313 (1965)
7. Oña, E., Cano-de la Cuerda, R., Sánchez-Herrera, P., Balaguer, C., Jardón, A.: A review of robotics in neurorehabilitation: towards an automated process for upper limb. J. Healthcare Eng. **2018**, 19 p. (2018)
8. Ravichandran, T., Wang, D., Heppler, G.: Simultaneous plant-controller design optimization of a two-link planar manipulator. Mechatronics **16**(3), 233–242 (2006)
9. Schmidt, H., Werner, C., Bernhardt, R., Hesse, S., Kruger, J.: Gait rehabilitation machines based on programmable footplates. J. Neuroeng. Rehabil. **4**(1), 2 (2007)
10. Valdez, S.I., Botello-Aceves, S., Becerra, H.M., Hernández, E.E.: Comparison between a concurrent and a sequential optimization methodology for serial manipulators using metaheuristics. IEEE Trans. Ind. Inform. **14**(7), 3155–3165 (2018)
11. Valdez-Peña, S.I., Hernández-Aguirre, A., Botello-Rionda, S.: Approximating the search distribution to the selection distribution in EDAs. In: Proceedings of the 11th Annual Conference on Genetic and Evolutionary Computation, GECCO 2009, pp. 461–468. ACM (2009). http://doi.acm.org/10.1145/1569901.1569965
12. Yano, H., Tanaka, N., Kamibayashi, K., Saitou, H., Iwata, H.: Development of a portable gait rehabilitation system for home-visit rehabilitation. Sci. World J. **2015**, 12 p. (2015)
13. Zavala, G., Nebro, A., Luna, F., Coello Coello, C.: A survey of multi-objective metaheuristics applied to structural optimization. Struct. Multidiscip. Optim. **49**(4), 537–558 (2014)
14. Zhang, C., Lan, B., Matsuura, D., Mougenot, C., Sugahara, Y., Takeda, Y.: Kinematic design of a footplate drive mechanism using a 3-DOF parallel mechanism for walking rehabilitation device. J. Adv. Mech. Des. Syst. Manuf. **12**(1), JAMDSM0017 (2018)

A Hybrid Multiobjective Optimization Approach for Dynamic Problems: Evolutionary Algorithm Using Hypervolume Indicator

Meriem Ben Ouada[1(✉)], Imen Boudali[2], and Moncef Tagina[1]

[1] COSMOS lab, ENSI, University of Manouba, Manouba, Tunisia
Meriem.BENouada@ensi.rnu.tn, moncef.tagina@ensi.uma.tn
[2] ENIT, University of Tunis ElManar, Tunis, Tunisia
Imen.Boudali@isi.utm.tn

Abstract. Real world problems are often dynamic and involve multiple objectives and/or constraints varying over time. Thus, dynamic multi-objective optimization algorithms are required to continuously track the moving of Pareto Front (PF) after any change. Recently, evolutionary algorithms have successfully solved dynamic single objective problems. However, few works have focused on multiobjective case through the common "detector technique" to detect/predict change in the fitness landscape, which is sometimes hard or impossible. In this paper, we propose a hybrid approach to tackle dynamic multiobjective optimization problems with undetectable changes. In this hybrid approach, a new local search and novel techniques of population selection and maintain diversity are combined within multiobjective evolutionary algorithm. Our proposed approach (Dynamic HV-MOEA) uses a hypervolume indicator as a performance metric in the local search and population selection techniques in order to accelerate convergence speed towards the Pareto Front. The population diversity is maintained through an efficient mutation strategy. The performances of our hybrid approach are assessed on various benchmark problems. Experimental results show the efficiency and the outperformance of Dynamic HV-MOEA in tracking pareto front and maintaining diversity in comparison with existing dynamic algorithms.

Keywords: Dynamic multiobjective optimization ·
Evolutionary algorithm · Undetectable changes · Hypervolume ·
Local search technique · Population selection strategy ·
Maintain diversity strategy

1 Introduction

In our modern life, many optimization problems involve multiple objectives varying over time, and so known as "Dynamic Multi-objective Optimization Problems" (DMOPs). Since DMOP is defined by merging Dynamic and Multiobjective optimization, the Pareto Front (PF) is unlikely to remain invariant whenever

© Springer Nature Switzerland AG 2020
A. M. Madureira et al. (Eds.): HIS 2018, AISC 923, pp. 208–218, 2020.
https://doi.org/10.1007/978-3-030-14347-3_21

a change occurs. Solving such problems require the optimization algorithm to not only converge towards a diverse PF but also to track it before a new change appears. By simulating the principles of biological evolution, several metaheuristics have been combined and applied for Dynamic Single Objective Optimization Problems (DSOPs) such as ant colony optimization, particle swarm optimization, etc. Nevertheless, Evolutionary algorithms (EAs) still receive the largest attention in handling DSOP and DMOP. One of the main challenge for adapting EAs to a dynamic environment is that they have to balance between convergence and diversity. In fact, the algorithm may lose its capacity to explore the search space when it converges to a partial front. Thereby, several techniques have been integrated into Multiobjective Evolutionary Algorithms (MOEAs) to increase or to maintain population diversity during the convergence process [5]. We distinguish diversity introduction approaches where diversity can be increased after a change detection by reinitializing the whole population [10,17] or by replacing some selected old solutions with other mutated ones [3,4,8,13]. Diversity can be also maintained during the run for preserving a constant level of exploration [2,20]. Two other categories of diversity enhancing approaches have been also used in literature. The first one is focused on exploiting past location of solutions and reusing it as the problem changes. The memory-based techniques [19,22] and the predictive methods [15,22] are instances of such category. The second one divides the main population into subpopulations with specific tasks to either search new optima or track the local ones in promising areas [10,14]. The main weakness of the majority of these approaches is that they have to detect or predict changes in fitness landscape. In fact, they employ the most common scheme named as "detectors technique", which re-evaluates some solutions and checks if their new fitness values are different from the previous ones. Nevertheless, such technique might miss changes because they might not occur in the same part of detectors. Therefore, it would be more appropriate to design future algorithms in DMOP without any detector technique to handle continuous changes. Thus, in this paper, we assume dynamic environment with undetectable changes. Moreover, these approaches are based on the most common static multiobjective algorithms that can be classified into four classes: (1) Scalarizing methods such as MOEA/D [21], (2) Parallel approaches such as VEGA [16], (3) Pareto dominance based approaches (e.g., NSGA II [7], SPEA II [23] and (4) indicator based algorithms (e.g., SMS-EMOA [6], FVMOEA [11]. Given the successful implementation of indicator based algorithms in the static case, we are interested in exploring their performance in solving DMOPs with undetectable changes. Motivated by this observation, we focused on a particular evolutionary algorithm that confirmed its superior performances over the classical MOEAs in static environment: The fast hypervolume-MOEA (FVMOEA) [11]. This recent algorithm quickly computes and updates the exact hypervolume contributions of each solution in the current PF and removes the irrelevant ones. Our basic idea is to propose a hybrid dynamic evolutionary approach based on hypervolume indicator that we called Dynamic HV-MOEA. In our approach, a hybridization of many techniques is performed for: handling the dynamic aspect of optimization problems, enhancing the solution quality, maintaining the diversity of the search process and improving the convergence speed. Thus, we employed the fast hypervolume method [11] to converge rapidly before the next change turns up. Moreover, we integrated a new local search technique to improve the quality of offspring solutions and to accelerate the convergence speed. Furthermore, we used an efficient maintain strategy inspired from the elitist and fast multi-objective genetic algorithm [8] in order to maintain the population diversity during the search process. In order to assess the performances of our Dynamic HV-MOEA, we performed an experimental study on several dynamic benchmark problems. The rest of this paper is organized as follows. Section 2 describes our Dynamic HV algorithm. Section 3 emphasizes the experimental studies on three benchmark problems with two objectives. Finally, the conclusion is given in Sect. 4.

2 Proposed Approach: Dynamic HV-MOEA

2.1 Basic Framework

Due to the dynamic effect of objective functions in DMOP, the PF may change over time. Thus, the optimization goal is to converge as quickly as possible to the current PF before new changes occur in the environment. The main challenge of the fast convergence, is that the optimization algorithm must maintain enough diversity to explore new areas of the space. Our Dynamic HV-MOEA is based on the fast hypervolume-MOEA (FVMOEA) which is a hypervolume indicator based approach [11]. This algorithm has shown good performances in handling the high time complexity of computing the exact HV contributions for each solution in PF. Equation (1) defines the calculation of HV contribution for solution a [11]:

$$HV(a) = HV(S) - HV(S \backslash \{a\}) \tag{1}$$

The time cost of (1) is high, because it requires to compute the HV for all solutions $HV(S)$ and the HV of the set excluding one solution $HV(S \backslash \{a\})$. To deal with this limitation, FVMOEA proposed a fast method to calculate the HV contributions by removing irrelevant solutions. Given the crucial role of time factor in our dynamic problem, we embedded this fast method in our approach to converge rapidly to the desired PF. Moreover, in order to improve solutions and ensure diversity whenever a change occurs, we defined a local search based on HV indicator and mutation strategy. For more details, our Dynamic HV-MOEA is described in Algorithm 1.

2.2 The Proposed Techniques

Offspring Generation. The offspring population can be considered as the largest part that can guide optimization closer to the PF and cover it with high quality solutions. Therefore, we produce an offspring population through a classical selection method and an evolutionary operator with a fixed size OS

Algorithm 1. DYNAMIC HV-MOEA

Input: *PS*, population size; *MaxGen*, maximum algorithm generation; *OS*, offspring number; *Period*, a fixed period of generations.

1 $Time = 0$; $CountGen = 0$
2 Initialize_Population(POP)
3 **while** *(CountGen < MaxGen)* **do**
4 $\quad$ $Time \leftarrow$ Adjust_Time($CountGen$) */Update Time after each Period/*
5 $\quad$ $POP \leftarrow$ Evaluate($POP,Time$)*/ Evaluate POP according to Time/*
6 $\quad$ $OFFS \leftarrow$ Offspring_Generation(POP,OS)
7 $\quad$ $OFFS\_LS \leftarrow$ LocalSearch_HV($\{OFFS\}$)
8 $\quad$ $UNION \leftarrow$ Merge($POP, OFFS \cup OFFS\_LS$)
9 $\quad$ $\{F_1, ..., F_n\} \leftarrow$ Ranking(UNION) */ Generate nondomination levels F_n /*
10 $\quad$ **if** *($|F_1|/PS$) < α* **then**
11 $\quad\quad$ MaintainDiv(UNION)
12 $\quad$ $POP \leftarrow \emptyset$, $i \leftarrow 1$
13 $\quad$ **while** *($|POP|+|F_i| < PS$)* **do**
14 $\quad\quad$ $POP \leftarrow$ Merge(POP,F_i)
15 $\quad\quad$ i++
16 $\quad$ $F_i \leftarrow$ SelectHV($F_i,PS-|POP|$)
17 $\quad$ $POP \leftarrow$ Merge(POP,F_i)
18 $\quad$ $CountGen$++
19 **return** *Non dominated set P*

(Algorithm 1, **line 6**). Contrary to others dynamic MOEA, we adjust the size of offspring solutions *OS* as the result of multiplying the value of the population size *PS* and the parameter *factor*. The aim of this value adjustment is to increase the probability of producing promising solutions that can survive during the optimization and guide the population with a high success rate of finding the true PS.

Local Search Based on HV Indicator. As noted in [12], the hypervolume (HV) metric is considered as a combined convergence-diversity metric. Moreover, it is characterized as a strictly monotonic performance. In other words, if a solution set S is better than a solution set S', the corresponding HV of S, noted as HV(S) is greater than HV(S'). Based on these features, we propose a local search method that improves the offspring population based on hypervolume indicator (Algorithm 2). The local search technique guides individuals of the current population to the nearest local optima that maximizes the hypervolume, since a better value of HV(S') depends on the contribution of solutions found in S' (Algorithm 2, **line 8**). Therefore, applying such technique on the subset of offspring population may drive the population towards the new promising search areas and thus accelerate the convergence speed to the desired PF. Through this method, a good distribution of the whole solution set is ensured since hypervolume metric takes into account the convergence and diversity of the population.

Algorithm 2. LOCALSEARCH_HV
___

Input: P_c, current population ; HV, HV value of P_c; $NTrial$, number of trials
1 $Count = 0$; $P' = \emptyset$
2 **while** *(Count $\leq$ NTrial)* **do**
3   **for** $i \leftarrow 0$ *to* $Size(P_c)$ **do**
4          $x \leftarrow$ Selection(P_c)
5          $x' \leftarrow$ Mutation(x)
6   Add(P',x');
7      $HV' \leftarrow$ MeasureHV(P');
8   **if** *(HV' > HV)* **then**
9          $P_c \leftarrow P'$
10   **else**
11         $P' \leftarrow P_c$
12     $Count++$
13 **return** P'

Moreover, the cost time may be reduced, because we neither need to compute hypervolume between each solution and its neighbor $HV(x, x')$, nor update values whenever a solution is added.

Maintaining Diversity. In each generation, the population diversity may decrease to a certain level which may lead the algorithm to lose its capacity to track the current front. Unlike the other methods that increase the diversity after each change detection, we use a metric to ensure an adaptive control of the population diversity (Algorithm 1, **line 10–11**). Once the diversity decreases to constant threshold α, we employ the mutation strategy that was proposed by [8] where a subset of *UNION* is replaced with better mutated solutions. Therefore, the proposed diversity strategy with a control metric has several advantages: First, it offers a better exploration of the search space since it is not based on any detection technique. Second, it provides an adaptive diversity maintain throughout the optimization process by using the collected information from the current population.

Population Selection Based on HV. The main idea of this method (Algorithm 1, **line 16**) is to eliminate the irrelevant solutions contributing to the PF with minimal hypervolume through new techniques of computing and updating HV as stated in [11]. The first one (Algorithm 3, **line 3–4**) measures the exact HV contribution values for each solution in the front F using a non-dominated worse set, which contains non-dominated solutions maximizing objective values between sol_i and $F \backslash \{ sol_i \}$. The second one determines also the non-dominated worse set but only with the (sol_{min})(**line 7–8**). Afterwards, it quickly updates the HV of each solution using $HV(sol_{min}, sol_i)$(**line 9**) instead of reinitializing and recalculating all HV values without (sol_{min})(**line 10**). Notice that the

---

Algorithm 3. SELECTHV

Input: F, current front ; RS, remain size;
1   R=Generate_ReferencePoint(F) /* R is the maximum objective values /*
2   **for** $i \leftarrow 0$ **to** $Size(F)$ **do**
3   W=NonDominated_Worse($sol_i, F\setminus \{sol_i\}$)
4      $HV(sol_i)= HV(sol_i, R)$-$HV(W, R)$
5   **while** $|F| > RS$ **do**
6   sol_{min}=MinHV($HV(sol_i)$)*/find the minimum of HV values /*
7   ws= worse(sol_{min}, sol_i)
8   W= NonDominated_Worse($ws, F\setminus \{sol_{min}, sol_i\}$)
9      $HV(\{sol_{min}, sol_i\})$=$HV(ws)$-$HV(W, R)$
10     $HV(sol_i)$=$HV(sol_i)$+$HV(\{sol_{min}, sol_i\})$
11   F'=$F\setminus\{sol_{min}\}$
12 **return** F'

---

combined use of the two new techniques may significantly reduce the time cost needed for measuring and updating HV solutions. This fact can greatly accelerate the whole algorithm in tracking the desired PF before a new change occurs.

3 Experimental Results and Discussion

The experiemnts of our algorithm are conducted by using the java-based framework jMetal [1] with some benchmarking problems. These problems come from FDA and dMOP suites [9,10]:

3.1 Parameter Settings

The parameters settings are represented as follows:

- Population size, PS = 100;
- Offspring number, OS = 20 × PS;
- Maximum number of generations, MaxGen = 1000;
- Selection method, Binary tournament [7];
- Crossover operator, SBX with a distribution index of 20;
- Mutation operator, Polynomial mutation with a distribution index of 20;
- The threshold, $\alpha = 0.3$;
- Ratio, the percentage of solutions taken for local search and diversity maintaining methods, Ratio = 10%;
- Independant runs time: Runs = 30
- Frequency of change, τ_t={5, 10, 25, 50}
- Severity of change, $\eta_t = \{1, 10\}$

3.2 Performance Indicators

The Inverted Generational Distance (IGD) [18] and the hypervolume ratio [12] are chosen as performance metrics to measure the convergence of the obtained pareto front. For the diversity, we have chosen the Maximum Spread metric (MS') [12,18]. In dynamic aspect, the mean of performance indicators can be calculated as follows:

$$\overline{P_{indicator}} = \frac{\sum_{i=0}^{NbChanges} P_{indicator}}{NbChanges} \tag{2}$$

Where $P_{indicator}$ is the performance indicator (IGD, HV or MS') computed for each change i and $NbChanges$ is the total number of changes over n generations.

3.3 Experimental Results

Impact of Severity and Frequency Parameters on the Performance of Dynamic HV-MOEA. First, we study the effect of the change severity on FDA1, dMOP1 and dMOP2 test problems. As illustrated in Table 1, when ($\eta_t = 10$), Dynamic HV-MOEA shows a good performance of adaption to slight changes. This may be explained by the high quality of solutions obtained before change appearance. By improving these solutions through the local search method and exploiting them whenever a change is occurred, Dynamic HV-MOEA converges rapidly to a diverse PF. When changes are important ($\eta_t = 1$), the algorithm performances remain stable and do not affect the adaptability capacity of Dynamic HV-MOEA. This observation may be explained by the good level of diversity that is maintained during the search process. This is due to the diversity strategy adjusted by a control parameter, the constant threshold α. Next, we study the effect of frequency change on test problems. Table 1 shows the $\overline{HV\,Ratio}$ of different τ_t values during 1000 generations. As presented in [8], a $\overline{HV\,Ratio}$ smaller than 94% is considered as a poor performance. Thus, by examining results, we can observe that with more frequent changes ($\tau_t = 5$), $\overline{HV\,Ratio}$ values reaches 1.0 for dMOP1 and dMOP2. This may reflect the best adaptive capacity of Dynamic HV-MOEA for short intervals. However, for FDA1, hypervolume ratio exceeds 1.0. This can be explained by a slightly larger hyperarea of the obtained PFs in comparison to true PFs. When $\tau_t > 5$, HV values decrease slightly but remain stable with ($\tau_t = 10$) and ($\tau_t = 25$) for all test problems. We also notice that HV results decrease for dMOP1 and dMOP2 mainly when $\tau_t = 50$, and affect their $\overline{IGD}$ values. Consequently, any decrease of the HVRatio is accompanied by an increase of IGD. For $\overline{MS'}$ values, Dynamic HV-MOEA maintains the best diversity among the different values of τ_t and η_t. It still converges quickly to a diversified PF, which can be explained by the fast selection based on HV indicator.

Table 1. $\overline{IGD}$, $\overline{HV\,Ratio}$ and $\overline{MS'}$ metrics for FDA1, dMOP1 and dMOP2 test problems over 1000 generations

(τ_t, η_t)		FDA1			dMOP1			dMOP2		
		IGD	HV Ratio	MS'	IGD	HV Ratio	MS'	IGD	HV Ratio	MS'
(5,10)	Med	3.03×10^{-4}	1.01	0.993	1.67×10^{-4}	1.00	0.999	3.56×10^{-4}	1.00	0.990
	IQR	3.5×10^{-6}	2.8×10^{-3}	4.3×10^{-4}	6.5×10^{-5}	1.4×10^{-2}	3.5×10^{-3}	5.9×10^{-6}	3.2×10^{-3}	4.1×10^{-4}
(10,10)	Med	1.45×10^{-4}	0.992	0.999	1.57×10^{-4}	0.996	1.00	1.58×10^{-4}	0.990	0.999
	IQR	2.9×10^{-7}	1.1×10^{-4}	1.7×10^{-4}	9.8×10^{-7}	8.7×10^{-3}	1.4×10^{-4}	1.1×10^{-6}	3.1×10^{-4}	8.7×10^{-5}
(25,10)	Med	1.43×10^{-4}	0.992	0.999	1.54×10^{-4}	0.996	0.999	1.55×10^{-4}	0.990	1.00
	IQR	3.4×10^{-7}	3.9×10^{-4}	2.6×10^{-4}	1.7×10^{-6}	1.1×10^{-2}	3.0×10^{-4}	2.0×10^{-6}	6.7×10^{-5}	1.8×10^{-5}
(50,10)	Med	1.43×10^{-4}	0.993	1.00	1.66×10^{-4}	0.990	1.00	1.69×10^{-4}	0.988	1.00
	IQR	3.7×10^{-7}	1.3×10^{-3}	2.8×10^{-4}	3.0×10^{-6}	4.5×10^{-3}	1.8×10^{-4}	3.8×10^{-6}	8.5×10^{-5}	2.6×10^{-5}
(25,1)	Med	1.43×10^{-4}	0.993	0.999	1.55×10^{-4}	0.993	0.999	1.55×10^{-4}	0.990	1.00
	IQR	4.1×10^{-7}	1.2×10^{-3}	3.5×10^{-4}	1.8×10^{-6}	1.6×10^{-2}	7.5×10^{-4}	1.4×10^{-6}	1.2×10^{-4}	7.5×10^{-5}

Med and IQR are the median and interquartile range of different values found over 1000 generations.

Effect of the Offspring Number on Dynamic HV-MOEA. In Dynamic HV-MOEA, the offspring population is produced within an offspring number OS, which is defined as $factor \times PS$ (Algorithm 1). To study the influence of this parameter in the performance of Dynamic HV-MOEA, it is tested with $OS = \{1, 5, 10, 20, 30\} \times$ PS. The $\overline{IGD}$ and $\overline{MS'}$ are evaluated on dMOP1 and dMOP2 with $\tau_t = 10$ and $\eta_t = 10$ over 1000 generations. From Table 2, we observe that the Dynamic HV-MOEA obtains higher performance with the increase of offspring number. In fact, when Dynamic HV-MOEA produces more offspring solutions, many elites solutions are preserved at each generation. Thus, our algorithm is maintaining a balance between convergence and diversity. However, the extreme case Factor = 30 does not make a significant improvement. It means that a large number of offspring solutions dominates the population and are reproduced for each generation, which may decrease the selection pressure.

Table 2. Impact of offspring number on the performance of Dynamic HV-MOEA for dMOP1 and dMOP2 test problems

Factor	dMOP1		dMOP2	
	$\overline{IGD}$	$\overline{HV\,Ratio}$	$\overline{IGD}$	$\overline{HV\,Ratio}$
1	2.43×10^{-3}	0.908	9.42×10^{-4}	0.962
5	1.87×10^{-4}	0.998	2.04×10^{-4}	0.996
10	1.62×10^{-4}	0.999	1.64×10^{-4}	0.999
20	1.57×10^{-4}	0.999	1.58×10^{-4}	0.999
30	1.57×10^{-4}	1.00	1.57×10^{-4}	1.00

Comparative Study. To further evaluate Dynamic HV-MOEA, we carried out a comparative study with four existing approaches that are based on evolutionary algorithms: MRP-MOEA [2], dCOEA [10], SGEA [17] and DNSGA-II-ADI [13]. We chose these approaches since they outperform the other existing dynamic multiobjective approaches in literature. Moreover, the first approach deals with undetectable changes while the last three ones employ change detector technique. These approaches were considered with the best parameter settings according to authors' simulation. Tables 3 and 4 report the detailed experimental results over 20 runs on dMOP1 and dMOP2. For dMOP1, Dynamic HV-MOEA reaches at the smallest $\overline{VD}$ among the four algorithms and obtains well distributed solutions with higher $\overline{MS'}$. This may reflects its better adaptability whenever a change is occurred in the shape of PF. When ($\tau_t = 25$), we can observe that DNSGA-II-ADI outperforms the other ones. Nevertheless, our algorithm is still able to find the most diversified PF despite its large value of $\overline{VD}$. We can also notice that Dynamic HV-MOEA produces better results for dMOP2 since it provides the smallest $\overline{VD}$ and the highest $\overline{MS'}$ for all other settings.

Table 3. Comparative results on dMOP1. Best performances are shown in bold.

(τ_t, η_t)		DNSGA-II-ADI		SGEA		dCOEA		MRP-MOEA		dynamic HV-MOEA	
		VD	MS'	VD	MS'	VD	MS'	VD	MS'	VD	MS'
(5,10)	Med	$5.36^{\times 10^{-3}}$	0.730	$2.64^{\times 10^{-2}}$	0.807	$8.2^{\times 10^{-3}}$	0.979	$9.1^{\times 10^{-3}}$	0.994	$\mathbf{3.41}^{\times 10^{-3}}$	**0.997**
	IQR	$9.6^{\times 10^{-4}}$	$1.3^{\times 10^{-1}}$	$8.1^{\times 10^{-3}}$	$1.4^{\times 10^{-1}}$	$3.5^{\times 10^{-3}}$	$1.5^{\times 10^{-2}}$	$1.0^{\times 10^{-3}}$	$1.71^{\times 10^{-5}}$	$2.2^{\times 10^{-4}}$	$3.3^{\times 10^{-3}}$
(10,10)	Med	$2.78^{\times 10^{-3}}$	0.922	$6.33^{\times 10^{-3}}$	0.872	$3.0^{\times 10^{-3}}$	0.990	$2.9^{\times 10^{-3}}$	**0.999**	$\mathbf{2.68}^{\times 10^{-3}}$	**0.999**
	IQR	$8.3^{\times 10^{-4}}$	$4.1^{\times 10^{-2}}$	$9.6^{\times 10^{-3}}$	$2.1^{\times 10^{-1}}$	$1.5^{\times 10^{-3}}$	$5.0^{\times 10^{-3}}$	$1.0^{\times 10^{-4}}$	$4.83^{\times 10^{-5}}$	$2.5^{\times 10^{-4}}$	$1.6^{\times 10^{-4}}$
(25,10)	Med	$\mathbf{8.56}^{\times 10^{-4}}$	0.986	$1.05^{\times 10^{-2}}$	0.827	$1.5^{\times 10^{-3}}$	0.991	$2.1^{\times 10^{-3}}$	**0.999**	$3.24^{\times 10^{-3}}$	**0.999**
	IQR	$6.6^{\times 10^{-5}}$	$2.9^{\times 10^{-2}}$	$1.1^{\times 10^{-2}}$	$1.6^{\times 10^{-1}}$	$4.0^{\times 10^{-4}}$	$7.3^{\times 10^{-3}}$	$2.0^{\times 10^{-4}}$	$4.53^{\times 10^{-5}}$	$5.4^{\times 10^{-4}}$	$3.1^{\times 10^{-4}}$

The VD metric measures the distance between the obtained PF and the true PF on the decision space.

Table 4. Comparative results on dMOP2. Best performance is shown in bold.

(τ_t, η_t)		DNSGA-II-ADI		SGEA		dCOEA		MRP-MOEA		dynamic HV-MOEA	
		VD	MS'	VD	MS'	VD	MS'	VD	MS'	VD	MS'
(5,10)	Med	$8.34^{\times 10^{-2}}$	0.911	$6.65^{\times 10^{-2}}$	0.888	$3.63^{\times 10^{-1}}$	0.989	$1.34^{\times 10^{-1}}$	0.952	$\mathbf{9.61}^{\times 10^{-3}}$	**0.990**
	IQR	$4.5^{\times 10^{-3}}$	$1.0^{\times 10^{-2}}$	$9.5^{\times 10^{-3}}$	$1.7^{\times 10^{-2}}$	$2.8^{\times 10^{-2}}$	$7.0^{\times 10^{-3}}$	$2.1^{\times 10^{-2}}$	$3.0^{\times 10^{-3}}$	$1.1^{\times 10^{-4}}$	$4.3^{\times 10^{-4}}$
(10,10)	Med	$1.11^{\times 10^{-1}}$	0.964	$2.88^{\times 10^{-2}}$	0.948	$1.73^{\times 10^{-1}}$	0.992	$7.0^{\times 10^{-2}}$	0.994	$\mathbf{2.26}^{\times 10^{-3}}$	**0.998**
	IQR	$3.8^{\times 10^{-3}}$	$5.6^{\times 10^{-3}}$	$4.7^{\times 10^{-3}}$	$1.1^{\times 10^{-2}}$	$1.7^{\times 10^{-2}}$	$5.1^{\times 10^{-3}}$	$4.0^{\times 10^{-3}}$	$8.2^{\times 10^{-3}}$	$2.4^{\times 10^{-5}}$	$2.6^{\times 10^{-5}}$
(25,10)	Med	$1.08^{\times 10^{-1}}$	0.993	$9.36^{\times 10^{-3}}$	0.961	$6.1^{\times 10^{-2}}$	0.994	$5.0^{\times 10^{-2}}$	0.994	$\mathbf{1.26}^{\times 10^{-3}}$	**1.00**
	IQR	$4.1^{\times 10^{-4}}$	$1.2^{\times 10^{-3}}$	$8.3^{\times 10^{-3}}$	$2.3^{\times 10^{-2}}$	$1.2^{\times 10^{-2}}$	$4.8^{\times 10^{-3}}$	$4.7^{\times 10^{-3}}$	$3.0^{\times 10^{-3}}$	$2.2^{\times 10^{-5}}$	$1.8^{\times 10^{-5}}$

4   Conclusion

In order to investigate the efficiency of our hybrid approach Dynamic HV-MOEA, we assessed its performances by using dynamic benchmark test problems. Experimental results have shown the ability of our approach to balance between convergence and diversity, since it tracks rapidly the PF after each change occurrence without losing its capacity in exploring the search space.

We also performed a comparative study of our approach Dynamic HV-MOEA with some successful existing approaches: MRP-MOEA, dCOEA, SGEA and DNSGA-II-ADI. According to the obtained results, our hybrid approach outperforms the existing ones in terms of convergence and diversity. These meaningful results were achieved through the hybridization of the following techniques: the fast hypervolume based method, local search technique for the evolutionary algorithm as well as the diversity maintain method. As future works, we intend to apply this approach to dynamic multiobjective real world problems such as the dynamic traffic regulation problem in public transportation systems.

References

1. Nebro, A.J., Durillo, J.J. (2013). http://jmetal.sourceforge.net
2. Azzouz, R., Bechikh, S., Ben Said, L.: A multiple reference point-based evolutionary algorithm for dynamic multi-objective optimization with undetectable changes. In: IEEE Congress on Evolutionary Computation (CEC 2014), pp. 3168–3175 (2014)
3. Azzouz, R., Bechikh, S., Ben Said, L.: A dynamic multi-objective evolutionary algorithm using a change severity-based adaptive population management strategy. Soft Comput. **21**, 885–906 (2015)
4. Azzouz, R., Bechikh, S., Ben Said, L.: Multi-objective optimization with dynamic constraints and objectives: new challenges for evolutionary algorithms. In: Proceedings of the 8th Annual Conference on Genetic and Evolutionary Computation (GECCO 2015), pp. 615–622 (2015)
5. Azzouz, R., Bechikh, S., Ben Said, L.: Dynamic multi-objective optimization using evolutionary algorithms: a survey. Recent Adv. Evol. Multi-objective Optim. **20**, 31–70 (2017)
6. Beumea, N., Naujoks, B., Emmerich, M.: SMS-EMOA: multiobjective selection based on dominated hypervolume. Eur. J. Oper. Res. **181**, 1653–1669 (2007)
7. Deb, K., Pratap, A., Agarwal, S., Meyarivan, T.: A fast and elitist multiobjective genetic algorithm: NSGA-II. IEEE Trans. Evol. Comput. **6**, 182–197 (2002)
8. Deb, K., Bhaskara Udaya Rao, N., Karthik, S.: Dynamic multi-objective optimization and decision-making using modified NSGA-II: a case study on hydro-thermal power scheduling. In: International Conference on Evolutionary Multi-Criterion Optimization (EMO 2007), pp. 803–817 (2007)
9. Farina, M., Deb, K., Amato, P.: Dynamic multiobjective optimization problems: test cases, approximations, and applications. IEEE Trans. Evol. Comput. **8**, 425–442 (2004)
10. Goh, C.K., Tan, K.C.: A competitive-cooperative coevolutionary paradigm for dynamic multiobjective optimization. IEEE Trans. Evol. Comput. **13**, 103–127 (2009)
11. Jiang, S., Zhang, J., Ong, Y.S., Zhang, A.N., Tan, P.S.: A simple and fast hypervolume indicator-based multiobjective evolutionary algorithm. IEEE Trans. Cybern. **45**, 2202–2213 (2015)
12. Helbig, M., Engelbrecht, A.P.: Performance measures for dynamic multi-objective optimisation algorithms. Inf. Sci. **250**, 61–68 (2013)
13. Liu, M., Zheng, J., Wang, J., Liu, Y., Jiang, L.: An adaptive diversity introduction method for dynamic evolutionary multiobjective optimization. In: IEEE Congress on Evolutionary Computation (CEC 2014), pp. 3160–3167 (2014)

14. Shang, R., Jiao, L., Ren, Y., Li, L., Wang, L.: Quantum immune clonal coevolutionary algorithm for dynamic multiobjective optimization. Soft Comput. **18**, 743–756 (2014)
15. Liu, R., Fan, J., Jiao, L.: Integration of improved predictive model and adaptive differential evolution based dynamic multi-objective evolutionary optimization algorithm. Appl. Intell. **43**, 192–207 (2015)
16. Schaffer, J.D.: Multiple objective optimization with vector evaluated genetic algorithms. In: Proceedings of the 1st International Conference on Genetic Algorithms, pp. 93–100 (1985)
17. Jiang, S., Yang, S.: A steady-state and generational evolutionary algorithm for dynamic multiobjective optimization. IEEE Trans. Evol. Comput. **21**, 65–82 (2016)
18. Jiang, S., Yang, S.: Evolutionary dynamic multiobjective optimization: benchmarks and algorithm comparisons. IEEE Trans. Cybern. **47**, 198–211 (2017)
19. Kundu, S., Biswas, S., Das, S., Suganthan, P.N.: Crowding-based local differential evolution with speciation-based memory archive for dynamic multimodal optimization. In: Proceedings of the 8th annual conference on Genetic and evolutionary computation (GECCO 2013), pp. 33–40 (2013)
20. Biswas, S., Das, S., Kundu, S., Patra, G.R.: Utilizing time-linkage property in dops: an information sharing based artificial bee colony algorithm for tracking multiple optima in uncertain environments. Soft Comput. **18**, 1199–1212 (2014)
21. Zhang, Q., Li, H.: MOEA/D: a multiobjective evolutionary algorithm based on decomposition. IEEE Trans. Evol. Comput. **11**, 712–731 (2007)
22. Peng, Z., Zheng, J., Zou, J., Liu, M.: Novel prediction and memory strategies for dynamic multiobjective optimization. Soft Comput. **19**, 2633–2653 (2015)
23. Zitzler, E., Laumanns, M., Thiele, L.: SPEA2: improving the strength pareto evolutionary algorithm. Technical report (2001)

Coarse Grained Parallel Quantum Genetic Algorithm for Reconfiguration of Electric Power Networks

Ahmed Adel Hieba$^{(\boxtimes)}$, Nabil H. Abbasy, and Ahmed R. Abdelaziz

Department of Electrical Engineering, Faculty of Engineering,
Alexandria University, Alexandria 21544, Egypt
ahmed.adelhieba@yahoo.com, nabil.abbasi@alexu.edu.eg

Abstract. In this paper, a Coarse Grained Parallel Quantum Genetic Algorithm (CGPQGA) is proposed to solve the network reconfiguration problem in distribution networks with the objective of reducing network losses, balancing load and improving the quality of voltage in the system. Based on the parallel evolutionary concept and the insights of quantum theory, we simulate a model of parallel quantum computation. In this frame, there are some demes (sub-populations) and some universes (groups of populations), which are structured in super star-shaped topologies. A new migration scheme based on penetration theory is developed to control both the migration rate and direction adaptively between demes and a coarse grained quantum crossover strategy is devised among universes. The proposed approach is tested on 33-bus distribution networks with the aim of minimizing the losses of reconfigured network, where the choice of the switches to be opened is based on the calculation of voltages at the system buses, real and reactive power flow through lines, real power losses and voltage deviations, using distribution load flow (DLF) program. Simulation results prove the effectiveness of the proposed methodology in solving the current challenges in this phase.

Keywords: Coarse grained · Quantum genetic algorithm ·
Network reconfiguration · Distribution load flow

1 Introduction

Network Reconfiguration in distribution systems (NRC) requires solving a complex combinatorial optimization problem with the purpose of reducing losses, balancing load and improving the quality of voltage in the system. NRC is realized by changing the status of sectionalizing switches. For several past decades, many approximation algorithms have been developed to find the acceptable near optimal solution of the problem. One of the NRC methods is genetic algorithm (GA). The extensive research in this area revealed that the computational time of Gas is costly and its global optimization ability is poor. As such, some researches began to join GA with other methods

A. R. Abdelaziz—Passed.

A. M. Madureira et al. (Eds.): HIS 2018, AISC 923, pp. 219–237, 2020.
https://doi.org/10.1007/978-3-030-14347-3_22

to provide a combination that would help to solve these problems. Recent developments in quantum mechanisms have shown that quantum computing can provide a dramatic advantage over classical computing for some algorithms. Therefore, it seems appropriate to consider how quantum parallelism can be applied to GA in order to provide the so called Quantum Genetic Algorithm (QGA). In this context, QGA is viable to greatly increase the production and preservation of good building blocks and thereby can dramatically improve the search process. This behavior can be attributed to "individuals" in the QGA which are actually the superposition of multiple individuals, where the effective statistical size of the population appears to be increased. However, there are some potential difficulties associated with the QGA. Some fitness functions may require "observing" the superimposed individuals, in a quantum mechanical sense. This would destroy the superposition of the individuals and ruin the quantum nature of the algorithm. This calls for a one-to-one fitness function that will also negate the advantages of the QGA, since it is not physically possible to exactly copy a superposition. As such, difficulties arise in both the crossover and reproduction stages of the algorithm. In this paper a brief background of GA and a quantum strategy is provided. Then the proposed coarse grained strategy applied to QGA is developed. In addition, a simple approach of optimization of power losses of a 33-bus network, based load flow calculations, to find a radial operating structure that minimizes the system power loss while satisfying operating constraints will be presented and analyzed.

2   Short Brief of GA

Genetic algorithms (GAs) are a subclass of evolutionary algorithms where the elements of the search space G are binary strings $(G = B^*)$ or arrays of other elementary types. As sketched in Fig. 1, the genotypes are used in the reproduction operations whereas the values of the objective functions $f \in F$ are computed on basis of the phenotypes in the problem space X which are obtained via the genotype-phenotype

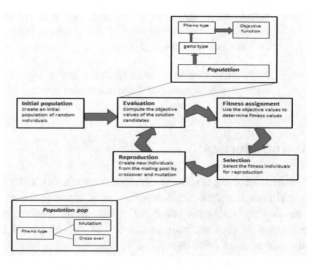

Fig. 1. Basic cycle of genetic algorithm

mapping "gpm" [1]. Among the methods which can give a global optimal solution is GA. Genetic algorithms use the principle of natural evolution and population genetics to search and arrive at a high quality near global solution. Due to the nature of the reconfiguration of electric power network design variables (Switches status), the required design variables are encoded into a binary string as a set of genes corresponding to chromosomes in biological systems [2].

The most important components in a GA consist of:

- Representation (definition of individuals)
- Evaluation function (or fitness function)
- Population
- Parent selection mechanism
- Variation operators (crossover and mutation)
- Survivor selection mechanism (replacement).

GAs differs from conventional algorithms in: 1 - GAs work with coding of parameters rather than the parameters themselves. 2 - GAs search from a population of points rather than a single point. 3 - GAs use only objective functions rather than additional information such as their derivatives. 4 - GAs use probabilistic transition rules, and not deterministic rules. These properties make GAs more robust, more powerful and less data-independent than many other conventional techniques. The theoretical foundation for GAs was first described by Holland [3], and was presented tutorially by Goldberg [4]. GAs provide a solution to a problem by working with a population of individuals; each representing a possible solution. Each possible solution is termed a 'chromosome'. New points of the search space are generated through GA operations, known as reproduction, crossover and mutation. These operations consistently produce fitter offspring through successive generations, which rapidly lead the search towards global optima [4].

3   Quantum Mechanism

The difference between a classical computing and a quantum computing lies basically in storing information with classical bits versus quantum q-bits, in addition to the quantum mechanical feature known as entanglement [5], which allows a measurement on some q-bits to affect the value of other q-bits.

A classical bit is in one of two states, 0 or 1. A q-bit can be in a superposition of the 0 and 1 states. This is often written as $\alpha^2|0> + \beta^2|1>$ where α^2 and β^2 are the probabilities associated with the 0 state and the 1 state. Therefore, the values α^2 and β^2 represent the probability of seeing a 0 (1) respectively when the value of the q-bit is measured. As such, the equation $\alpha^2 + \beta^2 = 1$ is a physical requirement [6]. The probability of measuring the answer corresponding to an original 0 bit is α^2 and the probability of measuring the answer corresponding to an original 1 bit is β^2.

Superposition enables a quantum register to store exponentially more data than a classical register of the same size (increase the search space). Whereas a classical register with N bits can store one value out of 2N, a quantum register can be in a superposition of all 2^N values. An operation applied to the classical register produces one result. An operation applied to the quantum register produces a superposition of all possible results. This is what is meant by the term "quantum parallelism" [7]. The quantum algorithm can be summarized as follows: First one produces a superposition and apply the desired functions, then, takes a Fourier transform of the superposition to deduce the commonalities, and finally, repeats these steps to pump up confidence in the information that was deduced from the transform. After that the feature of entanglement takes place, entanglement produces a quantum connection between the original superimposed q-bit and the final superimposed answer, so that when the answer is measured, collapsing the superposition into one answer or the other, the original q-bit also collapses into the value (0 or 1) that produces the measured answer. In fact, it collapses to all possible values that produce the measured answer [8].

There are some potential difficulties with the quantum mechanism. Some fitness functions may require "observing" the superimposed individuals in a quantum mechanical sense. This would destroy the superposition of the individuals and ruin the quantum nature of the algorithm. Another, more serious difficulty, is that it is not physically possible to exactly copy a superposition. This creates difficulties in both the crossover and reproduction stages of the algorithm. A possible solution for crossover is to use individuals consisting of a linked list rather than an array. The difficulty for reproduction is more fundamental. However, while it is not possible to make an exact copy of a superposition, it is possible to make an inexact copy. If the copying errors are small enough they can be considered as a "natural" form of mutation. Thus, those researchers who favor using only mutation may have an advantage in the actual implementation of a QGA.

4 CGQGA Strategy

4.1 Coarse Grained Parallel Model

The main idea of parallelism is to divide a large population into several sub populations. That evaluation process like several ways start simultaneously from different places for finding the optimal solution till covering the whole searching space [8]. In the coarse grained model which is called also island model, the whole population is divided into some independent sub-populations, and all of the sub-populations govern their own regions and accomplish their own evolution. Just like a lot of isolated islands, they occasionally communicate with neighborhood to exchange good information and that model is easily to accomplish and execute.

4.2 Population Structure

The NRC problem has a large amount of sub-populations; so many root nodes will be added to increase the corresponding cost. Therefore, a super star-shaped topology structure is adopted here, seen from Fig. 2. Sub-populations are divided into several groups, and each group is called a universe. The sub-population in each universe is called a 'deme' and uses the star-

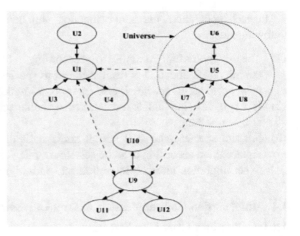

Fig. 2. Super star-shaped structure

shaped structure. In this frame, some universes arise, each of which does not contain a great deal of demes. In this way, the communication demand is relatively reduced and good information can be transferred among universes. Since it can make full use of all demes' information and overcome the shortcoming of huge cost, it is an excellent topology structure with good performance for the application NRC.

4.3 Communication Between Demes in One Universe

A newly proposed migration scheme based on penetration theory is used in this paper. The new strategy on how to migrate among demes is an available method to promote information communication and quicken algorithm's convergence speed. With this migration strategy, if the information changes in one deme, then the information changes immediately in other demes as well.

4.3.1 The Penetration Theory and Penetration Migration Strategy
The migration strategy directly comes from penetration theory, and we need not to define migration frequency, migration ratio or migration direction any more, which are replaced by a threshold value θ. Whether a migration is needed or not is decided by the threshold value. Here, $0 \leq \theta \leq max\{|fit_i - fit_j|\}$. Roughly speaking, when $\theta = 0$, it is more likely for individual information to transmit [9].

$$\lambda = \begin{cases} max\left\{1, \dfrac{|\Delta fit|}{max\{fit\ i, fit\ j\}}\right\} & \Delta fit > \theta \\ 0 & \Delta fit < \theta \end{cases}$$

Here, $\Delta fit = fit_i - fit_j$, and fit_i and fit_j are the best individual's fitness value in sub-population i, j, respectively.

In each generation, the above function will be calculated and executed in the following steps.

(1) The direction of migration is made according to the value of Δfit. If Δfit > 0, then the migration direction is from species i to species j, else if Δfit < 0, migration is from species j to species i, else Δfit = 0, do nothing.
(2) If λ > 0, λN individuals are selected by some selecting strategies to immigrate, else, do nothing.
(3) For the species who will accept some individuals, choose λN individuals to replace them according to some selection strategies. Thus we can use this strategy to set migration interval, migration rate and migration direction adaptively.

4.4   Information Communication Among Universes

4.4.1   Quantum Crossover Strategy

The selection of random individuals from two universes can be done by using quantum rotation gate, and then exchange optimal target to make individual's evolvement towards another universe's optimal direction, as follow:

(1) Randomly select one or more individuals from the current universe according to a given probability value;
(2) Evaluate these individuals to get their fitness value;
(3) Randomly select one universe, and then take the objective of the universe evolution as the aim for the evolution of the individuals above. If n (n > 1) universes are selected, then repeat the operation n times. Repeat the step 1–step 3 until all universes have experienced the quantum crossover operation.

4.4.2   Communication Period Among Universes

The typical communication period is 10%–20% amount of total iterative times. Because if communication among universes in small periods, it can help quickly spread best information and offer good instructions for population evolvement., but, it will increase the cost to communicate, and force some individuals to occupy the govern status and cut down the diversity of genes. That means It will disobey the rule of parallel searching and make populations convergence into local optimal solution. On the other hand, if communication among universes in large periods will lose the sense of migration, because it makes good genes hard to transmit and reduce the convergence speed.

5   A Network-Topology-Based Three-Phase Load Flow for Distribution Systems and Solution Techniques

Distribution load flow is a very important tool for the analysis of distribution systems and is used in operational as well as planning environments. Many real-time applications in the distribution automation system and distribution management system, such as network optimization, VAr planning, switching, state estimation and so forth, need the support of a robust and efficient power flow method. Such a power flow solution must be able to model the special features of distribution systems in sufficient detail.

If a line section (B_k) is located between Bus i and Bus j and Z_{ij} is the line impedance, two matrices, BIBC and BCBV, were developed based on the topological structure of distribution systems. The BIBC matrix is responsible for the relations between the bus current injections and branch currents while the BCBV matrix is responsible for the relations between the branch currents and bus voltages [10]. The corresponding variation of the branch currents, which is generated by the variation at the current injection buses, can be found directly by using the BIBC matrix. The corresponding variation of the bus voltages, which is generated by the variation of the branch currents, can be found directly by using the BCBV matrix. The relations between the bus current injections and bus voltages can be expressed as:

$$[\Delta V] = [BCBV][BIBC][I]$$
$$= [DLF][I]$$

The proposed algorithm is summarized as follows:

(1) *Input data.*
(2) *Form the BIBC matrix.*
(3) *Form the BCBV matrix.*
(4) *Form the DLF matrix.*
(5) *Iteration k = 0.*
(6) *Iteration k = k + 1.*
(7) *Solve for the three-phase power flow by using Equations.*

$$I_i^k = I_i^r(V_i^k) + jI_i^i(V_i^k) = \left(\frac{P_i + jQ_i}{V_i^k}\right)^*,$$

$$[\Delta V^{k+1}] = [DLF][I^k]$$

(8) If $\max_i\left(\left|I_i^{k+1}\right| - \left|I_i^k\right|\right) > tolerance$ go to (6).
(9) *Report and end.*

where $V^k i$ is the node voltage at the kth iteration;
$I^k i$ is the equivalent current injection at the kth iteration;
$I^r i$ and $I^i i$ are the real and imaginary parts of the equivalent current injection at the kth iteration, respectively.
DLF is the distribution load flow matrix.

6   The CGPQGA for NRC Problem of IEEE 33-Bus Network

In this section we introduce a CGPQGA to solve the NRC problem of IEEE-33 bus network. The crucial element in applying Coarse grained quantum Algorithm successfully for radial distribution system, in which the choice of the switches to be opened is based on the calculations of voltage at the buses, real and reactive power flowing through lines, real power losses and voltage deviation, using a simplified distribution load flow (SDLF) program. NRC is to develop an effective coding mechanism and a quantum rotation angle table. Meanwhile, to find solution with good quality, parallel evolution is necessary [14]. We separate a population into several demes with parallel architecture, and then carry out sequential QGA upon them with some new migration strategy. The solutions may get converged very early in the second generation therefore execution time is effectively small. Particular attention is paid to the relationship between electrical parameters of the distribution system and the mathematical parameters that influence the convergence properties of the algorithm. A 33-bus radial distribution test system is taken as a study system for performing the test. The results reveal the speed and the effectiveness of the proposed method for solving the problem.

6.1 NRC Problem

Reconfiguration of electrical networks is a combinatorial optimization problem and the exponentially large number of combinations is possible for large-scale problems. A CGPQGA based approach for distribution system loss minimum reconfiguration is proposed on this approach with the use of many objectives and constraints.

The search space for this problem is the set of all possible network configurations. Once the general layout of the distribution network is specified, the specific topology is determined by the status of each of the switches in the system. Switches which are normally open are called tie switches and normally closed switches are known as sectionalizing switches. Specifying the open/closed status of each switch completely characterizes the topology of the network. So if the total number of tied and section-alized switches in the system is i, the current configuration can be represented as a vector of individual switch states u, where

$$u = [u1, u2 \ldots us] \qquad ui \in \{0, 1\}, \ 1 \leq i \leq n_s,$$

where ui = 1 indicates that switch i is closed, and ui = 0 indicates that it's open. In order to calculate the cost function and check the constraints it is necessary to have complete information on the voltage magnitudes and angles at each bus. This infor-mation is included in the state variable x.

6.1.1 Mathematical Problem Formulation
Fundamental objective of reconfiguration of feeder is the reduction of all power losses as follow:

$$minimize \ Loss_P = \sum_{ij=1}^{N} Loss_{ij}^{line} \qquad (1)$$

And as mentioned before in the load flow technique and throw formulate the BIBC matrix, BCBV matrix, and DLF matrix, the solution for distributed load flow can be obtained by solving the next equation iteratively

$$I_i^k = I_i^r(V_i^k) + jI_i^i(V_i^k) = \left(\frac{P_i + jQ_i}{V_i^k}\right)^*,$$

$$[\Delta V^{k+1}] = [DLF][I^k]$$

$$V^{k+1} = V^k - \Delta V^k$$

Obtain voltage at the buses

$$V_j = V^{k+1} = V^k - \Delta V^k \tag{2}$$

V^{k+1} ... is the bus voltage after (k + 1) iterations
ΔV^k ... is change in bus voltage after successive iterations which is calculated from load flow algorithm
V^k ... is the bus voltage after (k) iterations.

Real power flow

$$P_{ij} = Real\left[\{(V_i - V_j)y_{ij}\}*\right] \tag{3}$$

P_{ij} ... line real power flowing between ith and jth buses
y_{ij} ... Admittance of the line between ith and jth buses.

Reactive power flow

$$Q_{ij} = Imag\left[\{(V_i - V_j)y_{ij}\}*\right] \tag{4}$$

Q_{ij} ... line reactive power flowing between ith and jth buses
PD_j ... Real power load at bus j.

Real Power Loss

$$Loss = \left\{V_{ss}\sum_{j\in ss}\left[(V_{ss} - V_j)y_{ss,j}\right] - \sum_{j=1}^{N}PD_i\right\} \tag{5}$$

V_{ss} ... main substation voltage
N ... number of buses in radial distribution system (RDS).

It must be observed that the limit of voltage busses must be within the range

$$V_{\min} < V_i < V_{\max}$$

Format of the configuration must be in radial No load-point interruption.

The previous equations used in single phase transmit ion line, and in case of the three phase let, $|Vi| = \left[|V_i^a|, |V_i^b|, |V_i^c|\right]^T$ and $|\theta_i| = \left[\theta_i^a, \theta_i^b, \theta_i^c\right]^T$ be the voltage magnitudes and angles respectively for phases a, b, and c at bus i. Given a three phase distribution network with a total of n buses, where bus 1 is the substation and buses 2, 3, …, n are the load buses, the state variable can be denoted by $x = \left[0_2, \ldots, 0n, |V_2|, \ldots, |V_n|\right]^T$.

Let the cost function, f(x, u), be the sum of the real power losses in each line

$$f(x, u) = \left[\sum_{i=1}^{n_1} Loss_i^{line}\right]$$

where $Loss_i^{line}$ represents the real power loss in line I, and n_i is the number of lines in the system. Stated in a different way, the total power loss in the system is the total power input to the system minus the total power delivered to the loads. Given the proper scaling, this cost function would give the number of money lost due to real power losses in the system. A more complete formulation might also include the cost of switching to configuration u from the current operating configuration.

In the proposed example of IEEE 33 bus network, it will be taken only loss of the lines for reconfiguration issue

$$f(x, u) = P_{input} - P_{delivered}$$

It is to be noted that not every configuration use is a reasonable solution to the network reconfiguration problem [11]. For example, if all of the switches were put in the open state and all bus voltages were set to zero, the real power losses in the system would also be zero, but a distribution system operated in this state would obviously cause the utility company to lose customers. So it is necessary to specify which states are feasible and which ones are not.

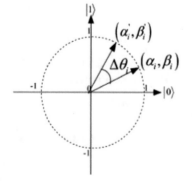

Fig. 3. Polar plot of the rotation gate for Q-bit individuals

6.2   The CGPQGA

Quantum Evolution Algorithm (QEA) has been introduced recently and gained much attention and wide applications for both function and combinatorial problems, the most important element in applying [12].

Quantum Evolution Algorithm successfully developed an effective coding mechanism and a quantum rotation angle table. Meanwhile, to find solution with good quality, parallel evolution is necessary. We separate a population into several demes with parallel architecture, and then carry out sequential QGA upon them with some new migration strategy (Fig. 3).

6.2.1 Q-Bit Individual

Instead of binary, numeric, or symbolic representation, qbit chromosome is adopted in our parallel quantum genetic algorithm as a new representation. The state of a qbit can be represented as $|\psi\rangle = \alpha|0\rangle + \beta|1\rangle$, where α and β are complex numbers that specify the probability amplitudes of the corresponding states. $|\alpha|^2$ is the probability that the qbit will be found in the "0" state and $|\beta|^2$ is the probability that the qbit will be found in the "1" state. Normalization of the state to unity guarantees $|\alpha|^2 + |\beta|^2 = 1$. A qbit may be in the "1", "0" states, or in any linear superposition of them. The advantage of qbit is that it can represent a linear superposition of solutions due to its probabilistic representation. For example, when there are two quantum bits, it can stay in adding states of four 00, 01, 10, 11.

The procedure of CGPQGA for NRC
Begin
t←0

1. *Initialization: PS (the initial population size of each sub-population), Pc (crossover probability), Pm (mutation probability), Ps (quantum crossover probability), ST (sampling times), Gen (iterative generation);*

2. *According to super star-shaped structure, randomly initialize s sub-populations in Q-bit representation PQ i(t), which are converted to the corresponding bus permutations in decimal code Di(t) (i = 1, 2, . . . , s). Use stochastic simulation technology to sample ST groups of processing time according to their probability functions. Evaluate each individual in Di(t) and then record the best minimum loss results. While (t < Gen) do*
   begin
   t←t +1

3. *for each sub-population i do begin*
   Select PQ i(t) from P Q i(t) based on fitness value, then apply crossover and mutation operations.
   if (catastrophe condition is satisfied)
   Perform catastrophe operation for P Q i(t) to generate P Q i(t +1).
   else
   Perform quantum rotation gate, for PQ i(t) to generate   PQ i(t +1).
   end
   end

4. *for demes in each universe do begin*
   Execute the migration strategy based on penetration theory.
   end

5. *for universes do begin*
   If (rand(1) < Ps)
   Execute the quantum crossover operator.
   end
   end

6. *Using stochastic simulation technology evaluated individuals of each sub-population and records the best minimum loss result.*
   end

7. *Output the global optimum minimum loss result.*
   End

If there is a system of m qbits, the system can represent 2 m states at the same time [14]. However, in the act of observing a quantum state, it collapses to a single state. So we define a Q-bit as the smallest unit of information, which consists of a pair of numbers $\begin{bmatrix} \alpha \\ \beta \end{bmatrix}$.

6.2.2   Quantum Rotation Gate

The state of a qbit can be changed by the operation mechanism named quantum gate. A quantum gate is a reversible gate and can be represented as a unitary operator U acting on the qbit basis states satisfying U + U = UU+, where U+ is the Hermitian transpose of U. There are several quantum gates, such as the Not gate, controlled NOT gate, rotation gate. The rotation gate U ($\Delta\theta$i) is applied to generate the probability amplitude of quantum states in Quantum evolution algorithm in order to maintain diversity of population, which is an important updating method in quantum evolution algorithm.

Where θi is rotation angle and it determines the rotation direction.

6.2.3   Genetic Operator

A. *Crossover Operator*

The crossover operator is important agent to avoid trap into local optimality and lead to the lack of diversity of genes. For this reason, cycle crossover is used in this paper which is appropriate to the quantum parallel theory of search till reach the optimum solution which is the main idea of parallel quantum theory.

Cycle crossover is used for chromosome with permutation encoding; it occurs by picking some cycles from one parent and the remaining cycles from the alternate parent, where the symbols of the first parent are divided into two subsets.

The first child is the copy of the first subset, and the second subset is copied to the second child. In addition, each subset of symbols preserves the exact positions from the same parent. Then the remaining symbols are copied to the child by preserving the order from the second parent [15].

$$P_c^+ = \begin{cases} \dfrac{P_c"\text{max}"}{1 + \dfrac{t}{t_{max}}} , & P_C^+ > P_C : "\text{min}" \\ P_c\text{max} , & P_C^+ < P_c"\text{min}" \end{cases}$$

where Pc_{min}, Pc_{max} are the minimum and maximum value of crossover probability and t_{max} is the algorithm's iterative times.

B. *Mutation Operator*

Executing mutation operator can effectively prevent premature convergence and improve searching ability in local space. In this part, we introduce the Not Gate as our mutation operator. The definition of Not Gate:

$$\begin{bmatrix} \alpha i \text{ after} \\ \beta i \text{ after} \end{bmatrix} = \begin{bmatrix} \beta i \\ \alpha i \end{bmatrix} \text{if } rand() < pm$$

where αi, βi are probability amplitudes before mutation, and they will turn to be "αi after" and "βi after". The details of the operation could be introduced below. Firstly, select the quantum individuals with mutation probability pm. Then, randomly generate a mutation position. Finally, exchange αi with βi to interconvert the probability of states "0" and "1".

C. *Selection Operator*

In order to obtain and maintain good performance of the fittest individuals, it is important to keep the selection competitive enough. It is no doubt that the fittest individuals have higher chances to be selected. In this paper, the 'roulette wheel selection' scheme is used [13], in which each string occupies an area of the wheel that is equal to the string's share of the total fitness

$$minimize \ Loss_P = \sum_{ij=1}^{N} Loss_{ij}^{line}$$

where P denotes the Power Loss of individual p at the current generation, and N no of buses.

D. *Catastrophe Operator*

To avoid premature convergence, a catastrophe operator is used in CGPQGA. The best solution until current generation does not change in some consecutive generations, and then we regard it to be trapped in local optimality and regenerate the initial generation randomly.

7   Illustrative Example

Based on the proposed algorithm, a computer program is implemented in MATLAB.A distribution system of IEEE 33-bus system, shown in Fig. 4, is tested by the proposed method. This system includes one transformer, four feeders, 32 branches, 5 tie lines, 32 buses and 37 switches. Substation voltage = 12.66 kV The minimum and maximum voltages are set at 0.92 and 1.0 p.u. Table 1 gives the necessary rearrangement of network data required to DLF in optimum case. The components of calculation of power loss of the distribution system is listed in Table 1. In addition, the running time is fast for application in an on-line system. For the 33 bus test system, the reconfiguration plan was obtained with CPU average time is 32.67 s on a 1.6 GHz Personal Computer. The reconfiguration problem attempts to determine the states of 37 switches for minimizing the system losses. Restated, the solution space contains 2^{37} possible combinations. The searching space is so large that most optimum algorithms cannot effectively solve the problem. Specially in time issue. The parameters of CGPQGA used in this system are described as follows: bit numbers of the solution fixed string: 37 bits, Deme number 8 (2 universes), population size 6, crossover probability $pc_{min} = 0.6$, $pc_{max} = 1$, mutation probability 0.1, sampling times 30, and communication period among universes 10. Consider a 3-phase balanced load with peak loads listed in below tables. In order to get performance of CGPQGA 30 runs performed, and due to random (stochastic) technique used we get 30 sampled value from the processing time optimal power losses solution 30 times, The results not had remarkable changes, each run time results are so close to each other with 100% probability to get optimal solution with average value (132.214 KW) illustrated in Table 3 with comparison between average values of different methods in [16–19], which are normal load flow method, on revisiting GA, and harmony search algorithm (HSA), PSO–ACO and Hybrid Genetic Algorithm Particle Swarm Optimization (HGAPSO) algorithm respectively. The worst voltage (p.u) in proposed method is 0.9413 (bus27) which is better than all mentioned before methods. Table 2 illustrates also the worst value of voltage (p.u) of different methods overall 30 run times (note: in [18] the value not mentioned as an average value) (Figs. 5 and 6).

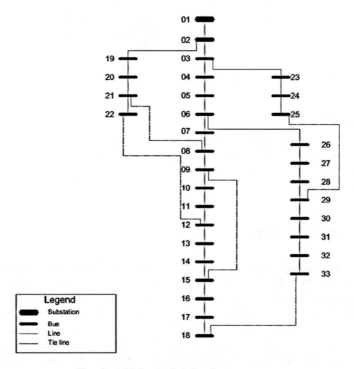

Fig. 4. A33-bus radial distribution system

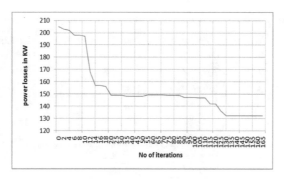

Fig. 5. Power losses vs. iterations

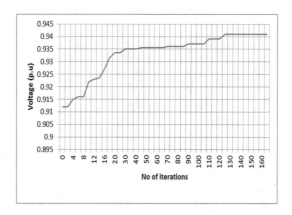

Fig. 6. Voltage deviation vs. iterations

8 Results for Reconfiguration Problem

The Reduction in the real power loss, and increased bus voltages are the merits shown by the method used. This can be understood by having a look on the Tables 3 and 4. Complete load flow solutions for optimal case. The CGPQGA algorithm was terminated after approximately 120 iterations that may be seen as big number of iterations but that achieved in the proposed method in very fast time due to the parallelism strategy and convergence to the optimum solution is too fast can be happened in second generation. The details of the results are presented in Tables 3 and 4 with comparison with different methods. The convergence pattern is presented in Fig. 4. The method was tested for 30 times and yields a good performance with 100% of find optimum solution.

Table 1. Load flow results (Currents, active and reactive power) for optimum case[*]

Line #	From bus	To bus	Line current (A)	Active power flow (kW)	Reactive power flow (kVAr)
1	1	2	207.208	3854.98	2404.88
2	2	3	134.679	2409.95	1691.92
3	3	4	32.0059	623.253	302.27
4	4	5	25.4082	502.128	221.697
5	5	6	22.3023	441.39	191.321
6	6	7	10.4064	200.061	100.201
8	8	9	20.0404	394.084	153.744
10	11	10	2.99602	60.0053	20.0018
11	12	11	5.50892	105.039	50.013
12	12	13	10.1473	180.53	115.457
13	13	14	6.8533	120.076	80.1004
15	15	16	14.2059	271.101	121.11
16	16	17	11.1929	210.65	100.78
17	17	18	8.18354	150.165	80.1334
18	2	19	67.7752	1333.15	646.909
19	19	20	63.2641	1240.89	604.753
20	20	21	58.675	1132.83	548.478
21	21	22	23.5252	438.893	244.174
22	3	23	98.8126	1669.87	1335.99
23	23	24	94.0907	1566.66	1276.96
24	24	25	72.9198	1122.81	1058.12
25	6	26	8.98807	180.107	70.0648
26	26	27	5.96543	120.058	45.0397
27	27	28	2.94466	60.0275	20.0243
29	29	30	46.0106	564.428	772.854
30	30	31	19.2828	361.205	171.212
31	31	32	11.2689	210.118	100.138
33	21	8	30.6048	599.704	259.363
34	9	15	17.0379	332.843	132.852
35	22	12	18.9177	347.716	202.618
36	18	33	3.47202	60.0181	40.0181
37	25	29	52.1995	688.515	846.941

Table 2. Voltage (p.u) results for optimum case

Bus #	Voltage (%)	Bus #	Voltage (%)	Bus #	Voltage (%)
2	99.7077	12	96.3079	23	96.4891
3	98.6985	13	96.0498	24	95.3589
4	98.5187	14	95.9704	25	98.0274
5	98.3703	15	99.5076	26	94.1959
6	98.0552	16	97.8245	27	94.1287
7	97.992	18	97.0155	28	94.7164
8	94.8519	19	96.2614	29	98.0015
9	94.7493	20	95.9246	30	97.9491
10	95.1435	21	96.2699	31	94.8566
11	95.3192	22	96.2784	32	94.537
33	97.9619				

Table 3. Results comparison between network configuration methods

	LOSS (KW)					Worst voltage (p.u)					Line switched out			
Case	Present	[16]	[17]	[18]	[19]	Present	[16]	[17]	[18]	[19]	Present	[16]	[17]	[19]
Optimal	132.214	158.24	139.55	138.06	140	0.9413 (bus27)	0.9388 (bus18)	0.9378 (bus31)	0.9342	0.9423	7, 9, 14, 28, 32	6, 9, 14, 32, 37	7, 9, 14, 37, 32	10, 7, 1, 4, 32, 3, 7

Table 4. Performance of the proposed method compared to other methods

Case	Loss reduction	Increment in worst voltage
Optimal	16.03 kW from [16]	Increase 0.00105 pu from best value
	7.33 kw from [17]	
	5.846 kW from [18]	
	7,786 KW from [19]	

9 Conclusion

This study presents a coarse grained Parallel Quantum Genetic Algorithm (CGPQGA) method for multi-objective programming to solve the reconfiguration in a distribution system. In reconfiguration problem, considered herein are to minimize the system losses, and constrained with bus voltage violations in conjunction with network constraint and renumbering the lines according to switches status. Merit of this method used is that it is very effective and solutions get conversed even in second generation there for the execution time of the proposed method is quite small. The above conclusions have been corroborated by the test results.

For the future work we suggest improve this method in aim of reach more better results like increase no of universes and demes, change the rate of rotation angle to produce processing time, and try to solve more complicated problems like reconfiguration and restoration of Distribution systems with Multiple DGs, also to repeat this method for different systems and different issues like network Maintenance Scheduling, network restoration after fault occurring, and Optimal Allocation of Capacitors in Distribution Systems and try to increase no of universes and demes aiming to increase search space for more long and complicated problems.

References

1. Maxwell, J.C.: A Treatise on Electricity and Magnetism, 3rd edn., vol. 2, pp. 68–73. Clarendon, Oxford (1892). Genetic Algorithms in Search, Optimization and Machine Learning
2. Sinclair, M.C.: Node-pair encoding genetic programming for optical mesh network topology design, Chap. 6. In: Telecommunications Optimization: Heuristic and Adaptive Techniques, pp. 99–114. Wiley (2000). In collection [450]
3. Holland, J.H.: Adaptation in Nature and Artificial Systems. The University of Michigan Press, Michigan (1975)
4. Goldberg, D.E.: Genetic Algorithms in Search, Optimization and Machine Learning. Addison-Wesley, Reading (1989)
5. Choy, C.K., Nguyen, H.Q., Thomas, R.: Quantum-inspired genetic algorithm with two searches supportive schemes and artificial entanglement. In: Proceedings of the 2014 IEEE Symposium on Foundations of Computational Intelligence, FOCI, Orlando, FL, USA, 9–12 December 2014, pp. 17–23 (2014)
6. Boghosian, B.M., Taylor IV, W.: Simulating quantum mechanics on a quantum computer. Phys. D **120**, 30–42 (1998). [CrossRef]
7. Deutsch, D.: Quantum theory, the Church-Turing principle, and the universal quantum computer. Proc. Roy. Soc. Lond. A **400**(1985), 97–1117 (1985)
8. Lahoz-Beltra, R.: Quantum Genetic Algorithms for Computer Scientists. Department of Applied Mathematics (Biomathematics), Faculty of Biological Sciences, 15 October 2016
9. Han, K.H., Park, K.H., Lee, C.H., Kin, J.H.: Parallel quantum-inspired genetic algorithm for combinatorial optimization problem. In: Proceedings of the 2001 IEEE Congress on Evolutionary Computation, Seoul, Korea, May 2001, pp. 1422–1429 (2001)
10. Carpaneto, E., Chicco, G., Roggero, E.: Comparing deterministic and simulated annealing-based algorithms for minimum losses reconfiguration of large distribution systems. In: IEEE Power Tech 2003 Conference, June 2003, p. 77 (2003)
11. Hayashi, Y., Matsuki, J.: Loss minimum configuration of distribution system considering N-1 security of dispersed generators. IEEE Trans. Power Syst. **14**, 636–642 (2004)
12. Han, K.K., Kim, J.H.: Quantum-inspired evolutionary algorithm for a class of combinatorial optimization. IEEE Trans. Evol. Comput. **6**(6), 580–593 (2002)
13. Bertsekas, D.P., Castanon, D.A.: Rollout algorithms for stochastic scheduling problems. J. Heuristics **5**(1), 89–108 (1999)
14. Erick, G.P.: Designing scalable multi-population parallel genetic algorithm. IlliGAL Report No. 98009. Illinois Genetic Algorithm Laboratory (1998)
15. Genetic Algorithm based Network Reconfiguration in Distribution Systems with Multiple DGs for Time Varying Loads SMART GRID Technologies, 6–8 August 2015

16. Thakur, T., Dhiman, J.: A new approach to load flow solutions for radial distribution system. In: IEEE PES Transmission and Distribution, pp. 1–6 (2006)
17. Wang, C., Gao, Y.: Determination of power distribution network configuration using non-revisiting genetic algorithm. IEEE Trans. Power Syst. **28**(4), 3638–3648 (2013)
18. Heidari, M.A.: Optimal network reconfiguration in distribution system for loss reduction and voltage-profile improvement using hybrid algorithm of PSO and ACO. In: 24th International Conference & Exhibition on Electricity Distribution, CIRED, 12–15 June 2017
19. Wazir, A., Arbab, N.: Analysis and optimization of IEEE 33 bus radial distributed system using optimization algorithm. J. Emerg. Trends Appl. Eng. (JETAE), **1**(2) (2016). ISSN 2518-4059. University of Engineering and Technology, Peshawar, Pakistan

PSO Evolution Based on a
Entropy Metric

E. J. Solteiro Pires[1]([✉]), J. A. Tenreiro Machado[2],
and P. B. de Moura Oliveira[1]

[1] INESC TEC – INESC Technology and Science (UTAD pole),
ECT–UTAD Escola de Ciêcias e Tecnologia,
Universidade de Trás-os-Montes e Alto Douro,
5000-811 Vila Rel, Portugal
{epires,oliveira}@utad.pt
[2] Department of Electrical Engineering, ISEP – Institute of Engineering,
Polytechnic of Porto, Rua Dr. António Bernadino de Almeida,
4249-015 Porto, Portugal
jtm@isep.ipp.pt

Abstract. Bioinspired search algorithms are widely used for solving optimization problems. The evolution progress isusially measured by the fitness value of the population fittest element. The search stops when the algorithm reaches a predetermined number of iterations, or when no improvement is achieved after some iterations. Usually, no information, behind the best global objective value, is fed into the algorithm to influence its behavior. In this paper, a entropy metric is proposed to measure the algorithm convergence. Several experiments are carried out using a particle swarm optimization to analyze the metric relevance. Moreover, the proposed metric is used to implement a strategy to prevent premature convergence to suboptimal solutions. The results show that the index is useful for analyzing and improving the algorithm convergence during the evolution.

Keywords: Particle swarm optimization · Entropy convergence

1 Introduction

Particle swarm optimization (PSO) is a search and optimization algorithm inspired on social species behavior. The PSO is a popular metaheuristic that has been successfully applied in a myriad engineering problems [12]. The PSO is inspired in the behavior of birds blocking and fish schooling [7]. Each bird or fish is represented by a particle with two components, namely by its position and velocity. A set of particles forms the swarm that evolves during several iterations giving rise to a powerful optimization method.

The algorithm evolution can be easily measured by monitoring the particle objective values. Indeed, when the performance over time is used, the best

© Springer Nature Switzerland AG 2020
A. M. Madureira et al. (Eds.): HIS 2018, AISC 923, pp. 238–248, 2020.
https://doi.org/10.1007/978-3-030-14347-3_23

objective value of the population is normally adopted. However, an advanced monitoring search process can be implemented using diversity measurements. Shi and Eberhart [18] discussed some diversity metrics for monitoring the PSO search process that can determine when particles are stucked in a minimum. They suggested that those metrics could be adopted to promote diversity mechanisms in the algorithm. Hendtlass [6] introduced random restarts in PSO. The particles can be dispersed by means of the positions that fit better than the average, at some iterations, when the swarm begins converging. It was shown that the random restarts provide promising results for deceptive problems. Zhang et al. [20] proposed a fast restarting PSO to preserve the population diversity and to provide better solutions. The algorithm reinitializes the local's best particle when the particle falls in a local optima. When the swarm stucks in a optima, the algorithm reinitializes the global best particle and all the local best particles. García Nieto and Alba [5] introduced a velocity term and a restarting mechanism for avoiding the early convergence by redirecting particles to promising areas in the search space. The particles are moved with regard to the best position after some iterations without improvement. Tatsumi et al. [19] proposed a PSO with two types of particles within multi-swarms. The algorithm restarts the position and velocity of inactive particles that are identified by their velocity. The results shown some improvements regarding the classic PSO. Pluhacek et al. [11] presented a hybrid PSO, where part of the population is restarted at certain moments during the run, based on the information obtained from the complex network analysis of the swarm.

The Shannon entropy has been applied in several fields, such as in communications, economics, sociology, biology among others [1], but its use in evolutionary computation it has been overlooked. Nonetheless, some examples as follows can be mentioned. Galaviz-Casas [4] studied the entropy reduction during the GA selection at the chromosome level. Masisi et al. [9] used the Renyi and Shannon entropies to measure the structural diversity of classifiers based in neural networks. The index is obtained by evaluating the parameter differences and the GA optimizes the accuracy of 21 classifiers ensemble. Myers and Hancock [10] predicted the GA behavior formulating appropriate parameter values. They suggested the population Shannon entropy for run-time performance measurement, and applied the technique to labeling problems. It was shown that populations with entropy smaller than a given threshold become stagnated and the population diversity disappears. Shapiro and Bennett [15,16] adopted the maximum entropy method to find out equations describing the GA dynamics.

In this paper, the dynamic and self-organization of particles along PSO algorithm iterations is analyzed. A Shannon entropy is proposed in order to promote the evaluation process towards the global optima. The study adopts several optimization functions and compares the performance of a standard and a modified PSO including a restarting process (PSOwR) based on the Shannon entropy.

Bearing these ideas in mind, the remaining of the paper is organized as follows. Sections 2 and 3 describe the standard PSO used in the experiments and some fundamental concepts about entropy, respectively. Section 4 presents the main motivation of using a restarting mechanism to promote algorithm convergence. Section 5 establishes five functions that are used to study the effect of random swarm initialization during PSO time evolution. Section 6 compares the standard PSO and the PSOwR. Finally, Sect. 7 outlines the main conclusions and discusses future work.

2   Particle Swarm Optimization

The PSO, proposed by Kennedy and Eberhart [7], has been successfully adopted to solve many complex optimization engineering applications. This optimization technique is inspired in the way swarms behave and its elements move in a synchronized way, both as a defensive tactic and foraging. An analogy is established between a particle and a swarm element. The particle movement is characterized by two vectors, representing its current position x and velocity v. Since 1995, several techniques emerged in order to improve the original PSO, namely by analyzing the tuning parameters [17] and by considering hybridization with other evolutionary techniques [8].

Algorithm 1 shows the original PSO. The algorithm begins by initializing the swarm randomly in the search space range. As it can be seen in the pseudo-code, were t and $t+1$ represent two consecutive iterations, the position x of each particle is updated during the iterations by adding a new velocity v term. This velocity is evaluated by summing an increment to the previous velocity value. The increment is influenced by two components representing the cognitive and the social knowledge.

The cognitive knowledge of each particle is obtained through the difference between its best position found so far b and the current position x. On the other hand, the social knowledge of each particle is included by taking the difference between the best swarm global position achieved so far g and its current position x. Both knowledge factors are multiplied by random uniformly generated terms ϕ_1 and ϕ_2, respectively.

The PSO is a optimization algorithm that proved to be efficient, robust and simple. However, if no care is taken, then the velocities may attain large values, particularly when particles are far away from local and global bests. Some approaches were carried out in order to eliminate this drawback. Eberhat et al. [3] proposed a clamping function (1) to limit the velocity, through the expression:

$$v_{ij}(t+1) = \begin{cases} v'_{ij}(t+1) & \text{if } |v'_{ij}(t+1)| < V_{\max j} \\ \frac{v'_{ij}(t+1)}{|v'_{ij}(t+1)|} V_{\max j} & \text{if } |v'_{ij}(t+1)| \geq V_{\max j} \end{cases} \tag{1}$$

where $v'_{ij}(t+1)$ is given by $v'_{ij}(t+1) = v_{ij}(t) + \phi_1 \cdot (b-x) + \phi_2 \cdot (g-x)$ for the parameter j of particle i at iteration $t+1$.

Later, a inertia weight was introduced [17] to control the velocity from exploding (2). The inertia weight ω is very important to ensure convergence behavior over evolution by adopting the equation:

$$v_{t+1} = \omega \cdot v_t + \phi_1 \cdot (b-x) + \phi_2 \cdot (g-x) \tag{2}$$

```
1  Initialize Swarm;
2  repeat
3  |   forall the particles do
4  |   |   calculate fitness f;
5  |   |   determine b
6  |   end
7  |   determine g;
8  |   forall the particles do
9  |   |   v_{t+1} = v_t + φ_1 · (b − x_t) + φ_2 · (g − x_t);
10 |   |   x_{t+1} = x_t + v_{t+1};
11 |   |   update b;
12 |   end
13 |   update g;
14 |   t = t + 1;
15 until stopping criteria;
```

Algorithm 1: Particle swarm optimization

Some studies also investigated the problem of how to determine the best inertia value [2] in order to obtain a more efficient PSO behavior.

3 Shannon Entropy

Several concepts and interpretations were used to express entropy, namely *disorder*, *mixing*, *chaos*, *spreading* and *information* [1]. For example, Boltzmann described entropy as the changing from ordered to disordered states. Guggenheim used entropy to describe the energy system diffusion from a small to a large system. In telecommunications, Shannon [14] defined as entropy, the information loss in the transmission of a given message throught the following equation:

$$H(X) = -K \sum_i p(x_i) \log p(x_i), \tag{3}$$

where $K \geq 0$ is a constant, often set to 1, that is used to express H in an unit of measure. Expression (3) considers a discrete random variable $x_i \in X$ characterized by the probability distribution $p(x_i)$.

4 Method Followed

The work is motivated by the need to understand the entropy signal during the PSO time evolution and to use it for improving its convergence. The Shannon entropy is adopted in the sequel considering different optimization functions. Since the PSO is stochastic, a large set of tests is executed to generate a representative statistical sample [13]. In a second phase, the entropy signal is used to influence the algorithm behavior, namely the reinizialization of the swarm (Fig. 1), along the execution for improving its convergence.

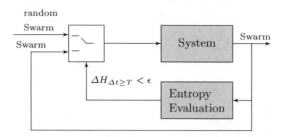

Fig. 1. PSO system with swarm reinitialization.

The index follows one interpretation of entropy. Indeed, entropy can express the "disorder" of a system energy, *i.e.* a measure of how spread out "particles" are within a system (search space). Taking this idea in mind, it is considered the distance d_i between swarm particles and the best global one. Each probability, p_i, is given by the distance of particle i to the best global particle over the maximum possible distance. Therefore, the probability is expressed as:

$$p_i = \frac{d_i}{d_{\max}} \tag{4}$$

where $d_{\max}$ is the maximum possible distance for a particle to the global best particle of the swarm. The Shannon particle diversity index, for a n swarm size, is based in quantification process and can be represented as:

$$H(X) = -\sum_{i=1}^{n} p_i \log p_i, \tag{5}$$

5   Test Functions

This section presents the optimization problems adopted during the PSO tests. The PSO algorithm adopts a real encoding scheme. The problems consists in minimizing 5 well known functions, with n parameters and f^* as the global optimum value (see Table 1), namely: Goldstein-Price's, Griewank, Rastrigin, Rosenbrock and Sphere represented in expressions (6–10), respectively [2], as described in the follow-up.

Goldstein-Price's function:

$$\begin{aligned}
f_1 = &\left(1 + (x_1 + x_2 + 1)^2 (19 - 14x_1 + 3x_1^2 - 14x_2 + 6x_1x_2 + 3x_2^2)\right) \\
&\left(30 + (2x_1 - 3x_2)^2 (18 - 32x_1 + 12x_1^2 + 48x_2 - 36x_1x_2 + 27x_2^2)\right)
\end{aligned} \tag{6}$$

with $n = 2$, $x_i \in [-2, 2]$, $i = \{1, 2\}$ and $\min(f_1) = f_1^*(0, -1) = 3$.
Griewank function:

$$f_2 = \frac{1}{4000} \sum_{i=1}^{n} x_i^2 - \prod_{i=1}^{n} \left[\cos \frac{x_i}{\sqrt{i}} + 1 \right] \tag{7}$$

Rastrigin function:

$$f_3(x) = \sum_{i=1}^{n} (x_i^2 - 10\cos(2\pi x_i) + 10) \tag{8}$$

Rosenbrock function:

$$f_4(x) = \sum_{i=1}^{n} \left[100(x_i + 1 - x_i^2)^2 + (x_i - 1)^2 \right] \tag{9}$$

Sphere function:

$$f_5(x) = \sum_{i=1}^{n} x_i^2 \tag{10}$$

6   Simulations Results

This section presents the PSO and PSOwR results for 5 experiments, each one considering one function described previously using the parameters listed in Table 1. The number of iterations are $N = 1500$ or $N = 15000$ for the {Goldstein-Price, Sphere} and {Griewank, Rastrigin, Rosenbrock} functions, respectively. The experiments consider $n_{pop} = 10$ particles for the Goldstein-Price's function and 50 for the others. For each experiment 101 runs are executed and the arithmetic mean, median, minimum and maximum values are taken to be used as the reference output. The interval range of the search space (see Table 1) varies according to the function used and the maximum distance to the global best particle varies with the range search. This variable, d_{max}, is used to evaluate the entropy.

Table 1. Functions parameters

No.	Function	N	n_{pop}	Parameters (n)	Range $(x_i \in)$	d_{max}	Optima
1	Goldstein-Price	1500	10	2	$[-2, 2]$	$2\sqrt{2}$	$f_1^*(0, -1) = 3$
2	Griewank	15000	50	30	$[-600, 600]$	$600\sqrt{2}$	$f_2^*(0, 0, ..., 0) = 0$
3	Rastrigin	15000	50	30	$[-5.12, 5.12]$	$-5.12\sqrt{2}$	$f_3^*(0, 0, ..., 0) = 0$
4	Rosenbrock	15000	50	30	$[-50, 50]$	$50\sqrt{2}$	$f_4^*(0, 0, ..., 0) = 0$
5	Sphere	1500	50	30	$[-50, 50]$	$50\sqrt{2}$	$f_5^*(0, 0, ..., 0) = 0$

The Goldstein-Price is the first function to be optimized. This problem is easy to solve and the algorithm converges in most cases. In Fig. 2(a) is verified that the median of the runs converges to the optimal value 3 as can be also seen in Table 2. The mean value is slightly higher due to some outlier run. It can be observed that the algorithm begins to converge about iteration 60. In iteration 200 all the particles are close together since the entropy index is too small, as

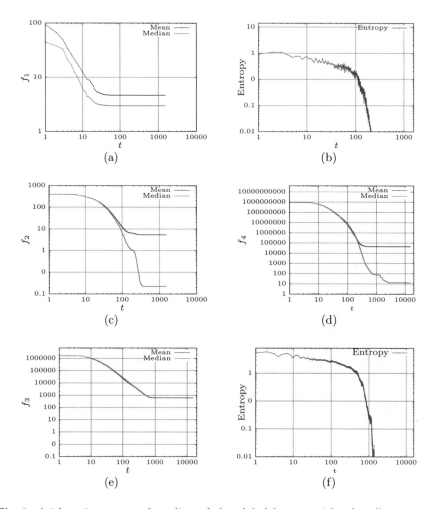

Fig. 2. Arithmetic mean and median of the global best particles for all runs using the functions: (a) Golstein-Price, (c) Griewank, (d) Rosenbrock, (e) Rastrigin. Swarm entropy of one PSO run along iterations for functions: (b) Golstein-Price, (f) Rastrigin.

shown in Fig. 2(b). It should be noted then when all the particles are in the same point, or are identical, the entropy is zero.

The entropy begins with a large value meaning that there is a large diversity among solutions. Then, along generations, the algorithm starts converging and the entropy decreases indicating that swarm is loosing diversity.

For the other functions the algorithm has the same behavior, $i.e.$ after some iterations the algorithm converges and the diversity is lost (see Figs. 2(c)–(e)). Some functions are more complex and need an higher number of iterations to converge, as revealed in Fig. 2(f), for the Rastrigin function.

Table 2. Results of the experiments

Function	Algorithm	Entropy	Minimum	Maximum	Median	Mean	Median restarts
Goldstein-Price	PSO	No	3.000	91.818	3.000	4.6814	
Goldstein-Price	PSOwR	Yes	3.000	3.000	3.000	3.0000	2
Griewank	PSO	No	3.8005e$-$10	90.893	0.022128	5.39380	
Griewank	PSOwR	Yes	2.0564e$-$10	1.4518	0.022141	0.16575	2
Rastrigin	PSO	No	268.77	1142.30	604.73	620.70	
Rastrigin	PSOwR	Yes	1.7494e$-$137	201.58	7.1770e$-$53	7.3180	23
Rosenbrock	PSO	No	1.6733e$-$5	2.5002e+5	12.6360	44745.0	
Rosenbrock	PSOwR	Yes	2.1765e$-$8	2.4897e+5	4.3733	2931.7	24
Sphere	PSO	No	2.6190e$-$9	2500.000	3.5704e$-$6	321.78	
Sphere	PSOwR	Yes	1.4227e$-$9	34.026	6.5980e$-$6	1.2458	2

In order to prevent premature convergence to sub-optimal solutions and, consequently, having the PSO stagnated (*i.e.* with all particles very near to each other) the entropy index is included to restart the swarm. If the entropy variation, $\Delta H = \max_{t=t-T}^{t} |H(t) - H(t-1)|$, is lower than $\epsilon < 0.01$ during $T = 500$ iterations, *i.e.* $\Delta H_{\Delta t \geq T} < \epsilon$, then the swarm is restarted. Therefore, all particles and its local bests are reinitialized, except the global best of the swarm that remains the same.

To analyze the benefits of this approach, all simulation were performed considering identical conditions, but incorporating the restarting mechanism. Figure 3 shows the obtained results. In the Goldstein-Price's function, all runs converged to the global optimum (Table 2). Indeed, all runs converge by introducing only 2 swarm re-initializations. Figure 3(b) shows the evaluation for one run, where the re-initializations took place at generations $t = 636$ and $t = 1162$. When the algorithm converges to a sub-optimum it is useless to continue executing it. Usually, it is better to re-initialize the PSO than wasting time to continue, since the particles can not generate new movements outside the vicinity of the search zone. This behavior is present in all functions: the algorithm reaches best values when the restarting mechanism based on the entropy is used. In particular with the Rastrigin function, represented in Fig. 3(e), the swarm is re-initialized 23 times since the number of iterations is longer, see Fig. 3(f).

Table 2 sums up the results obtained in the experiments. The first two columns identify the optimized function and the algorithm used. Column 3 indicates the algorithm used: standard PSO or PSOwR taking account the Entropy. Columns 4 and 5 indicate the minimum and maximum values found by the 101 runs, respectively. Columns 6 and 7 show the arithmetic and mean values. Finally, column 8 indicates the median restarts value occurred in each PSOwR experiment. It can be observed that, in general, the restarting mechanism improves significantly the PSO.

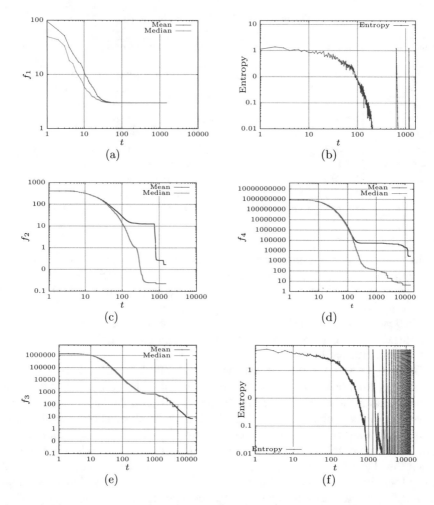

Fig. 3. Arithmetic mean and median of the global best particles for all runs with entropy restarting using the functions: (a) Golstein-Price, (c) Griewank, (d) Rosen-brock, (e) Rastrigin. Swarm entropy of one PSOwR run along iterations for functions: (b) Golstein-Price, (f) Rastrigin.

7 Conclusions

In this paper an entropy based index was proposed to measure the diversity of the swarm. The diversity measurement takes into account the global best position. To observe this mechanism, several function were optimized using the standard PSO. In a second phase, the PSO was run and the swarm was initialized every time the starting mechanism detected that the swarm is stagnated. The proposed index is used to identify stationary stages. Therefore, if the PSO is not able to improve, the mechanism re-initializes the swarm to help it moving from the present local optimum towards more promising areas.

The results show that the entropy index could measure the population diversity. Moreover, when the swarm is stagnated is better to reinitialized the swarm than to continue its execution with the same particles.

References

1. Ben-Naim, A.: Entropy and the Second Law: Interpretation and Misss-Interpretationsss. World Scientific Publishing Company (2012)
2. den Bergh, F.V., Engelbrecht, A.P.: A study of particle swarm optimization particle trajectories. Inf. Sci. **176**(8), 937–971 (2006)
3. Eberhart, R., Simpson, P., Dobbins, R.: Computational Intelligence PC Tools. Academic Press Professional Inc., San Diego (1996)
4. Galaviz-Casas, J.: Selection analysis in genetic algorithms. In: Coelho, H. (ed.) Progress in Artificial Intelligence - IBERAMIA 98. Lecture Notes in Computer Science, vol. 1484, pp. 283–292. Springer, Lisbon (1998)
5. García-Nieto, J., Alba, E.: Restart particle swarm optimization with velocity modulation: a scalability test. Soft Comput. **15**(11), 2221–2232 (2011)
6. Hendtlass, T.: Restarting particle swarm optimisation for deceptive problems. In: 2012 IEEE Congress on Evolutionary Computation, pp. 1–9, June 2012
7. Kennedy, J., Eberhart, R.C.: Particle swarm optimization. In: Proceedings of the 1995 IEEE International Conference on Neural Networks, Perth, Australia, vol. 4, pp. 1942–1948. IEEE Service Center, Piscataway (1995)
8. Løvbjerg, M., Rasmussen, T.K., Krink, T.: Hybrid particle swarm optimiser with breeding and subpopulations. In: Proceedings of the Genetic and Evolutionary Computation Conference (GECCO 2001), pp. 469–476. Morgan Kaufmann, San Francisco, 7–11 July 2001
9. Masisi, L., Nelwamondo, V., Marwala, T.: The use of entropy to measure structural diversity. In: IEEE International Conference on Computational Cybernetics, ICCC 2008, pp. 41–45 (2008)
10. Myers, R., Hancock, E.R.: Genetic algorithms for ambiguous labelling problems. Pattern Recogn. **33**(4), 685–704 (2000)
11. Pluhacek, M., Viktorin, A., Senkerik, R., Kadavy, T., Zelinka, I.: Extended experimental study on PSO with partial population restart based on complex network analysis. Log. J. IGPL, jzy046 (2018)
12. Reyes-Sierra, M., Coello, C.A.C.: Multi-objective particle swarm optimizers: a survey of the state-of-the-art. Int. J. Comput. Intell. Res. (IJCIR) **2**(3), 287–308 (2006)
13. Seneta, E.: A tricentenary history of the law of large numbers. Bernoulli **19**(4), 1088–1121 (2013)
14. Shannon, C.E.: A mathematical theory of communication. Bell Syst. Tech. J. **27**, 379–423, 623–656 (1948)
15. Shapiro, J., Prügel-Bennett, A., Rattray, M.: A statistical mechanical formulation of the dynamics of genetic algorithms. In: Lecture Notes in Computer Science, vol. 865, pp. 17–27. Springer-Verlag (1994)
16. Shapiro, J.L., Prügel-Bennett, A.: Maximum entropy analysis of genetic algorithm operators. In: Lecture Notes in Computer Science, pp. 14–24. Springer-Verlag (1995)
17. Shi, Y., Eberhart, R.: A modified particle swarm optimizer. In: The 1998 IEEE International Conference on Evolutionary Computation Proceedings, IEEE World Congress on Computational Intelligence, pp. 69–73 (1998)

18. Shi, Y., Eberhart, R.: Monitoring of particle swarm optimization. Front. Comput. Sci. China **3**(1), 31–37 (2009)
19. Tatsumi, K., Yukami, T., Tanino, T.: Restarting multi-type particle swarm optimization using an adaptive selection of particle type. In: 2009 IEEE International Conference on Systems, Man and Cybernetics, pp. 923–928, October 2009
20. Zhang, J., Zhu, X., Wang, W., Yao, J.: A fast restarting particle swarm optimizer. In: 2014 IEEE Congress on Evolutionary Computation (CEC), pp. 1351–1358, July 2014

Hybridizing S-Metric Selection and Support Vector Decoder for Constrained Multi-objective Energy Management

Jörg Bremer[(⊠)] and Sebastian Lehnhoff

University of Oldenburg, 26129 Oldenburg, Germany
{joerg.bremer,sebastian.lehnhoff}@uni-oldenburg.de

Abstract. An adequate Orchestration of individually configured and operated distributed energy resources within the smart grid demands for multi-objective optimization with integrated constraint handling. Usually, no closed form description is available for the individual operation constraints of many renewable energy units. Possible operation and restrictions can merely be assessed indirectly by simulation. Thus we hybridized a support vector based constraint modeling technique with a standard multi-objective evolution strategy based on S-metric selection. The proposed method is capable of handling individual, local objectives like individual cost minimization as well as global objectives like minimizing deviations from the joint energy product. A simulation study evaluates the applicability and effectiveness of the proposed approach and the impact of different parametrizations is analyzed.

Keywords: SMS-EMOA · Support vector decoder · Constrained multi-objective optimization · Virtual power plant

1 Introduction

In order to allow for a transition of the current central market and grid control structure of today's electricity supply system towards a decentralized smart grid, an efficient management of numerous distributed energy resources (DER) becomes more and more indispensable. To enable small and individually operated energy devices to responsibly take over control tasks, pooling of different DER is necessary in order to gain enough potential and flexibility. An established concept for such pooling is the virtual power plant (VPP) [14,15]. Orchestration of such groups of energy units is done by different scheduling procedures that frequently involve multi-objective optimization. Here, we go with the example of predictive scheduling [18].

Predictive scheduling [18] describes an optimization problem for day-ahead planning of energy generation in VPPs, where the goal is to select a schedule for each energy unit – from an individual search space of feasible schedules with respect to a future planning horizon – such that as a global objective the

A. M. Madureira et al. (Eds.): HIS 2018, AISC 923, pp. 249–259, 2020.
https://doi.org/10.1007/978-3-030-14347-3_24

distance to a target power profile for the VPP is minimized (e. g. a product from an energy market). This is a multi-objective problem. Further objectives like cost minimization, maximization of residual flexibility (for later planning periods) or environmental impact are to be achieved concurrently [2].

The difficulty of this optimization tasks lies in finding feasible solutions as every participating energy units has its own technical restrictions situationally depending on individual process integration. Usually, no closed form description of the feasible region (or the individual constraints) exists. Feasibility of a solution can be checked with the help of individually parameterized simulation models. For the single objective case, decoder approaches have been developed [3] for handling these individual feasible regions during optimization.

Here, we extend the decoder approach to the multi-objective case by integration in a S-metric based evolution strategy [9].

The rest of the paper is organized as follows. We start with a brief recap on constrained multi-objective optimization and the support vector decoder. We then describe the hybridization with the S-metric selection approach and demonstrate applicability and appropriateness with different simulation studies.

2   Related Work

In optimization problems with more than one and at least two conflicting objectives, Pareto optimization has become an appropriate means for solving [4]. As improving on one objective degrades each conflicting one, multi-objective optimization deals with finding a set of Pareto optimal solutions as trade-off between opposing solutions. Different algorithms have been designed to find an approximation to the Pareto-optimal set $M = \{x \in \mathbb{S} | \nexists x^* \in \mathbb{S} : x^* \prec x$ for a set of objective functions $f_{1,...,n} : \mathbb{S} \to \mathbb{R}$ defined on some search space $\mathbb{S}$ [21]; and with $x \prec x^*$ denoting that x dominates x^*, i. e. all objective values of x are better than x^*. Different algorithms have been proposed [21]; among them are evolutionary algorithms [9,22], genetic algorithms including the famous NSGA-II [5], or swarm-based approaches [8].

In 2005, [9] proposed an evolution strategy based on S-metric selection, the SMS-EMOA. It has been designed for covering maximum hypervolumes with a minimum number of necessary points. SMS-EMOA is based on the hypervolume measure or S-metric [22] that measures the size of the dominated space, i. e. the Lebesgue measure of hypercubes between non-dominated points and a reference point. With SMS-EMOA, no reference point has to be chosen a priori. To achieve this, SMS-EMOA combines ideas from established methods like NSGA-II [5, 12]. The implementation of SMS-EMOA makes it easy to be hybridized with decoders.

For constraint-handling in multi-objective optimization two general concepts are usually applied [17]. Either a penalty [16,19] is added to each objective function degrading constraint violating solutions or the definition of Pareto-dominance is extended to take into account constraint violation [6,17]. Introducing a penalty term changes the objective function and as in multi-objective optimization the impact on different objectives has to be balanced, a too weak set of penalties may lead to infeasible solutions whereas a too strong impact leads to poor distributions of solutions [17].

Nevertheless, all approaches for constraint integration so far need a closed form description of constraints; i. e. the problem is defined as

$$\begin{aligned}
\text{minimize} : \ &f_i(\boldsymbol{x}), \ i = 1, 2, \ldots, n \\
\text{s.t.} : \ &g_k(\boldsymbol{x}) \geq 0, \ k = 1, 2, \ldots, m_g \\
&h_\ell(\boldsymbol{x}) = 0, \ \ell = 1, 2, \ldots, m_h \\
&x_j \in [x_{L_j}, x_{U_j}].
\end{aligned} \tag{1}$$

Constraints are given as a set of (possibly non-linear) in-equalities and equalities as well as a box-constraint demanding all parameters being from a specific range. In the smart grid domain, often no closed form descriptions of constraints are available.

We here consider rather small, distributed electricity producers that are supposed to pool together with likewise distributed electricity consumers and prosumers (like batteries) in order to jointly gain more degrees of freedom in choosing load profiles. In this way, they become a controllable entity with sufficient market power. In order to manage such a pool of DER, the following distributed optimization problem has to be frequently solved: A partition of a demanded aggregate schedule has to be determined in order to fairly distribute the load among all participating DER. A schedule $\boldsymbol{x}$ is a real valued vector with each element x_i denoting the amount of energy generated or consumed during the ith time interval withing the planning horizon Optimality usually refers to local (individual cost) as well as to global (e.g. environmental impact) objectives in addition to the main goal: Resemble the wanted overall load schedule as close as possible.

In [3], a support vector decoder has been introduced to cope with individual constraints of different types of energy units. Constraints may be technically rooted like the state of charge of attached batteries or thermal buffer stores or be economically soft rooted or be due to individual preferences. The basic idea is to start with a set of feasible schedules derived from a simulation model of the energy unit [3]. This sample is used as a stencil for the region (the sub-space in the space of all schedules) that contains the feasible schedules. The sample is used as a training set for a support vector based machine learning approach [20] that derives a geometrical description of the sub-space that contains the given data; in our case: the feasible schedules. Given a set of data samples, the inherent structure of the region (the phase space of the energy unit) is derived as follows: After mapping the data to a high dimensional feature space by means of an appropriate kernel, the smallest enclosing ball in this feature space is determined. When mapping back this ball to data space, it forms a set of contours (not necessarily connected) enclosing the given data sample.

In data space, the set of alternatively feasible schedules of a unit is represented as pre-image of a high-dimensional ball $\mathcal{S}$ in feature space. Figure 1 shows this situation. This representation has some advantageous properties. Although

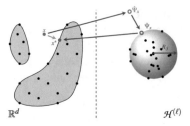

Fig. 1. General support vector model and decoder scheme for constraint handling.

the pre-image might be some arbitrary shaped non-continuous blob in $\mathbb{R}^d$, the high-dimensional representation is still a ball and thus geometrically easier to handle (right hand side of Fig. 1). The relation is as follows: If a schedule is feasible, i.e. can be operated by the unit without violating any constraint, it lies inside the feasible region (grey area on the left hand side in Fig. 1). Thus, the schedule is inside the pre-image (that represents the feasible region) of the ball and thus its image in the high-dimensional representation lies inside the ball. An infeasible schedule (e. g. x in Fig. 1) lies outside the feasible region and thus its image $\hat{\Psi}_x$ lies outside the ball. But, we know some relations: the center of the ball, the distance of the image from the center and the radius of the ball. Hence, we can move the image of an infeasible schedule along the difference vector towards the center until it touches the ball. Finally, we calculate the pre-image of the moved image $\tilde{\Psi}_x$ and get a schedule at the boundary of the feasible region: a repaired schedule x^* that is now operable by the respective energy unit. No mathematical description of the original feasible region or of the constraints is needed to do this. More sophisticated variants of transformation are e. g. given in [3]. Formally, a decoder function γ

$$\gamma : [0,1]^d \to \mathcal{F}_{[0,1]} \subseteq [0,1]^d \tag{2}$$
$$x \mapsto \gamma(x)$$

transforms any given (maybe in-feasible) schedule (scaled to $[0,1]^d$) into a feasible one. Thus, the scheduling problem is transformed into an unconstrained formulation when using a decoder:

$$\delta\left(\sum_{i=1}^{d} s_i \circ \gamma_i(x'_i), \zeta\right) \to \min, \tag{3}$$

where γ_i denotes the decoder of unit i that produces feasible schedules $x' \in [0,1]^d$ and s_i scales these schedules entrywise to correct power values resulting in schedules that are operable by that unit.

Technically, scaling can also be integrated into the decoding process by combining both functions. Thus, for the rest of the paper we refer with γ to a decoder function that maps an infeasible schedule into the feasible region and scales it appropriately to the rated power of the respective energy unit. Please note that this constitutes only a single objective solution and multi-objective scenarios so far have to combine different objectives to a single one by a weighted aggregation what is not possible in case of a mixture of global and local objectives.

3   Hybrid Multi-objective Scheduling Algorithm

We extended this approaches by integrating the decoder into a multi-objective algorithm. For this purpose, we chose an S-metric based approach due to its ability to approximate the Pareto front with a rather equally spread solution set, an important aspect in VPP automation.

Algorithm 1. Basic scheme of the SMS-EMOA used here (cf. [9]).

$P^{(0)} \leftarrow$ randomPopulation()
$t \leftarrow 0$
while $t <$ max iterations **do**
　　$o \leftarrow$ mutate $\circ$ crossover$(P^{(t)})$
　　$\{R_1, \ldots, R_k\} \leftarrow$ fast-nondominated-sort$(P^{(t)} \cup \{o\})$
　　$p \leftarrow \arg\min_{s \in R_k}[\Delta_S(s, R_k))]$
　　$P^{(t+1)} \leftarrow P^{(t)} \{p\}$
　　$t \leftarrow t + 1$
end while

Using S-metric selection for evolutionary multi-objective algorithms (SMS-EMOA) has first been proposed by [9]. The S-metric is based on the hypervolume encapsulated by the set of non-dominated solutions and a reference point [22] and can thus be described as the Lebesgue measure Λ of the union of hypercubes defined by the reference point x_r and the set of non-dominated points m_i [4,9]:

$$S(M) = \Lambda(\bigcup\{a_i | m \in M\}). \tag{4}$$

This metric constitutes an unary quality measure by mapping a solution set to a single value in $\mathbb{R}$: the size of the dominated space [10]. As it is desirable to have a large S-metric value for solution sets in multi-objective optimization, [10] first used this measure in a Simulated Annealing approach and [9] developed an evolution strategy (SMS-EMOA) based on this measure. We hybridized the latter with the decoder approach for flexibility modeling and constraint-handling in multi-objective energy management.

For integration, two types of objective had to be considered. In [11] constraints have been identified on different locality levels. The same holds true for objectives in a VPP. Objectives on a global as well as on a local level have to be integrated. Objectives on a global level have to be achieved jointly. An example is given by the minimization of the deviation of the aggregated joint schedule from a given product schedule that has to be delivered as contracted. These objectives can only be achieved with joint effort. In contrast local objectives like individual cost minimization are also to be integrated. Although evaluation can only be performed locally (individual cost), help from other to be able to choose a cheaper schedule is often necessary to achieve the goal. In the following we denote with f local and with F global objectives.

Algorithm 1 shows the basic idea of SMS-EMOA. The algorithm repeatedly evolves a population of μ solutions. in each iteration, first a new solution is generated and added to the population. Subsequently, a selection process is started to find the worst individual in the solution which is then removed from the population. Thus, the number of individuals stays constant from a steady state perspective. Selection is done by first issuing a fast non-domiated sort after [5]. In this way, the Pareto fronts are ranked and from the front with the lowest rank the individual with the lowest contribution to the hypervolume (measured by the S-metric) is removed. This process is repeated until some stopping criterion – e. g. a number of maximum objective evaluations – is met. As criterion that measures the contribution of a solution candidate to the hypervolume,

$$\Delta_S(s, R_k) = S(R_k) - S(R_k \backslash s) \tag{5}$$

is used (cf. Algorithm 1). $S(R_i)$ denotes the hypervolume of the ith front and the contribution of s is determined by the difference between the Lebesque measure of the set R_i with and without s.

For hybridizing with the decoder approach we need two solution representations. A genotype representation

$$X = \begin{pmatrix} x_1 \\ \vdots \\ x_n \end{pmatrix}, \ x_i \in [0,1]^d \tag{6}$$

is mutated and recombined by crossover. The modified solution has than to be mapped to a feasible solution for selection. The phenotype representation

$$M = \begin{pmatrix} \gamma_1(x_1) \\ \vdots \\ \gamma_n(x_n) \end{pmatrix} \tag{7}$$

contains in each row a schedule that is feasible and thus operable by the respective energy unit. Feasibility is ensured by mapping with the individual decoder γ_i that is individually trained for each energy unit.

For applying the fast non-dominance-sort as introduced in [7] and S-metric selection we extend the definition of dominance by integrating the decoder set:

$$y \prec y^* \equiv \forall i = 1, \ldots, d : y_i < y_i^* \tag{8}$$

by setting

$$y = \begin{pmatrix} f_{1,1}(\gamma_1(\boldsymbol{x}_1)) + \cdots + f_{1,n}(\gamma_n(\boldsymbol{x}_n)) \\ f_{2,1}(\gamma_1(\boldsymbol{x}_1)) + \cdots + f_{2,n}(\gamma_n(\boldsymbol{x}_n)) \\ \vdots \\ f_{m,1}(\gamma_1(\boldsymbol{x}_1)) + \cdots + f_{m,n}(\gamma_n(\boldsymbol{x}_n)) \\ F_1(\boldsymbol{M}) \\ \vdots \\ F_\ell(\boldsymbol{M}) \end{pmatrix} \qquad (9)$$

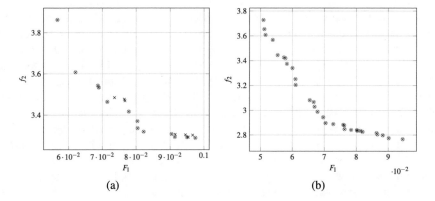

(a)   (b)

Fig. 2. Resulting Pareto front approximations after 5000 and 50000 iterations for a two-objective load scheduling problem (10 CHP) with f_1: error between target and aggregated load profile; and f_2: state of charge deviation. A cross denotes non-dominated solutions.

In this way, variations of the previous solutions in the solution set are produced by applying variation operators to the genotype. In this way the selection and crossover operators merely have to obey an easy to integrate box constraint that ensures that each value of a solution candidate is kept within $[0, 1]$. No further constraints have to be integrated. Thus, the problem formulation can be regarded as constraint-free. Constraint-handling is introduced by using decoder functions that abstract from individual capabilities or technical constraints of energy units. The set of decoders ensures that selection is done on feasible solutions only and thus that the solution set approaches a Pareto front without any knowledge about controlled energy units.

4   Results

For evaluation, we used a model that has already been used in several studies and projects [1,3,11,13]. This model comprises a micro CHP with 4.7 kW of rated electrical power (12.6 kW thermal power) bundled with a thermal buffer store.

Constraints restrict power band, buffer charging, gradients, minimum on and off times, and satisfaction of thermal demand. Thermal demand is determined by simulating losses of a detached house (including hot water drawing) according to given weather profiles. If not otherwise stated, random selection, Gaussian mutation and uniform crossover have been used as standard operators for the SMS-EMOA part.

In a first experiment, we approximated the Pareto front of a scheduling problem with two objectives. The first objective f_1 evaluates the goodness of fit between the desired target load schedule and the aggregated schedule as mean absolute percentage error: $\delta_{MAPE} = \delta(\boldsymbol{x}, \boldsymbol{\zeta}) = \frac{1}{d} \sum_{i=1}^{d} \left| \frac{\zeta_i - x_i}{\zeta_i} \right|$, with d denoting the length of the planning horizon and thus the dimension (number of time intervals) of the schedules. The second objective f_2 evaluates the deviation from a desired individual state of charge of the attached buffer store at the end of the energy delivery period. In this way, we integrated a global (F_1) and an individual objective (f_2). The group of energy units comprises 10 co-generation plants. Figure 2(a) shows the approximated front after 5000 iterations. Almost all solutions are already non-dominated. Figure 2(b) shows the situation after 50000 iterations. All solutions are marked by an ×, all solutions that are non-dominated and thus member of the front are additionally marked by a ○.

Figure 3 shows an example result from a three objective scheduling problem. As a third objective, the minimization of the thermal losses has been integrated. Thermal losses occur by incomplete insulation if a thermal buffer store attached to a co-generation plant is charged too early in advance.

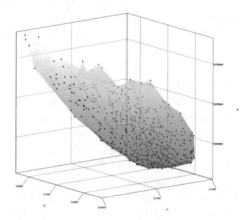

Fig. 3. Resulting Pareto front of a three-objective scheduling problem after 250000 iterations.

Next, we scrutinized the convergence behavior. Figure 4 shows the relationship of the (growing) hypervolume that is covered by a solution set and the number of iterations. We tested different mutation operators. A constant mutation rate is tested against a sigmoidal shrinking one with different evaluation

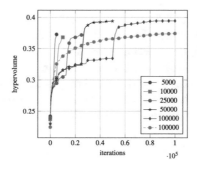

Fig. 4. Convergence behavior as growing hypervolume.

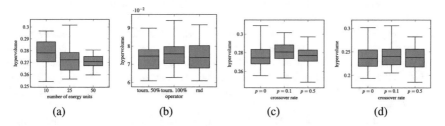

Fig. 5. Sensitivity of the achieved hypervolume to the number of energy units in the VPP (a), selection operator (b), and crossover rate for different objectives (b) and (d).

budgets. Obviously, in the long run, the sigmoidal one yields the better final results, whereas for smaller budgets the constant rate performs better.

Finally, we analyzed the sensitivity of the choice of the selection operator and the crossover rate to the achieved hypervolume with a fixed budget of objective evaluations. We also analysed the impact of different VPP sizes (number of energy units). Figure 5 shows the result. For the VPP size case only the variability can be compared. Obviously, larger VPP lead to more robust results according to the growing flexibility (due to more individual energy units); this has also been observed in single objective cases. For the selection operator, the tournament selection operator succeeds, but also consumes the most computation time. The crossover rate only has an impact for scheduling scenarios with longer (96 dimensions) planning horizons Fig. 5(c). In the 16-dimensional case Fig. 5(d), the differences are marginal.

5 Conclusion

Orchestration of pooled distributed energy resources for joint control demands multi-objective optimization for finding appropriate schedules for each unit. The optimization problems are of a constrained nature, but a closed form description of constraints is usually not available due to the individual operation environment of different energy units. Thus, a support vector decoder technique for surrogate description of individual feasible regions and systematic generation of feasible solutions has been hybridized with an established multi-objective evolution strategy: the S-metric selection EMOA.

A simulation study has shown the effectiveness of the hybrid approach with good approximations of the Pareto front that can be achieved with a reasonable amount of objective evaluations due to the integrated decoder that reduces the search space significantly to the feasible region.

References

1. Bremer, J., Rapp, B., Sonnenschein, M.: Encoding distributed search spaces for virtual power plants. In: IEEE Symposium Series on Computational Intelligence (SSCI 2011), Paris, France, April 2011
2. Bremer, J., Sonnenschein, M.: Automatic reconstruction of performance indicators from support vector based search space models in distributed real power planning scenarios. In: Horbach, M. (ed.) Informatik 2013, 43. Jahrestagung der Gesellschaft für Informatik e.V. (GI), Informatik angepasst an Mensch, Organisation und Umwelt, 16–20 September 2013, Koblenz. LNI, vol. 220, pp. 1441–1454. GI (2013)
3. Bremer, J., Sonnenschein, M.: Constraint-handling for optimization with support vector surrogate models - a novel decoder approach. In: Filipe, J., Fred, A.L.N. (eds.) ICAART 2013 - Proceedings of the 5th International Conference on Agents and Artificial Intelligence, Barcelona, Spain, 15–18 February 2013, vol. 2, pp. 91–100. SciTePress (2013)
4. Coello, C.A.C., Lamont, G.B., Veldhuizen, D.A.V.: Evolutionary Algorithms for Solving Multi-objective Problems (Genetic and Evolutionary Computation). Springer, Heidelberg (2006)
5. Deb, K., Pratap, A., Agarwal, S., Meyarivan, T.: A fast and elitist multiobjective genetic algorithm: NSGA-II. IEEE Trans. Evol. Comput. **6**(2), 182–197 (2002)
6. Deb, K.: Multi-objective Optimization Using Evolutionary Algorithms. Wiley, New York (2001)
7. Deb, K., Agrawal, S., Pratap, A., Meyarivan, T.: A fast elitist non-dominated sorting genetic algorithm for multi-objective optimization: NSGA-II. In: Schoenauer, M., Deb, K., Rudolph, G., Yao, X., Lutton, E., Merelo, J.J., Schwefel, H.P. (eds.) Parallel Problem Solving from Nature PPSN VI, pp. 849–858. Springer, Heidelberg (2000)
8. Durillo, J.J., García-Nieto, J., Nebro, A.J., Coello, C.A., Luna, F., Alba, E.: Multi-objective particle swarm optimizers: an experimental comparison. In: Proceedings of the 5th International Conference on Evolutionary Multi-criterion Optimization, EMO 2009, pp. 495–509. Springer, Heidelberg (2009)
9. Emmerich, M., Beume, N., Naujoks, B.: An EMO algorithm using the hypervolume measure as selection criterion. In: Coello Coello, C.A., Hernández Aguirre, A., Zitzler, E. (eds.) Evolutionary Multi-criterion Optimization, pp. 62–76. Springer, Heidelberg (2005)
10. Fleischer, M.: The measure of pareto optima applications to multi-objective meta-heuristics. In: Fonseca, C.M., Fleming, P.J., Zitzler, E., Thiele, L., Deb, K. (eds.) Evolutionary Multi-criterion Optimization, pp. 519–533. Springer, Heidelberg (2003)

11. Hinrichs, C., Bremer, J., Sonnenschein, M.: Distributed hybrid constraint handling in large scale virtual power plants. In: IEEE PES Conference on Innovative Smart Grid Technologies Europe (ISGT Europe 2013). IEEE Power & Energy Society (2013)

12. Knowles, J., Corne, D., Fleischer, M.: Bounded archiving using the lebesgue measure. In: 2003 Congress on Evolutionary Computation, CEC 2003, vol. 4, pp. 2490–2497. IEEE Computer Society, USA (2003)

13. Neugebauer, J., Kramer, O., Sonnenschein, M.: Classification cascades of overlapping feature ensembles for energy time series data. In: Proceedings of the 3rd International Workshop on Data Analytics for Renewable Energy Integration (DARE 2015). Springer (2015)

14. Nieße, A., Beer, S., Bremer, J., Hinrichs, C., Lünsdorf, O., Sonnenschein, M.: Conjoint dynamic aggrgation and scheduling for dynamic virtual power plants. In: Ganzha, M., Maciaszek, L.A., Paprzycki, M. (eds.) Federated Conference on Computer Science and Information Systems - FedCSIS 2014, Warsaw, Poland, September 2014

15. Nikonowicz, L.B., Milewski, J.: Virtual power plants - general review: structure, application and optimization. J. Power Technol. **92**(3), 135–149 (2012)

16. Smith, A., Coit, D.: Penalty Functions. In: Handbook of Evolutionary Computation, p. Section C5.2. Oxford University Press and IOP Publishing, Department of Industrial Engineering, University of Pittsburgh, USA (1997)

17. Snyman, F., Helbig, M.: Solving constrained multi-objective optimization problems with evolutionary algorithms. In: Tan, Y., Takagi, H., Shi, Y., Niu, B. (eds.) Advances in Swarm Intelligence, pp. 57–66. Springer, Cham (2017)

18. Sonnenschein, M., Lünsdorf, O., Bremer, J., Tröschel, M.: Decentralized control of units in smart grids for the support of renewable energy supply. Environ. Impact Assess. Rev. (2014, in press)

19. Srinivas, N., Deb, K.: Multiobjective optimization using nondominated sorting in genetic algorithms. Evol. Comput. **2**(3), 221–248 (1994)

20. Tax, D.M.J., Duin, R.P.W.: Support vector data description. Mach. Learn. **54**(1), 45–66 (2004)

21. Zhou, A., Qu, B.Y., Li, H., Zhao, S.Z., Suganthan, P.N., Zhang, Q.: Multiobjective evolutionary algorithms: a survey of the state of the art. Swarm Evol. Comput. **1**(1), 32–49 (2011)

22. Zitzler, E., Thiele, L.: Multiobjective optimization using evolutionary algorithms–a comparative case study. In: Eiben, A.E., Bäck, T., Schoenauer, M., Schwefel, H.P. (eds.) Parallel Problem Solving from Nature–PPSN V, pp. 292–301. Springer, Heidelberg (1998)

A Hybrid Recommendation Algorithm to Address the Cold Start Problem

Licínio Castanheira de Carvalho[1]([⊠]), Fátima Rodrigues[1,2], and Pedro Oliveira[1]

[1] Institute of Engineering, Polytechnic of Porto (ISEP/IPP), Porto, Portugal
{1120533, mfc, pmdso}@isep.ipp.pt
[2] Interdisciplinary Studies Research Center (IRSC), Porto, Portugal

Abstract. In this age where there are a lot of available data, recommender systems are a way to filter the useful information. A recommender system's purpose is to recommend relevant items to users, and to do that, it requires information on both, data from users and from items, to better organize and categorize both of them.

There are several types of recommenders each best suited for a specific purpose, and with specific weaknesses. Then there are hybrid recommenders, made by combining one or more types of recommenders in a way that each type suppresses or at least limits the weaknesses of the other types. A very important weakness of recommender systems occurs when the system doesn't have enough information about something and so, it cannot make a recommendation. This problem known as a Cold Start problem is addressed in this study.

There are two types of Cold Start problems: those where the lack of information comes from a user (User Cold Start) and those where it comes from an item (Item Cold Start). This article's main focus is on User Cold Start problems. A novel approach is introduced that combines clients' segmentation with association rules. Although the proposed solution's average precision is similar to other main Cold Start algorithms it is a simpler approach to most of them.

Keywords: Recommender system · Data mining · Hybrid system · Cold start user

1 Contextualization

In the current age of information, with ever-increasing amounts of user-generated content, a problem arises linked to the need to filter all this content. This problem occurs because, although users having access to more information can be beneficial, the process of choosing which information is relevant becomes increasingly more complex as the alternative sources grow, which creates a demand for approaches that allow users to distinguish relevant from irrelevant information in the most efficient possible way.

This demand has prompted the interest in personalized recommender system's because the use of recommender systems is not only limited to making suggestions or helping in the decision process but also to filtering content.

© Springer Nature Switzerland AG 2020
A. M. Madureira et al. (Eds.): HIS 2018, AISC 923, pp. 260–271, 2020.
https://doi.org/10.1007/978-3-030-14347-3_25

Recommender systems are software tools and techniques that give suggestions and their main purpose is giving them to users, so that they can make better decisions. A more in-depth way to characterize them would be as systems that use past opinions of members of a community to assist users in the same community in finding content that may be of their interest, usually from a very large set of alternatives.

A Collaborative Filtering (CF) recommender system focuses on the similarity between users. It puts its emphasis on the assumption that two users have similar interests if they acquired the same item in the past and that they will have similar tastes in the future. In this type of recommendation, filtering items from a large set of alternatives is done collaboratively between the preferences of the different users.

However, this approach has been known to reveal two major types of problems: sparsity and scalability [1]. Data sparsity is a problem that occurs when only a small part of the available items is rated by the users, which leads to the amount of ratings in the rating matrix being insufficient for the system to make accurate predictions [2, 3]. Data scalability comes from situations when a system has millions of users or items, making the use of traditional Collaborative Filtering algorithms impractical. According to Twitter, the solution for these situations requires the scaling of recommendations through the use of clusters of machines [4].

A Content-Based recommender system focuses on the similarity between the items and user profiles when making a recommendation. It analyses a set of items rated by a user, as well as ratings given by that same user, to create a profile to be used in future recommendations of items for users that fit that same profile.

This type of system recommends items that are similar to those that users with the same profile have shown interest for in the past. This means that it needs a measure with which to calculate the similarity between different items. This calculation is done by taking into account the features associated with the compared items and is matched with the user's historical preferences. This recommendation approach does not take into account the user's neighborhood preferences, which means it doesn't require the item preferences of a large user group to increase its recommendation accuracy as it only considers the user's past preferences and item features.

However, this type of system also has limitations. One of them lies in the fact that it causes overspecialized recommendations, which means that it only recommends items very similar to those that the user already knows that exist [1]. Another problem of this type of recommender system lies with its limited content analysis. This is due to the fact that if two distinct items have the same features, they are considered exactly the same to the system. For example, a good article and a poor article would be indistinguishable to the system if they both used the same terms [5].

Knowledge-based recommenders, or constraint-based recommendation systems are used in specific situations, where the users' purchase history is smaller. This type of recommender takes into account item features, user preferences and recommendation criteria when trying to make recommendations. The user preferences are defined by explicitly asking the user in question. And the model accuracy is determined based on how useful the recommended item is to the user.

Hybrid Systems are combinations of various features of distinct recommender systems with the purpose of building a more robust system. The combinations of various recommender systems can be used to remove the disadvantages of one system while using the advantages of another system, thus making the system more robust. The most common combination of recommenders is that of a CF recommender system with other recommender systems, in an attempt to avoid problems as Cold-Start, sparseness or scalability problems [6].

There have also been successful attempts at tackling the problems of CF systems by adding external information to the rating information. However this type of approach, besides the difficulty inherent to the acquisition of external information has limitations such as making the recommender system less flexible or the difficulty inherent to the acquisition of external information [2].

There are several different approaches when building a Hybrid recommender system: Parallelized, which run the different recommenders separately and then combine their results; Pipelined, which run the recommenders in sequence, with the output of one being the input of the next; and Monolithic Hybrid Systems, which integrate the different approaches in the same algorithm (some examples of this are feature combination and feature augmentation).

2   Background

2.1   Cluster Analysis

Clustering, or cluster analysis, is the process that involves the grouping of objects in a way that objects grouped in one cluster are the most similar between themselves and objects in different clusters are as different as possible from each other.

Clustering is an unsupervised learning method, which means that it does not have response variables to predict, and that it instead tries to find patterns within the dataset made available. It identifies a set of clusters, each representing a different category, used to describe the data. Among several clustering algorithms, the most popular in the area of Recommender Systems are K-means and Self-Organizing Maps (SOM) algorithms [1].

K-means Clustering [7] is an iterative clustering algorithm that takes as input k, in which k is the number of clusters to be formed from the data. It then makes partitions of a set of n items into the k clusters. To achieve clustering, the K-means algorithm requires two steps: first, it randomly chooses center points for each of the k clusters, assigning each data point to whichever cluster center it is closer to; second, it moves the centroid (i.e. the center of the cluster in the previous iteration), choosing its new position by calculating the mean position of all data points in its cluster. It then repeats these two steps until all data points are grouped and the mean of the data points of each cluster (i.e. the centroid) does not change.

The K-Medoids algorithm is a clustering algorithm related to the K-Means algorithm and the medoidshift algorithm [8]. Like K-Means, this algorithm is partitional, which means it breaks the input dataset up into groups, or clusters.

This algorithm minimizes the sum of the dissimilarities between the points labelled to be in a cluster and a point designated as the center of that cluster, while also choosing data points as centers, also known as medoids or exemplars.

The K-Medoids algorithm is easy to understand, more robust to noise and outliers when compared to K-Means and has the added benefit of having an observation serve as the exemplar for each cluster, however both run time and memory are quadratic (i.e. $O(n^2)$).

2.2   Association Rule Mining

This data mining technique finds all association rules that have values above a user-specified minimum support and minimum confidence threshold levels.

Given a set of transactions, with each transaction containing a set of items, an association rule has the form "X=>Y" (i.e. X implies Y), where X and Y are two sets of items [9].

The support and confidence thresholds are used to find the most interesting sets of rules. While the support measure refers to the frequency with which the set of items appears in the dataset, the confidence is a measure indicative of how many times the rule was found to be true (i.e. when a transaction has the set of items X, the set of items Y also occurring).

Another very important measure is the lift, which can be used to find out if the sets of items in a rule are dependent of one another. If the lift value equals one (1), then the two sets of items are independent of each other, but if it is lower than one, then they are dependent of each other and the respective rule can be useful.

3   Related Work

Cold Start (CS) problems come from the data sparsity limitations that CF recommender systems have with recommendations for new users or new items. This has led many researchers to try and solve these types of situations.

The papers here presented show some work on those fields and were selected due to having interesting approaches, being recent and/or using the "MovieLens" dataset, the same dataset used in this study, which allows for a better comparison with the proposed solution.

[10] propose an algorithm that predicts a user's preference in a certain context, using the past experiences of his neighbors in that same context. The algorithm creates contextual profiles for the users, based on users' preferences in different contexts and it's based on the k-Nearest Neighbors approach, using the Pearson Correlation Coefficient to predict the user's preferences. The authors used two sets of data for the evaluation of their approach: 4,500 movie metadata from "The Movie Database" crossed with more movie metadata from the "Internet Movie Database"; and contextual user data retrieved from 200 anonymous users from a portuguese university.

The evaluation metrics used by the authors to evaluate the algorithm's performance were the Average Precision (AP) and Mean Average Precision (MAP).

The recommendations were generated 10 times, with $k = 5$, meaning the number of recommendations made were five for each time. The evaluation was made in four different context types (i.e. contexts), and once without a context, with 20 users representing new users, meaning they all had little or no information in their user profile model. The results for every context had an average precision well above the run without a context, and had a variation from 30-70%. The mean average precision was then used for each of the different contexts, having a variation between 0.412 and 0.640, while the scenario without a context had a Mean Average Precision of 0.224.

[11] propose a method to get better recommendations for users without their history information (i.e. new users). The proposed method works in three phases: the first phase consists of the computation of the similarity matrix of the users. The authors use the Vector Cosine-based similarity measure to calculate the user similarities, due to the fact that it can measure the similarity between two users without using rating information. In the second phase, the algorithm uses clustering to divide the users into different groups. The clustering method used is the breadth first searching method of graph theory. And, in the third phase, the average rating of each item of all users in the same group is calculated, and for each group, the top n items with the highest average rating are recommended.

The data used for the evaluation of this method came from the "MovieLens" dataset, from where they extracted the data of 100 users. This data was then randomly partitioned into training and testing sets, with two different ratios: 60% training and 40% testing for the first; and 80% training and 20% testing for the second. The evaluation metrics used by the authors are mean absolute error (MAE), root mean square error (RMSE) and precision. The precision of the authors' method was 0.651.

In this paper [12] propose the use of a similarity measure that reduces the impact of a User Cold Start problem on recommenders. The proposed measure involves the creation of a linear combination for a group of simple similarity measures, with its respective weights being obtained through a neural learning optimization process. The authors decided on using a leave-one-out cross validation when evaluating the recommender's results, due to the low amount of items voted on by Cold Start users, in order to have the highest possible amount of training items.

The data used in the evaluation process came from the "MovieLens 1M" dataset. The evaluation metrics used by the authors in this paper were the mean absolute error (MAE), coverage, precision and recall.

The authors created two experiments, with the first one being divided into two scenarios: one for checking the results of the MAE and coverage for the range of neighborhoods between 100 and 200; and another for checking the values of precision and recall for a range of number of recommendations between 2 and 20.

The precision metric has overall better results for the proposed measure, which led the authors to define the neighborhood size for the second experiment as 700, with the number of recommendations being 10 and the range of votes cast by Cold Start users being between 2 and 20. The authors also found that the recall metric also has overall better results for the proposed metric. In the second experiment the proposed measure performs better overall, having lower MAE values and higher precision and recall.

4 Dataset

The data that is to be used to train and test the solution is from the "MovieLens" website, a website for movie rating and recommendation, and is provided by Group-Lens, that own a repository with various-sized datasets of "MovieLens"[1]. "MovieLens" has several datasets, each with a different number of ratings, movies and users. For the training and testing of the solution, the "MovieLens 100K" was used. As its name refers, "MovieLens 100K" is a dataset with 100,000 ratings. It has 1,682 movies and 943 users (Fig. 1).

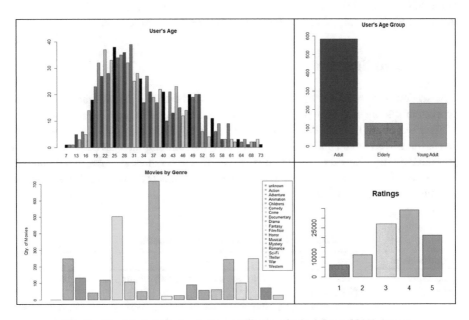

Fig. 1. Graphical analysis of some attributes of MovieLens 100K dataset

The ages with most users are between twenty and thirty. In a way to help in detecting the patterns in the user data, the users' age was converted into age groups: Child, Teenager, Young Adult, Adult and Elderly. According to the gender attribute distribution, 71% of the users are male, while 29% are female. The movie distribution according to the release year, starts in 1922 and ends in 1996, but most of the films date from 1993 onwards. There are nineteen distinct film genres. The genre most common is Drama, followed by Comedy. Regarding users' ratings young adult users seem to be more active than elderly users. They represent 24.81% of all system users but are responsible for 26.54% of all ratings, while elderly users represent 13.26% but they made only 11.52% of ratings. The adult users' values don't differ much, with a value of

---

[1] https://grouplens.org/datasets/movielens/.

61.94%. Similarly to the users' age group, the users' gender doesn't differ much from the users' demographic information, with 26% of ratings having been made by female users and 74% by male users. The user ratings' distribution is in the range 1 to 5, with 1 being the worst evaluation and 5 the best. The two lowest tiers of evaluation have the lowest percentage of ratings: around 6% of the ratings have a value of 1, while around 11% have a value of 2. Ratings with a value of 4 are the most common, with around 34% of the ratings, followed by those with a value of 3, that represent 27% of the ratings. And the highest tier of evaluation (i.e. a rating value of 5) represents around 21% of all ratings.

The attributes Release Year, Rating Year, Years since Release and Novelty were not used in the approaches presented in this chapter but were kept as future work could involve using them in the clustering and recommendation processes.

5   Proposed Solution

This section introduces the final solution as well as the various approaches used in the process of finding a hybrid algorithm capable of dealing with the CS problem.

It is important to reiterate that the algorithm is split according to its two functions, clustering and recommending. This means that, the solution and the other approaches presented in this chapter will be divided in the same manner. The nomenclature identifying the approaches shows the differences between each of them and its meanings are shown in Table 1.

Table 1. Characteristics of the approaches

Nomenclature	Characteristics
Clustering AR-RG	- Uses clustering to filter the user data when creating the association rules (AR) model - The user data attributes chosen for the clustering process include user demographics and the users' average rating per movie genre
Simple AR	- Uses all users' data when creating the association rules (AR) model
Eval. By movie title	- Comparison of the recommendations with the test set uses the movies' titles
Eval. By movie genre	- Comparison of the recommendations with the test set uses the movies' genres

The design of the solution has to take into account that it is a hybrid recommender, which means its purpose is to be a more robust recommender system. It also has to be able to deal with CS users. These challenges lead to the creation of two separate components: one to find and classify the different user types (i.e. a User Categorization); and another one to make the recommendations for users in CS situations (i.e. a Recommendation component).

5.1 User Categorization Component

The User Categorization component takes all the available data relating to the users and applies a Clustering algorithm to it, so that it can find the different groups of users there are. The Clustering algorithm used for this component is the K-medoids algorithm [13] due to its efficacy when dealing with categorical data, unlike K-means. This component uses the users' demographic data, as well as the mean of all the ratings of the user per movie genre, to find and create the optimal number of clusters. The users' demographic data gets a greater weight than the remaining attributes to show their importance in the definition of the users' clusters.

The distance metric used when calculating the dissimilarity matrix was the Gower distance, mainly because the data used has mixed attributes, which made the use of other metrics impractical. The 'NbClust' R package [14] was used for calculating the number of clusters for the model. In order to find the ideal number of clusters it applies thirty different indices and proposes the best clustering scheme from the different results obtained. As can be seen in Fig. 2 the ideal number of clusters was six.

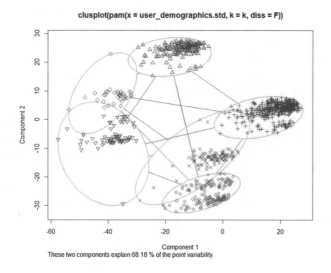

Fig. 2. Clusters of the Proposed Solution with weights, for k = 6

After each user's clusters are found, the pertaining information is saved, so it can be used by the Recommendation component.

5.2 Recommendation Component

When the recommendation component is trying to recommend items to a user in a CS situation, it finds the users most similar to that user and all of their ratings. It then uses an association rule algorithm to find the associations between all the rated items and recommends the genres of the top n movies returned by the algorithm. Using the movies' genres instead of their titles was a way to fight the lack of information inherent to users in a CS situation. The Apriori algorithm [9] is the association rule algorithm used by this component to find new items to recommend.

This component starts by getting the CS user's demographic information as well as the cluster it belongs to. If the solution uses clustering to filter its data, it then checks the CS user's cluster and gets the demographic information of all other users with the same cluster and all the movies rated by those same users. With the list of every movie rated by the users of the same cluster as the CS user in question, the algorithm proceeds to get every rating of those same users to derive the association rules from that data. The value chosen for the confidence threshold was 80% in order to select only the rules more reliable. For the support threshold it is chose one that attempts to get a number of association rules closest to the number of distinct movies in the dataset (i.e. around 1,600).

6   Evaluation and Experiments

The proposed solution's model evaluation uses the leave-one-out validation which was chosen because the solution will be evaluated CS user by CS user and, as each CS user has 19–20 ratings, the training and testing sets will be using the five-fold cross validation method, with one fold used for testing and the remaining four for training.

When evaluating the performance of the proposed solution, the evaluation metrics used to compare it to the other approaches are the accuracy, precision, recall and F1 measure.

Three alternative approaches were created, each with small differences from the proposed solution: the first, "Clustering AR-RG, eval. by Movie Title", uses clustering to filter the users when creating the association rules but its evaluation is made using the movies' titles, rather than their genres; and the two remaining approaches don't use clustering to filter the users when creating the association rules, with the difference between them being the evaluation method – "Simple AR, eval. by Movie Genre" evaluates its performance using the movies' genres, while "Simple AR, eval. by Movie Title" uses the movies' titles.

The following Table 2 shows the results obtained with these four approaches.

Table 2. Evaluation metrics per approach

Method	Accuracy	Precision	Recall	F1 Measure
Proposed solution	0.996	0.459	0.892	0.585
Clustering AR-RG, eval. by movie title	0.992	0.035	0.121	0.052
Simple AR, eval. by movie genre	0.996	0.457	0.837	0.571
Simple AR, eval. by movie title	0.993	0.0171	0.065	0.026

The proposed solution's results are overall better than the remaining approaches, with the exception of "Simple AR, eval. by Movie Genre", that has only slightly lower results than the proposed solution.

The proposed solution was also compared to the papers presented in the Related Work section, with their results being as follows:

When comparing the proposed solution's results with those from [10], the proposed solution shows higher precision values for the worst performance (i.e. 0.34, compared to 0.20), its best and average performances are lower than the approach from [10]. Although the approach's best performance (i.e. 0.68) is only slightly better than the proposed solution's (i.e. 0.61), the difference in the average value is much higher (i.e. 0.46, compared to 0.64).

Using the Welch two sample t-test, the conclusion of this comparison is that there is not enough information to safely assume that one approach is better than the other, even though the approach in [10] has better results. Other metrics such as the recall and the F1 Measure should be used to better compare the two methods. As seen in the previous section, the statistical test results also show that the performance of both approaches is likely equivalent.

The proposed solution's average precision is 46%, which is lower than the approach in [11] that has a precision of 61%. However the authors of [11] don't give much information other than this value for the precision, making it difficult to better compare it with the proposed approach. The recall and the F1 Measure should be used to better compare these methods.

When comparing the proposed approach with the approach by [12] it is important to note that users are considered CS users in the same situation for both approaches (i.e. those with twenty or less ratings available) and that the approach in [12] uses a bigger version of the MovieLens dataset, the "MovieLens 1M". The use of a larger dataset gives a better chance for finding patterns in the data and get overall better results. This study has the same average precision as the proposed solution, but a lower recall. The F1 Measure is slightly higher for [12] which means it is slightly better than the proposed solution.

7   Conclusions

The CS user situation is a serious problem in recommenders because can lead to the system's loss of new users. This increases the importance of recommenders capable of dealing with this problem.

The algorithm documented in this report aims to deal with this situation and its development required the implementation of two components: one for clustering, that assignes users to the groups that best characterized them; and one for association rules, that uses the users in the same cluster as the target for the recommendations, looking for the best movies to recommend.

The clustering component uses the users' demographic data and the mean rating of each user for all movies with a specific movie genre. When creating the clusters a higher weight was given to the demographic attributes due to their importance when characterizing a user.

The association rules component filters the data according to the cluster of the user to which the system is trying to recommend. However, this filtering process increases the runtime of the algorithm, which was found when comparing the algorithm's runtime with the runtime of its variant that does not use clustering.

Another limitation of the algorithm comes from the method that chooses the support threshold and returns the pruned association rules, which is heavy on the processing power and has a considerable runtime.

The algorithm here presented shows that clustering has marginal improvements when used to filter the information for the creation of association rules. The solution shows a marginally higher precision but it has a higher recall value, which means it was able to make more interesting recommendations to its users. It also shows a higher F1 Measure value, which implies a better performance.

Although this algorithm does not have overall better results than the others state of the art methods, it is a simpler approach than most of them and besides it is able to get better results in some of the evaluation metrics such as recall.

Further improvements on the solution could include a method that makes better recommendations when given a set of movie genres (e.g. using movies' popularity or novelty).

References

1. Park, D.H., Kim, H.K., Choi, I.Y., Kim, J.K.: A literature review and classification of recommender systems research. Expert Syst. Appl. **39**(11), 10059–10072 (2012). https://doi.org/10.1016/j.eswa.2012.02.038
2. Shambour, Q., Lu, J.: An effective recommender system by unifying user and item trust information for B2B applications. J. Comput. Syst. Sci. **81**(7), 1110–1126 (2015). https://doi.org/10.1016/j.jcss.2014.12.029
3. Wu, D., Zhang, G., Lu, J.: A fuzzy preference tree-based recommender system for personalized business-to-business e-services. IEEE Trans. Fuzzy Syst. **23**(1), 29–43 (2015). https://doi.org/10.1109/TFUZZ.2014.2315655
4. Gupta, P., Goel, A., Lin, J., Sharma, A., Wang, D., Zadeh, R.B.: WTF: the who-to-follow system at twitter. In: Proceedings of the 22nd International Conference on World Wide Web
5. Adomavicius, G., Tuzhilin, A.: Toward the next generation of recommender systems: a survey of the state-of-the-art and possible extensions. IEEE Trans. Knowl. Data Eng. **17**(6), 734–749 (2005)
6. Lu, J., Wu, D., Mao, M., Wang, W., Zhang, G.: Recommender system application developments: a survey. Decis. Support Syst. **74**, 12–32 (2015). https://doi.org/10.1016/j.dss.2015.03.008
7. Hartigan, J.: A Clustering Algorithms. Wiley, New York (1975)
8. Jain, A.K.: Data clustering: 50 years beyond K-means. Pattern Recogn. Lett. **31**(8), 651–666 (2010)
9. Agrawal, R., Imielinski, T., Swami, A.N.: Mining association rules between sets of items in large databases. In: Proceedings of the 1993 ACM SIGMOD International Conference on Management of Data, pp. 207–216 (1993)
10. Otebolaku, A.M., Andrade, M.T.: Context-aware personalization using neighborhood-based context similarity. Wirel. Pers. Commun. **94**(3), 1595–1618 (2017). https://doi.org/10.1007/s11277-016-3701-2
11. Yanxiang, L., Deke, G., Fei, C., Honghui, C.: User-based clustering with top-n recommendation on cold-start problem (2013). https://doi.org/10.1109/ISDEA.2012.381

12. Bobadilla, J., Ortega, F., Hernando, A., Bernal, J.: A collaborative filtering approach to mitigate the new user cold start problem. Knowl. Based Syst. - KBS **26**, 225–238 (2011). https://doi.org/10.1016/j.knosys.2011.07.021
13. Kaufman, L., Rousseeuw, P.J.: Finding Groups in Data: An Introduction to Cluster Analysis. Wiley, New York (1990)
14. Charrad, M., Ghazzali, N., Boiteau, V., Niknafs, A.: NbClust: an R package for determining the relevant number of clusters in a data set. J. Stat. Softw. **61**, 1–36 (2014)

A Decision-Support System
for Preventive Maintenance in Street
Lighting Networks

Davide Carneiro[1,2(✉)] ⓘ, Diogo Nunes[1], and Cristóvão Sousa[1,3] ⓘ

[1] CIICESI/ESTG, Polytechnic Institute of Porto, Porto, Portugal
{dcarneiro,8140365,cds}@estg.ipp.pt
[2] Algoritmi Center/Department of Informatics, University of Minho, Braga, Portugal
[3] INESC TEC, Porto, Portugal

Abstract. An holistic approach to decision support systems for intelligent public lighting control, must address both energy efficiency and maintenance. Currently, it is possible to remotely control and adjust luminaries behaviour, which poses new challenges at the maintenance level. The luminary efficiency depends on several efficiency factors, either related to the luminaries or the surrounding conditions. Those factors are hard to measure without understanding the luminary operating boundaries in a real context. For this early stage on preventive maintenance design, we propose an approach based on the combination of two models of the network, wherein each is representing a different but complementary perspective on the classifying of the operating conditions of the luminary as normal or abnormal. The results show that, despite the expected and normal differences, both models have a high degree of concordance in their predictions.

Keywords: Preventive maintenance · Public lighting · Distributed Random Forest

1 Introduction

Street lighting is nowadays an intrinsic component of a city's environment and its role go far beyond that of illuminating. It contributes, directly, for traffic and citizens saffety [12]. Additionally, street lighting also influences the perception of safety and well-being, in what regards to pedestrians point-of-view [8,12]. Nonetheless, there is only weak statistically significant evidence that street lighting impacts the level of crime or road casualties [8], independently of the street lighting adaptation strategies used (e.g. switch off, part-night lighting, dimming, white light) [10].

Finally, street lighting also influences the comfort of the citizens. Many factors can be considered when studying lighting performance and comfort, including correlated color temperature, mesopic vision illuminance, dark adaption or

© Springer Nature Switzerland AG 2020
A. M. Madureira et al. (Eds.): HIS 2018, AISC 923, pp. 272–281, 2020.
https://doi.org/10.1007/978-3-030-14347-3_26

color perception [6]. Moreover, luminaires with different color temperatures will have difference performances in these factors. The combination of these factors is so complex that some municipalities are even staging so-called atmospheres (through different light configurations) in certain areas of their cities to foster a sense of community and safety [1].

In the last years, solid state lighting (LED luminaries) with electronic drivers are gradually replacing the conventional High-Pressure Sodium (HPS) lamp-based luminaries. Among other advantages, such as decreased power consumption and increased lifetime, these luminaries (sometimes deemed *intelligent*) allow the acquisition of operational real-time data as well as data characterising the luminaire location and setup [7], pushing the public lightning management into a new paradigm of the internet of things (IoT) [5]. All these factors increase the potential use cases and applications of modern public lighting networks.

Considering the new opportunities offered by the technological shift on street lighting devices, we have been concerned, both on smart energy efficiency and maintenance. However, in this paper we discuss about maintenance of luminaries networks. Unlike traditional luminaries, the LED ones are being exposed to artificial and intended operating variations, in order to decrease their overall consumption, which may influence the pre-established maintenance plans. Tradicional approaches, where maintenance is carried out in-place, with human technicians visiting each luminary either preventively or reactively, are no longer feasible as further explored in Sect. 2. The continuous update of maintenance plans and early warning mechanisms are needed to overcome the new challenges posed by the new street lighting setups.

Accordingly, conducting new preventive maintenance design approaches within intelligent decision support systems for public lighting control, must address the operating characteristics of each luminary and of the environment in which it is located. The result of this novel approach is a model that is sensitive even to slight changes in the operation of each luminary (such as increased power consumption or working temperature), allowing for a more efficient and effective preventive maintenance. Maintenance action plans also become more context-aware and fine-grained, as they can be carried out only on the flagged luminaries. This paper provides some background, and shares the details o how was developed a model about the operation of 305 luminaries. The model is to be used in the preventive maintenance of those luminaries.

2   Background

The operation of an LED luminary is significantly different from other technologies, such as HPS. LEDs do not usually fail when they reach their lifetime. Instead, lumen output gradually decreases over time. This is known as Lumen Depreciation Factor (LDF) and it consists of one the must important light loss factors. As a convention, manufacturers lifetime are typically set by a decrease in lumen output of 30%. Thus, the expected lifetime of LED street lights is usually 10 to 15 years, which is two to four times the lifetime of HPS lights.

For obvious reasons, the failing of an HPS lamp is easily detectable: no light is being emitted. It is, however, hard to predict since HPS lights maintain their luminescence fairly well, with 90% still available halfway through their lifespan. LED lights, on the other hand, gradually loose luminous flux over time. Whereas the risk for a luminary to fail is smaller, it is also more difficult to determine its lifetime. This generally requires to measure a set of indicators affecting LED efficiency. These indicators range from lamp related factors, luminaries related factores and changes on the location of the luminaries [3,9].

When considering illuminated street furniture maintenance, there are two possible approaches: (i) **planned preventive maintenance**, whose objective is to regularly visit assets (e.g. luminaries, control gear, circuit breakers) to undertake routine maintenance activities. Regarding the lamps' lifetime, preventive maintenance advocates the replacement of all lamps after a given period of time, which is the lamp's estimated lifetime. If this lifetime is estimated by default, lamps that are still in good condition will be disposed. If it is estimated in excess, lamps that are not longer working as efficiently as projected will be in use; (ii) **reactive maintenance**, in which assets are only visited when they are known to have failed. Our stance is that reactive maintenance is no longer suited for LED lights since they do not *fail* in the traditional sense. However, we also believe that traditional preventive maintenance schemes can be improved, namely through a continuous acquisition and analysis of the operating behaviour of each luminary. Hence, we propose an automated approach to improve preventive maintenance, by creating a model that considers different aspects of the luminaries' operation as well as their individual characteristics.

3   Data Acquisition

The data used in this study was collected from a network of 305 luminaries manufactured by Arquiled (ARQUICITY R1 model), in a public lighting network located in a Portuguese municipality. Several variables describing the operation of the luminary are gathered at 5-m intervals, including instant voltage (V), luminary temperature (°C), instant power (W), accumulated energy (Wh), uptime (s) or dimming (0%–100%). These data were collected over a period of 4 months, between September 5th 2017 and January 3rd 2018.

Aiming at studying the influence of weather on luminary efficiency, data from a local weather station was also collected. These data include air and dew temperature (°C), humidity (%), wind speed (m/s), wind direction (degrees), wind gust (m/s), pressure (mbar), solar irradiance (W/m^2) and rain (mm/h). Thus, for each instance of data collected from each luminary there is an associated weather data.

An initial cleaning of the dataset was carried out in which instances with missing data in one or more variables were removed. The resulting dataset contains 2.407.672 instances, describing more than 200.000 h of operation of the 305 luminaries and the corresponding weather information.

A visual analysis of the data was then carried out. While many insights were acquired about the operation of the network, we focus on the operation differences that are observed between the luminaries. This is important as the operation conditions of each luminary will influence its lifetime. For instance, the LED chip temperature has an inverse proportion with the LED lifetime. Temporary or prolonged working temperature (for example due to a defective heat sink) may cause a complete failure of the LED, a temporary or permanent decrease in light output (lamp lumen depreciation - LLD) or changes in color temperature [4].

Figure 1 shows how working temperature varies among luminaries (data extracted from 65 luminaries, selected randomly), with some specific luminaries working at significantly higher temperatures. These luminaries are in an increased risk of failure and/or are likely to have a shorter lifetime. Figure 2 also shows differences between luminaries in what concerns power consumption.

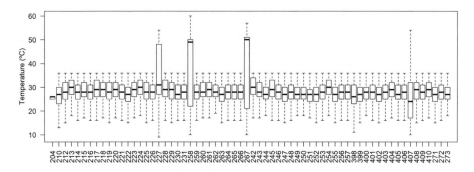

Fig. 1. Distribution of working temperature for 65 of the luminaries (21%, selected randomly) over the 4 months of data collection (outliers removed for clarity).

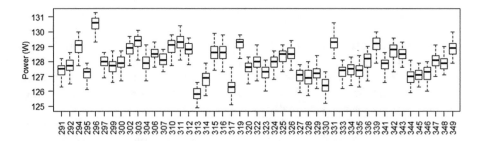

Fig. 2. Distribution of power consumption for 47 of the luminaries (15%, selected randomly) over the 4 months of data collection (outliers removed for clarity).

While, to some extent, these differences may be attributed to the physical properties of each chip or some manufacturing defect, it is not the purpose of this paper to explain these differences or understand how and why they happen.

Rather, we acknowledge the existence of these differences and we propose a method for the autonomous detection of strange operating behaviours, that takes into consideration what is *normal* for all the luminaries as well as for each specific luminary.

4   Data Labelling

On pursuing this paper's goals, there was the need to train a model to classify the luminaries' operation as being *normal* or *abnormal*. Accordingly, a labelled dataset was necessary to be used by a Machine Learning algorithm for inferencing purposes. The dataset, in its original form, was not labelled. That is, luminaries provide no information regarding the nature of their operation.

Domain labelling could be achieved wit the help of Human experts, however, and given the size of the dataset, this approach was not feasible. We did, however, ask a Human expert to formalize the rules according to which such labelling could be carried out. Thus, based on the normative values of the luminary provided by the manufacturer, the expert defined the acceptable boundaries for each variable. The expert also identified moments in which the operation of the lighting network was not normal (such as turning on the luminaries during the day for maintenance purposes). Finally, the first 5 min of data from each luminary were also labelled as abnormal since, according to the expert, there is a small warm-up period in which the values read may vary in abnormal ways.

This labelling resulted in a total of 64.272 instances marked as abnormal (2.67% of the data). While this relatively inexpensive approach solves, at least in part, the problem, we believe that it is not enough. Specifically, it does not consider fluctuations of variables that differ significantly from the normal operation but that are still between the range of acceptable values. For instance, a temperature of 35 °C falls within the acceptable range for this variable. However, it may be seen as abnormal if observed on a luminary that usually operates at 30°. This approach does not capture such subtleties. Hence, it was carried out a second labelling task. Four new variables (labels) were added to each instance of the dataset, three of which describe if the operation of the luminary is considered normal concerning the variables *accumulated energy*, *power* and *temperature*, respectively. The fourth variable is a logical conjunction of the first three, that is, if the three variables are considered normal, the fourth variable has the same value. This fourth variable represents, under this new approach, whether the luminary is operating normally or not.

Each of the three first labels is determined as follows. For each luminary and each variable (accumulated energy, power and temperature), the interquartile range (IQR) is calculated. Then, all the values that lie more than one and a half times the length of the IQR are labelled as not normal, i.e., values below $Q_1 - 1.5 * IQR$ or above $Q_3 + 1.5 * IQR$. While different boundaries could be used, this particular one was suggested by John Tukey (who first proposed boxplots) and is commonly used in these diagrams as a demarcation line for outliers [11].

As a result, 55.839 instances are labelled as abnormal considering *accumulated energy*, 121.831 are labelled as abnormal considering *temperature* and 2.612

are labelled as abnormal considering *power*. Finally, 161.415 instances are globally marked as abnormal (when any of the three labels is abnormal), which represents 6.70% of the data. This is, as expected, slightly higher than the instances labelled by the previous method.

Of all the instances, 55.428 are marked by both approaches as representing abnormal operating conditions (2.3% of the data). It is also interesting to note that all the instances that were labelled as abnormal by the first approach were also labelled as such by the second approach.

5   Model Training

This section describes the training of two models for predicting the normality of the operation of each luminary in the network. The first was trained to predict the label which was assigned using rules (which will be referred as the rule-based model) while the second was trained to predict the label assigned using the IQR (which will be referred as the iqr model).

To train the models, a Distributed Random Forest (DRF) algorithm was used. This is a learning algorithm that generally results in an accurate classifier and that deals efficiently with large datasets [2]. A DRF generates a group of classification trees (forest), rather than a single tree. Each of the trees is trained using a randomly chosen subset of rows and columns. The prediction result is the average prediction of all trees. Thus, and since in the original dataset the numbers 1 (depicting a normal operation) and 0 (depicting an abnormal operation) were used for labelling, both models produce as output a value between 0 and 1.

In the training of each model, all variables mentioned in Sect. 3 were considered, including meteorological ones. While meteorological variables should not prove very relevant in the first model, in which a "blind" labelling of the data was carried out, they may be relevant in the second model. Consider, for example, the positive correlation between air temperature and the luminary's working temperature, deviating it from its normal operational state.

Both models were trained for the same dataset using 5-fold cross-validation, but with different response variables (each predicts one of the two labels generated through rules or through the iqr approach). Fifty trees were trained for each model, each with a maximum depth of 20 levels. Each tree was trained with 60% of the variables, selected randomly. Both models are able to perform very accurate predictions. Table 1 depicts some cross validation metrics for each model and Fig. 3 shows the evolution of the logloss metric during the training of each model. Logloss is a classification loss function that quantifies the accuracy of a classifier by penalizing false classifications.

Figure 4, on the other hand, shows the relative importance of each variable in each model, when predicting the normality of the operation of a luminary. The uptime (which measures the time spent since turning on the luminary) and the temperature of the luminary, are the two most relevant variables in both models, although with different relative importance.

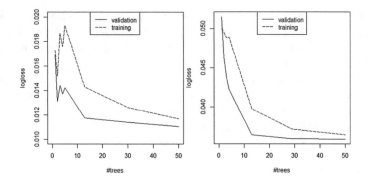

Fig. 3. Evolution of the logloss with the increasing number of trees in both the rule-based (left) and iqr (right) models.

Table 1. Cross validation metrics for each model.

Model	Accuracy	Precision	Recall	AUC
Rule-based	0.9999	1.0	1.0	1
iqr	0.9993	0.9996	0.9996	0.999929

It is interesting to note that, in the iqr model, the top 3 of the most relevant variables includes both the temperature of the luminary as well as the environmental temperature. It is also interesting to note that the device_id (the unique identifier of each luminary) is the fourth most important variable in the iqr model (while it is the second least important variable in the rule-based model).

This reveals an important characteristics about the iqr model: it captures both the influence of the surrounding environment and of the individual characteristics of each luminary. For this reason, we are more inclined to use such a model than the rule-based one, which is based on a blindly labelled dataset that does not take into consideration the context of the data.

6   Results

Section 5 described the training of two models for predicting if the operation of a luminary, in a certain context, can be considered normal. That is, within normal parameters according to the specifications of the manufacturer and within normal parameters according to each luminary's history. The data used for training the first model (called rule-based) was labelled using a set of expert-defined rules, while the data used for the second model (called iqr) was labelled using a statistical approach (4).

In this Section we analyse these two models with the goal to assess their differences and their concordance and, eventually, support the selection of one or the other for use in the creation of a Decision-Support System for preventive maintenance. To this end, 199.402 instances of data were used. These instances

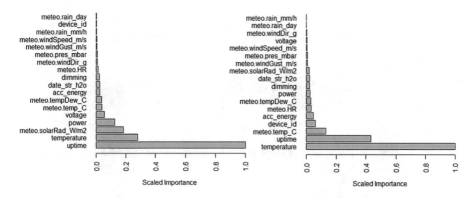

Fig. 4. Scaled relative importance of each variable in both the rule-based (left) and iqr (right) models.

result form the cleaning (as described in Sect. 3) of a set of 200.000 instances extracted from the initial dataset and not used in the training phase.

Each model predicted on each of these instances of data, each producing a value between 0 and 1 for each instance. A threshold of 0.5 was used to distinguish between *normal* and *abnormal*.

The rule-based model classified 193.924 of the instances as *normal* (97.25%) while the iqr model did so for 186.827 (93.69%). This could point out either that there are instances that are neglected by the rule-based model or that the iqr model has a high false-positive rate. We are inclined to the first hypothesis since the rule-based model only considers normative values, whereas the iqr model embodies additional aspects (such as what is the *normal* operation of each luminary).

In the rule-based model, 99.88% of the predictions are what we call high-confidence predictions, that is, predictions in which the output is higher than 0.85 (high confidence on a normal operation) or lower than 0.15 (high confidence on an abnormal operation). This value is of 99.60% for the iqr model: a slightly lower value (which may be attributed to the fuzzier nature of the labelled data) but very high nonetheless.

It is also important to analyse what we call the concordance of the models, that is, the extent to which both models agree in their predictions. According to the results, there are 190.857 instances of data (95.71%) in which both models agreed (either on a normal or abnormal operation). This indicates a very high concordance of the models, which increases our confidence on the use of both.

Figure 5 depicts how the predictions of each model relate to each other. Specifically, it shows that when both models agree, their confidence on the prediction is also higher as values are significantly less scattered in the agreement quadrants, especially when both predictions are positive.

The correlation between both predictions could also be used to quantify this concordance. When considering all data, the correlation is of 0.58 which is not particularly high, due to the influence of the instances in which there is no agreement. When considering only instances in which there is agreement, correlation increases to 0.99 and decreases to −0.95 in the cases of disagreement.

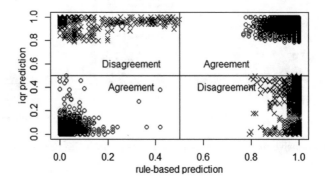

Fig. 5. Concordance of the predictions of both models: the quadrants of agreement contain 95.71% of the instances.

The DSS for preventive maintenance uses both models. Indeed, and as argued in Sect. 7, we believe that both provide a different but valid and relevant view on the problem. Thus, the predictive service that results from this work was developed as a weighted sum of the results of both models. The same threshold of 0.5 is applied to the result of this weighted sum, to decide between *normal* and *abnormal*. Moreover, the weight is defined by the client of the service in the moment of its use. Thus, the client can determine how much weight should be attributed to each model, obtaining in result a classification that is more rigid (based on rules and normative values) or more context-aware and based on each luminary's characteristics.

7   Discussion and Conclusions

In the previous sections we detailed the methodology followed in the development of two models for classifying the operation of a luminary as normal or abnormal. We critically analysed both models. The iqr model is especially interesting: it captures the influence of the surrounding environment as well as the individual characteristics of each luminary. We tend to consider iqr more suitable for the context of the problem, and for that reason, should have a higher weight. However both models have, in our opinion, similar scientific relevance, mainly deu to two reasons: (i) first, the rule-based model classifies nearly more 4% of the instances as *normal* than the iqr model. We argue that there are instances of *abnormal* operation that are ignored by the first model and that are detected by the second, especially those instances that constitute small but statistically significant variations from the regular operation of a specific luminary. However, both views/approaches have their merits; (ii) second, let us consider a case in which a luminary is consistently operating in an abnormal manner, since the first moment. For such a luminary, what is generally considered *abnormal* would be considered *normal*. Hence the need to also consider the rule-based approach.

Finally, we would like to point out two indicators that support our confidence on the resulting model. First, the results of the training, namely the very high AUC, show that this type of data are relatively easy to distinguish. Second, in both models more than 99% of the predictions have values above 0.85 or below 0.15, which indicates high confidence. All these results support the claim that this may be a valid approach to characterize the operation of a public lighting network for the purpose of continuous preventive maintenance.

Acknowledgments. This work is co-funded by Fundos Europeus Estruturais e de Investimento (FEEI) through Programa Operacional Regional Norte, in the scope of project NORTE-01-0145-FEDER-023577.

References

1. Bille, M.: Lighting up cosy atmospheres in denmark. Emot. Space Soc. **15**, 56–63 (2015)
2. Breiman, L.: Random forests. Mach. Learn. **45**(1), 5–32 (2001)
3. Budzyński, L., Zajkowski, M.: Automatic measurement system for long term LED parameters, Wilga, Poland, p. 96620L, September 2015
4. Cheng, H.H., Huang, D.S., Lin, M.T.: Heat dissipation design and analysis of high power led array using the finite element method. Microelectron. Reliab. **52**(5), 905–911 (2012)
5. Chiang, M., Zhang, T.: Fog and IoT: an overview of research opportunities. IEEE Internet Things J. **3**(6), 854–864 (2016)
6. Jin, H., Jin, S., Chen, L., Cen, S., Yuan, K.: Research on the lighting performance of led street lights with different color temperatures. IEEE Photonics J. **7**(6), 1–9 (2015)
7. Ouerhani, N., Pazos, N., Aeberli, M., Muller, M.: IoT-based dynamic street light control for smart cities use cases. In: 2016 International Symposium on Networks, Computers and Communications (ISNCC), pp. 1–5. IEEE (2016)
8. Peña-García, A., Hurtado, A., Aguilar-Luzón, M.: Impact of public lighting on pedestrians' perception of safety and well-being. Saf. Sci. **78**, 142–148 (2015)
9. Royer, M.: Lumen maintenance and light loss factors: consequences of current design practices for LEDs. LEUKOS **10**(2), 77–86 (2014). http://www.tandfonline.com/doi/abs/10.1080/15502724.2013.855613
10. Steinbach, R., Perkins, C., Tompson, L., Johnson, S., Armstrong, B., Green, J., Grundy, C., Wilkinson, P., Edwards, P.: The effect of reduced street lighting on road casualties and crime in england and wales: controlled interrupted time series analysis. J. Epidemiol. Commun. Health **69**(11), 1118–1124 (2015)
11. Tukey, J.W.: Exploratory Data Analysis, vol. 2. Pearson, Reading (1977)
12. Yao, X., Guo, J., Ren, C., Wang, X.: The influence of urban road lighting on pedestrian safety. Int. J. Eng. Innov. Res. **7**(2), 136–138 (2018)

Improving the Research Strategy in the Problem of Intervention Planning by the Use of Symmetries

Mounir Ketata$^{(\boxtimes)}$, Zied Loukil, and Faiez Gargouri

MIRACL, University of Sfax, Sfax, Tunisia
mounir.isims@gmail.com, zied.loukil@gmail.com,
faiez.gargouri@gmail.com

Abstract. The problem of interventions planning allows assigning technicians to geographically dispersed clients to serve them for the resolution of one or more incidents. This problem can be a relatively complex task, and is usually done manually in most existing IT management software since it must be ensured that each intervention is assigned to the most competent technician and many constraints must be respected such as availability and skill for each technician and customer priorities.

There is the problem of planning interventions in the maintenance management processes or after-sales services, especially when there is a need to respect quality standards, for example ITIL standard which is a repository of good practice and which improves the quality of information systems and the quality of assistance.

Several works have been developed in this context, for example by trying to turn the problem of intervention planning into a vehicle routing problem (VRP), or to transform it into a CSP. Also, filtering rules have been proposed in the literature to optimize the resolution time. However, even with these proposed rules, the resolution time remains very important.

In this paper, we propose to improve the CSP and COP models for the problem of intervention planning and we propose other filtering rules with the use of techniques based on the elimination of symmetries to detect equivalent solutions to optimize the resolution time.

Keywords: ITIL · Intervention planning · CSP · COP · Symmetry techniques

1 Introduction

The management of interventions nearby customers is an important or even decisive factor in the quality of the after-sales services of IT companies.

Several standards exist to regulate this area, such as the ISO/IEC 20000 standards [1] for IT services, based on the ITIL repository [2]. Several software programs have been developed to automate the processes related to this standard, however, among these software, it is rare to find one who does the planning of the interventions because of the great complexity of this problem.

© Springer Nature Switzerland AG 2020
A. M. Madureira et al. (Eds.): HIS 2018, AISC 923, pp. 282–293, 2020.
https://doi.org/10.1007/978-3-030-14347-3_27

Some research has tried to solve the problem of planning interventions by declarative methods by transforming it into a vehicle routing problem (VRP), a constraint satisfaction problem (CSP) or a constraints optimization problem (COP) [3] to resolve them using constraint solvers. In addition, in [4], the authors proposed filtering rules that allowed reducing significantly the search time with eliminating from domains of variables all values that cannot lead to solutions. However, for COPs, we note that despite this contribution, the search time for an optimal solution is still too long.

In this paper we propose an improved modeling of the problem of planning interventions in CSP and COP form and we also propose other filtering rules based on symmetries to improve the search time of solutions, especially in COPs.

2   Problem of Intervention Planning

The problem of intervention planning involves assigning technicians to geographically dispersed customers to provide them with the services they need (repair, maintenance, installation, etc.). In addition, the intervention must be done in a period of time already planned.

The intervener assigned for an intervention must have the skills required by this intervention and must be available to perform the intervention from start to finish. The availability of the intervener must also coincide with that of the client. Some data must also be taken into account when solving the problem of interventions planning such as:

- The time required to go from one client to another
- The time interval representing the availability time whether for the customer or the intervener
- The priority of each client.

Now, we will Model the problem of intervention planning into a CSP. We start with the definition of a CSP.

A CSP (Constraint Satisfaction Problem) is defined as a triplet (X, D, C), with X the set of variables, D the domain of values for each variable and C the set of constraints imposed on these variables.

Solving a CSP involves finding an assignment for all variables in their respective domains that respects all constraints of the problem [5, 6].

The CSP is an NP-Complete problem, and thanks to this property, there are NP-Difficult problem-solving techniques, including the problem of interventions planning, in which the problem to be solved is transformed into an equivalent CSP and we then use the constraint solvers for the resolution.

The modeling of the problem of interventions planning in the form of CSP begins with the specification of a set of constants that represents the data of the problem, then the set of variables and their respective domains and finally the constraints that link these variables.

Inputs of Problem
We start by defining the constants that represent the initial data of this problem and which we need to present our contribution in this paper and the other proposed constants are definined in [4]:

- NbClt: The number of clients.
- Availab(clt$_i$): the availability of a client clt$_i$ $\forall i \in \{1, \ldots, \text{NbClt}\}$, which will be represented by an interval (of time) or the union of several intervals,
- Needed_Skills(clt$_i$): the skills required for an intervention to serve a client clt$_i$ $\forall i \in \{1, \ldots, \text{NbClt}\}$.
- NbInterv: The total number of technicians in the problem.
- Availab(interv$_i$): the availability of a technician interv$_i$ $\forall i \in \{1, \ldots, \text{NbInterv}\}$
- Earned_Skills(interv$_i$): the skills of a technician interv$_i$ $\forall i \in \{1, \ldots, \text{NbInterv}\}$.
- Max_dur: The maximum duration that a mission may have for any technician.

Variables and Their Domains
Solving a problem of planning interventions is assigning to each customer a technician who starts at a precise time. For this reason, the variables of the problem are:

- The technician assigned to each client,
- The time when the technician arrives at the targeted client to start the intervention.

  Formally:

$$X = \{(\text{Aff}_i, S_i)\} \text{ such as } i \in \{1, \ldots, \text{NbClt}\} \tag{1}$$

  With:

- Aff$_i$ is the technician assigned to the client clt$_i$ so, $\forall i \in \{1, \ldots, \text{NbClt}\}$, $D(\text{Aff}_i) = \{1, \ldots, \text{NbIntervnants}\}$
- S$_i$ is the start time of the intervention for a client clt$_i$. So $\forall i \in \{1, \ldots, \text{NbClt}\}$, $S_i = [0, \text{Max\_dur}]$.

Constraints
The first constraint concerns the skills of the affected technician which must include those necessary for the intervention. The second constraint indicates that the total duration of the mission of any technician interv$_i$ can't exceed the maximum duration set for missions. Finally, the third constraint indicates that for every client clt$_j$, the intervention must be carried out by the affected technician Aff$_i$ in the period of time when both of them are available. All these constraints are detailed in [4].

Search for Optimal Solution
When we solve a CSP, we obtain the first solution that respects all the problems' constrains. However, in the intervention planning problem, it would be better to define quality criteria ad to find the best solution respecting them. For this reason, it would be interesting to transform the CSP to a COP by adding an objective function which is detailed in [4].

3   Filtering Rules for the Problem of Interventions Planning

3.1   Filtering Rules in CSPs

In the resolution of a CSP, the assignment of values to variables sometimes does not provide solutions. For this reason, it is better to reduce the values domains of these variables by eliminating values that can't lead to solutions.

For the problem of intervention planning, several filtering rules have been proposed in [4] to reduce the search space. Despite these filtering rules, we note that the resolution time remains very important, especially for COPs where we do not stop at the first solution found but rather we seek the optimal solution.

For this reason, we propose in this paper other filtering rules that allow verifying the existence of symmetrical solutions, which mean assignments of value for variables that can lead to the same solution. They are eliminated and thus reduce the search time of the optimal solution.

3.2   Use of Symmetries in CSPs and COPs

Definition
Either a CSP (X,D,C), a permutation δ between two variables x_i and x_j of X that are in the same domain denoted $\delta (x_i) = x_j$ is a relation that simultaneously change in all the constraints of C all the occurrences of x_i in x_j and all those of x_j in x_i. We obtain a new CSP $(X, D, C') = \delta (X, D, C)$
A permutation δ is symmetry for the CSP (X, D, C) if and only if $\delta (X, D, C) = (X, D, C)$. In other words, the set of constraints does not change by permuting x_i and x_j.

Use of Symmetries
Constraint satisfaction problems often have symmetries [7] which are bijections transforming assignments into other equivalent assignments in the sense of consistency. The symmetries have the effect of increasing the search time of the solutions. On other hand, there are symmetry elimination's techniques developed for constraint programming that aim to improve the efficiency of the resolution methods by reducing the search time for solutions.

- **As filtering rules for CSPs:**

  We can use symmetries [8] as filtering rules thanks to this property: if we reduce the domain of a variable X_i by eliminating certain values, they must be eliminated from the domain of the variable X_j symmetrical to X_i.

  Indeed, it is easily demonstrable that since these values cannot be assigned to X_i, they cannot be assigned to X_j.

- **To detect equivalent solutions:**

The symmetries in a problem form a group and allow the definition of a set of equivalence classes between the symmetric assignments. The elimination of symmetries [9] consists in limiting the exploration of the search space to a single representative for each equivalence class. Once a solution is found, the symmetries can be applied to quickly derive its symmetrical solutions. In addition, the areas of the search space can be ignored if they are symmetric or dead-end already explored.

Symmetries can be used to search for equivalents solutions [10, 11]: Sol_1 represent a solution for a CSP, then for any pair of variables (X_i, X_j) such that X_i and X_j are symmetrical, if the assignments of X_i and X_j are exchanged, a solution Sol_2 of the CSP equivalent to the first Sol_1 will be obtained.

Application of Symmetries in the Problem of Intervention Planning

Interveners who will serve geographically dispersed customers may have similar skills. For this reason, interveners with the same skills are grouped into groups of symmetries.

$$\forall i, j \in \{1, \ldots, \text{NbInterv}\}, \text{Earned\_Skills}(\text{interv}_i) \triangle \text{Earned\_Skills}(\text{interv}_j) = \{\emptyset\}$$
$$\Rightarrow \{\text{interv}_i, \text{interv}_j\} \subseteq E \tag{2}$$

- E = Symmetry group that includes interveners who have the same skills.

To check the validity of the concept of symmetries for the problem of interventions planning, we first have to check that the variables are all in the same value domains. For that reason we have:

Aff_i represents the technician who will serve the client i so:

- $\forall i \in \{1, \ldots, \text{NbClt}\}, D(Aff_i) = \{1, \ldots, \text{NbIntervnants}\}$

Aff_j represents the technician who will serve the client j so:

- $\forall j \in \{1, \ldots, \text{NbClt}\}, D(Aff_j) = \{1, \ldots, \text{NbIntervnants}\}$

So $D(Aff_i) = D(Aff_j)$
S_i it is the time of beginning intervention at the Client i therefore:

- $\forall i \in \{1, \ldots, \text{NbClt}\}, S_i = [0, \text{Max\_dur}]$.

S_j it is the time of beginning intervention at the Client j therefore:

- $\forall j \in \{1, \ldots, \text{NbClt}\}, S_j = [0, \text{Max\_dur}]$.

So, $D(S_i) = D(S_j)$
So, all variables are in the same value domains.

After the verification of the variables, it is necessary to prove that the symmetry is valid for all the constraints.

The first stakeholder skills constraint requires that the affected intervener Aff_i must have the necessary skills to serve the client i. If we switch the assigned intervener with another who is symmetrical and has the same skills, then the constraint remains valid. For that reason we have:

$$\text{NeededSkills} \subseteq \text{EarnedSkills}(\text{Aff}_i) \Leftrightarrow \text{NeededSkills} \subseteq \text{EarnedSkills}(\text{interv}_j)$$
$$\Leftrightarrow \text{NeededSkills} \subseteq \text{EarnedSkills}(\text{interv}_k) \quad (3)$$

With:

$$\text{EarnedSkills}(\text{Aff}_i) \, \Delta \, \text{EarnedSkills}(\text{interv}_j)$$
$$= \{\emptyset\} \wedge \text{EarnedSkills}(\text{interv}_j) \, \Delta \, \text{EarnedSkills}(\text{interv}_k) = \{\emptyset\} \quad (4)$$

- Earned_Skills(Aff_j): the skills of an intervener affected to client i
- Earned_Skills(interv_j): the skills of an intervener interv_i $\forall j \in \{1,\dots,\text{NbInterv}\}$.
- Earned_Skills(interv_k): the skills of an intervener interv_k $\forall k \in \{1,\dots,\text{NbInterv}\}$.

The second constraint indicates that the total duration of the mission of an intervener cannot exceed the maximum duration defined for the missions; this constraint remains valid with another intervener symmetrical to the first one. If we switch both of the two intervener, the constraint must be satisfied. For this reason we have:

$$\forall i \in \{1,\dots,\text{NbInterv}\}, \text{Tot\_dur}(\text{interv }_i) < \text{Max\_dur} \Leftrightarrow$$
$$\forall j \in \{1,\dots,\text{NbInterv}\}, \text{Tot\_dur}(\text{interv }_j) < \text{Max\_dur} \quad (5)$$

- interv_i and interv_j are two symmetrical intervener with the same skills.

The last constraint requires that for each client, the intervention must be performed by the intervener at the same time in which both of them (client and intervener) are available. This constraint remains valid if the intervener affected is exchanged by another who is symmetrical and has the same skills.

Illustrative Example
In this part we present an illustrative example that illustrates how we can find symmetrical solutions in the problem of intervention planning.

For that reason, we consider in our example 3 clients who can request a hardware or software intervention and 3 interveners with different skills.

Needed_Skills(CLT1) = Software,   Needed_Skills(CLT2)   and   Needed_Skills (CLT3) = Hardware

Earned_Skills(TECH1) = Software,   Earned_Skills(TECH2)   and   Earned_Skills (TECH3) = Hardware

Two solutions are obtained:

- First solution: CLT1 served by TECH1, CLT2 served by TECH2 and CLT3 served by TECH3
- Second solution: CLT1 served by TECH1, CLT2 served by TECH3 and CLT3 served by TECH2.

We can consider that the two solutions are symmetrical since the CLT2 and CLT3 can be served by either TECH1 or TECH2 since both of them have the same skills.

The symmetries have the effect of increasing exponentially the resolution time of constraint satisfaction problems. The filtering rules that have been proposed in [4] applied to the proposed CSP and COP model for problem of intervention planning can significantly reduce the resolution time but this time can be improved by eliminating the symmetrical solutions. In the next part of the paper we detail the techniques for eliminating symmetric solutions.

4   Techniques for Eliminating Symmetric Solutions for the Problem of Intervention Planning

In the context of constraint programming, the resolution of problem can leads to symmetric multiple case explorations that significantly affects the resolution time. This is also valid for the problem of intervention planning that is modeled as a CSP and COP.

The most important application of symmetry in constraint programming and especially in our problem of intervention planning is the "Breaking symmetries" [5] in order to reduce the time needed to find solutions. The goal of symmetry breaking is to never explore two research states that are symmetrical, because we know that the result in both cases must be the same. There are three approaches for breaking symmetries in constraint programming:

- The first approach is to reformulate the problem so that it has a reduced amount of symmetry.
- The second is to add symmetry breaking constraints before start the search,
- The final approach is to dynamically eliminate symmetry during the search, by adapting an appropriate search procedure.

In the problem of intervention planning we note that intervener who will serve dispersed client may have similar skills levels which can result during the resolution of problem many symmetrical solutions which mean that if we switch two intervener with the same skills so the customer can be served by either the first or the other one.

If we have:

$$\text{Aff}_i = \text{interv}_j \land \forall j, k \in \{1, \ldots, \text{NbInterv}\}, \text{Earned\_skills}(\text{interv}_j)$$
$$= \text{Earned\_skills}(\text{interv}_k) \Rightarrow \text{Aff}_i = \text{interv}_k \tag{6}$$

Our contribution is then to go through the list of interveners to extract those who have the same skills. During the exploration of the search tree we eliminate from the list of affected interveners to serve client i every intervener who is symmetrical with another intervener who is already affected to serve the same client. Two interveners are symmetrical if and only if the first has the same skills with the second.

Client(interv$_j$) represents the customer list assigned to the intervener i.

$$
\begin{aligned}
\text{si}(\text{Aff}_i = \text{interv}_j \;\wedge\; \forall j, k \\
\in \{1, \ldots, \text{NbInterv}\}, \text{Earned\_skills}(\text{interv}_j) \;\triangle\; \text{Earned\_skills}(\text{interv}_k)) = \{\emptyset\}) \\
\Rightarrow i \not\subset \text{Client}(\text{interv}_k)
\end{aligned}
$$

$$(7)$$

5 Implementation in a Constraint Solver

The CSP model of the intervention planning problem that has been improved in the second part of this paper has been implemented in a "choco" constraint solver. In this context, our problem is modeled declaratively with specifying the variables and constraints to satisfy.

To achieve the approach proposed in this paper, which consists of using the symmetric assignment elimination process, we have followed the following approach:

From the first solution obtained during the resolution, the algorithm starts the search for a strictly better solution in which the cost of the objective function must be strictly lower than the cost of the objective function of the first solution.

FctObj1, FctObj2: Objective functions [4].

For that reason we have added a constraint which concretizes the improvement of the second objective function compared to the first.

$$\text{Cost}(\text{FctObj2}) < \text{Cost}(\text{FctObj1})$$

Then we must proceed to the elimination of the symmetrical solutions which have the same costs. The final solution obtained after filtering operations will be saved.

To concretize our approach, we propose an algorithm. For that we need a set of variables. A solver is an object to solve a CSP problem. This solver must be alimented with a model object in which it's declared the variables of the problem and their respective domains and the constraints that must be respected at each assignment of values for the variables

We need two solver objects S1 and S2 and a model object M:

- S1: it is a solver object with the model of the intervention planning problem proposed in this paper.
- S2: it's a solver object initially empty but during the exploration of the search tree the constraints that concretize the cost improvement of the objective function will be added to the model object M which will be assigned to the solver object S2.
- M: It's a model object that models the problem.

Then we must proceed to the elimination of symmetries which mean the solutions with the same costs. As long as checking of feasibility for the current solution is satisfied, this procedure is repeated and the next solution is passed.

- C: it's a constraint object.
- ElTech: It is a table of size NbInterv * NbInterv which represents the symmetry between the interveners.
- addConstraint(): allows to add constraint to the model object.
- lt: operator less then.
- getVar(var): allows to get variable named var.
- getValue(): allows to get value of a variable.
- removeVal(value): allows to remove a value from the domain of a variable.
- read(model): allows reading the model.
- solve(): allows solving problem.
- isfeasible: test if the current solutions is feasible or not.

ALGORITHM: Elimination of symmetries

```
VARIABLES
S1: Solver;S2: Solver; M: Model;
FctObj1: Variable;FctObj2: Variable;
BEGIN
S2← ∅
if(S1.isFeasible()){
do{
//----- Add the objective function improvement constraint
M.addConstraint(lt(FctObj2, FctObj1));
//----- Eliminate symmetries from the domain of M -------
    for(int i=1;i< NbClt;i++)
      for(int j=1;j< NbClt;j++)
        for(int k = 0; k < NbInterv-1; k++)
        {for(int l = 0; l < NbInterv-1; l++){
//---If the technicians have the same skills
if (S1.getVar(Intervenant[i]).getVal()==k &&
S1.getVar(Intervenant[j]).getVal()==l && ElTech[k][l]==1)
          { Intervenant[i].removeVal(l); }}}
    S2.read(M);S2.minimize(true);S2.solve();
    if(S2.isFeasible()){S1=S2;}
    S1.nextSolution();}
  while (S2.isFeasible());}
```

6   Experimental Results

6.1   Experimentation Field

We will try to plan interventions for single and several days. For that reason, we have fixed for all instances of the problem 10 IT skills and 5 technicians with different skills.

Our approach which consists on eliminating symmetries works the first solution is found, for that reason this approach is not efficient for CSPs when the aim is to find one valid solution. So, we implemented it for the COP model of the problem of intervention planning.

We will test the improved COP model, without elimination of symmetries and then apply these techniques to the same instance to compare the results. We used Choco library [12] for the test of all instances.

6.2   Single Day vs Multiple Day Planning

We started by planning interventions for a single day and then for several days. We test our approach of elimination symmetries from problem of intervention planning for single and multiple days planning with and without eliminations of symmetries.

COP with vs Without Filtering Rules vs Symmetries Elimination

For the COPs, the goal it's to find solution which optimizes the objective function mentioned above while respecting all the other constraints. Thanks to the proposed filtering rules which consist on the elimination of symmetries, we can see in Figs. 1 and 2 that it is possible to plan interventions for more clients. We can see also that elimination of symmetries is useful for improvement of research time needed to the resolution of problem.

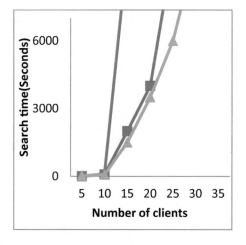

Fig. 1. COP solving time for a single day

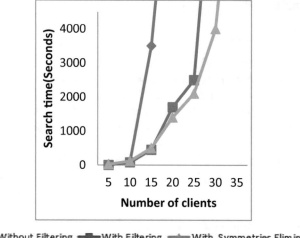

Fig. 2. COP solving time for multiple days

Results Analysis

For COPs, after testing instances without any filtering rules, we conclude that interventions planning can be scheduled for up to 15 clients. With the filtering rules proposed in [4] in addition with which have proposed in this paper we conclude that intervention planning can be scheduled for 30 clients and more.

With single day planning interventions we can see in Fig. 1, without elimination symmetries, it is hard to plan interventions for 15 clients or more, but thanks to filtering rules, it is possible to solve problems with 20 and 25 clients. For multiple days planning we can see in Fig. 2 that it is possible to plan interventions for more than 30 clients. We can notice that the resolution of problem by using the techniques of symmetries elimination takes more time at the beginning of the resolution compared to that using the techniques of filtering proposed in [4], this it is due to a set of pretreatment which consists of grouping technicians with the same skills into groups and the elimination of symmetrical assignments, but after a certain time there is a considerable reduction of the resolution time.

From these results, we can also conclude that in the majority of COP cases, filtering rules which consist on the elimination of symmetries can be useful. They can reduce up to 20% of the search time consumed by the model with the filtering rules proposed in [4].

7   Conclusion

In this paper, we have introduced a new model for the intervention planning problem in the form of CSP, COP. We have also proposed new filtering rules which are based on the detection of symmetrical solutions. Our approach has automatically detected a large number of symmetries of variables. To eliminate the symmetries, an algorithm has been proposed in which constraints have been introduced to eliminate them.

The practical interest of our approach has been shown on many series of instances. Indeed, the number of resolved instances has been significantly improved, demonstrating the robustness and effectiveness of this approach. To summarize, our results confirm that eliminating symmetries is a significant improvement for the resolution times required by the CSP solver "CHOCO" used to solve the intervention planning problem represented as a COP.

References

1. Quesnel, J.: Comprendre ITIL V3 Normes et meilleurs pratiques pour évolue r vers ISO 20000. Editions ENI (2010)
2. TSO: ITIL® Service Operation. ISBN 9780113313075 (2011)
3. Ghrab, I., Ketata, M., Loukil, Z., Gargouri, F.: Using constraint programming techniques to improve incident management process in ITIL. In: proceedings of International Conference on Artificial Intelligence and Pattern Recognition (AIPR 2016) (2016)
4. Ketata, M., Loukil, Z., Gargouri, F.: Filtering techniques for the intervention planning problems. In: International Conference on Natural Computation, Fuzzy Systems and Knowledge Discovery (ICNC-FSKD 2017) (2017)
5. Tsang, E.: Foundations Constraint Satisfaction. Academic Press, Cambridge (1993)
6. Apt, K.: Principles of constraint programming. Cambridge University Press, Cambridge (2003). ISBN 0-521-82583-0
7. Benhamou, B.: Study of symmetry in constraint satisfaction problems. In: PPCP-94, Orcas Island, Washington, pp. 246–254, May 1994
8. Benhamou, B., Saïdi, M.R.: Elimination des symétries locales durant la résolution dans les CSPs
9. Allignol, C.: Suppression des symétries en programmation par contraintes, Toulouse, Septembre 2006
10. Kelsey, T., Linton, S., Roney-Dougal, C.: New developments in symmetry breaking in search using computational group theory
11. Lecoutre, C., Tabary, S.: Des symétries locales de variables aux symétries globales. Inria-00291567, June 2008
12. Fages, J.-G., Lorca, X., Prud'homme, C.: Choco solver user guide (2016)

Towards a Hybrid System for the Identification of Arabic and Latin Scripts in Printed and Handwritten Natures

Karim Baati[1,2]([✉]) and Slim Kanoun[3]

[1] REGIM-Lab.: REsearch Groups on Intelligent Machines,
National Engineering School of Sfax (ENIS), University of Sfax,
BP 1173, 3038 Sfax, Tunisia
karim.baati@enis.tn
[2] HESTIM School of Engineering and Management, Casablanca, Morocco
[3] MIRACL Laboratory, University of Sfax, Sfax, Tunisia
slim.kanoun@gmail.com

Abstract. This paper deals with the problem of script identification. Our aim is to use a global approach as a first step towards a hybrid one in order to resolve the problem of differentiation between the Arabic and the Latin scripts in printed and handwritten natures. In our proposed system, features are extracted using three tools namely, Gabor filters, Gray-Level Co-occurrence Matrices (GLCMs) and wavelets. The highest correct identification rate is obtained with GLCMs (84.75%). This result could be assessed as satisfactory when considering the simplicity of the approach and the complexity of the treated problem.

Keywords: Script identification · Global approach · Gabor filters · Gray-level co-occurrence matrices · Wavelets

1 Introduction

Nowadays, documents generated in whole the world and especially in international administrative environments encompass different scripts with both printed and handwritten natures. Consequently, current research in the field of document image analysis aims at conceiving and implementing systems which can automatically differentiate a set of scripts in order to select the convenient recognition system for each textual entity.

Three main approaches can be considered to conceive these systems. Indeed, we distinguish the systems based on a global approach, the systems based on a local approach and the systems relying on a hybrid approach.

Systems based on a global approach view the text block as being only one entity and hence do not invoke other analysis from text lines, words or connected components. To do that, these systems use features based on textural analysis and they assume that the text block to be identified is normalized (height and

© Springer Nature Switzerland AG 2020
A. M. Madureira et al. (Eds.): HIS 2018, AISC 923, pp. 294–301, 2020.
https://doi.org/10.1007/978-3-030-14347-3_28

width are equals) and uniformed (same spaces between the lines and the words) [1–3].

Systems relying on a local approach analyze the document image at the level of a list of connected components (like lines, words and characters) and hence require segmentation of the document as a preprocessing step [4,5].

Systems based on a hybrid approach look for establishing a script differentiation strategy that rests on all information available in the three principal levels of a textual entity script to identify, namely the text block, the text line or the word and the connected component. Obviously, this strategy combines the global and the local approaches [6,7].

Compared to the local approach which requires a finer preliminary segmentation, the global strategy is often considered as simpler and it is often deemed as a necessary step before calling a hybrid approach. However, the global approach was never used to resolve a script identification problem which combines the Arabic script and the handwritten nature. In this work, our goal consists in using this approach with the aim to resolve the problem of differentiation between the Arabic and the Latin scripts in printed and handwritten natures. The ensuing difficulty of this problem chiefly stems from the cursive nature of the Arabic and the handwritten Latin.

The rest of this paper is organized as follows. Section 2 starts with a survey of systems based on the global approach. Then, Sect. 3 provides details of our proposed system which calls the global approach by investigating three tools for feature extraction, namely, Gabor filters, Gray-Level Co-occurrence Matrices (GLCMs) and wavelets. Section 4 is devoted to show and discuss the experimental results obtained on a dataset formed of 800 homogeneous text blocks. Finally, Sect. 5 presents some concluding remarks and suggests some directions for future work.

2   A Survey of Systems Based on the Global Approach

The survey of existing systems based on the global approach reveals that very few work concerned the Arabic script [8] and the handwritten nature [9]. Moreover, no existing work has combined the identification of the Arabic script and the handwritten nature in the same identification problem.

On the other hand, we note that the correct identification rates obtained with systems based on the global approach and dealing with a differentiation problem, are very good to excellent. These rates are shown in Table 1.

3   Proposed System

As in most systems using the global approach [1,3], our system is based on three main steps: preprocessing, feature extraction and classification.

Table 1. Correct identification rates of different systems based on the global approach

Reference	Scripts and nature	Rate
[1]	- Chinese, Korean, English, Greek, Russian, Persian and Malayalam - Printed	96.7%
[2]	- Oriental and Euro-American - Printed	93.31%
[3]	- Latin, Chinese, Japanese, Greek, Cyrillic Hebrew, Farsi and Sanskrit - Printed	99%
[11]	- Roman, Devanagari, Bangla and Telugu - Handwritten	91.6%
[12]	- 7 Indian scripts and English - Printed	100%
[13]	- Chinese, Korean, Japanese and English - Printed	98.76%
[10]	- 10 Indian scripts - Printed	97.11%

3.1   Preprocessing

Preprocessing is a necessary step in the global approach [1,3]. In our system, this step consists in extracting from each document of the data set, a normalized and uniformed text block. To create such block, we chose a size of 512×512 pixels. It is a size which guarantees that a sufficient amount of information is available for the texture feature extraction. To form the content of each block, we start by extracting the first line of the document. Afterward, words of this line are put in the normalized block with the same spacing and without exceeding the predefined size (words or connected components which are outside, are removed). Lastly, the line obtained in the normalized block is duplicated throughout the block by maintaining the same space between lines.

An example of the preprocessing of a printed Arabic document is shown in Fig. 1.

Fig. 1. Preprocessing of a printed Arabic document

3.2   Feature Extraction

Features are extracted through three texture tools: Gabor filters, GLCMs and wavelets. In the following, we present how features are extracted from each tool and what are the numerical values selected for different parameters.

– **Gabor filters**

The energy of the output of a bank of Gabor filters is often considered as a good feature to identify the script of a document image [13]. In a two-dimensional context, the Gabor filter is described by a pair of real filters defined by:

$$h_e = g(x, y)cos(2\Pi f(xcos\theta + ysin\theta))$$
$$h_o = g(x, y)sin(2\Pi f(xcos\theta + ysin\theta)) \tag{1}$$

where x and y are the spatial coordinates, f and θ respectively the frequency and the orientation of the Gabor filter. The function $g(x, y)$ is the two-dimensional Gaussian function defined by:

$$g(x, y) = exp[-\frac{1}{2}(\frac{x^2}{\sigma_x{}^2} + \frac{y^2}{\sigma_y{}^2})] \tag{2}$$

where σ_x and σ_y are the frequencies of the Gaussian envelope along the principal axes, typically with $\sigma_x = \sigma_y$.

In our system, we fixed $\sigma_x = \sigma_y = 1.5$ and we used 3 frequency values ($f = 8$, 16 and 32). For each frequency, filtering is performed at 8 orientations values spaced equidistantly between 0 and $\frac{\Pi}{4}$. This results in 24 output text blocks (8 from each frequency) from which the texture features are extracted as follows: the mean and the standard deviation of each output text block are calculated which yields 48 features per input text block.

– **Gray-level co-occurrence matrices**

GLCMs are used to represent the pairwise joint statistics of the pixels of an image. For a gray-scale image quantized to R discrete levels, such matrices contain R x R elements. When the document is binary, the GLCM extracted is of size 2×2. If the GLCM is moreover diagonal, three features can be extracted in this case. By varying the values of d and θ which are respectively the linear distance in pixels and the angle between them, features are multiplied [3]. In our system, 5 distances ($d = 1, 2, 3, 4$ and 5) and 4 angles ($\theta = 0$, 45, 90 and 135) are chosen. This leads to a total of 60 features per input test block.

– **Wavelets**

A discrete, two dimensional form of the wavelet transform can be defined by:

$$A_j = [H_x * [H_y * A_{j-1}]_{\downarrow 2,1}]_{\downarrow 1,2}$$
$$D_{j1} = [G_x * [H_y * A_{j-1}]_{\downarrow 2,1}]_{\downarrow 1,2}$$
$$D_{j2} = [H_x * [G_y * A_{j-1}]_{\downarrow 2,1}]_{\downarrow 1,2}$$
$$D_{j3} = [G_x * [H_y * A_{j-1}]_{\downarrow 2,1}]_{\downarrow 1,2} \tag{3}$$

where A_j and D_{jk} are the approximation and detail coefficients at each resolution level j, H and G are the low and high pass filters, respectively, and $\downarrow x, y$ represents downsampling along each axis by the given factors.

The energies of each detail band of this transform can be calculated by:

$$E_{jk} = \frac{\sum_{m=1}^{M} \sum_{n=1}^{N} D_{jk}(m,n)}{MN} \qquad (4)$$

These energies can be extracted as features giving a total of $3J$ features where J is the total number of decomposition levels used in the transform [3]. In our system, we chose $J = 12$ which leads to a total of 36 features per input text block.

3.3   Classification

The k-nearest neighbors is often considered as a simple and effective classifier. For this reason, it has been selected to make decision within our system.

On the other hand, the choice of the value of k mainly depends on the treated application. In our case, we found that the best results are obtained with $k = 5$.

4   Experiments and Results

In order to assess our suggested system, we constituted a data set of 800 homogeneous text documents involving 200 of each class: printed Latin, printed Arabic, handwritten Arabic and handwritten Latin. This data set is then subdivided in two data sets: a first for the training and a second for the test including each one 100 documents of each class.

Tables 2, 3 and 4 exhibit the correct identification rates obtained by our system using features extracted respectively from Gabor filters, GLCMs and wavelets. In these tables, the abbreviations CI, C, PA, PL, HA and HL respectively represent Correct Identification rate, Confusion rate, Printed Arabic, Printed Latin, Handwritten Arabic and Handwritten Latin.

The analysis of the results presented in Tables 2, 3 and 4 reveals that the identification rate in the case of the global approach depends on the tool used for features extraction. For our system, the highest correct identification rate is obtained with GLCMs (84.75%). This rate could be considered as satisfactory if we take into account both the simplicity of the method (which does not require any preliminary segmentation) and the complexity of the treated problem (mainly coming from the cursive nature of the Arabic and the handwritten Latin).

Table 2. Results of script identification based on features extracted by Gabor filters

Script and nature	% CI	% C	Confusion matrix			
			PA	PL	HA	HL
PA	69%	31%	69	1	16	14
PL	60%	40%	18	60	0	22
HA	91%	9%	0	0	91	9
HL	81%	19%	1	0	18	81
Mean	75.25%	24.75%				

Table 3. Results of script identification based on features extracted by GLCMs

Script and nature	% CI	% C	Confusion matrix			
			PA	PL	HA	HL
PA	86%	14%	86	4	8	2
PL	95%	5%	3	95	1	1
HA	83%	17%	0	7	83	10
HL	75%	25%	0	10	15	75
Mean	**84.75%**	15.75 %				

Table 4. Results of script identification based on features extracted by wavelets

Script and nature	% CI	% C	Confusion matrix			
			PA	PL	HA	HL
PA	71%	29%	71	1	17	11
PL	80%	20%	14	80	4	2
HA	88%	12%	3	1	88	8
HL	90%	10%	0	0	10	90
Mean	82.25%	17.75%				

5   Conclusion and Prospects

In this paper, we aimed at establishing a first system of differentiation between the Arabic and the Latin scripts in printed and handwritten natures based on the global approach. A highest rate of correct identification of 84.75% is obtained with GLCMs.

Even if this outcome could be considered as acceptable, it remains insufficient when compared with rates obtained with other systems based on the same approach [3,12]. Thus, some improvements should be considered for the suggested system.

On the one hand, possible improvements could reinforce the current approach by using a larger data set or either recalling other texture tools such as Log Gabor filters [14].

On the other hand, enhancements could affect the used approach by adding some local features to the global ones. Obviously, among major challenges in such a hybrid strategy [15,16] is how to perform an efficient decision-making [17,18] from the available heterogeneous attributes [19].

References

1. Tan, T.N.: Rotation invariant texture features and their use in automatic script identification. IEEE Trans. Pattern Anal. Mach. Intell. **20**(7), 751–756 (1998)
2. Tao Y., Tang Y.Y.: Discrimination of Oriental and Euramerican scripts using fractal feature. In: International Conference on Document Analysis and Recognition (ICDAR), pp. 1115–1119 (2001)
3. Busch, A., Boles, W.W., Sridharan, S.: Texture for script identification. IEEE Trans. Pattern Anal. Mach. Intell. **27**, 1720–1732 (2005)
4. Elgammal, A.M., Ismail, M.A.: Techniques for language identification for hybrid Arabic-English document images. In: International Conference on Document Analysis and Recognition (ICDAR), pp. 1100–1104 (2001)
5. Fan, K., Wang, L., Tu, Y.: Classification of machine-printed and handwritten texts using character block layout variance. Int. J. Pattern Recognit. **31**(9), 1275–1284 (1998)
6. Desai, A.A.: Support vector machine for identification of handwritten Gujarati alphabets using hybrid feature space. CSI Trans. ICT **2**(4), 235–241 (2015)
7. Saeed, K., Tabedzki, M.: A new hybrid system for recognition of handwritten-script. Int. J. Comput. **3**(1), 50–57 (2014)
8. Gaddour, H., Kanoun, S., Vincent, N.: A new method for Arabic text detection in natural scene image based on the color homogeneity. In: International Conference on Image and Signal Processing, pp. 127–136 (2016)
9. Jemni, S.K., Kessentini, Y., Kanoun, S., Ogier, J.M.: Offline Arabic handwriting recognition using BLSTMs combination. In: IAPR International Workshop on Document Analysis Systems (DAS), pp. 31–36 (2018)
10. Joshi G.D., Garg S., Sivaswam J.: Script identification from indian documents. In: IAPR International Workshop on Document Analysis Systems (DAS), pp. 255–267 (2006)
11. Singhal V., Navin N., Ghosh D.: Script-based classification of hand-written text documents in a multilingual environment. In: Workshop on Parallel and Distributed Simulation, pp. 47–54 (2003)
12. Manthalkar, R., Biswas, P.K.: An automatic script identification scheme for Indian languages. IEEE Trans. Pattern Anal. Mach. Intell. **19**(2), 160–164 (1997)
13. Pan W.M., Suen C.Y., Bui T.D.: Script identification using steerable gabor filters. In: International Conference on Document Analysis and Recognition (ICDAR), pp. 883–887 (2005)
14. Hajian, A., Ramli, D.A.: Sharpness enhancement of finger-vein image based on modified un-sharp mask with log-gabor filter. Proc. Comput. Sci. **126**, 431–440 (2018)

15. Baati, K., Hamdani, T.M., Alimi, A.M.: Hybrid Naïve possibilistic classifier for heart disease detection from heterogeneous medical data. In: International Conference on Hybrid Intelligent Systems, pp. 235–240 (2013)
16. Baati, K., Hamdani, T.M., Alimi, A.M.: A modified hybrid Naïve possibilistic classifier for heart disease detection from heterogeneous medical data. In: International Conference on Soft Computing and Pattern Recognition, pp. 353–35 (2014)
17. Baati, K., Hamdani, T.M., Alimi, A.M., Abraham, A.: A new classifier for categorical data based on a possibilistic estimation and a novel generalized minimum-based algorithm. J. Intell. Fuzzy Syst. **33**(3), 1723–1731 (2017)
18. Baati, K., Hamdani, T.M., Alimi, A.M.: Diagnosis of lymphatic diseases using a Naïve Bayes style possibilistic classifier. In: IEEE International Conference on Systems, Man and Cybernetics (SMC), pp. 4539–4542 (2013)
19. Baati, K., Hamdani, T.M., Alimi, A.M., Abraham, A.: A new possibilistic classifier for heart disease detection from heterogeneous medical data. Int. J. Comput. Sci. Inf. Secur. **14**(7), 443–450 (2016)

Social Media Chatbot System - Beekeeping Case Study

Zine Eddine Latioui, Lamine Bougueroua$^{(\boxtimes)}$, and Alain Moretto

AliansTic, Efrei Paris, Villejuif, France
{zine.eddine.latioui, Lamine.bougueroua,
Alain.moretto}@efrei.fr

Abstract. The aim of this paper is to present an innovative way for assisting beekeepers during the process of taking care of their apiary based on text mining and deep learning. To reach this goal, we propose an innovative social media *Chatbot* called *ApiSoft*. This system is able to extract relevant information by processing data from different sources like social media, web, data provided by expert and our applications embedded on the beekeepers' smartphone. Once data are collected, *ApiSoft* can send alerts, information and pieces of advice about the state of apiaries to all subscribers according to their specific interests. We believe that this approach will not only lead to a better monitoring of production but will also allow an enhanced monitoring of the sector at regional and national level.

Keywords: Text mining · Deep learning · Social media data · Chatbot

1 Introduction

Social Networks are indisputably popular nowadays and show no sign of slowdown. According to the Kepios study [1], the number of active users of social networks increased by 13% in 2017 to reach 3.3 billion users in April 2018. For example, Facebook attracts more than 2.2 billion users a month. Penetrating ever more aspects of our daily life, they become not only a considerable threat for our privacy, but also an encompassing tool for analyzing opinions, habits, trends and some would even say – thoughts.

In the current growth of artificial intelligence, machine learning and natural language processing, driven by new technological possibilities, it is possible to automate the analysis of vast amounts of publicly published data.

Text Mining and Social Network Analysis have become a necessity for analyzing not only information but also the connections across them. The main objective is to identify the necessary information as efficiently as possible, finding the relationships between available information by applying algorithmic, statistical, and data management methods on the knowledge. The automation of sentiment detection on these social networks has gained attention for various purposes [2–4].

Twitter is a social network that allows the user to freely publish short messages, called Tweets via the Internet, instant messaging or SMS. These messages are limited to 140 characters (more exactly, NFC normalized codepoints [5]). With about 330

A. M. Madureira et al. (Eds.): HIS 2018, AISC 923, pp. 302–310, 2020.
https://doi.org/10.1007/978-3-030-14347-3_29

million monthly active users (as of 2018, Twitter Inc.), Twitter is a leading social network, which is known for its ease of use for mobile devices (90% of users access the social network via mobile device). Note also, the more mature age of Twitter users: the most represented age group is 35–49 years old. Twitter known by the diversity of content, as well as its comprehensive list of APIs offered to developers.

With an average of 500 million messages sent per day, the platform seems ideal for live tracking opinions on various subjects. Furthermore, the very short format messages facilitate classification since short messages rarely discuss more than one topic. However, embedded links, abbreviations and misspellings complicate the automated interpretation. Facing these challenges is becoming increasingly important for Economic and Market Intelligence in order to successfully recognize trends and threats.

We are interested in a particular use case; it concerns the sector of beekeeping. The importance of honeybees in agriculture has gained public attention in recent years, along with wide news coverage of their decline. Growing numbers of people are becoming concerned about the plight of honeybees. According to *FranceAgrimer* [6], the average age of beekeepers in France is 42 years old.

The beekeeping industry faces many problems related to the health of bee colonies concomitant with a general decline in production. Scientific work focused on understanding these phenomena has allowed identifying several causes and establishing concrete elements to guide the decisions of beekeepers.

Due to pollution and climate change, experts confirmed that the massive destruction of marine creatures could occur very soon. Many terrestrial species are also exposed to the same fate. The death of the honeybees' population is considered to be the most important warning to humanity, as they play a more vital role in pollination (this is an important step in horticulture and agriculture).

In this paper, we will present an original method in order to help beekeepers take care of their apiary. Once consistent data from different sources are collected, a deep learning AI is run to give beekeepers advices trough a smart Chatbot.

This paper is structured as follows. In the first place, we discuss the state of art then we describe the utilities and functionalities of our smartphone application in Sect. 3. In Sect. 4, we talk about text mining and data aggregation before concluding and giving some perspectives.

2   Related Work

This idea is inspired by previous work in the domain of machine learning especially in deep learning. We could find a lot of interesting deep neural net applied to speech recognition needed in our application.

We can mention, for example, [7–9], however we will be using Baidu deep speech 2 [10] because its performance was proven over two different languages (English and Mandarin). Since the approach is highly generic, it can quickly be applied to extra languages. Therefore, we suppose that it will be suitable for other languages like French, which is our main language.

Text mining is an active research area when it comes to web and social media data, as mentioned in [11, 12], people on social media such as Twitter don't pay much attention to their spelling or grammatical construction faults, as a result preprocessing these data is a prerequisite towards a valid mining system.

In this regards we intend to do as in [13], so we will be using an n-gram model to correct the spelling and grammar faults, we will replace the emoticons with their meaning and also we will remove punctuation, symbols, links, hashtags, targets, replace opinion phrases and idioms with their true meaning. We will also filter the language used in the tweet. Since we want to use machine-learning techniques to deal with this problem we first need to extract feature from the preprocessed tweets.

Different type of features are used in the literature. We note, for example, the use of an n-gram model to link words with their negative and positive probabilities.

3   System Architecture

The core of our system is a Chatbot capable of sending valuable information to subscribed beekeepers by collecting and extracting data from blocks and from different sources. The collected data will help training machine learning models capable of generating advice or alerts to users in human understandable language.

Figure 1 shows the structure of our proposed model we can see that this system is composed of different parts; it consists of four major components:

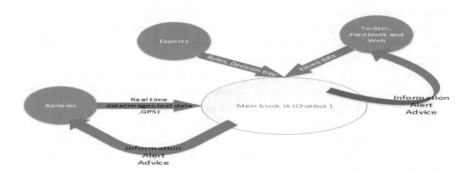

Fig. 1. General model structure.

- Apiary (smartphone application),
- Social media and web mining,
- Expertise rules,
- Chatbot.

(i)   We use our application to collect real-time information from beekeepers that uses our application to control their apiaries.

(ii) We apply text-mining approach to extract data from social media such as Twitter and we filter it to keep the most relevant information.

(iii)   At this stage, we use expert's knowledge to form some sort of rules. We thus create an expert agent.

(iv)   The Chatbot ApiSoft is the main component of our system. Its mission is to address beekeepers with eventual alerts or valuable information based on data provided from all other parts of the system.

A. Smartphone application (Apiaries)

The reason for creating this application is to help beekeepers manage their apiaries. At the same time, we can use data collected by this application to feed the Chatbot. For example, if there is a certain bee disease that has been located in a geographic area, then the Chatbot can notify all nearby beekeepers about that disease.

In this application, we propose a speech recognition system inspired by Baidu deep speech 2 [10]. Figure 2 presents the structure of deep neural network we use. The first levels consist of convolutional layers (the core building block of a Convolutional Network that does most of the computational heavy lifting) since they have proved being capable of extracting pertinent features.

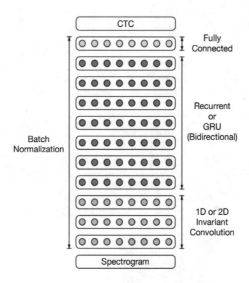

Fig. 2. Deep neural network used for speech recognition [10].

After the convolutional step, we implemented bidirectional recurrent layers using Gated Recurrent Units (GRUs). GRU units are preferred to Long Short-Term Memory networks (LSTM) since they consume much less energy and are easier to train compared to LSTMs. We finish with a one fully connected layer. Furthermore, we have used Connectionist Temporal Classification (CTC) loss function [14] to train our model. For performance reasons, we choice to use WrapCTC [10]. The most challenging part in this deep learning approach was to find data for training since deep learning required more training data then traditional approaches.

With this Speech Recognition (SR) system, users are capable of wearing their beekeeper outfit and yet communicating commands and information about their bee-hives to this application over their voice, at the same time. They do not need to touch their phones so they are not bothered in their movements. One additional utility of this application consists in processing images data such in order to detect and count Varroa mites out of photos taken by a smartphone.

B. Social media and web mining

The block, described in Fig. 3, is responsible for collecting topic related data from the web particularly from social media like Twitter or Facebook. They represent major sources of data. Because of its data type and the number of users, we use a python client interface to connect to twitter API to search for tweets on a specific topic. For example, we can search for tweets containing words like apiary, honey, beekeeping, etc.

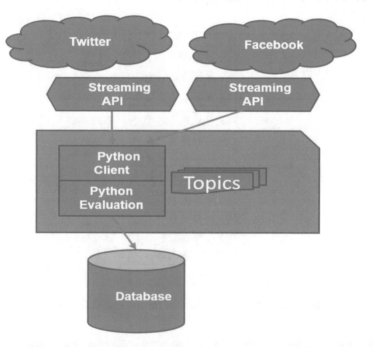

Fig. 3. Proposed structure for collecting and mining social media network data.

Twitter is proved to be a great source of information [15, 16] since it contains a large community and only handles short text data, which make it ideal of text mining techniques.

We perform an evaluation of the collected tweets, in which we will be using a probabilistic n-gram model to indicate the polarity of the tweet. This mechanism will allow us to extract knowledge or patterns and store it in our database so that we can use it to generate advice or alerts. Chatbot communicates all information to users.

C. Expertise rules

In this part, we take bee experts knowledge to form a set of rules. We use these rules to build a decision tree. We uses this tree on information provided from other blocks, which can help us generate meaningful warnings and assistance.

We came to choose a decision tree since despite it does not require too much data; it is capable of handling multi output problems. Besides, the cost of implementing such a tree is logarithmic which a gain in complexity.

First we begin by creating a web survey addressed to bee experts, to collect their knowledge in an easily usable format, after that we pass to train the model using C5.0 which is a sophisticated data mining tool involving if-else- rules [17].

D. Chatbot

This is the main component of our system. It uses all information and approaches accumulated from other blocks, to generate human comprehensive text.

To create a system capable of generating human understandable content we use a sequence-to-sequence neural net. We came to this choice as these models provide versatility compared to traditional machine learning methods that required a fixed output length. The generated message is then broadcasted to all concerned beekeepers, so that it can help them while they maintain their apiary.

4   Results and Discussion

In this study, we are more specifically interested in the voice recognition part. First, we speak about *hotword* detection, which consists in detecting specific words in the speech. In order to select the suited system we tested two methods *Snowboy* detection and *CMU Sphinx*.

Snowboy is a personalized real-time deep neural network capable of identifying *hotword* from continues speech. The advantage of this library is that it uses insignificant computing power that makes it capable of working even in Raspberry Pi of the first generation.

The second toolkit is *CMU Sphinx*, which is mainly a full speech recognition system. It also offers a special mode that makes it able to detect targeted words.

We tested the performance between these two libraries by a dataset containing words like hey google, *Apisoft* etc. We have used confusion matrix (shown in figure Fig. 4) as a metric of performance evaluation.

We can clearly see that in term of *hotword* detection *Snowboy* is clearly a better approach with 95% of accuracy compared to 64% for CMU Sphinx.

For speech recognition, we have also tested two approaches: *CMU sphinx* and an end-to-end deep neural network.

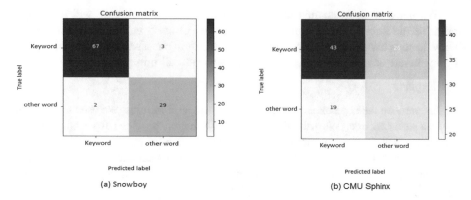

Fig. 4. Confusion matrix

The first approach is based on the classical pipeline of a typical speech recognition system which involving an acoustic model. In the case of *CMU Sphinx*, they have used a Hidden Markov model capable of converting a set of audio features into a list phonemes. The second component is a dictionary that allows doing the mapping between the phonemes the words. The last component is the language model that allows minimizing vocabulary and grammatical errors.

The second approach is a deep recurrent neural network that we trained based on deep speech. We have used 50 h of training data (mainly from VoxForge dataset plus books reading samples). Since we do not really have a large amount of data, we just keep the network as simple as possible with three convolutional layers followed by a bidirectional neural network.

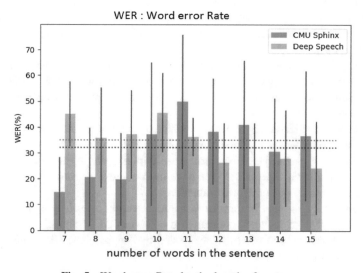

Fig. 5. Word error Rate by the length of sentences

We did a comparison of the two above described methods. Figure 5 shows the WER (Word error Rate) according to the length of sentences for the two systems. We can see that CMU Sphinx, in general, is slightly better than the end-to-end DNN approach (32% compared to 35% for the neural network). We think that this is due to the quantity of data we have since the end-to-end DNN approaches generally requires a larger dataset to train. However, we will interpret the data problem, considering we have some collaborators that are willing to give us the necessary data to train a <10% WER Speech recognition system based on this approach.

5 Conclusion

In this paper, we have presented a new architecture capable of helping beekeepers to control their apiary by communicating to users' notifications about the state of apiaries in their respective areas and inform them if there is a phenomenon to consider.

Our method, based on machine learning and text mining techniques, tries to build a Chatbot that collects data from various sources such as social media, experts and real data (different types of sensors in hives) collected by smartphones applications. We think this approach will continue to improve as this system continues to work and training.

At last, we expect that this system will give valuable aids in the beekeeping process. We also hope that our solution will contribute to the improvement of the quantity and quality of honey production and will participate in the prevention against the extinction of honeybees.

References

1. Kepios: Digital in 2018, essential insights into internet, social media, mobile, and ecommerce use around the world, April 2018. https://kepios.com/data/. Accessed 2 Nov 2018
2. Ghiassi, M., Skinner, J., Zimbra, D.: Twitter brand sentiment analysis: a hybrid system using n-gram analysis and dynamic artificial neural network. Expert Syst. Appl. **40**(16), 6266–6282 (2013)
3. Zhou, X., Tao, X., Yong, J., Yang, Z.: Sentiment analysis on tweets for social events. In: Proceedings of the 2013 IEEE 17th International Conference on Computer Supported Cooperative Work in Design, CSCWD 2013, 27–29 June 2013, pp. 557–562 (2013)
4. Salathé, M., Vu, D.Q., Khandelwal, S., Hunter, D.R.: The dynamics of health behavior sentiments on a large online social network. EPJ Data Sci. **2**(4), 1–12 (2013). https://doi.org/10.1140/epjds16
5. Sriram, B., Fuhry, D., Demir, E., Ferhatosmanoglu, H., Demirbas, M.: Short text classification in Twitter to improve information filtering. In: Proceedings of the 33rd international ACM SIGIR Conference on Research and Development in Information Retrieval, 19–23 July 2010, pp. 841–842 (2010). https://doi.org/10.1145/1835449.1835643
6. Proteis+: Audit economique de la filière apicole française. Final report. FranceAgriMer (2012)

7. Graves, A., Navdeep, J.: Towards end-to-end speech recognition with recurrent neural networks. In: International Conference on Machine Learning, pp. 1764–1772 (2014)
8. Maas, A., Xie, Z., Jurafsky, D., Ng, A.: Lexicon-free conversational speech recognition with neural networks. In: Proceedings of the 2015 Conference of the North American Chapter of the Association for Computational Linguistics: Human Language Technologies, pp. 345–354 (2015)
9. Chan, W., Jaitly, N., Le, Q., Vinyals, O.: Listen, attend and spell: a neural network for large vocabulary conversational speech recognition. In: 2016 IEEE International Conference on Acoustics, Speech and Signal Processing (ICASSP), pp. 4960–4964. IEEE (2016)
10. Amodei, D., Ananthanarayanan, S., Anubhai, R., Bai, J., Battenberg, E., Case, C., Casper, J., et al.: Deep speech 2: end-to-end speech recognition in english and mandarin. In: International Conference on Machine Learning, pp. 173–182 (2016)
11. Parikh, R., Movassate, M.: Sentiment analysis of user-generated twitter updates using various classification techniques. CS224 N Final Report, 118, 4 June 2009
12. Gokulakrishnan, B., Priyanthan, B., Ragavan, T., Prasath, N., Perera, A.: Opinion mining and sentiment analysis on a twitter data stream. In: 2012 International Conference on Advances in ICT for Emerging Regions (ICTer), 12 December 2012, pp. 182–188. IEEE (2012)
13. Kharde, V., Sonawane, P.: Sentiment analysis of twitter data: a survey of techniques. arXiv preprint arXiv:1601.06971, 26 January 2016
14. Graves, A., Fernández, S., Gomez, F., Schmidhuber, J.: Connectionist temporal classification: labelling unsegmented sequence data with recurrent neural networks. In: Proceedings of the 23rd International Conference on Machine Learning, pp. 369–376. ACM (2006)
15. Barbosa, L., Feng, J.: Robust sentiment detection on twitter from biased and noisy data. In: Proceedings of the 23rd International Conference on Computational Linguistics: Posters, pp. 36–44. Association for Computational Linguistics (2010)
16. Bifet, A., Frank, E.: Sentiment knowledge discovery in twitter streaming data. In: International Conference on Discovery Science, pp. 1–15. Springer, Heidelberg (2010)
17. Hou, S., Hou, R., Shi, X., Wang, J., Yuan, C.: Research on C5. 0 algorithm improvement and the test in lightning disaster statistics. Int. J. Control Autom. 7(1), 181–190 (2014)

Development of an Intelligent Diagnosis System for Detecting Leakage of Circulating Fluidized Bed Boiler Tubes

Yu-Hyun Kim[1], In-Kyu Jeong[1], Jae-Young Kim[1], Jae-Kyo Ban[2], and Jong-Myon Kim[1,2(✉)]

[1] School of IT Convergence, University of Ulsan, Ulsan, South Korea
dbgus115@naver.com, jeonginkeyu@gmail.com,
kjy7097@gmail.com
[2] ICT Safety Convergence Center, University of Ulsan, Ulsan, South Korea
{jagurben, jmkim07}@ulsan.ac.kr

Abstract. The detection of leaks in circulating fluidized bed boiler tubes is an important issue in thermal power plants, because leaks lead to enormous economic and social losses. To address this issue, a time-based maintenance (TBM) method has been employed in power plants, but it cannot detect leakage of a tube during operation. Instead, a condition-based maintenance (CBM) method is required to detect unexpected leakage of a boiler tube in real time. This paper proposes an acoustic emission (AE)-based diagnostic system for detecting leakage of a circulating fluidized bed boiler tube using a data acquisition (DAQ) system and support vector machines (SVMs) for data acquisition, data analysis, and leakage detection. Experimental results show that the proposed diagnosis system perfectly detects leakage of a boiler tube on the simulation testbed.

Keywords: Support vector machine · Condition-based maintenance · Boiler tube · Acoustic emission

1 Introduction

Thermal power generation uses turbines to convert heat generated by burning fuel to rotational kinetic energy, which is then converted to electric energy. Thermal power plants have various advantages over other power plants. They can be constructed irrespective of topography, and they can reduce cost loss by reduction of transmission cost and low construction cost. Because of these advantages, thermal power generation accounts for the majority of electricity production [1].

In a circulating fluidized bed boiler, water is converted into steam within about 30,000 tubes at high temperature and high pressure. However, deterioration of the tubes leads to increasingly frequent shutdowns of circulating fluidized bed boilers, leading to enormous economic and social losses [1]. Power plants are implementing time-based maintenance (TBM) to reduce these losses. TBM is a post-maintenance method in which maintenance is performed even though the equipment can operate normally, or maintenance is performed after the equipment is defective. However, time-based

© Springer Nature Switzerland AG 2020
A. M. Madureira et al. (Eds.): HIS 2018, AISC 923, pp. 311–320, 2020.
https://doi.org/10.1007/978-3-030-14347-3_30

maintenance is unnecessarily costly. Therefore, there is a need to apply standardized condition-based maintenance (CBM) to make early predictions of unexpected leakage of boiler tubes and to establish a maintenance plan [2].

There are six components of condition-based maintenance: data acquisition (DA), data manipulation (DM), state detection (SD), health assessment (HA), prognosis assessment (PA), and advisory generation (AG). Data acquisition, data manipulation, and state detection are important initial steps in condition-based maintenance. Therefore, this paper proposes a diagnostic system for circulating fluidized bed boiler tubes involving data acquisition and diagnostic devices that are used for the data acquisition, data manipulation, and state detection steps of a standardized condition-based maintenance process [2]. The data acquisition step consists of sensor selection and data acquisition device selection. During the sensor selection, it is important to select a sensor appropriate to the situation. In this paper, an acoustic emission sensor was chosen to diagnose early defects in circulating fluidized bed boiler tubes. Acoustic emission sensors are used for early detection of defects in pipes, storage tanks, etc. [3] We developed a data acquisition device using the Texas Instruments DSP and ADC to fabricate a noise filter circuit and a power distribution circuit. We used Bluetooth wireless communication technology to communicate with the diagnostic device [4].

The data manipulation and state detection phases are performed on diagnosis devices consisting of a small embedded single-board computer and an LCD screen for mobility, operator convenience, and efficiency. The data manipulation phase involves signal analysis, during which the characteristics of the time domain and the frequency domain are extracted for the acquired signal. The extracted features are used as input data for the diagnosis algorithm in the state detection step. That is, the data manipulation step is a preprocessing step for applying the algorithm in the next step. Facility diagnosis occurs during the state detection step; after signal analysis is completed in the data manipulation stage, a diagnosis algorithm is needed to diagnose the condition of the equipment. In particular, support vector machines (SVM) are suitable for use in small embedded devices because of the small amount of calculation required [5].

Finally, we develop a diagnostic system for condition-based maintenance of circulating fluidized bed boiler tubes based on acoustic emission using SVM. To verify the system's performance, a simulated test bed consisting of an actual circulating fluidized bed boiler tube was fabricated, and a pinhole defect in the tube was artificially generated. We then tested the ability of the system described in this paper to accurately identify fault conditions in circulating fluidized bed boiler tubes.

2   Background

2.1   Condition-Based Maintenance Method

Condition-based maintenance is a diagnosis technology that uses sensors to monitor the status of equipment or mechanical systems and to identify signs of failure. Based on the acquired data, the system develops a model according to the state using machine learning techniques and diagnoses the failure of the facility. This makes it possible to respond more quickly to accidents or breakdowns in areas where worker access is difficult [2].

2.2   Acoustic Emission Techniques

Acoustic emission is a phenomenon in which solids emit sound during deformation or destruction via elastic waves. In addition, since small defects and small amounts of deformation cause acoustic emissions before the material is destroyed, it is possible to detect and predict defects and destruction of materials and structures by analyzing the patterns in acoustic emission occurrence [3].

2.3   Support Vector Machine

The SVM is based on minimizing the structural risk to minimize the probability of misclassifying data with unknown probability distributions. In particular, SVM is suitable for small, embedded devices because of the small amount of computation involved. In this paper, we apply the widely used Gaussian radial basis function (RBF) kernel in the kernel function model for SVM learning to map input features to infinite feature space [5].

3   Proposed Diagnosis System

3.1   Wireless Data Acquisition Device

3.1.1   Configuration of Wireless Data Acquisition Device
The wireless data acquisition device is composed of two layers. On the first level there is a 24 V power cable, a power switch, and a power distribution circuit to supply power to each module. On the second level, there is a two-channel BNC signal filter circuit, a Bluetooth module, and a TI DSP (TMS320F28377D) to control all functions of the acquisition device. Figure 1 shows the configuration of the wireless data acquisition device.

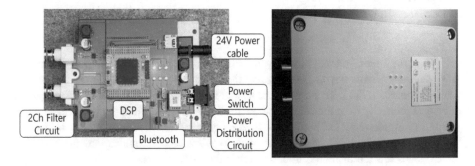

Fig. 1. DAQ device configuration.

3.1.2   Firmware Algorithm of Wireless Data Acquisition Device
Figure 2 is a flow chart illustrating the firmware algorithm of the data acquisition device developed in this paper. The device waits for Bluetooth communication when first powered on. When the data acquisition device is paired with the diagnosis device, they communicate with each other through six protocol commands, as shown in Table 1.

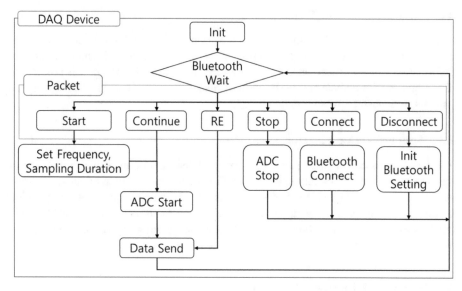

Fig. 2. DAQ device firmware algorithm.

Table 1. Packet functions of DAQ device.

Packet	Function
Start	Set frequency (500 kHz) and sampling duration (1 s), start ADC, send data
Continue	Start ADC, send data
RE	Re-send data
Stop	Stop ADC
Connect	Connect with diagnosis device
Disconnect	Init Bluetooth setting

When the "start" packet is received from the diagnostic device, the DSP sets the ADC sampling frequency to 500 kHz and the ADC sampling duration to 1 s. After the setup is complete, the DSP starts the ADC and sends the acquired data to the diagnosis device. When the "continue" packet is received, the acquired data is transmitted without changing the ADC settings. It then repeats the above routine until it receives a "stop" packet. When data loss occurs due to wireless communication, the acquired ADC values are retransmitted after receiving the "RE" packet.

3.2   Diagnosis Device

3.2.1   Configuration of Diagnosis Device

The diagnosis device is shown in Figs. 3 and 4. The diagnosis system uses Odroid, a single board computer (SBC), to operate the diagnosis algorithm, and Odroid-VU7, a touch LCD, for the operator's convenience. First, an acoustic emission signal is acquired from the data acquisition device and transmitted to the diagnosis device using

Bluetooth, which is a wireless communication technology. The diagnosis device diagnoses the data through the SVM algorithm and displays it on the LCD screen.

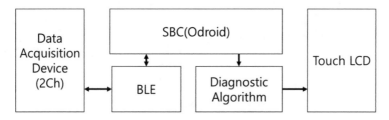

Fig. 3. Diagnosis architecture.

Fig. 4. Diagnosis device.

3.2.2 Diagnosis Algorithm

In this paper, the SVM machine learning technique is used, and an overall flow chart for the diagnosis algorithm is shown in Fig. 5. First, data acquired by the data acquisition device are received using Bluetooth. Then, 21 features are extracted from the acquired data, as shown in Table 2. The extracted features are either incorporated in SVM model generation or used as inputs to the SVM model. In this paper, we have generated an SVM model that distinguishes between the steady state and the defective state of a circulating fluidized bed boiler tube. To verify the performance of the generated model, test data were acquired in each state and the diagnosis accuracy was confirmed.

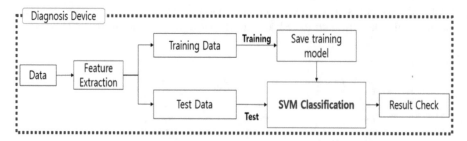

Fig. 5. Diagnostic algorithm.

Table 2. Features of time domain and frequency domain.

Num	Feature	Num	Feature	Num	Feature
Time domain					
1	Peak	2	RMS	3	Kurtosis
4	Crest factor	5	Impulse factor	6	Shape factor
7	Skewness	8	SMR	9	Margin factor
10	Peak to peak	11	Kurtosis factor	12	Entropy
13	Energy	14	Clearance factor	15	Normalize5
16	Normalize6	17	Shape2		
Frequency domain					
1	Center frequency	2	RMS frequency		
3	Root variance frequency	4	Frequency spectrum energy		

3.2.3   Configuration of Diagnosis Program

The configuration of the diagnosis program is shown in Fig. 6. Table 3 shows the function of each part. In the "Bluetooth Settings" panel, the connection state with the data acquisition device can be confirmed. In the "Diagnosis Settings" panel there are options for "Training," "Model," and "Save Path." "Training" specifies the path of the SVM model to be created and specifies the class name; data acquired are then learned in the specified class. "Model" selects the path of the saved SVM model. "Save Path" sets the path to save the acquired data.

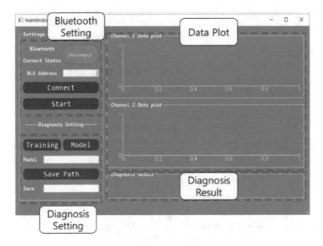

Fig. 6. Overview of diagnosis program.

Table 3. Functions of diagnosis program.

Part	Function
Bluetooth setting	Communication connection, start data acquisition
Diagnosis setting	Training, set model path and data storage path
Data plot	Display acquired ADC data graph
Diagnosis result	Display diagnosis result

3.2.4   Flowchart of Diagnosis Program

The flowchart for the program is shown in Fig. 7. First, the wireless connection with the data acquisition device is started with the UI. After confirming the connection with the data acquisition device, the user chooses to use either the training mode or the stored model. In training mode, the training process is performed and the diagnosis result is displayed based on the data acquired by the data acquisition device using the generated SVM model. Conversely, if a stored model is selected, the diagnosis result is displayed for the data acquired by the data acquisition device using the stored model.

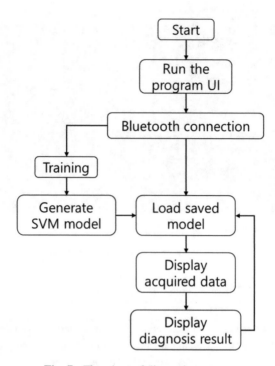

Fig. 7. Flowchart of diagnosis program.

4   Experimental Environment and Results

4.1   Experimental Environment

In this study, an actual circulating fluidized bed boiler tube test bed was fabricated. As shown in Fig. 8, the waveguide was attached to the membrane of the boiler water wall tube to allow for testing. In addition, to create the 0.6 mm pinhole condition, the solenoid valve was manufactured as a switch type so that defects could be made at a desired point. The acoustic emission sensor was attached to the waveguide as shown in Fig. 8 and connected to the data acquisition device. We made 30 measurements using 500 kHz sampling frequency and 1 s sampling duration for each of the normal state and the 0.6 mm pinhole condition, and this was used as training data to generate the SVM model. Then, to verify the performance of the generated model, 20 measurements were made using a sampling frequency of 500 kHz and sampling duration of 1 s for each of the normal state and the 0.6 mm pinhole state, and the diagnosis accuracy was confirmed.

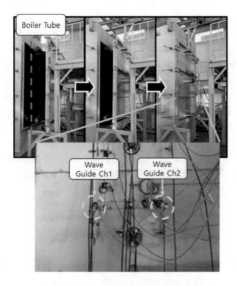

Fig. 8. Boiler tube testbed.

4.2 Experiment Results

As shown in Table 4, 100% diagnostic accuracy was achieved for both the normal state and the 0.6 mm pinhole. The results of diagnosing defects for pinhole are shown in Fig. 9.

Table 4. Diagnosis accuracy results.

State	Diagnosis accuracy
Normal	100%
Pinhole	100%

Fig. 9. Photo of the diagnosis device and program.

5 Conclusions

In this paper, an intelligent diagnosis system for detecting leakage of circulating fluidized bed boiler tubes was developed and tested. The proposed intelligent diagnosis system includes a two-channel DAQ and a diagnosis algorithm to detect leakage of boiler tubes. Experimental results showed that the proposed diagnosis system perfectly detects boiler tube leaks. In the future, we will increase the number of different sizes of leaks, test in various environments, and improve the diagnosis system.

Acknowledgment. This work was supported by the Korea Institute of Energy Technology Evaluation and Planning (KETEP) and the Ministry of Trade, Industry & Energy (MOTIE) of the Republic of Korea (No. 20161120100350).

References

1. Lee, S.B., Roh, S.M.: Developing an early leakage detection system for thermal power plant boiler tubes by using acoustic emission technology. J. Korean Soc. Nondestr. Test. **36**(3), 181–187 (2016)
2. Kang, T., Han, S.-W., Lee, J.-H., Yoon, D.-B., Park, J.-H.: Development of CBM (condition-based maintenance) based abnormal condition diagnosis of centrifugal pump. In: Conference of the Korean Society for Noise and Vibration Engineering, pp. 318–320 (2016)
3. Kim, D.-H., Lee, S.-B., Kim, Y.-H., Won, J.-H., Son, Y.-C., Cha, Y.-J.: Pulverizer condition monitoring system using acoustic emission technique in thermal power plant. J. Korean Soc. Nondestr. Test. **37**(6), 435–442 (2017)
4. Kim, Y.-H., Jeong, I.-K., Kim, J.-M.: Development of embedded based real time wireless data acquisition system for diagnosis of industrial equipment faults. J. Korean Soc. Eng. Art Sci. **15**(1), 115–116 (2017)
5. Park, S., Kim, K.-J., Lee, J.-S., Lee, S.-R.: Red tide prediction using neural network and SVM. J. Inst. Electron. Eng. Korea **48**(5), 39–45 (2011)

Optimizing Dispatching Rules
for Stochastic Job Shop Scheduling

Cristiane Ferreira$^{(\boxtimes)}$, Gonçalo Figueira, and Pedro Amorim

INESCTEC, Faculdade de Engenharia da Universidade do Porto, Porto, Portugal
`cristiane.m.ferreira@inesctec.pt`

Abstract. Manufacturing environments commonly present uncertainties and unexpected schedule disruptions. The literature has shown that in these environments simple and fast dynamic dispatching rules are efficient sequencing methods. However, most of the works in the automated designing of these rules have considered deterministic processing times. This work aims to design dispatching rules for problem settings similar to the ones found in real environments such as uncertain processing times and sequence-dependent setup times. We use Genetic Programming to generate efficient rules for stochastic job shops with setup times. We show that the generated rules outperform benchmark dispatching rules, specially in settings with high setup time levels.

Keywords: Dynamic job shop scheduling · Stochastic scheduling · Genetic Programming · Dispatching rules

1 Introduction

Dynamic scheduling refers to environments where the schedule of jobs is subject to unexpected conditions. In dynamic job shops, the sequence of operations must be continuously re-optimized to adapt to the unexpected deviations [2]. In this work we aim to design efficient heuristics for job shops that incorporate real-world characteristics such as uncertainty in the processing times and setup times.

The dynamic version of the Job Shop Scheduling Problem (JSSP) is defined on a set of n jobs $\{1, ..., j, ..., n\}$ and a set of m machines $\{1, ..., i, ..., m\}$. Each job j requires one operation to be processed by each machine i, in a given processing time p_{ij}. The sequence of machines to visit may be different among jobs and each machine can only process one operation at a time. The objective is to find a schedule of jobs that minimizes the makespan, i.e., the maximum completion times [14].

Dispatching rules have been shown to be a promising approach for dynamic job shop scheduling [6]. They are simple and fast sequencing methods, used to prioritize jobs waiting to be processed. Blackstone et al. [1] and Haupt et al. [4] review the literature on rule-based sequencing methods for the dynamic JSSP.

© Springer Nature Switzerland AG 2020
A. M. Madureira et al. (Eds.): HIS 2018, AISC 923, pp. 321–330, 2020.
https://doi.org/10.1007/978-3-030-14347-3_31

Lawrence and Sewell [10] compare the performance of seven simple rules and exact methods for dynamic job shop scheduling. They show that the dynamic heuristics outperform static optimum schedules even with moderate levels of uncertainty in processing times. As uncertainty increases the performance of the dispatching rules converges to that of the exact methods, reinforcing the use of simple scheduling methods in dynamic stochastic environments.

Despite the fact that several rules present good performance in general, the existence of a single rule that performs best in all settings is unlikely. This encourages studying methods that combine several dispatching rules and automated heuristic design [2], i.e. a process to explore the "heuristic search space" instead of the solution search space [15].

Genetic Programming (GP) [8] has been widely used for designing efficient dispatching rules and heuristics for scheduling problems. In the last decade there has been a increasing attention to the use and improvement of this technique [5,12]. A recent survey on GP is found in Nguyen et al. [13] and it indicates that most of the works have considered deterministic processing times.

The vast majority of works have not considered aspects found in real-world manufacturing environments. As far as we know, the authors Karunakaran et al. [7] are the only to deal with dynamic job shop and stochastic processing times. But their work does not include setup times between the operations.

This work analyses the influence of two frequent aspects of real job shops, uncertainty in processing times and the sequence dependent setup times between operations in the same machine. We adapt the instance set used by Lawrence and Swell and extend their work by automating the rules design with GP.

The reminder of this paper is organized as follows: The Genetic Programming algorithm and its parameters are described in Sect. 2. In Sect. 3 we present the experiments design together with the computational results. We also discuss some insights regarding instances characteristics and the performance evaluation of rules. Finally the paper is concluded with Sect. 4.

2   Methodology

In general, evolutionary strategies evolve solution candidates for a problem. In Genetic Programming (GP) individuals are computer programs, which can be interpreted or executed to produce a solution to the original problem. Thus, the fitness value of each individual corresponds to how well its able to perform a specific computational task.

GP algorithms commonly represent individuals as symbolic expression trees. This is due to the fact that in GP the final structure and size of the program is not known in advance. The trees are composed by *terminal* nodes and *functions*. The former include parameters of the problem, which are operated by the latter.

Figure 1 illustrates an individual, represented by a parsing tree and the corresponding dispatching rule. In the example 2, a, b and c are terminals, possibly related the attributes of the operation. The function set is composed by the

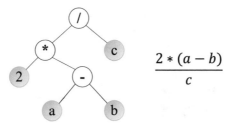

$$\frac{2 * (a - b)}{c}$$

Fig. 1. A GP individual tree example and its corresponding mathematical expression

mathematical operators $*$, $/$ and $-$. Details on the specific implementation of GP in this work will be given next.

In this work an individual represents a rule to prioritize awaiting operations. Each rule results in a sequence of operations which will define the schedule for the job shop. The fitness of an individual is the average performance measure of many schedules built by sorting operations according to the corresponding dispatching rule.

2.1 Breeding

Breeding is achieved via crossover and reproduction. The crossover operation is standard: it picks two individual trees, selects a node in each tree, and swaps the two sub-trees rooted by those nodes. The reproduction operation simply copies the individuals to be part of the population in the next generation.

The crossover and reproduction operations are chosen according to a likelihood probability. Both processes use tournament selection for selecting individuals. According to the *Tournament Size t*, it returns the best individual from a set of t individuals randomly selected.

The breeding process pick certain kinds of nodes with different probabilities. For example, one can state to pick non-terminals 90% of the time and terminals 10% of the time.

2.2 Function and Terminal Sets

The function set is composed by the four basic arithmetic operations, the conditional *If* and the logical operators. The terminal set is based on the seven rules from the work of Lawrence and Sewell. We also incorporated relevant attributes not considered by them such as setup times. Table 1 describes the complete function and terminal sets.

2.3 Fitness Evaluation

The primary objective of the fitness function is to guide the evolution process through efficient dispatching rules. Therefore a dynamic job shop simulator was implemented for measuring their performance.

Table 1. Function set and terminals of GP

Function set	Description
+	Addition
-	Subtraction
*	Multiplication
/	Protected division
If	Conditional
&	Logical operator *And*
‖	Logical operator *Or*
$\geq$	Logical operator *Greater than or equal to*
$\leq$	Logical operator *Less than or equal to*
Terminal set	**Description**
PT[a]	Expected processing time of current operation
−PT[a]	The negative of the expected processing time of current operation
OR[a]	Number of operations remaining for the current job
WF	Work following: Sum of the expected processing time of all remaining operations of the current job
SWF[a]	Successive work following: Sum of the expected processing time of all remaining operations (excluding the current operation)
SD[a]	Difference between current expected processing time and the expected processing time on the next machine
TR[a]	Right tail of the distribution of the remaining processing time of the current job
FCFS[a]	First-come first-served (Of jobs arrival time)
QSize	Queue size of current machine
RQSize	Sum of the queue size of the remaining machines for the current job
ST	Setup time between the last operation performed by the current machine and the current operation
n	Total number of jobs in the problem instance
m	Total number of machines in the problem instance

[a] Dispatching rule used by Lawrence and Sewell [10]

The simulation builds the schedule by sequencing the operations on a machine's queue every time a new job arrives. Once an operation finishes processing, its actual processing time is used to update the schedule, the job moves to the next machine and the dispatching rule is called again. When all operations are executed, the final makespan is then returned.

The simulator receives as input a dispatching rule and a job shop problem instance and returns the respective makespan C. Given the lowerbound $C*$ for a problem instance, the *relative performance* is $(C/C*)$. The overall performance of an individual i is estimated by $RelPer_i$, the average *relative performance* of the schedules generated by using the corresponding rule. Smaller values of $RelPer_i$ indicate better performances of individual i.

Another aspect considered by the fitness function is the size of trees. Large trees may adapt well for specific problem instances, but lose performance when applied outside of the training set. This is known as over-fitting [9]. By promoting short trees, the final result might not only be more elegant, but also have more generalizability.

The most common way of keeping trees short is to limit them to a certain depth. In addition, we defined a parsimony strategy to smoothly penalize the number of nodes. There is a penalty function $h(i, g)$ that depends on the size of the tree $|T_i|$ and the number of generations g and is given by $|T_i| * g^{1.2}$. As we do not want to prematurely prune good branches, which can be critical for the evolution process, h exponentially increases with g.

At each generation all individuals from the population are evaluated by the fitness function with the objective of preserving the best ones for the next generations. Considering the pursuit of efficient rules and the avoidance of very large trees, the fitness $f(i, g)$ of individual i in generation g is calculated as follows:

$$f(i, g) = 1/(RelPer_i + h(i, g) * 10^{-8})$$

Both the power of 1.2 and the coefficient of 10^{-8} were obtained by conducting some preliminary tests.

3   Computational Experiments

The GP algorithm was coded in Java and uses the evolutionary computation library ECJ [11]. All experiments ran on a personal computer using an Intel(R) Core(TM) i7 CPU with 2.8 GHz and 12 GB of RAM. The parameters calibration for the GP were defined in preliminary tests. We consider 100 generations and a population of 200 individuals. We use a tournament size of 3 and the maximum depth of the tree was set to 17. The breeding probabilities are 0.9 and 0.1 for the crossover and reproduction, respectively. At last, the probability of selecting a terminal and a non-terminal node were set to 0.1 and 0.9, respectively. The reported rules are the best found after 10 runs.

In the following we detail the instance generation and the design of the computational experiments.

3.1   Experimental Design

The basic problem set is the same as that used by Lawrence and Sewell, which is composed by 53 deterministic job-shop benchmark instances. Problem sizes range from 6 machines and 6 jobs, to 15 machines and 20 jobs. For all instances, all jobs visit all machines.

Lawrence and Sewell introduced uncertainty into the problem set by generating stochastic versions of each instance. Given the processing time of an operation p_{ij} and a coefficient of variation (cv), the actual processing times were found by a gamma distribution with mean $E[p_{ij}]$ and standard deviation $\sigma_{ij} = E[p_{ij}] \times cv$. The expected processing times $E[p_{ij}]$ are the same as in the deterministic version. The coefficient of variation cv determines the uncertainty level (in the deterministic case, $cv = 0$). In this work we use the same distribution with cv in five different levels: 0.2, 0.4, 0.6, 0.8 and 1.0.

As the objective of the study is to approximate the instances characteristics to the ones found in real-world settings, we included sequence-dependent setup times in the basic problem set. These were sampled from a gamma distribution with mean and standard deviation equal to the mean and standard deviation of the instance processing times, multiplied by a factor ϕ (setup time level). For each deterministic instance in the basic problem set there are four variants generated with different setup time levels ($\phi = 0.25, 0.5, 0.75, 1$).

Training Set. In order to perform a fair evaluation of the resulting rule, we do not use the complete instance set in the GP training. The objective is to evaluate if training the GP with a fraction of the instances is sufficient for it to design good rules for the complete set. Thus the fitness function uses the stochastic version of the benchmark instances with $cv = 0.6$ and setup time levels (ϕ) of 0.25 and 1. The average relative performance refers to 10 replications of each instance. As the benchmark instances compose a set of 53 problems, when evaluating an individual the fitness function calls the simulator $53 \times 2 \times 10 = 1040$ times.

3.2   Results and Discussion

To evaluate the quality of the final rules, we report the (*RelPer*) values on the testing set. The latter is composed by 100 replications of each instance type. The lowerbound $C*$ is the optimal value of the corresponding deterministic instance, provided by Gonçalves and Resende [3].

Existing Rules. The first set of experiments aimed to evaluate the performance of the seven existing benchmark dispatching rules in the instances with setup times. For the sake of brevity we present only the results for $\phi = 0, 0.5$ and 1.

Table 2 presents the average *RelPer* for all uncertainty levels and the number of instance types in which each rule presented the best average results (# *Inst.*). On both indicators, the three rules stand out: MOR, LTR and MWF. Given these results we decided to select the best three rules for the comparison with those evolved by GP.

Evolved Rules. We performed three experiment sets with GP which are explained in the following.

Evolved Rule R1 - As observed in Table 2, the benchmark rules MOR, LTR and MWF present superior performance than the remaining ones for most of the instances with setup times. The first experiments with GP aimed to analyse if there is a combination of these rules suitable to perform better than any of

Table 2. Results comparison of benchmark rules

Rule	Description	$\phi = 0$		$\phi = 0.5$		$\phi = 1$	
		RelPer	# Inst.	RelPer	# Inst.	RelPer	# Inst.
SPT	Shortest Processing Time	1.34	1	1.79	2	2.25	5
LPT	Longest Processing Time	1.48	0	1.92	0	2.38	0
LSD	Largest Successive Difference	1.30	6	1.79	6	2.23	3
MOR	Most Operations Remaining	**1.29**	**3**	**1.74**	**6**	**2.20**	**13**
LTR	Longest Tail Remaining	**1.27**	**16**	**1.70**	**12**	**2.16**	**11**
FCFS	First-come First-served	1.31	5	1.79	1	2.25	1
MWF	Most Work Following	**1.25**	**22**	**1.68**	**23**	**2.15**	**20**

them separately for the complete instance set. We thus ran GP using a terminal set composed only by {OR, TR, SWF}. This resulted in following rule:

$$SWF/TR$$

Evolved Rule R2 - In the subsequent experiments we included all the remaining terminals listed in Table 1 except the setup time (ST). The objective was to verify if simple rules with a poor performance could help producing efficient schedules when combined with other terminals. The result of GP is given below:

$$-OR \times ((PT \times n) + 2 \times PT + 2 \times SD)$$

Evolved Rule R3 - In addition to the terminals considered by R2, the last experiment set also considered the setup time terminal (ST). The evolved rule is:

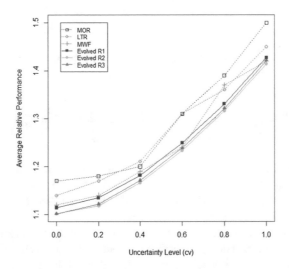

Fig. 2. Average relative performance of evolved rules in benchmark instances without setup time

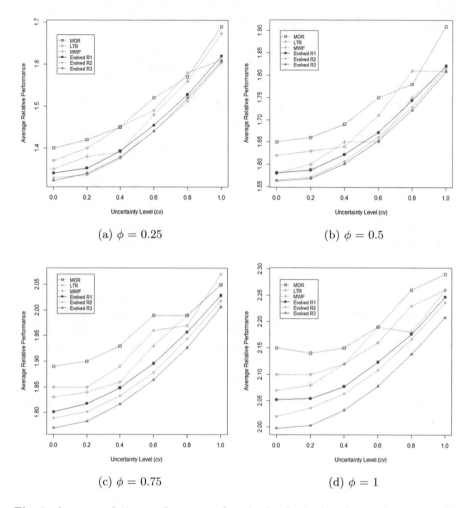

Fig. 3. Average relative performance of evolved rules in benchmark instances with different setup time levels (ϕ)

$$\frac{\frac{SWF}{ST+PT\times(1-1/(m+RQSize+n))} + m + RQSize + \frac{(2\times(m+RQSize)+PT+(SWF\times m))}{m\times(RQSize+n)}}{m \times (RQSize + ST)}$$

Let us start by analysing the results of the instances without setup times. The chart in Fig. 2 presents the average *RelPer* for all uncertainty levels. As reported in the work of Lawrence and Sewell, the rules tend to deteriorate their performance as the uncertainty level increases.

Observe that GP was able to find an efficient rule from the combination of the best simple benchmark rules. R1 performes better than MOR, LTR and MWF for most of the cases, except when $cv = 0.6$, where MWF is able to outperform it. This is actually surprising, given that this was the cv value used in training.

Regarding the rules R2 and R3, the comparison of theirs relative performances shows that both resulted in better schedules than all the other rules. Although R3 includes a terminal that is not used in these instance set (ST), its results are competitive. In addition, given that instances with $\phi = 0$ were not present in the training set, it is interesting one that all the three evolved rules were able to perform so well on those instances.

Figure 3 presents the results obtained for the instances with setup times. Again the average *RelPer* is greater for larger values of *cv*. As the lowerbound used to calculate it was provided by instances without setup times, the relative performance also increases with the level of ϕ.

The overall results show that in general the three evolved rules performed clearly better than the benchmark ones. The difference in performance becomes greater as the value of ϕ increases. This was expected since the training set included setup times.

Comparing the evolved rules, we notice that the results from R2 and R3 are always better than those produced by R1. We also observe that for low levels of setup time (Fig. 3a) R2 and R3 have similar results. As ϕ increases R3 becomes better. Finally, Fig. 3d shows meaningful differences between the results found by R3 and all the other rules.

4   Conclusions and Future Work

In this work we applied Genetic Programming for designing efficient dispatching rules for stochastic job shops. Our main contribution is the analysis of the processing times uncertainty together with setup times between operations in the same machine. Benchmark instances were adapted to incorporate these characteristics.

The proposed method was able to find rules that performed better than the benchmark for the complete instance set. Moreover we observed that the performance of the evolved rules was clearly better for higher setup time levels.

As future research we aim to test the evolved rules in real problem instances and incorporate other characteristics found in real-world environments, such as due dates. Further analysis may also include other scheduling performance measures such as mean and total tardiness. Finally, one may adapt the method to deal with flexible job shops and other variants of scheduling problems.

Acknowledgments. This work is financed by the ERDF - European Regional Development Fund through the Operational Programme for Competitiveness and Internationalisation - COMPETE 2020 Programme, and by National Funds through the Portuguese funding agency, FCT - Fundação para a Ciência e a Tecnologia, within project SAICTPAC/0034/2015- POCI-01-0145-FEDER-016418.

References

1. Blackstone, J.H., Phillips, D.T., Hogg, G.L.: A state-of-the-art survey of dispatching rules for manufacturing job shop operations. Int. J. Prod. Res. **20**(1), 27–45 (1982)
2. Dominic, P.D.D., Kaliyamoorthy, S., Kumar, M.S.: Efficient dispatching rules for dynamic job shop scheduling. Int. J. Adv. Manuf. Technol. **24**(1), 70–75 (2004)
3. Gonçalves, J.F., Resende, M.G.C.: Biased random-key genetic algorithms for combinatorial optimization. J. Heuristics **17**(5), 487–525 (2011)
4. Haupt, R.: A survey of priority rule-based scheduling. Oper.-Res.-Spektrum **11**(1), 3–16 (1989)
5. Hildebrandt, T., Heger, J., Scholz-Reiter, B.: Towards improved dispatching rules for complex shop floor scenarios: a genetic programming approach. In: Proceedings of the 12th Annual Conference on Genetic and Evolutionary Computation, GECCO 2010, pp. 257–264. ACM, New York (2010)
6. Holthaus, O., Rajendran, C.: Efficient dispatching rules for scheduling in a job shop. Int. J. Prod. Econ. **48**(1), 87–105 (1997)
7. Karunakaran, D., Mei, Y., Chen, G., Zhang, M.: Evolving dispatching rules for dynamic job shop scheduling with uncertain processing times. In: 2017 IEEE Congress on Evolutionary Computation (CEC), pp. 364–371 (2017)
8. Koza, J.R.: Genetic programming as a means for programming computers by natural selection. Stat. Comput. **4**(2), 87–112 (1994)
9. Kronberger, G.K.: Symbolic regression for knowledge discovery - bloat, overfitting, and variable interaction networks. Ph.D. thesis, Johannes Kepler University, Linz, Austria (2010)
10. Lawrence, S.R., Sewell, E.C.: Heuristic, optimal, static, and dynamic schedules when processing times are uncertain. J. Oper. Manag. **15**(1), 71–82 (1997)
11. Luke, S.: ECJ then and now. In: GECCO (2017)
12. Nguyen, S., Mei, Y., Xue, B., Zhang, M.: A hybrid genetic programming algorithm for automated design of dispatching rules. Evol. Comput. 1–31 (2018, preprint). https://doi.org/10.1162/evco_a_00230
13. Nguyen, S., Mei, Y., Zhang, M.: Genetic programming for production scheduling: a survey with a unified framework. Complex Intell. Syst. **3**(1), 41–66 (2017)
14. Pinedo, M.L.: Scheduling - Theory, Algorithms and Systems, 3rd edn. Springer, Heidelberg (2008)
15. Shahzad, A., Mebarki, N.: Data mining based job dispatching using hybrid simulation-optimization approach for shop scheduling problem. Eng. Appl. Artif. Intell. **25**(6), 1173–1181 (2012)

A Simple Dual-RAMP Algorithm for the Uncapacitated Multiple Allocation Hub Location Problem

Telmo Matos$^{(\boxtimes)}$ [ID], Fábio Maia, and Dorabela Gamboa [ID]

CIICESI – Center for Research and Innovation in Business Sciences
and Information Systems, School of Management and Technology,
Polytechnic of Porto, Porto, Portugal
{tsm, dgamboa}@estg.ipp.pt, fabio7maia@gmail.com

Abstract. This paper presents a Dual-RAMP algorithm for the solution of the multiple allocation hub location problem (UMAHLP). This approach combines information of a lagrangean relaxation procedure with subgradient optimization on the dual side with primal-feasible solutions on primal side, that are obtained by a simple improvement method. The overall performance of the proposed algorithm was tested on standard Australian Post (AP) and Civil Aeronautics Boarding (CAB) instances, comprising 192 test instances. The effectiveness of our approach has been proven by comparing our results with other state-of-the-art algorithms.

Keywords: Hub location problem · Primal-dual algorithm ·
Lagrangean relaxation · RAMP

1 Introduction

In this paper we tackle the Uncapacitated Multiple Allocation Hub Location Problem (UMAHLP) in which the objective is to choose the nodes that will act as hubs and the optimal assignment of other nodes to the selected hubs. This problem assumed a great focus lately and gave rise to many algorithms applied to different variants of the problem (some surveys on HLP can be found in [1, 9, 10]).

The first linear programming formulation for the HLP with and without capacity constraints was proposed by Campbell [5], where the author addressed single and multiple allocation. Klincewicz [13] presented an algorithm that applies the dual ascent and dual adjustment procedure together with a branch-and-bound method for UMAHLP. Later, Mayer and Wagner [15] implemented some improvements to the dual ascent procedure, considering the aggregated (considering the combination of $i - j - k - m$) model formulation despite the disaggregated one (replacing the combination of node pair (i, j) and hub pair (k, m) by single variables h and l, respectively) leading to a reduced number of constraints in the constraint set and reducing the computation time of the branch-and-bound procedure. Kratica *et al.* [14] developed a genetic based algorithm using the cache technique. The main goal of this technique is to store the "genetic code" of the visited solutions to avoid returning to the same solution.

© Springer Nature Switzerland AG 2020
A. M. Madureira et al. (Eds.): HIS 2018, AISC 923, pp. 331–339, 2020.
https://doi.org/10.1007/978-3-030-14347-3_32

Later in 2007, Cánovas *et al.* [6] proposed a new heuristic based on dual ascent, also using a branch-and-bound technique. The dual ascent procedure is initiated with preprocessing to improve the solution, following the feasibility of each solution. Next, the dual adjustment procedure takes place and the algorithm continues in the primal side with an exact method. The algorithm's performance was tested on a standard benchmark instance set providing results to instances up to 120 nodes. In the next year, Camargo *et al.* [4] presented an algorithm based on benders decomposition [3] to solve the problem. This approach divides the problem into two simple problems: a higher-level problem (or master problem), which determines the chosen hubs and a lower-level problem, (known as subproblem) that defines the allocation of nodes to the chosen hubs. Contreras *et al.* [7] proposed an exact algorithm to solve large instances for this problem, performing tests with instances up to 500 nodes. The algorithm is based on benders decomposition with the inclusion of several features such as reduction tests, and a heuristic procedure incorporating two distinct phases (namely the estimation and intensification phase), which aims to construct an initial solution and to generate feasible solutions covering the sets of open hubs obtained in the previous phase, respectively.

More recently, Mokhtar *et al.* [16] also used the benders procedure for the UMAHLP. A modified version of the benders procedure is presented with different tuning parameters and with subproblems reformulation. These subproblems are then solved using a minimum cost network flow algorithm resulting in a more effective benders algorithm with a smaller number of benders iterations and therefore reduced running times. The performance of the algorithm is compared only with other benders algorithms ([3] and [7]) and for AP and CAB instances, taking in average around two thirds less computational time (with the same solutions quality) then the other algorithms with the same technique.

2   Problem Description

UMAHLP is a well-known combinatorial optimization problem belonging to the class of the NP-Hard problems [17]. This problem can be described as follows. Consider the complete graph $G = (N, A)$, where N is the set of nodes $N = \{1, 2, \ldots, n\}$, that correspond to origins/destinations as well as potential hub locations. Let w_{ij} be the flow between i and j. For each node $i \in N$, let f_i the fixed set-up cost of hub i. The distance between nodes i and j is assumed to satisfy the triangle inequality and is denoted by d_{ij}. We will use these distances as a measure of the per unit flow transportation costs along the links of the graph. These distances are weighted by some discount factors, denoted χ, α and δ, to represent the collection, transfer and distribution costs per unit of flow, respectively. The objective consists in choosing the set of nodes to be established as hubs, while minimizing the total cost of assigning all the non-hubs to the chosen hubs. The total cost of routing the flow along the path $i - j - k - m$ (these are the paths between origin destination pairs, where i and j represents the origin and destination, respectively, and k and m are the hubs to which i and j are allocated, respectively) is given by:

$$F_{ijkm} = w_{ij}\left(\chi d_{ik} + \alpha d_{km} + \delta d_{mj}\right) \tag{1}$$

and for each pair $i, k \in N$ the following sets of binary decision variables are defined by:

$$Z_k = \begin{cases} 1 & \text{if node } k \text{ is a hub;} \\ 0 & \text{otherwise.} \end{cases} \tag{2}$$

Variable Z_k denotes the establishment or not of a hub at node k, when $i = k$. An additional set of binary variables will be defined. These variables indicate if there is flow through each link of the graph. For each $i, j, k, m \in N$ the fraction of flow is defined by X_{ijkm}.

The mathematical formulation for UMAHLP is:

$$\text{UMAHLP} = min \sum_{k \in N} f_k Z_k + \sum_{i \in N} \sum_{j \in N} \sum_{k \in N} \sum_{m \in N} F_{ijkm} X_{ijkm} \tag{3}$$

$$s.t. \sum_{k \in N} \sum_{m \in N} X_{ijkm} = 1 \; \forall i, j \in N \tag{4}$$

$$\sum_{m \in N} X_{ijkm} \leq Z_{ik} \forall \, i, j, k \in N \tag{5}$$

$$\sum_{k \in N} X_{ijkm} \leq Z_{jm} \forall \, i, j, m \in N \tag{6}$$

$$Z_k \in \{0, 1\} \forall \, k \in N \tag{7}$$

$$X_{ijkm} \geq 0 \forall \, i, j, k, m \in N \tag{8}$$

Constraints (4) assure that every single node is assigned to one hub. Constraints (5) and (6) ensure that flow is only sent via open hubs. Constraints (7) and (8) are integrality and non-negativity constraints. The objective function (3) represents the total cost of hub establishments plus total cost of routing the flow along the path $i - j - k - m$. Being uncapacitated and multiple allocation, means that the hubs do not have capacity limits and each node can be assigned to more than one hub, respectively. Unlike the single allocation, where a node is assigned to only a hub, this version is more difficult to solve, involving much more variables in its formulation.

3   RAMP Algorithm for the UMAHLP

The Relaxation Adaptive Memory Programming (RAMP) emerged in 2005 by Rego [19], combining fundamental principles of mathematical relaxation with con-cents of adaptive memory programming techniques, with the objective of incorporating information obtained by primal and dual solutions spaces. RAMP comprises two levels of sophistication, namely Dual-RAMP and Primal-Dual RAMP. At the first level of sophistication (Dual-RAMP or simply RAMP), this framework explores more intensively the dual side, restricting the primal side interaction to the projection of dual solutions to the primal solutions space and to the improvement of these solutions.

Higher levels of sophistication (Primal-Dual RAMP or simply PD-RAMP) allow a more intensive exploration of the primal side, incorporating the simple level, the Dual-RAMP, with more complex memory structures. Several combinatorial optimization problems have already been solved by RAMP applications, producing excellent results, in some cases with new best-known solutions. Some examples of RAMP approaches with different levels of sophistication are the capacitated minimum spanning tree [20], the linear ordering problem [13], the resource constrained project scheduling problem [21], or the capacitated single allocation hub location problem [18], among others.

We propose a Dual-RAMP algorithm for the UMAHLP that uses a Lagrangean Relaxation on the dual side and an improvement method on the primal side. With this simple level of sophistication, the algorithm explores and combines dual and primal solutions spaces achieving excellent results for standard data instances. Below, (Fig. 1) the RAMP model for UMAHLP is presented.

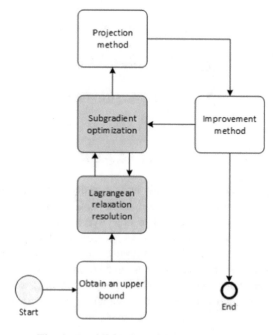

Fig. 1. Dual-RAMP model for UMAHLP.

The algorithm starts by getting a local search procedure to construct a feasible solution to get a good starting solution to lagrangean relaxation (LR). Next, for each iteration of the algorithm, LR is solved and a projection method is used to project the solution onto the primal solutions space and then it is improved by an improvement method. After the solution is improved, the algorithm alternates to dual solution space, completing one iteration of the LR. The algorithm stops when it reaches one of the four stopping criteria:

- The agility parameter is less than 0.005;
- The norm reaches the value 0;
- The maximum number of iterations is reached;
- The difference between the upper bound and lower bound is less than 1.

3.1   Dual Method

In our design, the dual phase relies on the exploration of the dual solutions space with subgradient optimization to solve the dual problem obtained by the Lagrangean Relaxation. At each iteration of the subgradient optimization, a solution for the relaxed problem is obtained, then a Projection Method projects it to the primal solutions space and an Improvement Method tries to improve it. Specifically, if we relax constraints (5) and (6), we obtain the following optimization problem:

$$
\begin{aligned}
L(\lambda) \ = min &\sum_{i \in N} \sum_{j \in N} \sum_{k \in N} \sum_{m \in N, m=k} (F_{ijk} + \lambda_{ijk}) \\
+ &\sum_{i \in N} \sum_{j \in N} \sum_{k \in N} \sum_{m \in N, m \neq k} (F_{ijkm} \lambda_{ijk} + \lambda_{ijm}) + \sum_{k \in N} f_k (Z_k - \sum_{i,j \in N} \lambda_{ijk}) \\
& s.t. (4), (7) \ and \ (8)
\end{aligned}
$$

$$(9)$$

The problem given by Eq. (9) can be separated in two subproblems, denoted as $LY(\lambda)$ and $LX(\lambda)$, for Y space variables and for X space, respectively.

Regarding the $LY(\lambda)$, the subproblem is:

$$
LY(\lambda) = min \sum_{k \in N} Z_k \left(f_k - \sum_{i,j \in N} \lambda_{ijk} \right)
$$

$$(10)$$

s.t.

$$
\sum_{k \in N} Z_k \geq 1
$$
$$(7)$$

$$(11)$$

that can be solved by inspection $\left(min \left\{ k : f_k - \sum_{i,j \in N} \lambda_{ijk} \right\} \right)$ and form $LX(\lambda)$ we get:

$$
\begin{aligned}
LX(\lambda) \ = min &\sum_{i \in N} \sum_{j \in N} \sum_{k \in N} \sum_{m \in N, m=k} (F_{ijkm} + \lambda_{ijk}) \\
+ &\sum_{i \in N} \sum_{j \in N} \sum_{k \in N} \sum_{m \in N, m \neq k} (F_{ijkm} \lambda_{ijk} + \lambda_{ijm})
\end{aligned}
$$
$$(12)$$
$$s.t. \ (4) \ and \ (8)$$

For the previous problem, it is not as simple as solving the $LY(\lambda)$, especially for large datasets. For each node pair (i, j), it is necessary to evaluate the minimum cost considering only one hub k or hub pair (k, m) and the associated lagrangean multipliers.

The relaxed problem $L(\lambda)$, can be obtained by the sum of the two subproblems, one for each space variables $(LY(\lambda) + LX(\lambda))$. The lower bound computation, Z_D will be $max_\lambda L(\lambda)$. After solving the dual problem, a subgradient optimization method will be used. At each iteration, the solution for the relaxed problem $L(\lambda)$ is obtained, and it is projected to the primal solution space through a projection method and it is improved by an improvement method. Let the lower bound be $Z_D = 0$, and starting with an initial value for the initial upper bound Z_{UB} (that we will see in Primal Phase) and letting the lagrangean multipliers be $\lambda_\lambda = 0$. The agility parameter π is initialized with the value 2, where this parameter is divided by 2 every 30 consecutive iterations without improving Z_D and it is reset to the original value every 100 iterations to decrease the step size. The step size (Δ) is calculated as $\Delta = \frac{\pi\left(Z_{UB}-L(\lambda)\right)}{||\delta||}$. For a given vector λ, let $z(\lambda)$ and $x(\lambda)$ represent the optimal solution to $L(\lambda)$. A subgradient $\delta(\lambda)$ of $L(\lambda)$, is given by

$$\delta(\lambda) = \left(\sum_m X_{ijkm}(\lambda) + \sum_{m,m\neq k} X_{ijmk}(\lambda) - Z_k(\lambda)\right) \tag{13}$$

Finally, to determine the set of the lagrangean multipliers λ that maximizes the function $L(\lambda)$, the subgradient method requires generating a new sequence of multipliers, one for each iteration of the Lagrangean Relaxation, given by $\lambda^{iter+1} = \lambda^{iter} + \Delta\delta(\lambda)$.

3.2   Primal Method

The LR procedure provides a dual solution from which it is possible to build a primal feasible solution, through a projection method. This method is very simple and takes in consideration that once the hubs are defined, a feasible primal solution is obtained.

In each iteration of the LR procedure, the projection method projects a dual solution to the primal solution space, and the resulting solution is improved by an improvement method (IM) based on Tabu Search [11, 12]. The IM starts with an admissible initial solution S_0, with the objective function value vs_0. The initial size of the tabu list $ts = p/2$, and p is the number of hubs in the solution returned by the greedy method. The maximum number of iterations without improving the best solution found so far depends on the number of nodes of the problem. After several tests performed and taking in consideration the computational time, $\sqrt{numNodes}$ was used as the maximum number of iterations. For each iteration, this improvement method checks if there is any move that improves the current solution. If so, the move is applied, and the current solution is updated. If there are no moves that improve the current solution, a hub is randomly chosen to be closed. The size of tabu list is dynamic, that is, it will be changed according to whether the algorithm finds a current better solution or the best so far. Its size varies between 1 and $p+2$. The tabu list influences the choice of the next possible move, since only moves outside the tabu list are permitted. If the move is tabu, it will only be considered if it improves the best solution found so far (the common aspiration criterion).

4   Computational Results

The performance of the RAMP algorithm was evaluated on a standard AP (Australia Post) dataset introduced by Ernst and Krishnamoorthy [8] and obtained in Beasley's OR-Library [2] and CAB (Civil Aeronautics Boarding) data benchmark proposed by O'Kelly [17]. The AP benchmark includes 184 instances, of which 28 are standard instances available by Beasley [2] and are referred as AP-1, 84 were proposed by Cánovas et al. [6] (where the author has 3 sets of 28 asymmetric instances) referred as AP-2 and 72 were proposed by Contreras et al. [7] referred as AP-3. The CAB benchmark consists of 8 instances, two instances proposed by O'Kelly [17] and the remaining six are variants of standard instances, proposed by Cánovas et al. [6].

The algorithm was coded in C programming language and run on an Intel Pentium I7 2.40 GHz (only one processor was used) with 8 GB RAM under Ubuntu operating system. The Dual-RAMP algorithm was compared with the state-of-the-art algorithms for the solution of the UMAHLP. The best-known approaches are the dual ascent combined with the branch and bound method proposed by Cánovas et al. [6] (DA-BB), genetic algorithm proposed by Katrica et al. [14] (GA) and benders decomposition heuristic proposed by Contreras et al. [7] (BD).

For all results tables, "instances" stands for the designation of the dataset and "OF" is the number of optimal/best-known solutions found. The value of the column "gap" was computed as $(UB - Z*)/UB*100$ ($Z*$ is the optimal/best-known solution and UB is the value of the upper bound obtained). The "cpu" column is the computational time (in seconds) needed to achieve the total running time. The "b-k" is the best-known solution present in literature and "Nb-k" is the new best-known solution found by RAMP algorithm.

Table 1 shows the results for AP benchmark proposed by Beasley [2], Cánovas et al. [6] and by Contreras et al. [7]. Analyzing the table, the RAMP algorithm achieved the optimal solutions for all instances, while GA cannot reach the optimal solution in one instance in AP-1 and three instances in AP-2. Regarding the AP-3 dataset and comparing our algorithm with BD, RAMP algorithm achieved 70 out of 72 optimal/best-known solutions in a short computational time while BD fail to obtain the optimal solutions for four instances. Although computational time is not comparable, we choose to present this result.

Table 1. Aggregated results for the AP data instances (Beasley, Cánovas and Contreras) and comparing RAMP, GA and BD algorithms.

Instances	GA			BD			RAMP		
	of	gap	cpu	of	gap	cpu	of	gap	cpu
AP-1	27/28	0.030	11.61	–	–	–	28/28	0.0	3.92
AP-2	81/84	0.003	9.11	–	–	–	84/84	0.0	2.44
AP-3	–	–	–	68/72	0.010	9.50	28/28	0.0	11.21

For the last table, Table 2 we can see that our algorithm was able to find all optimal/best-known solutions. Comparing with DA-BB algorithm that managed to achieve in average 0.921% of deviation regarding the best-known solution, RAMP achieved all best-known solution and for one specific instance, it managed to obtain an even better result.

Table 2. Results for the CAB data instances and comparing RAMP vs DA-BB algorithm.

Instances	b-k	DA-BB		RAMP		Nb-k
		gap	cpu	gap	cpu	
25La	390369	2.020	0.000	0.000	0.056	–
25L1	196903	0.000	0.000	0.000	0.041	–
25L5	234523	2.600	0.000	0.000	0.034	–
25L9	256691	0.380	0.000	0.000	0.034	–
25Ta	484591	1.950	0.000	0.000	0.041	–
25T1	258577	0.420	1.000	−0.001	0.042	258557
25T5	279144	0.000	0.000	0.000	0.034	–
25T9	284852	0.000	0.000	0.000	0.033	–
Average		0.921	0.125	0.000	0.150	–

For the CAB instances, RAMP algorithm managed to improve the best-known solutions for instance 25T1. For this instance, it found a new solution with value 258557 as indicated in the table below in column "Nb-k".

5   Conclusions

The Hub Location Problem (HLP) has been extremely studied by the scientific community due to its complexity and the vast real-world applications, motivating many authors to present state-of-the-art algorithms. A Dual-RAMP algorithm to solve the UMAHLP is proposed, combining lagrangean relaxation with subgradient optimization in the dual side with an improvement method on the primal side.

Numerous tests were performed to access the effectiveness of our algorithm. For the 192 standard instances used, the RAMP algorithm successful achieved excellent results in very reduced time outperforming all other best approaches in literature. In fact, for all instances shown, the proposed algorithm achieved all optimal/best-known solution and found a new best-known solution for one instance in the CAB dataset.

Once again, the RAMP approach was able to efficiently solve a complex optimization problem. The use of primal-dual exploration techniques, that use adaptive memory in metaheuristics such as tabu search, are extremely efficient when applied to such problems. The application of this framework to other complex optimization problems is expected to obtain results of the same quality as we obtained for the UMAHLP.

References

1. Alumur, S., Kara, B.Y.: Network hub location problems: the state of the art. Eur. J. Oper. Res. **190**(1), 1–21 (2008)
2. Beasley, J.: OR-library: distributing test problems by electronic mail. J. Oper. Res. Soc. **65**, 1069–1072 (1990)
3. Benders, J.F.: Partitioning procedures for solving mixed-variables programming problems. Numer. Math. **4**(1), 238–252 (1962)
4. de Camargo, R.S., et al.: Benders decomposition for the uncapacitated multiple allocation hub location problem. Comput. Oper. Res. **35**(4), 1047–1064 (2008)
5. Campbell, J.F.: Integer programming formulations of discrete hub location problems. Eur. J. Oper. Res. **72**(2), 387–405 (1994)
6. Cánovas, L., et al.: Solving the uncapacitated multiple allocation hub location problem by means of a dual-ascent technique. Eur. J. Oper. Res. **179**(3), 990–1007 (2007)
7. Contreras, I., et al.: Benders Decomposition for large-scale uncapacitated hub location. Oper. Res. **59**(6), 1477–1490 (2011)
8. Ernst, A.T., Krishnamoorthy, M.: Solution algorithms for the capacitated single allocation hub location problem. Ann. Oper. Res. **86**, 141–159 (1999)
9. Farahani, R.Z., et al.: Hub location problems: a review of models, classification, solution techniques, and applications. Comput. Ind. Eng. **64**(4), 1096–1109 (2013)
10. Fernandez, E.: Locating Hubs: an overview of models and potential applications. (2013)
11. Glover, F.: Tabu search—part I. ORSA J. Comput. **1**(3), 190–206 (1989)
12. Glover, F.: Tabu search—part II. ORSA J. Comput. **2**(I), 4–32 (1990)
13. Gamboa, D.: Adaptive memory algorithms for the solution of large scale combinatorial optimization problems. Ph.D. thesis. Instituto Superior Técnico, Universidade Técnica de Lisboa (2008). (In Portuguese)
14. Kratica, J., et al.: Genetic algorithm for solving uncapacitated multiple allocation hub location problem. Comput. Informatics. **24**(4), 415–426 (2005)
15. Mayer, G., Wagner, B.: HubLocator: an exact solution method for the multiple allocation hub location problem. Comput. Oper. Res. **29**(6), 715–739 (2002)
16. Mokhtar, H. et al.: A new Benders decomposition acceleration procedure for large scale multiple allocation hub location problems. In: International Congress on Modelling and Simulation, pp. 340–346 (2017)
17. O'Kelly, M.E.: A quadratic integer program for the location of interacting hub facilities. Eur. J. Oper. Res. **32**(3), 393–404 (1987)
18. Matos, T., Gamboa, D.: Dual-RAMP for the capacitated single allocation hub location problem. In: Gervasi, O., et al. (eds.) Proceedings of the 17th International Conference on Computational Science and Its Applications – ICCSA 2017, Part II, Trieste, Italy, 3–6 July 2017, pp. 696–708. Springer (2017)
19. Rego, C.: RAMP: a new metaheuristic framework for combinatorial optimization. In: Rego, C., Alidaee, B. (eds.) Metaheuristic Optimization via Memory and Evolution: Tabu Search and Scatter Search, pp. 441−460. Kluwer Academic Publishers (2005)
20. Rego, C., et al.: RAMP for the capacitated minimum spanning tree problem. Ann. Oper. Res. **181**(1), 661–681 (2010)
21. Riley, R.C.L., Rego, C.: Intensification, diversification, and learning via relaxation adaptive memory programming: a case study on resource constrained project scheduling. J. Heuristics (2018)

Solving Flexible Job Shop Scheduling Problem Using Hybrid Bilevel Optimization Model

Hajer Ben Younes, Ameni Azzouz[(✉)], and Meriem Ennigrou

Institut Supérieur de Gestion, SMART Laboratory, Université de Tunis,
Tunis, Tunisia
ameni.azzouz@isg.rnu.tn

Abstract. Flexible Job Shop Problem (FJSP) has an important significance in both fields of production management and combinatorial optimization. This problem is decomposed into two sub-problems: the assignment problem and the scheduling problem. Following this structure, we consider in this work the FJSP as a bilevel problem. For that, we are interested to solve this problem with bilevel optimization method in which the upper level optimizes the assignment problem and the lower level optimizes the scheduling problem. Therefore, we propose, for the first time, an hybrid bilevel optimization model named HB-FJSP based on both exact and approximate methods to solve the FJSP in order to minimize the makespan. The computational results confirm that our model HB-FJSP provides better solutions than other models.

Keywords: Flexible Job Shop Problem · Makespan · Hybridization · Bilevel optimization · Genetic Algorithm

1 Introduction

Nowadays, the manufacturing industry faces new challenges like an increasing product varieties, higher product customization, shorter lead-times and reduced product life cycles. Scheduling is one of the most known problems in manufacturing and service industries. The Job Shop scheduling Problem (JSP) is considered as the most active research field with this area. It consists of a set of different machines that perform a set of jobs. We are focused on a generalization of the classical JSP named Flexible Job Shop problem (FJSP) where each operation can be performed by more than one machine. According to [6], two types of approaches have been used in the literature to solve the FJSP. The first category called hierarchical approach. This approach reduces the complexity of the FJSP by dividing the problem into two sub-problems: the routing and the sequencing sub-problems. Firstly, we solve the assignment of operations to their eligible machines and then we find the feasible schedule by solving the sequencing problem. Alternatively, in the second category called integrated approach these latter are made concurrently.

© Springer Nature Switzerland AG 2020
A. M. Madureira et al. (Eds.): HIS 2018, AISC 923, pp. 340–349, 2020.
https://doi.org/10.1007/978-3-030-14347-3_33

For the hierarchical approach, [6] was the first who used this decomposition for the FJSP. He solved the assignment problem using some dispatching rules and the sequencing problem using a tabu search metaheuristic. In [17], the authors proposed a hierarchical solution approach for the multi-objective FJSP where they used the particle swarm optimization algorithm to assign operations on machines and simulated annealing algorithm to schedule operations on each machine. In [8], the authors developed a hierarchical approach used simulated annealing algorithm to solve the FJSP with overlapping in operations in order to minimize the makespan.

The integrated approach is much more difficult to solve than the hierarchical approach, but in general achieves better results. In [9], the authors proposed a particle swarm optimization (PSO) algorithm to solve the FJSP in order to minimize the makespan time. The different functions of the proposed PSO that are proposed to evolve concurrently the assignment and scheduling problems. In [4], an integrated approach for the FJSP with sequence-dependent setup times using a variable neighborhood search algorithm is proposed in order to minimize the makespan time and mean tardiness. They used a neighborhood structure related to sequencing problem and a neighborhood structure related to assignment problem in order to generate the neighboring solutions. In [18], the authors developed a memetic algorithm to solve the multi-objective FJSP. They incorporated a tabu search as local search algorithm into the Nondominated Sorting Genetic Algorithm II (NSGA- II) in which the local search was applied to every chromosome generated by the genetic algorithm. [3] introduced a hybrid algorithm based on a genetic algorithm combined with variable neighborhood search to solve the FJSP with sequence dependent setup time.

Following the structure of our considered problem, we are interested to a special case of hierarchical architecture which is the bilevel optimization approach. This latter was first introduced by von Stackelberg in 1934 [16]. It is inspired by a hierarchical game composed of two players known as Stackelberg game that has a leader (upper level) and a follower (lower level). Several researchers have considered a great attention for the bilevel optimization problems among them we cite, [12] where the authors formulated the vehicle routing problem as a bilevel optimization problem where the upper level consists in assigning the set of customers to the vehicles and in the lower level the decision maker finds the optimal routes of these assignments. The decision maker of the first level, once the cost of each routing has been calculated in the second level, estimates which assignment is the better one to choose. In [7], the authors proposed a bilevel program for solving the knapsack problem where the leader (upper level) controls the knapsacks capacity and the follower (lower level) solves the knapsack problem including the capacity set by the upper level. [10] formulated the assembly job shop scheduling problem as a bi-level programming approach where the project manager is considered as the upper level (the leader) aims to minimize the earliness and tardiness of completed jobs and the shop floor manager is considered as the lower level aims to minimize the average shop floor throughput time. The interest of using the bilevel optimization is that it facilitates the search for solutions and improves the search space to explore more promising areas in both levels.

In this paper, we propose for, the first time, a hybrid bilevel optimization algorithm based on both exact and approximate methods to solve the FJSP where the upper level optimizes the assignment problem using a genetic algorithm and the lower level optimizes the sequencing problem via exact method. The main objective of this research is to design a new bilevel hybrid algorithm in order to improve the resolution efficiently of this problem. This paper is organized as follows. Section 2 defines the considered problem. Section 3 describes our proposed model HB-FJSP. Section 4 shows and discuss the performance of our HB-FJSP on a set of benchmark problems. Conclusions and some future works are presented in Sect. 5.

2   FJSP Formulation

The FJSP can be formulated as follows: let $J = J_1, J_2, ..., J_n$ be a set of n jobs performed on a set of m machines noted as $M = M_1, M_2, ..., M_m$. Each job consists of a sequence of n_i operations $(O_{i,1}, O_{i,2}, ..., O_{i,n_i})$. Each operation $O_{i,j}$ requires one machine from a set of eligible machines. The execution of each operation j of a job i on a specific machine requires a fixed processing time noted as $p_{i,j,k}$. In this paper, the objective is to minimize the makespan, and the main assumptions are described as follows:

- All machines and jobs are available at time zero.
- Machines are independent from each other.
- Jobs are independent from each other.
- At a given time, one machine can handle at most one operation.
- There are precedence constraints among the operations of the same job.
- Each operation may be performed on one machine out of a set of capable machines without preemption.
- Transportation times of jobs and setup time of machines are not considered.
- Same priority for all the jobs.

3   Proposed HB-FJSP

In this paper, we consider the FJSP as a bilevel problem in which the upper level optimizes the assignment problem and the lower level optimizes the sequencing problem. We solve the upper level by a Genetic Algorithm (GA) and the lower level by an exact method. We detail the two optimization levels in the following:

3.1   Upper Level Optimization Procedure

In this model, we opt to use an evolutionary algorithm in order to optimize the assignment of operations to their eligible machines. The step-by-step procedure of the upper level GA is described as follows:

Step 1 (Encoding solution): Our proposed representation is designed as a binary matrix [2].

Step 2 (Initial population): Generate firstly an initial population. In order to increase the quality of the first population, we generate feasible solutions from the beginning, i.e. we fix for each operation the available machines where that can be performed on them and then we assign randomly this operation to exactly one of these machines.

Step 3 (Selection): In this step, we randomly select K individuals from the population in order to maintain the diversity of solutions.

Step 4 (Crossover): We have adapted the crossover operator "order 1" to our GA [15]. This operator generates offspring by choosing randomly two positions X and Y in Parent1. The part between these two positions is copied to the offspring1. Then, starting from the position $Y + 1$ of Parent2 and filled the rest of elements of the offspring1. To generate the offspring2 just repeat the same steps starting with Parent2.

Step 5 (Mutation): We select a random operation to which we assign a new machine. This new machine is selected from the available machines.

Step 6 (Evaluation): We combine both upper level solutions (the assignment in our case) and lower level solutions (the sequencing in our case) and we evaluate them based on the lower level optimization procedure.

Step 7 (Stop criterion): If the stopping criterion is met then return the best solution found by the upper level; otherwise, return to Step 3.

3.2 Lower Level Optimization Procedure

In order to optimize the operations' sequencing of all jobs according to the passed assignment upper solution, we give in the following the used parameters and the decision variables used in our lower level optimization procedure. We used IBM ILOG CPLEX CP Optimizer to implement the considered model.

The proposed mathematical model is defined as follows:

Minimize C_{max}

$$
\begin{cases}
S_{ijk} \geq C_{i\prime j\prime k\prime} - (Y_{ iji\prime j\prime k}) \times L & \forall i < i\prime, \forall j \in O_i, \forall j\prime \in O_{i\prime}, \forall k \in M_j \cap M_{j\prime}, \ (1) \\
S_{i\prime j\prime k\prime} \geq C_{ijk} - (1 - Y_{iji\prime j\prime k}) \times L & \forall i < i\prime, \forall j \in O_i, \forall j\prime \in O_{i\prime}, \forall k \in M_j \cap M_{j\prime}, \ (2) \\
\displaystyle\sum_{k \in M_j} S_{ijk} \geq \sum_{k \in M_j} C_{i,j-1,k} & \forall i \in J, \forall j \in O_i - \{O_{if_i}\}, \ (3) \\
C_i \geq \displaystyle\sum_{k \in M_j} C_{i,O_{il_i},k} & \forall i \in J, \ (4) \\
C_{max} \geq C_i & \forall i \in J, \ (5) \\
S_{ijk} \geq 0 & \forall i \in J, \forall j \in O_i, \forall k \in M_j, \\
C_{ijk} \geq 0 & \forall i \in J, \forall j \in O_i, \forall k \in M_j, \\
Y_{iji\prime j\prime m} \in \{0,1\} & \forall i < i\prime, \forall j \in O_i, \forall j\prime \in O_{i\prime}, \forall k \in M_j \cap M_{j\prime}, \\
C_i \geq 0 & \forall i \in J.
\end{cases}
$$

Indices and sets	Description
i	Jobs $(i,\ i' \in J)$
j	Operations $(j,\ j' \in O)$
k	Machines $(k \in M)$
J	The set of jobs
M	The set of machines
O	The set of operations
O_i	Ordered set of operations of job i $(O_i \subseteq O)$, where O_{if_i} is the first and O_{il_i} is the last element of O_i
M_j	The set of alternative machines on which operation j can be processed $(M_j \subseteq M)$
$M_j \cap M_{j'}$	The set of machines on which operations j and j' can be processed
Parameters	**Description**
t_{ijk}	The processing time of operation O_{ij} on machine k
L	A large number
Decision variables	**Description**
X_{ijk}	1, if machine k is selected for operation O_{ij}; 0, otherwise
S_{ijk}	The starting time of operation O_{ij} on machine k
C_{ijk}	The completion time of operation O_{ij} on machine k
$Y_{iji'j'k}$	1, if operation O_{ij} precedes operation $O_{i'j'}$ on machine k; 0, otherwise
C_i	The completion time of job i
C_{max}	Maximum completion time over all jobs (makespan)

The constraints (1) and (2) guarantee that the operations O_{ij} and $O_{i'j'}$ must not be performed simultaneously on any machine in the set of available machines. Constraint (3) assures the precedence relationships between the operations of the same job. Constraint (4) defines the completion time of the final operations of each job. Constraint (5) determines the makespan. The last constraints define the nature of decision variables.

3.3   HB-FJSP Basic Scheme

According to the two procedures described above, we present in the following the architecture of our hybrid bilevel model. In fact, a leader decision level focuses on the assignment of jobs to machines, while a follower is interested by the operations' sequencing. The Fig. 1 describes the flowchart of the proposed model based on hybrid bilevel optimization model.

Our considered model, as can be seen in Fig. 1, contains two levels of decision making: the upper level and the lower level. The upper level's (leader) objective is to affect each operation to one of the available machines and the lower level's (follower) objective is to find the best scheduling of operations in order to minimize the leader's objective function. In our model, the upper level is optimized with a GA where, first, we generate randomly the population. After that, we choose a set of random solutions for reproduction phase. According to

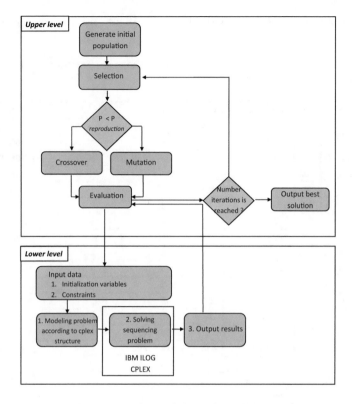

Fig. 1. Flowchart of the proposed model

a probability, the selected individual performs crossover or mutation. To solve
the sequencing problem, the lower level will be launched in order to evaluate
the set of solutions using CPLEX. Then, the output results return to the upper
level. These steps will be repeated until a certain number of iteration is reached.
Finally, we determine the best solution among all the solutions according to their
fitness.

It is important to note that the main motivation behind the design of FJSP
as a bilevel model is the possibility to well accelerate the diversity and the
convergence of the search by exploring more job assignment in the upper level
search space and then concentrate on one fixed assignment at each way, in order
to find the corresponding optimal job sequencing.

4   Experimental Study

In order to evaluate the performance of the proposed model, this section describes
the computational experiments. Our proposed HB-FJSP has been implemented
using the integrated development environment (IDE) Eclipse and IBM CPLEX
Optimization Studio V12.8 run on an Intel(R) Core(TM) i3-2328M 2.20 GHz
machine (in 64 bits mode) and 4 GB of RAM memory.

Table 1. The results of *Cmax* obtained on the HUdata benchmarks of Hurink et al. [11] set edata.

Instances	n	m	LB	UB	Exact approach	N2-1000	MAS-GATS	GATS+HM	MACROG-FJSP	HB-FJSP	
										Cmax	*GAP*(%)
La01	10	5	609	609	609	618	621	609	609	**609**	0
La02	10	5	655	655	655	656	704	655	655	**655**	0
La03	10	5	550	554	567	566	578	567	573	**550**	0
La04	10	5	568	568	568	578	607	568	579	**568**	0
La05	10	5	503	503	503	503	524	503	503	**503**	0
La06	15	5	833	833	833	833	833	-	-	**833**	0
La07	15	5	762	765	765	778	813	-	-	**762**	0
La08	15	5	845	845	845	845	860	-	-	**845**	0
La09	15	5	878	878	878	878	891	-	-	**878**	0
La10	15	5	866	866	866	866	873	-	-	**866**	0
La11	20	5	1078	1103	1106	1106	1129	-	-	1103	2.31
La12	20	5	960	960	960	960	960	-	-	**960**	0
La13	20	5	1053	1053	1053	1053	1053	-	-	**1053**	0
La14	20	5	1123	1123	1123	1123	1137	-	-	**1123**	0
La15	20	5	1111	1111	1111	1121	1229	-	-	**1111**	0
La16	10	10	892	915	915	924	1019	892	896	**892**	0
La17	10	10	707	707	707	757	773	707	707	**707**	0
La18	10	10	842	843	843	864	914	843	867	**842**	0
La19	10	10	796	796	799	813	937	804	806	799	0.37
La20	10	10	857	864	857	919	886	857	863	863	0.70

The benchmark problems used the 20 instances of edata taken from HUdata which is introduced by [11]. To evaluate the obtained results and to compare the performance of our model we use two kinds of metrics: the makespan which denotes the total completion time of all the jobs and the GAP which is obtained as follows: $GAP = Sol_{model} - Sol_{algo}/Sol_{algo}$ where Sol_{model} denotes the objective function of our model and Sol_{algo} indicates the objective function of the lower bound. In our experiment, we tested different values for our parameters, and computational experience proves that the following values more effective for the GA:

- Population size: 100 individuals;
- Crossover: probability 0.8;
- Mutation: probability 0.2;
- Stopping criterion: Number of iterations.

We compare our proposed model HB-FJSP with an exact approach proposed by [5] which is based on constraint programming. Moreover, we compare our results against four approximate approaches; the first one presented by [11] where the authors developed a tabu search algorithm denoted as N2-2000 based

Table 2. The results of CPU(s) obtained on the HUdata benchmarks of Hurink et al. [11] set edata.

Instances	N2-1000	GATS+HM	HB-FJSP
La01	6.1	24.64	9.97
La02	39.2	4.65	18
La03	3.5	10.67	35
La04	39.4	22.13	32
La05	5.7	10.22	9
La06	107.8	–	15
La07	119.3	–	29
La08	115	–	19
La09	129.7	–	20
La10	87.3	–	12
La11	276.1	–	26
La12	271.1	–	21
La13	176.3	–	26
La14	266.2	–	25
La15	195.9	–	69
La16	12.5	73.14	57
La17	13.0	116.58	54
La18	7.3	34.98	60
La19	11.4	36.88	61
La20	9.2	70.36	60

on a neighborhood structure. The second one proposed by [1], which is a multi agent model based on a hybridization between genetic algorithm and tabu search, called MAS-GATS model, for the FJSP. The third one presented by [14] where the authors developed a hybrid metaheuristics based on genetic algorithm combined with tabu search in a holonic multi agent model (GATS+HM). The last one established by [13], denoted as MACROG-FJSP, where the authors developed an hybrid model based on chemical reaction optimization with greedy algorithm.

As can be seen in Table 1, our HB-FJSP outperforms the exact approach in 5 out of 20 problems, N2-1000 in 12 out of 20 problems, MAS-GATS in 17 out of 20 problems, GATS+HM in 3 out of 10 instances and MACROG-FJSP in 5 out of 10 instances. For the comparison with the literature lower bound (LB), the HB-FJSP attains the optimal solutions in 17 out of 20 instances. It is evident from this table that HB-FJSP keeps its robust performance and generates the better GAP values in different problem sizes. Therefore, the results indicate the successful hybridization of GA and an exact method. This is due to the fact that the GA has guaranteed a better assignment whereas the exact method has assured a better scheduling of all the jobs.

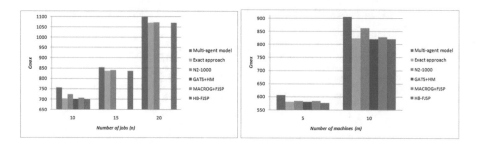

Fig. 2. Variation of $Cmax$ VS. Number of jobs (n) and Number of machines (m)

To further evaluate the performance of our hybrid bilevel model, we describe the obtained CPU time in Table 2. We remark that our HB-FJSP does not take too much time against the other algorithms. As can be seen, our model outperforms the N2-1000 in 12 of 20 instances with 60%. Moreover, it outperforms also the GATS+HM in 5 of 10 instances with 50%.

Figure 2 shows the interaction between the performance of our proposed model and the problem sizes for Hurink's benchmark. We state that our HB-FJSP keeps its robust performance in different problem sizes.

5   Conclusion

In this paper, we proposed a new hybrid bilevel model for the FJSP. Our model is composed of two levels. The GA is used to optimize the assignment problem i the upper level and an exact method is considered to solve the scheduling problem in order to minimize the makespan in the lower level. Our HB-FJSP demonstrate its effectiveness against several methods existing in the literature, and returned better objective function for the majority of test problems. For future work, it will be interesting to solve a multi-criteria FJSP, such as earliness and tardiness. Another interesting research direction which consider additional constraints, such as maintenance requirements in order to more reflect real situations.

References

1. Azzouz, A., Ennigrou, M., Jlifi, B.: Diversifying TS using GA in multi-agent system for solving flexible job shop problem. In: 12th International Conference on Informatics in Control, Automation and Robotics (ICINCO), vol. 2, pp. 94–101 (2015)
2. Azzouz, A., Ennigrou, M., Said, L.B.: Flexible job-shop scheduling problem with sequence-dependent setup times using genetic algorithm. In: 18th International Conference on Enterprise Information Systems, vol. 2, pp. 47–53 (2016)
3. Azzouz, A., Ennigrou, M., Said, L.B.: A hybrid algorithm for flexible job-shop scheduling problem with setup times (2017)

4. Bagheri, A., Zandieh, M.: Bi-criteria flexible job-shop scheduling with sequence-dependent setup times variable neighborhood search approach. J. Manuf. Syst. **30**(1), 8–15 (2011)
5. Behnke, D., Geiger, M.J.: Test instances for the flexible job shop scheduling problem with work centers. Helmut-Schmidt-University, Logistics Management Department, Hamburg, Germany (2012)
6. Brandimarte, P.: Routing and scheduling in a flexible job shop by tabu search. J. Ann. Oper. Res. **22**, 158–183 (1993)
7. Brotcorne, L., Hanafi, S., Mansi, R.: A dynamic programming algorithm for the bilevel knapsack problem. Oper. Res. Lett. **37**(3), 215–218 (2009)
8. Fattahi, P., Jolai, F., Arkat, J.: Flexible job shop scheduling with overlapping in operations. Appl. Math. Model. **33**(7), 3076–3087 (2009)
9. Girish, B.S., Jawahar, N.: A particle swarm optimization algorithm for flexible job shop scheduling problem. In: 2009 IEEE International Conference on Automation Science and Engineering, pp. 298–303 (2009)
10. Huang, G.Q., Lu, H.: A bilevel programming approach to assembly job shop scheduling. In: 2009 International Conference on Computers & Industrial Engineering, pp. 182–187 (2009)
11. Hurink, J., Jurisch, B., Thole, M.: Tabu search for the job-shop scheduling problem with multi-purpose machines. Oper.-Res.-Spektrum **15**(4), 205–215 (1994)
12. Marinakis, Y., Migdalas, A., Pardalos, P.M.: A new bilevel formulation for the vehicle routing problem and a solution method using a genetic algorithm. J. Global Optim. **38**(4), 555–580 (2007)
13. Marzouki, B., Driss, O.B., Ghedira, K.: Multi agent model based on chemical reaction optimization with greedy algorithm for flexible job shop scheduling problem. Procedia Comput. Sci. **112**, 81–90 (2017)
14. Nouri, H.E., Belkahla, O., Ghédira, D.K.: Solving the flexible job shop problem by hybrid metaheuristics-based multiagent model. J. Indus. Eng. Int. **14**(1), 1–14 (2017)
15. Pezzella, F., Morganti, G., Ciaschetti, G.: A genetic algorithm for the flexible job-shop scheduling problem. Comput. Oper. Res. **35**(10), 3202–3212 (2008)
16. Stackelberg, H.V.: Marktform und gleichgewicht. Julius Springer, Vienna, Austria (1934)
17. Xia, W., Wu, Z.: An effective hybrid optimization approach for multi-objective flexible job-shop scheduling problems. Comput. Indus. Eng. **48**(2), 409–425 (2005)
18. Yuan, Y., Xu, H.: Multiobjective flexible job shop scheduling using memetic algorithms. IEEE Trans. Autom. Sci. Eng. **12**(1), 336–353 (2015)

Deep Reinforcement Learning as a Job Shop Scheduling Solver: A Literature Review

Bruno Cunha[1(✉)], Ana M. Madureira[1], Benjamim Fonseca[2], and Duarte Coelho[1]

[1] Interdisciplinary Studies Research Center, School of Engineering, Polytechnic of Porto (ISEP/IPP), Porto, Portugal
{bmaca,amd,1130652}@isep.ipp.pt
[2] University of Trás-os-Montes and Alto Douro, Vila Real, Portugal
benjaf@utad.pt

Abstract. Complex optimization scheduling problems frequently arise in the manufacturing and transport industries, where the goal is to find a schedule that minimizes the total amount of time (or cost) required to complete all the tasks. Since it is a critical factor in many industries, it has been, historically, a target of the scientific community. Mathematically, these problems are modelled with Job Shop scheduling approaches. Benchmark results to solve them are achieved with evolutionary algorithms. However, they still present some limitations, mostly related to execution times and the difficulty to generalize to other problems.

Deep Reinforcement Learning is poised to revolutionise the field of artificial intelligence. Chosen as one of the MIT breakthrough technologies, recent developments suggest that it is a technology of unlimited potential which shall play a crucial role in achieving artificial general intelligence.

This paper puts forward a state-of-the-art review on Job Shop Scheduling, Evolutionary Algorithms and Deep Reinforcement Learning. It also proposes a novel architecture capable of solving Job Shop Scheduling optimization problems using Deep Reinforcement Learning.

Keywords: Job Shop scheduling · Deep Reinforcement Learning · Machine learning · Optimization · Evolutionary algorithm

1 Introduction

Job Shop scheduling is one of the most important industrial activities, and can be found in fields such as transportation (workforce planning), distribution and manufacturing (tasks allocation and resource assignments) [1,2]. Hence, there has always been an industry demand for the optimization of real-world Job Shop scheduling problems, which has attracted the attention of the scientific community.

© Springer Nature Switzerland AG 2020
A. M. Madureira et al. (Eds.): HIS 2018, AISC 923, pp. 350–359, 2020.
https://doi.org/10.1007/978-3-030-14347-3_34

A Job Shop can be described as a location in which a number of general purpose work stations exist and are used to perform a variety of tasks (i.e. the "jobs"). The Job Shop problem is characterized by the different jobs that are processed on the available machines. Each job consists of a sequence of tasks and the order in which they must be processed. Each task can only be performed on a specific machine. The critical decision is how to schedule the tasks on the machines in order to minimize the length of the schedule, i.e. the time it takes since the jobs start until all jobs are completed. This is where optimization is required, since the planning of such schedules is one of the hardest known combinatorial optimization problems [3]. In computer theory terms, not only is this an *NP-hard* problem, it is actually one of the worst known *NP-hard* problems [4].

The contemporary approaches to solve a Job Shop scheduling problem that present the best benchmark results are based on metaheuristics. The main advantage of using these methods is the fact that they can achieve good solutions with a low computational effort; i.e. they shall be seen as a mean to obtain not the optimal, but a "good enough" solution, to be used in an acceptable time. Some of these optimization techniques have been created using nature as an inspiration and, as such, are usually known as evolutionary algorithms. One of the most used and well-known evolutionary techniques is the Genetic algorithm, inspired by the process of natural selection.

Even if a Job Shop scheduling problem can be solved using these techniques, there is still a lot of room for improvements. Some of the current issues are the difficulty to generalize to other problems, the execution times and the required effort to develop these methods (i.e. modelling and codification) [5,6].

Looking at the computer science landscape of the latest years, machine learning - or, more specifically, deep learning - is the emerging technique that should be applied to this type of problem. Deep learning is revolutionizing almost every computer science field, obtaining extraordinary results and taking advantage of the advanced hardware we posses today [7]. While deep learning contains multiple branches of investigation (and is, by itself, a branch of machine learning), one of the most promising is its combination with reinforcement learning (RL), known as Deep Reinforcement Learning (Deep RL). Selected as one of the MIT 10 breakthrough technologies in 2017 [8], recent developments suggest that Deep RL is a field of unlimited potential which shall play a crucial role in achieving artificial general intelligence.

Considering that we have one of the worst known computer science problems and that current solutions still present some flaws, our main goal is to take advantage of the latest developments in deep learning to present a novel, modern method able to solve Job Shop scheduling problems. This shall be done using Deep RL, taking into account the benefits already brought to other fields.

The remaining sections are organized as follows: Sect. 2 starts by providing an overview of the Job Shop scheduling problem, followed by a review of the fields of evolutionary algorithms and Deep RL; Sect. 3 puts forward the proposed Deep RL architecture to solve Job Shop scheduling problems; and, at last, Sect. 4 contains the final conclusions and puts forwards ideas for future works.

2   Literature Review

2.1   Job Shop Scheduling

The scheduling function, in a manufacturing organization, is integrated in the global management decision level, considering the diverse organizational levels, their functional perspective or the integration of functions and business processes [9]. Most real manufacturing scheduling problems can be described as dynamic and extended versions of the classic Job Shop scheduling combinatorial optimization problem. In practice, scheduling environment tends to be dynamic, i.e. new jobs can arrive at unpredictable intervals, machines can breakdown, jobs can be cancelled and due dates, release dates or priorities can change [10].

In the last decades, there have been advances in research and application of metaheuristics based approaches, with special concern on real-world problems solving. However, most studies have focused on developing specific aspects of optimization either for static or deterministic scenarios. Several contributions have been proposed in literature to address different classes of manufacturing systems subject to unexpected and imponderable events, such as machine failure; rush orders; job cancellation; due date modification (postpone or advance) delay in the arrival of raw materials or components; and changes in job priority [11–13].

Dios and Framinan presented, in 2016 [14], a collection of all publicly available computer-based manufacturing scheduling tools. In total, 99 instances were discovered, starting from the year 1988. From the study of the most well-known metaheuristic-based manufacturing scheduling tools, some general insights are:

- Most tools are designed for a specific scheduling model/problem;
- The incorporation of human expertise into manufacturing scheduling tools is disputed and few systems allow a proper incorporation of the user knowledge - Recent contributions on how to achieve this are presented in [15];
- Pragmatically, it is impossible to test algorithms from one system into another one;
- Most systems attempt to design a "master" algorithm that can beat all the literature existing ones, instead of making available a collection of algorithms (such as different metaheuristics for the user to select [16]);
- Almost all authors consider the rescheduling problem. However, most approaches are relatively simple and/or driven by user inputs [14];
- User interfaces are not homogeneous, which can be crucial in the acceptance of any system by its target audience;
- Surprisingly, very few systems have as an optimization objective the makespan (i.e. performing all tasks on the least time possible);
- 50% of systems have the goal of fulfilling all orders before a specific date; meaning that they are not primarily concerned with the optimization of a plan, stopping when any feasible solution is achieved.

In sum, the area of computer-based manufacturing tools for optimization is very active. However, these tools still have difficulties in real world situations and, hence, human intervention is required to maintain real-time adaptation and optimization. There is an increasing interest and exploration on decision support systems that make use of autonomous entities to handle problems in complex manufacturing systems through a self-organized, automatic behaviour [17].

2.2 Evolutionary Algorithms

A Job Shop scheduling problem can be described as a combinatorial optimization problem. It is one of the worst known *NP-hard* problems (as previously explained). As such, it is not possible to obtain optimal solutions in acceptable times for large problems. This happens for two major reasons:

- The complexity of the problem makes it so that resolution to optimality is impracticable;
- Many problems must deal with a dynamic reality, meaning that when the optimal solution is reached the characteristics of the problem have changed already.

Exact methods can explore all available solutions for a problem in order to find the optimal solution [18]; however, they typically involve algorithms of exponential complexity, meaning they cannot be used on large problems – which is what most organizations have to deal with on a daily basis. Approximation methods have the goal of finding good solutions in an acceptable period [19]. When the optimal solution to the problem is not known or impossible to determine, these methods can consider the best solution known so far.

A metaheuristic is an approximation method. A term coined by Glover [20], it refers to a combination of methodologies that assume the existence of other algorithms that, at a lower level, are capable of generating feasible solutions to a problem and calculating the value of an objective function (for those solutions). The metaheuristic is responsible for coordinating the cooperation of the many lower level algorithms, guiding them during the solution search. The main advantage of using metaheuristics is the fact that they can achieve good solutions with a low computational effort; i.e. they shall be seen as a mean to obtain not the optimal, but a "good enough" solution, to be used in an acceptable time. Some examples of popular metaheuristic are *Genetic algorithms, Simulated Annealing, Tabu Search, Ant Colony Optimization, Particle Swarm Optimization* and *artificial bee colonies* [18,19].

Metaheuristics are also known as optimization techniques inspired in nature; as such, they are often called "evolutionary algorithms". The parallel between the search algorithms and what can be found in nature is quite harmonious; e.g. the Genetic algorithm is inspired by Darwin's Origin of Species, where the best individuals have the best chance of surviving [21].

Metaheuristics can be split in two main categories: local search based heuristics and population based heuristics. Nonetheless, some common concepts are shared between both metaheuristic categories: the representation/codification of solutions, the existence of an objective, fitness function and the capacity to deal with multiple constraints. Local search-based heuristics are based on the search for a single solution through a process that covers the solution space looking for the local/global optimum. This search usually follows well-defined rules (e.g. Tabu search) and may contain stochastic characteristics (e.g. Simulated Annealing). Population based heuristics can be described as an iterative process that improves a group (the "population") of solutions. A new population is generated and integrated on the current one based on a selection mechanism, and this goes on until the stopping criteria is met. Population metaheuristics can be split in two major groups: evolutionary computation and swarm intelligence. The Genetic algorithm metaheuristic is one of the most used evolutionary techniques, while the particle swarm optimization is one of the most famous swarm intelligence approaches. The algorithms based on evolutionary computation are stochastic and iterative, and do not guarantee convergence. The iterative process is completed by reaching a predefined maximum number of generations (i.e. stopping criteria) or by obtaining an acceptable solution. These algorithms operate on a set of individuals (usually designated by population). Each individual represents a potential solution to the given problem. A solution is obtained via a coding/decoding mechanism. Initially, the population is generated randomly and each individual is assigned a value through a fitness function. This value, known as the aptitude value [22], is a measure of its quality relative to the problem under consideration and is used to guide the search process.

2.3   Deep Reinforcement Learning

RL, at its core, is defined by the process of an agent learning the best actions based on environmental rewards. These are the two principal components of a RL system (i.e. the agent and the environment), but there are other relevant pieces that shall be defined. The state describes the current situation of the environment (e.g. for a chess program, the state is the positions of all the pieces on the board). The action is what an agent can do in each state (e.g. the available chess positions to move a piece). The reward is the result of taking an action on a state. "Reward" is an abstract concept that aims to represent the feedback from the environment of the agent actions, and is inspired in the scientific investigation of the human behaviour conducted in psychology [23]. A reward can be positive or negative; positive corresponding to good actions and negative representing punishments. The policy is known as the strategy of an agent. It is used to determine the next action that agent takes based on the current state. There are two types of policies: on-policy, which is built around learning the value based on an action derived from the current policy, and off-policy, based on obtaining the action from another policy (e.g. greedy). The value is the expected long-term return of an action based on a policy (instead of the short-term feedback of the reward). The Q-value is a variation of the value, but also considers the state; i.e. the long-term return of the current state after taking an action based on the agent's policy.

Based on the aforementioned components, it is easy to describe the main cycle of a simple RL algorithm: in the beginning, the agent knows what is the current state of the environment. Based on that state and its current knowledge, the agent performs an action. Then, the environment will return a new state and the reward of the action taken. With that reward, the agent updates its beliefs. This keeps on going until the environment signals the end, usually known as the terminal state.

To update its beliefs, the agent needs a learning algorithm. Considering an optimal system, the agent would have a model of all the transitions between state and action pairs (known as model-based systems). However, this is simply impractical due to the huge number of actions and states variations. Model-free systems are built around trial-and-error methods, removing the requirement to store all the state and action combinations. A brief explanation of the main learning algorithms, all model-free due to performance and historic achievements [24], is presented below.

Q-learning, proposed by Watkins and Dayan [25], is a RL algorithm based on the bellman equation, where the next action to take is chosen to maximise the next state Q-value. Hence, Q-learning is an off-policy approach. Fundamentally, Q-learning can be de-constructed in a two-dimensional array of action-state pairs that contain the probability of selecting the action on that state. When each pair is experienced, the affected probabilities will be updated. Q-learning avoids the exploration-exploitation problem [26] by using a common learning-rate solution. In spite of Q-learning being very powerful, it has some limitations regarding the capacity to generalize and to known the value of unseen states; i.e. it has difficulty in deciding how to proceed in completely new states [24].

Deep Q-Network (DQN) is the deep learning evolution of Q-learning. The DQN approach replaces the action-state matrix by neural networks (hence the deep learning). This neural network is able to provide an estimate of the Q-value; it receives the current value and outputs the corresponding one of taking each action. The first application of this technique was famously achieved by the DeepMind team, using it to play Atari games [27]. The training cycle of this neural network is based on the squared error between target and output Q-values; and also contains some key techniques, developed in [27]: experience replay (explained on the next chapter) and a separate target network.

Even if DQNs are successful, its action space is discrete. This is prohibitive in some environments (e.g. robotic physical tasks) where the action space is continuous. For these situations, a possible approach is to use a Deep deterministic policy gradient (DDPG) algorithm. It relies on its actor-critic architecture [28] where an actor decides the best action for a specific state and a critic is used to evaluate the policy that is being estimated by that actor. DDPG takes advantage of Q-learning and DQN advances, such has the aforementioned experience replay and separate target networks. The main issue with DDPG is the rare option to perform exploration of unknown actions; although some workarounds have already been proposed and experimented with [29].

3 Proposed Architecture

Due to the complexity and dimension of the Job Shop scheduling problem, existing solutions still present some flaws, as explained in detail in Sect. 2.1. Hence, an argument can be made for the exploration of the latest developments in deep learning to present a novel, modern method able to solve this type of problem. From the review on Deep RL (presented in Sect. 2.3) and the benefits it has brought to other fields, it is quite clear that these techniques have an immense potential, calling for, at the very minimum, to be experimented with. Worst case scenario, the results of the experience are not near what we can achieve today (with metaheuristics) will serve as a useful case study for the scientific community. However, if successful, it may revolutionize the scheduling system landscape.

The proposed architecture consists in the utilisation of a Deep RL system that receives, as input, a Job Shop scheduling problem and outputs an optimised valid solution. A diagram is presented in Fig. 1.

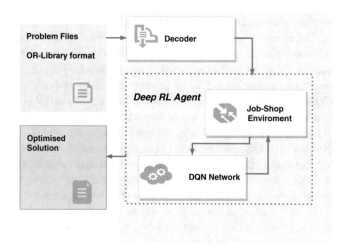

Fig. 1. Diagram of the proposed architecture.

To handle the input, we propose the adoption of the OR-Library format, developed by Beasley [30]. Amongst other benefits, the adoption of this standard format enables a faster and less cumbersome test procedure since benchmark optimal solutions have been published with it. Also, it shall enable the scientific community to better reproduce this proposal. A case could be made for the adoption of the Taillard format [31]; however, in practice, it is mostly used for Flow-shop problems. The output of the valid solution by the system shall specify, for each machine, the order and the start times of all the tasks that need to be processed. Nonetheless, the most interesting and innovative component is the Deep RL agent. The proposed architecture employs a DQN network. The two main reasons are the ability to gain experience via replays and the separate neural network to estimate the Q-value.

Experience replay is, in concept, the storage of an agent's experiences to be later used, in random batches, to train the network; i.e. to increase the likelihood of performing well in the scheduling problem.

The neural network itself is an attempt at estimating the "magical" formula that would calculate the exact reward of taking a specific action. Unfortunately, there is no such formula or, otherwise, all Job Shop scheduling problems would be solved. Since neural networks are universal function estimators, one can be used to estimate the magical formula. In sum, the deep neural network will try to predict the (long-term) quality of taking each action given the current state.

In order to enable this degree of training and interaction, it is fundamentally necessary to develop an environment where a Deep RL agent can interact with Job Shop scheduling problems. This makes it necessary to define state and action sets and to devise a formula that based on a state and an action calculates the next state and a reward. The rewards shall be calculated based on the progression of the overall completion of the plan (i.e. how many tasks remain), amongst other specific scheduling criteria (e.g. the total amount of machines' idle times).

To move forward the artificial intelligence field, researchers usually tend to follow two paths: the development of novel techniques and the attempts to improve current results on specific problems. The architecture proposed in this paper does not fall far from the tree. It takes advantage of cutting-edge machine learning techniques to solve a decades-old computer science problem. As such, the importance of the proposed architecture is that, if it proves successful, it will have the capacity to revolutionize multiple industries across various domains and, hence, sustaining the belief that reinforcement learning shall play a crucial role in achieving artificial general intelligence [8].

4   Conclusions and Future Work

This paper presents a literature review on Job Shop scheduling problems, evolutionary algorithms and Deep RL. Job Shop scheduling problems are present in many industries, being one of the fields where optimization is most necessary. This review presents existing applications, how they are being solved today and the issues that still exist. Evolutionary algorithms are a group of techniques that can be used whenever an optimal solution can not be found in a practical time. Usually known as metaheuristics, there is a wide array of robust approaches, from artificial bee colonies to the more famous genetic algorithm. The main advantage and shortcomings of using these techniques were discussed, and a detailed explanation of its underlying mechanisms was also presented.

Deep RL is one of the most exciting fields with a huge potential. This paper puts forward the concept of RL and its characteristics, such as states, actions, environments and rewards, and an explanation on how to incorporate deep learning techniques into classic RL. It is also presented a thorough review and comparison of several techniques that can be used to implement a Deep RL system.

All of these topics are relevant so that one can understand the current landscape of Job Shop scheduling problems, the current techniques that are being used today (including the existing limitations), and the most promising technologies that can be applied to it. With that in mind, this paper includes a proposal that takes advantage of cutting-edge developments to create an architecture capable of solving Job Shop scheduling optimization problems. Essentially, it consists on the application of a Deep RL agent that shall learn how to optimize Job Shop scheduling problems. The learning mechanism of the agent is based on experimentation, on the proposed environment, using a DQN approach.

Future works shall consist, foremost, in the development of a system that puts in practice the proposed architecture. After that, the Deep RL agent shall undergo the training phase, where it learns how to handle Job Shop scheduling problems. Whenever the system becomes ready and operational, it shall be deployed into real-world, non-academic problems. The architecture itself will require further refinements. Some examples are the formal definitions, after multiple experimentation and fine-tuning phases, of the reward function and the (hyper) parameters of the Deep RL networks. At last, there shall be a presentation of the obtained results; including, as expected, a thorough discussion and analysis of the required steps to achieve the final architecture.

References

1. Casavant, T.L., Kuhl, J.G.: A taxonomy of scheduling in general-purpose distributed computing systems. IEEE Trans. Softw. Eng. **14**(2), 141–154 (1988)
2. Ahire, S., Greenwood, G., et al.: Workforce-constrained preventive maintenance scheduling using evolution strategies. Decis. Sci. **31**(4), 833–859 (2000)
3. Sonmez, A.I., Baykasoglu, A.: A new dynamic programming formulation of (n x m) flowshop sequencing problems with due dates. Int. J. Prod. Res. **36**(8), 2269–2283 (1998)
4. Yamada, T., Yamada, T., Nakano, R.: Genetic algorithms for job-shop scheduling problems. In: Modern Heuristic for Decision Support, pp. 474–479 (1997)
5. McKay, K.N., Safayeni, F.R., Buzacott, J.A.: Job-shop scheduling theory: what is relevant? Interfaces **18**(4), 84–90 (1988)
6. Lawler, E.L., Lenstra, J.K., et al.: Sequencing and scheduling: algorithms and complexity. Handb. Oper. Res. Manag. Sci. **4**, 445–522 (1993)
7. LeCun, Y., Bengio, Y., Hinton, G.: Deep learning. Nature **521**(7553), 436 (2015)
8. Knight, W.: Reinforcement Learning: 10 Breakthrough Technologies 2017 - MIT Technology Review (2017)
9. Pinedo, M.L.: Scheduling: Theory, Algorithms, and Systems, 4th edn. Springer, New York (2008)
10. Madureira, A., Pereira, I., Falcão, D.: Dynamic adaptation for scheduling under rush manufacturing orders with case-based reasoning. In: International Conference on Algebraic and Symbolic Computation (2013)
11. Lee, Y.H., Kumara, S.R.T., Chatterjee, K.: Multiagent based dynamic resource scheduling for distributed multiple projects using a market mechanism. J. Intell. Manuf. **14**, 471–484 (2003)
12. Ouelhadj, D., Cowling, P., Petrovic, S.: Utility and stability measures for agent-based dynamic scheduling of steel continuous casting. In: 2003 IEEE International Conference on Robotics and Automation, vol. 1 (2003)

13. Goren, S., Sabuncuoglu, I., Koc, U.: Optimization of schedule stability and efficiency under processing time variability and random machine breakdowns in a job shop environment. Naval Res. Logistics **59**, 26–38 (2012)

14. Dios, M., Framinan, J.M.: A review and classification of computer-based manufacturing scheduling tools. Comput. Industr. Eng. **99**, 229–249 (2016)

15. Cunha, B., Madureira, A., et al.: Evaluating the effectiveness of Bayesian and neural networks for adaptive scheduling systems. In: 2016 IEEE Symposium Series on Computational Intelligence (SSCI), pp. 1–6. IEEE, December 2016

16. Madureira, A., Gomes, S., et al.: Prototype of an adaptive decision support system for interactive scheduling with metacognition and user modeling experience. In: Sixth World Congress on Nature and Biologically Inspired Computing (NaBIC) (2014)

17. Madureira, A., Pereira, I., Cunha, B.: Specification of an architecture for self-organizing scheduling systems. In: Madureira, A., Abraham, A., Gamboa, D., Novais, P. (eds.) ISDA 2016, vol. 557, pp. 771–780. Springer, Heidelberg (2017). https://doi.org/10.1007/978-3-319-53480-0_76

18. Talbi, E.G.: Metaheuristics: From Design to Implementation. Wiley, Hoboken (2009)

19. Gonzalez, T.: Handbook of Approximation Algorithms and Metaheuristics. Chapman & Hall, London (2007)

20. Glover, F.: Future paths for integer programming and links to artificial intelligence. Comput. Oper. Res. **13**(5), 533–549 (1986)

21. Holland, J.H.: Adaptation in Natural and Artificial Systems, p. 183. University of Michigan Press, Ann Arbor (1975)

22. Rangel-Merino, R.M.A., López-Bonilla, J.L.: Optimization Method Based on Genetic Algorithms. Apeiron **12**(4), 393–408 (2005)

23. Ludvig, E.A., Bellemare, M.G., Pearson, K.G.: A primer on reinforcement learning in the brain: psychological, computational, and neural perspectives. In: Computational Neuroscience for Advancing Artificial Intelligence: Models, Methods and Applications, pp. 111–144. IGI Global (2011)

24. Degris, T., Pilarski, P., Sutton, R.: Model-free reinforcement learning with continuous action in practice. In: 2012 American Control Conference (ACC), pp. 2177–2182. IEEE (2012)

25. Watkins, C.J.C.H., Dayan, P.: Q-learning. Mach. Learn. **8**(3–4), 279–292 (1992)

26. Kaelbling, L.P., Littman, M.L., Moore, A.W.: Reinforcement learning: a survey. J. Artif. Intell. Res. **4**, 237–285 (1996)

27. Mnih, V., Kavukcuoglu, K., et al.: Playing Atari with Deep Reinforcement Learning. ArXiv e-prints, no. 1312.5602 (2013)

28. Lillicrap, T.P., Hunt, J.J., et al.: Continuous control with deep reinforcement learning, September 2015

29. Plappert, M., Houthooft, R., et al.: Parameter Space Noise for Exploration (2017)

30. Beasley, J.E.: OR-Library: distributing test problems by electronic mail. J. Oper. Res. Soc. **41**(11), 1069–1072 (1990)

31. Taillard, E.: Benchmarks for basic scheduling problems. Eur. J. Oper. Res. **64**(2), 278–285 (1993)

Improving Nearest Neighbor Partitioning Neural Network Classifier Using Multi-layer Particle Swarm Optimization

Xuehui Zhu[1], He Zhang[1], Lin Wang[1(✉)], Bo Yang[1,2], Jin Zhou[1],
Zhenxiang Chen[1], and Ajith Abraham[3]

[1] Shandong Provincial Key Laboratory of Network Based Intelligent Computing,
University of Jinan, Jinan 250022, China
wangplanet@gmail.com
[2] School of Informatics, Linyi University, Linyi 276000, China
[3] Machine Intelligence Research Labs (MIR Labs),
Scientific Network for Innovation and Research Excellence, Auburn, USA

Abstract. Nearest neighbor partitioning (NNP) method has been proved to be an effective method to enhance the quality of neural network classifiers. However, there are many cluster shapes in NNP, which results in a large number of local optimal solutions in the searching space by the traditional particle swarm optimization (PSO) algorithm. Therefore, the multi-layer particle swarm optimization (MLPSO) is introduced to increase the diversity of searching groups through increasing the number of layers, thereby improving the performance when facing with large scale problems. In this study, we adopt the combination of multi-layer particle swarm optimization and nearest neighbor partitioning to solve the local optimal problem caused by multi-cluster shapes in the optimization of NNP. Experimental results show that this method improves the performance of classifier.

Keywords: Classification · Nearest neighbor partitioning ·
Neural network · Multi-layer particle swarm optimization

1 Introduction

Classification is a kind of supervised learning, which knows the labels of training samples in advance and separates the samples that belong to different categories of labels by mining. Once the classification model is obtained, this model will be used to predict which category the sample belongs to. Some classification methods contain Bayesian networks [1], gene expression programming [2] and neural networks [3–6]. In these classification techniques, neural network has been shown that it can be close to nonlinear functions arbitrarily, and neural network classifiers have been successfully applied to classification tasks [7–9].

In the neural network classifier, nearest neighbor partitioning (NNP) [10] has been proved to be an effective classification method and its performance is

© Springer Nature Switzerland AG 2020
A. M. Madureira et al. (Eds.): HIS 2018, AISC 923, pp. 360–369, 2020.
https://doi.org/10.1007/978-3-030-14347-3_35

superior to other classification methods. However, NNP has a variety of cluster shapes, resulting in a large number of local optimal solutions in the searching space by the traditional particle swarm optimization (PSO) [11] algorithm. Its biggest drawback is that it often leads to premature convergence to a local minima.

In this paper, the multi-layer particle swarm optimization is introduced to improve the performance of nearest neighbour partitioning, and named by the combination of multi-layer particle swarm optimization and nearest neighbor partitioning (MLPSO+NNP). MLPSO increases the number of layers for improving search ability in the solution space. For the multi-layer searching strategy, the upper layer guides the lower layer to thoroughly search for the solution space, and the best positions are located in each of the best potential layer areas at the same time. MLPSO is able to jump out of the local optimum by means of the cross-layer information exchange of the whole population.

The rest of the paper is organized as follows. Section 2 illustrates the related works. Section 3 explains the basic methods and the MLPSO+NNP. Section 4 explores the results of the experiment. Finally, Sect. 5 gives the conclusions.

2   Related Works

For traditional neural network classifier, it contains binary classification and multiple classification problem, respectively. For a binary classification, it predicts the probability that the sample is the "true" class. For the multiple classification problem, it classifies the samples based on the category with the highest probability output value.

For the above situation, the total of the output probabilities for individual cannot be equal to one. Due to sample class labels are mutually exclusive, each sample can only be allocated for one label, which signifies that the total of these individual probabilities ought to be one. Therefore, Bridle [13] applied the SoftMax activation function to normalize the output of the neural network so that the total of individual probabilities are always equal to one. Dietterich and Bakiri [14] applied an error correction output code (ECOC) method to improve the performance of inductive learning programs on multiple types of problems. In addition, Wang et al. [15] proposed a floating centroid method (FCM) using centroid clustering [16] that removes the constraint of fixed centroid and improves the performance of classifier. However, this method cannot produce flexible decision boundaries when classifying. Thus, nearest neighbor partitioning (NNP) [10] was proposed to solve this problem and obtained good performance.

Researchers made a lot of contributions for improving neural network classifiers based on optimization algorithms. Xiong et al. [17] adopted Improved Genetic Algorithms (MGA) and improved operators of Simple Genetic Algorithms (SGA) to construct neural network classifier, which achieved good performance. Zhou et al. [18] adopted genetic algorithms to construct and optimize a new BP neural network classifier, which achieved good performance. Lin et al. [19] adopted real-coded genetic algorithm (RCGA) to improve the classification performance of Polynomial Neural Networks (PNN).

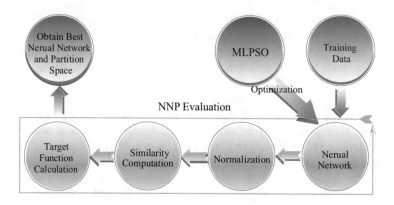

Fig. 1. The training precess of NNP.

3   Methodology

In this section, the basic methods are introduced first, including multi-layer particle swarm optimization and nearest neighbor partitioning. Then, the MLPSO+NNP is described in detail. In addition, the process of the algorithm and the training model are given.

3.1   Basic Methods

Multi-layer Particle Swarm Optimization. In MLPSO [12], particles search solution space in multiple layers. A *subswarm* called a *swarmparticle* in the lower layer can also be considered as a particle directly in the upper layer. The swarmparticles search for optimal solutions on a small scale in the lower layer and large scale in the upper layer. In addition, the swarmparticles in the lower layer not only obtain information from the upper layer, but also obtain information from the current layer. This multi-layer searching strategy enables swarmparticles search the solution space completely.

MLPSO includes two versions, global version called global MLPSO and local version called local MLPSO. Although the local MLPSO has advantages with the increase of problem dimensions, the global MLPSO is more suitable for solving the local optimal problems caused by various cluster shapes of NNP. Therefore, we adopt global version to optimize the neural network. A few swarmparticles will be defined in each layer as follows:

$$SP = (SP_1, SP_2, \ldots, SP_m) \tag{1}$$

where m stands for the number of layers, SP_i stands for the number of swarm-particles at layer $i - 1$ dominated by a swarmparticle at layer i. For example, $SP_2 = 4$ implies that a swarmparticle at layer 3 contains 4 swarmparticles at layer 1. Since the particle in the first layer is inseparable, SP_1 is always 1. Then

$$SPN = \prod_{i=1}^{m} SP_i \qquad (2)$$

where SPN is the total number of particles. The velocity update formula for particles follows the accumulation of information from all the layers. The formula for global MLPSO is as follows:

$$\begin{cases} v^i = \omega v^i + \sum_{j=1}^{m} \phi_j r(pbest_j^i - x^i) & \phi_j = \frac{\varphi}{m} \\ x^i = x^i + v^i \end{cases} \qquad (3)$$

where i stands for the current particle, v^i and x^i are the velocity and position of particle, respectively. ω is the random inertia weight and its range is $[0, 1]$, ϕ is the summary of the acceleration constant, $pbest_j$ stands for the optimal solution found by the swarmparticle in each layer j.

Algorithm 1. Algorithm of MLPSO+NNP

---

1 Initialize the velocity and position of particles;
2 **while** *maximum generations has not been achieved* **do**
3   **for** *each particle* **do**
4   Decode the position of each particle to a neural network;
5   **for** *each sample in the data set* **do**
6   Map the current samples to a partition space;
7   **end**
8   **for** *mapped points in the partition space* **do**
9   Normalize the points using Z-score;
10   **end**
11   **for** *normalized points in the partition space* **do**
12   Constrain points to hypersphere using (4);
13   **end**
14   **for** *constrained points in the partition space* **do**
15   Calculate the summation of within-class similarities and the summation of similarities between classes using (5);
16   **end**
17   Compute the fitness value of current particle using (6);
18   **end**
19   Return the fitness values of all particles;
20   Update the best position for all of swarmparticles at every layer;
21   **for** *each particle* **do**
22   Update the velocity and position using (3);
23   **end**
24   Regenerate particles and enter the next generation;
25 **end**
26 Return the optimal neural network and distribution of the training points.

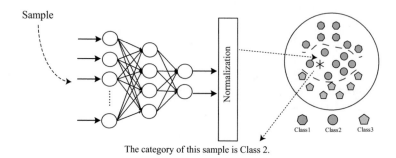

Fig. 2. An example of an unknown sample classification.

Nearest Neighbor Partitioning. The biggest advantage of NNP [10] is that it can create flexible decision boundaries in a partition space where it categorizes the sample. In NNP training process, the particles are initialized first, and each particle corresponds to a set of weights of the neural network. The training points are mapped into a partition space via the neural network and then the new data is normalized with Z-score [20]. Then, the normalized points are constrained to a hypersphere, and the formula is defined as follows:

$$g_1(\mathbf{x}) = \frac{\mathbf{x}(1 - e^{-|\mathbf{x}|/2})}{|\mathbf{x}|(1 + e^{-|\mathbf{x}|/2})} \tag{4}$$

where $\mathbf{x}$ stands for the normalized points. After that, the similarity of any two points based on the constraint formula is

$$s(\mathbf{x_1}, \mathbf{x_2}) = 2 - |g_1(\mathbf{x_1}) - g_1(\mathbf{x_2})| \tag{5}$$

where $\mathbf{x_1}$ and $\mathbf{x_2}$ stand for the two normalized points. By calculating within-class similarity and similarity between classes. The final optimization target function is

$$F = \sum_{i=1}^{n} \{\omega(\mathbf{x}_i)(S_{nonself}(\mathbf{x}_i) - \alpha S_{self}(\mathbf{x}_i))\} \tag{6}$$

where $\omega(\mathbf{x}_i)$ stands for the weight of current point and α is a adjustment coefficient. $S_{nonself}(\mathbf{x}_i)$ is the sum of similarities between different classes. $S_{self}(\mathbf{x}_i)$ is the sum of similarities between the same classes. After that, an optimal neural network is found.

3.2   MLPSO+NNP

Different from conventional neural network classifier, nearest neighbour partitioning has plenty of nearest optimal solutions on account of its flexibility on points distribution. As far as this point is concerned, the complexity of optimization has to confront multimodal landscape in searching space, which impedes its performance. As an effective variant of particle swarm optimization, multilayer particle swarm optimization is able to exploit local optimums simultaneously at different scales while exploring unknown regions. Therefore, the MLPSO is adopted to aid the training process of nearest neighbour partitioning.

MLPSO runs through the whole training process of NNP. It makes NNP jump out of the local optimum during the optimization process via a cross-layer searching strategy. MLPSO mainly optimizes the connection weights, which correspond to the position of the particles. The optimization aims to choose the position of the particle with the smallest fitness value, i.e., a set of weights. Then we get the neural network composed of this set of weights.

To evaluate the connection weight values, the particles are first decoded into a set of parameters of neural network and the training data is mapped to a partition space via the neural network. In this space, the optimization follows the principle that keeps points from same class together and keeps points from the different class away from each other. Then, the fitness value of each particle is obtained via three steps, which are Z-score normalization, similarity computation by Eq. 5, and fitness computation by Eq. 6. After optimization, we will get a set of parameters of the optimal neural network and the distribution of training sample points in the partition space. Figure 1 shows the optimization training process of the MLPSO+NNP. To clearly describe the entire algorithm, the pseudo-code is presented in Algorithm 1.

Table 1. Characteristics of data set

Data sets	No. of samples	No. of features	No. of classes
Haberman	306	3	2
Ionosphere	351	34	2
Fertility	100	9	2
WBCD	569	30	2
Iris	150	4	3
Vertebral column (VC)	310	6	3
Seeds	210	7	3
Wine	178	13	3
User knowledge modeling (UKM)	403	5	4
Vehicle	846	18	4
Breast tissue (BT)	106	9	6
Glass	214	9	6
Zoo	101	16	7

When classifying an unknown sample, it is first mapped by an optimal neural network, then the mapped point is normalized by using Z-score and then is constrained to hyperspheres. Finally, the unknown sample is classified by using the weighted k-nearest neighbors (KNN) [21] on the basis of the distribution of training points in the partition space. Figure 2 shows an example of an unknown sample classification.

4   Experiments

In order to prove the effectiveness of MLPSO+NNP, we select 13 data sets from the UCI database to test its performance. Table 1 explains the attributes of these data sets. Due to different variables for the data set often have different units and degrees of variation. In order to eliminate the effect of dimensionality, and the size and magnitude of the variation of the variables themselves, the data is standardized by using min-max normalization [10]. By eliminating the dimensional relationship between variables, which makes the data comparable.

Generalization accuracy (GA) [10] and average F-measure (Avg.FM) [10] are selected as measurement standards to evaluate the performance of MLPSO+NNP. GA refers to the proportion of correctly classified samples to the total number of samples. The higher GA indicates the better performance among the classification methods. Avg.FM is a comprehensive evaluation index, which reflects the quality of the classifier.

To obtain a reliable and stable model, tenfold cross-validation is used in the experiment. It aims to divide the data set into ten subsets, and take nine subsets as training and take remaining subset as testing. The procedure will be repeated for ten times. Then ten results are averaged as an estimate of the accuracy of the algorithm.

Table 2. Results of accuracy on all data sets

	Traditional [10]	SoftMax [10]	ECOC [10]	FCM [10]	NNP [10]	MLPSO+NNP
Haberman	72.81 (±4.43)	N/A	N/A	**73.75** (±3.67)	71.56 (±11.17)	70.63 (±4.70)
Ionosphere	88.06 (±5.41)	N/A	N/A	94.17 (±4.80)	93.89 (±3.88)	**94.72** (±2.05)
Fertility	78.18 (±9.77)	N/A	N/A	**83.64** (±3.83)	79.09 (±13.59)	80.91 (±9.04)
WBCD	93.13 (±5.14)	N/A	N/A	94.90 (±3.15)	95.60 (±1.91)	**95.79** (±1.89)
Iris	94.67 (±6.89)	96.67 (±5.67)	93.33 (±6.29)	97.33 (±4.66)	**99.33** (±2.11)	**99.33** (±2.11)
VC	77.74 (±3.86)	84.19 (±5.15)	79.35 (±7.78)	85.16 (±6.31)	85.48 (±6.32)	**87.74** (±6.94)
Seeds	93.33 (±6.02)	92.86 (±7.19)	94.29 (±7.38)	95.24 (±5.02)	96.19 (±4.92)	**96.67** (±3.21)
Wine	97.78 (±3.88)	98.33 (±3.75)	97.78 (±2.87)	98.89 (±2.34)	98.89 (±2.34)	**99.44** (±1.76)
UKM	92.38 (±5.12)	91.67 (±6.66)	93.33 (±2.46)	95.95 (±1.96)	95.24 (±2.97)	**96.90** (±2.52)
Vehicle	81.74 (±1.98)	**83.26** (±2.14)	79.88 (±3.05)	79.19 (±3.78)	83.02 (±2.46)	77.56 (±1.46)
BT	67.86 (±11.79)	72.14 (±14.85)	65.00 (±12.80)	73.57 (±10.13)	71.43 (±14.29)	**74.29** (±9.04)
Glass	65.27 (±8.56)	61.94 (±13.91)	59.04 (±15.59)	**66.58** (±20.54)	66.11 (±9.91)	61.67 (±19.92)
Zoo	94.00 (±5.84)	94.00 (±5.84)	94.00 (±5.84)	93.33 (±7.03)	94.67 (±5.26)	**95.33** (±4.50)

A three layers feedforward neural network with 20 hidden nodes is used in this experiment. In NNP, the range of adjustment coefficient α is set from 10^{-2} to 10^1, the nearest neighbors L in training is set from 1 to 50, and the nearest neighbors K in testing for KNN is set from $\{1, 3, ..., 25\}$. In MLPSO, the

Table 3. Results of average F-measure on all data sets

	Traditional [10]	SoftMax [10]	ECOC [10]	FCM [10]	NNP [10]	MLPSO+NNP
Haberman	59.42 (±8.02)	N/A	N/A	55.93 (±9.61)	64.41 (±13.54)	**66.39** (±4.99)
Ionosphere	86.18 (±6.42)	N/A	N/A	93.44 (±5.43)	93.21 (±4.31)	**94.10** (±2.33)
Fertility	55.72 (±14.67)	N/A	N/A	57.66 (±17.11)	64.93 (±21.57)	**71.80** (±10.51)
WBCD	92.73 (±5.13)	N/A	N/A	94.53 (±3.24)	95.28 (±2.00)	**95.51** (±2.00)
Iris	94.57 (±7.01)	96.60 (±5.78)	93.22 (±6.40)	97.26 (±4.82)	**99.33** (±2.13)	**99.33** (±2.13)
VC	72.50 (±5.72)	79.98 (±6.31)	73.59 (±9.61)	81.21 (±7.77)	82.11 (±7.60)	**85.26** (±8.16)
Seeds	93.32 (±6.01)	92.85 (±7.15)	94.26 (±7.43)	95.18 (±5.08)	96.16 (±4.96)	**96.63** (±3.26)
Wine	97.85 (±3.69)	98.33 (±3.77)	97.78 (±2.88)	98.88 (±2.36)	98.97 (±2.16)	**99.44** (±1.77)
UKM	92.73 (±5.03)	85.22 (±14.79)	93.26 (±2.98)	96.11 (±2.03)	94.88 (±3.44)	**96.72** (±2.89)
Vehicle	81.88 (±2.14)	**83.28** (±2.19)	80.10 (±3.12)	78.54 (±4.17)	82.92 (±2.56)	77.12 (±1.54)
BT	62.88 (±10.34)	66.14 (±17.03)	60.13 (±11.00)	69.78 (±11.70)	65.94 (±16.12)	**69.86** (±10.01)
Glass	60.74 (±12.57)	51.73 (±16.27)	52.83 (±19.89)	55.52 (±20.84)	**67.30** (±12.87)	64.85 (±18.21)
Zoo	86.88 (±12.98)	86.64 (±13.27)	86.88 (±12.98)	86.45 (±13.54)	87.61 (±12.14)	**91.60** (±7.81)

searching layers m is set from 4 to 7, the summarization of the acceleration constant ϕ is set from [3.0, 4.5], the SP is set from 2 to 5, Max Generation is set to 1000, $Vmax$ is set to 0.4. For the fair comparison of experiments for different methods, the number of the population is also set to 20.

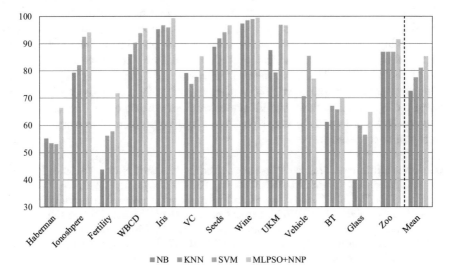

Fig. 3. Comparison with other classification methods on F-measure. (Results for NB, KNN, SVM can be found in the ref. [10].)

Table 2 displays the results of the accuracy on selected data sets. The results are analyzed that MLPSO+NNP achieves better accuracy than the NNP method on most of the data sets. Compared with the other classification methods, MLPSO+NNP has better performance on Ionosphere, WBCD, VC, Seeds, Wine,

UKM, BT and Zoo. It can also be seen that MLPSO+NNP is tied with NNP first on Iris data set. Moreover, the smallest standard deviation is yielded through MLPSO+NNP on most data sets, which indicates MLPSO+NNP is rather stable. Table 3 shows the results of average F-measure for different methods. From this table, the results further demonstrate that MLPSO+NNP has good performance in F-measure, and which exceeds the F-measure of other classification methods on eleven data sets.

The proposed method is also compared with other classification methods, e.g., K-Nearest Neighbor (KNN), Naive Bayes (NB) and Support Vector Machines (SVM), which are outside the field of neural networks. Figure 3 explains the results of the F-measure. The last item represents the average F-measure of each method on the 13 data sets. It can be seen that MLPSO+NNP is more superior to the other three methods. Therefore, The proposed method has higher performance than other methods.

5   Conclusion

In this study, the combination of multi-layer particle swarm optimization and nearest neighbor partitioning is proposed to deal with the local optimal problem caused by multi-cluster shapes in the optimization of NNP. Furthermore, MLPSO is used to optimize the parameters of neural network. It employs a multi-layer searching strategy and jumps out of the local optimum by means of the cross-layer information exchange of the whole population, which improves the searching performance of the particle population. To prove the combination of the MLPSO+NNP, we selected thirteen well-known data sets to verify its performance. Compared with the other classification methods, the results of the proposed method show significant advantages in accuracy and average F-measure. In future works, other experiments will be performed in evaluating the quality of this method.

Acknowledgments. This work was supported by National Natural Science Foundation of China under Grant No. 61573166, No. 61572230, No. 61872419, No. 61873324, No. 81671785, No. 61672262. Science and technology project of Shandong Province under Grant No. 2015GGX101025. Project of Shandong Province Higher Educational Science and Technology Program under Grant no. J16LN07. Shandong Provincial Key R&D Program under Grant No. 2016ZDJS01A12, No. 2016GGX101001. Taishan Scholar Project of Shandong Province, China.

References

1. Heckerman, D., Wellman, M.P.: Bayesian networks. Commun. ACM **38**(3), 27–31 (1995)
2. Wang, L., Yang, B., Wang, S., Liang, Z.: Building image feature kinetics for cement hydration using gene expression programming with similarity weight tournament selection. IEEE Trans. Evol. Comput. **19**(5), 679–693 (2015)

3. Yeh, I.C.: Modeling of strength of high-performance concrete using artificial neural networks. Cem. Concr. Res. **28**(12), 1797–1808 (1998)
4. Misra, B.B., Dehuri, S., Dash, P.K., Panda, G.: A reduced and comprehensible polynomial neural network for classification. Pattern Recogn. Lett. **29**(12), 1705–1712 (2008)
5. Wang, L., Orchard, J.: Investigating the evolution of a neuroplasticity network for learning. IEEE Trans. Syst. Man Cybern. Syst. **PP**(99), 1–13 (2017)
6. Chen, C.P., Liu, Z.: Broad learning system: an effective and efficient incremental learning system without the need for deep architecture. IEEE Trans. Neural Netw. Learn. Syst. **29**(1), 10–24 (2018)
7. Gutierrez, P.A.: Neuro-logistic models based on evolutionary generalized radial basis function for the microarray gene expression classification problem. Neural Process. Lett. **34**(2), 117–131 (2011)
8. Avci, E.: An expert target recognition system using a genetic wavelet neural network. Appl. Intell. **37**(4), 475–487 (2012)
9. Yu, Q., Yan, R., Tang, H., Tan, K.C., Li, H.: A spiking neural network system for robust sequence recognition. NIEEE Trans. Neural Netw. Learn. Syst. **27**(3), 621–635 (2016)
10. Wang, L., Yang, B., Chen, Y., Zhang, X., Orchard, J.: Improving neural-network classifiers using nearest neighbor partitioning. IEEE Trans. Neural Netw. Learn. Syst. **28**(10), 2255–2267 (2017)
11. Wang, L., Yang, B., Orchard, J.: Particle swarm optimization using dynamic tournament topology. Appl. Soft Comput. **48**, 584–596 (2016)
12. Wang, L., Yang, B., Chen, Y.: Improving particle swarm optimization using multilayer searching strategy. Inf. Sci. **274**(8), 70–94 (2014)
13. Bridle, J.S.: Probabilistic interpretation of feedforward classification network outputs, with relationships to statistical pattern recognition. In: Neurocomputing, pp. 227–236. Springer, Berlin (1990)
14. Dietterich, T.G., Bakiri, G.: Solving multiclass learning problems via error-correcting output codes. J. Artif. Intell. Res. **2**(1), 263–286 (1995)
15. Wang, L., Yang, B., Chen, Y., Abraham, A., Sun, H., Chen, Z., Wang, H.: Improvement of neural network classifier using floating centroids. Knowl. Inf. Syst. **31**(3), 433–454 (2012)
16. Zhou, J., Chen, L., Chen, C.L.P., Zhang, Y., Li, H.X.: Fuzzy clustering with the entropy of attribute weights. Neurocomputing **198**, 125–134 (2016)
17. Xiong, Z.Y., Liu, D.Q., Zhang, Y.F.: Constructing classifier of neural networks using improved genetic algorithms. Comput. Appl. **25**(01), 31–34 (2005)
18. Zhou, W., Xiong, S.: Optimization of BP neural network classifier using genetic algorithm. Energy Procedia **11**, 578–584 (2013)
19. Lin, C.T., Prasad, M., Saxena, A.: An improved polynomial neural network classifier using real-coded genetic algorithm. IEEE Trans. Syst. Man Cybern. Syst. **45**(11), 1389–1401 (2015)
20. Booth, H.S., Maindonald, J.H., Wilson, S.R., Gready, J.E.: An efficient Z-score algorithm for assessing sequence alignments. J. Comput. Biol. A J. Comput. Mol. Cell Biol. **11**(4), 616–625 (2004)
21. Yu, Z., Liu, Y., Yu, X., Pu, K.Q.: Scalable distributed processing of K nearest neighbor queries over moving objects. IEEE Trans. Knowl. Data Eng. **27**(5), 1383–1396 (2015)

Adaptive Sequence-Based Heuristic for the Two-Dimensional Non-guillotine Bin Packing Problem

Óscar Oliveira[(✉)] and Dorabela Gamboa

CIICESI-IPP, Felgueiras, Portugal
{oao, drg}@estg.ipp.pt

Abstract. We consider the non-guillotine bin packing problem in which a set of items must be packed into the minimum number of identical bins. We present a simple and fast heuristic that iteratively creates a new sequence of items that defines the packing order used to generate the new cutting plan. The new sequences retain, adaptively, characteristics of the previous sequences for search intensification and diversification. Computational experiments of the effectiveness of this approach are presented and discussed.

Keywords: Two-dimensional · Rectangular · Non-guillotine · Bin packing · Heuristics

1 Introduction

In this study, we address the problem in which it is intended to pack m items into the minimum number of identical bins. We consider the two-dimensional (2D) case where items and bins are characterized by their length and height. The items cannot be rotated and must be packed orthogonally into the bins. Following the typology of Wäscher et al. [1], the problem addressed is classified as Single Bin Size Bin Packing Problem (SBSBPP).

Several approaches were presented to solve this problem. Berkey and Wang [2] studied the performance of heuristics proposed in the literature for the Open Dimensional Problems (ODP, see Wäscher et al. [1]) adapted for the SBSBPP. The ODP considers that the height of the bin is infinite, and the objective is to minimize the packing height.

Martello and Vigo [3] showed that the Continuous Lower Bound $\left(LB = \left\lceil \sum_{i=1}^{m} l_i h_i / LH \right\rceil\right)$ for the SBSBPP has a worst-case performance ratio of $\frac{1}{4}$ and presented new lower bounds that are used in a Branch-and-Bound algorithm.

Lodi et al. [4] presented heuristics to solve the 2D (non-)oriented (non-)guillotine SBSBPP and the Unified Tabu Search Framework that is adaptable for each specific problem uniquely changing the inner heuristic to explore the neighborhood.

Boschetti and Mingozzi in [5] presented new lower bounds and in [6] presented a heuristic approach. The proposed heuristic, denoted as HBP, generates solutions considering the current allocation method and pricing rule. At the end of each iteration,

A. M. Madureira et al. (Eds.): HIS 2018, AISC 923, pp. 370–375, 2020.
https://doi.org/10.1007/978-3-030-14347-3_36

the value of the items is updated, and a new solution is generated. The complete heuthe heuristic considers two allocation method and four pricing rules based on area, length, width, and perimeter.

Faroe *et al.* [7] presented a heuristic based on Guided Local Search (GLS, see Voudouris and Tsang [8]) for the 3D SBSBPP and evaluate this approach also for the 2D case. This heuristic starts with an upper bound, obtained by a greedy heuristic, and iteratively decreases the number of available bins searching for a feasible solution with this new value. This process is repeated until the time limit is reached or the optimality is guaranteed through a bounding scheme.

Monaci and Toth [9] presented the Set-Covering Heuristic (SCH) formulating the problem as a set-covering problem. This heuristic first generates a large set of columns through heuristic procedures that will define the set-covering instance to be solved by means of a Lagrangean-based heuristic.

Parreño *et al.* [10] presented a Greedy Randomized Adaptive Search Procedure (GRASP, see Feo and Resende [11]) with Variable Neighborhood Descent (VND, see Hansen and Mladenović [12]) for the 2D and 3D SBSBPP.

Blum and Schmid [13] presented a heuristic, denoted as EA-LGFi, in which an evolutionary approach is used to generate the input sequences to the LGFi heuristic proposed by Wong *et al.* [14] in order to perform the placement.

This paper is organized as follows. Section 2 describes the proposed algorithm. Computational experiments are presented in Sect. 3, and in Sect. 4 conclusions and future work directions are provided.

2 Adaptive Sequence-Based Heuristic (ASH)

We propose a multi-start (see Martí *et al.* [15]) heuristic that iteratively creates a new sequence of items that defines the packing order to generate a new cutting plan.

Starting with the base sequence of items (S_{base}) ordered by non-increasing area, at each iteration, a new sequence $(S_{current})$ is generated that will be used to create a new cutting plan. This heuristic iterates until either the optimality is guaranteed (using the Continuous Lower Bound LB as reference) or a maximum number of iterations has been performed.

We generate new sequences based on the algorithm proposed by Lesh *et al.* [16]. Starting with an empty set of items $(S_{current})$, we add to this set, with a probability of α, one element at a time from the base sequence (S_{base}) until all elements of S_{base} are in $S_{current}$.

The items are packed, following the ordering defined in $S_{current}$, one at a time, into the empty rectangular space (ERS) that have space to pack it inside and that have the lowest area of all the available ERS present in the current opened bin. We use the Difference Process proposed by Lai and Chan [17] to keep track of all ERS that are created during the packing of items. A new bin is opened when no more items can be packed inside the current opened bin, if any.

If the ordering of $S_{current}$ generates the best cutting plan found so far, the current sequence will replace the current base and the current probability α is reset to its minimum value (α_{mim}). Otherwise, meaning that $S_{current}$ did not led to a new best solution, S_{base} is not altered and α is incremented $(\alpha = \min(\alpha + \alpha_{inc}, \alpha_{max})$.

The reset of the value of the current probability (α) to its minimum will generate in the next iteration a sequence that is very similar to S_{base} intensifying the search of neighbourhoods considered promising, while incrementing α allows to diversify the search space generating sequences with incrementally more differences to the base sequence.

3   Computational Results

The proposed heuristic was implemented in C and the tests were run on a computer with an Intel Core i7-4800MQ at 2.70 GHz with 8 Gb RAM and operating system Linux Ubuntu 18.04.

Each instance was run one time generating at most 2000 cutting plans with α_{min}, α_{max}, and α_{inc} set to 0.1, 0.65, and 0.002, respectively.

We tested our algorithm on a set of instances, usually referred to as CLASS, generated by Berkey and Wang [2] (CLASS 1 to 6) and by Martello and Vigo [3] (CLASS 7 to 10). This set is divided into 10 classes of 50 instances each. Each class contains 5 subsets composed of 10 instances with an identical number of m items to pack (with $m \in [20, 40, 60, 80, 100]$). The bin dimensions range from 10×10 to 300×300.

Table 1 presents the total number of bins needed to solve all the instances of the dataset CLASS by the heuristics HBP (Boschetti and Mingozzi [5]), GLS (Faroe et al. [7]), SCH (Monaci and Toth [9]), GRASP/VND (Parreño et al. [10]) and EA_LGFi (Blum and Schmid [13]).

Table 1. Comparison with literature approaches

Algorithm	HBP	GLS	SCH	GRASP/VND	EA_LGFi
Number of bins	7275	7284	7243	7241	7239

Observing Table 1, the best results are clearly obtained by SCH, GRASP/VND and EA_LGFi. The results obtained by these best approaches and by ASH solving Berkey and Wang instances are given in Table 2, and in Table 3 the results solving Martello and Vigo instances.

Tables 2 and 3 present in the first column the class number, the second column denotes the number of items of each subset, then for each of the heuristics the number of bins and the average time, in seconds, needed to solve each subset.

Table 4 presents a summary of the results obtained, denoting the total number of bins for the 500 instances of the CLASS dataset, and the average and maximum time to solve a subset.

ASH could not provide better results than those obtained by the other three approaches, but it is extremely attractive considering that the number of bins is acceptable for the time required.

Table 2. Results for Berkey and Wang instances

CLASS	m	SCH		GRASP/ VND		EA_LGFi		ASH	
		Bins	t (s)	Bins	t (s)	Bins	t (s)	Bins	t (s)
1	20	71	0.06	71	0.00	71	0.00	71	0.01
	40	134	2.42	134	0.00	134	0.00	134	0.02
	60	200	7.26	200	4.50	200	0.01	200	0.04
	80	275	4.63	275	1.50	275	0.00	275	0.07
	100	317	5.21	317	0.00	317	0.00	317	0.08
2	20	10	0.06	10	0.00	10	0.00	10	0.00
	40	19	0.67	19	0.00	19	0.00	19	0.00
	60	25	0.07	25	0.00	25	0.00	25	0.00
	80	31	0.07	31	0.00	31	0.00	32	0.01
	100	39	0.79	39	0.00	39	0.00	39	0.00
3	20	51	0.07	51	0.00	51	0.02	51	0.01
	40	94	2.66	94	3.00	94	0.01	94	0.02
	60	139	6.21	139	4.60	139	0.27	140	0.06
	80	189	8.80	189	4.10	189	20.68	192	0.08
	100	223	12.80	223	4.90	224	26.17	225	0.11
4	20	10	0.06	10	0.00	10	0.00	10	0.00
	40	19	0.07	19	0.00	19	0.00	19	0.00
	60	25	6.15	25	3.00	23	12.18	25	0.02
	80	32	10.35	31	1.90	31	0.00	32	0.03
	100	38	4.72	38	1.50	37	0.00	38	0.03
5	20	65	0.06	65	0.00	65	0.00	65	0.00
	40	119	1.98	119	0.00	119	0.03	119	0.02
	60	180	1.93	180	1.50	180	0.14	181	0.06
	80	247	20.66	247	9.00	247	0.03	247	0.09
	100	282	18.50	282	5.20	284	27.33	287	0.11
6	20	10	0.06	10	0.00	10	0.00	10	0.00
	40	17	6.85	17	3.00	17	0.03	18	0.03
	60	21	0.66	21	0.10	21	0.00	22	0.01
	80	30	0.23	30	0.00	30	0.00	30	0.00
	100	34	6.29	34	3.00	32	0.58	34	0.00

Although not directly comparable the computational time required by the ASH is extremely low, and noteworthy that the gap between the average and maximum time is very tight when compared with the other approaches (e.g., 1.77 s of average time to 27.33 s of maximum time needed to solve a subset with the EA_LGFi) providing in this manner a heuristic with a predictable execution time.

Although the other approaches can be seen as simple heuristics to the more experienced researcher or practitioner, ASH is much simpler to implement and requires less parametrization.

Table 3. Results for Martello and Vigo instances

CLASS	m	SCH		GRASP/VND		EA_LGFi		ASH	
		Z	Time (s)	Z	Time (s)	Z	Time (s)	Z	Time (s)
7	20	55	0.13	55	0.00	55	0.00	55	0.01
	40	111	3.02	111	3.00	111	0.01	112	0.03
	60	158	8.85	159	4.50	159	0.00	159	0.06
	80	232	54.79	232	12.00	232	0.00	232	0.08
	100	271	25.06	271	3.10	271	0.01	273	0.10
8	20	58	0.06	58	0.00	58	0.03	58	0.01
	40	113	0.96	113	1.50	113	0.00	113	0.03
	60	162	9.05	161	4.20	161	0.02	162	0.06
	80	224	11.60	224	1.60	224	0.00	226	0.09
	100	279	47.13	278	6.10	277	0.25	278	0.00
9	20	143	0.06	143	0.00	143	0.00	143	0.01
	40	278	0.07	278	0.00	278	0.00	278	0.03
	60	437	0.07	437	0.10	437	0.00	437	0.07
	80	577	0.08	577	0.00	577	0.00	577	0.10
	100	695	0.11	695	0.00	695	0.00	695	0.00
10	20	42	0.12	42	0.00	42	0.02	43	0.01
	40	74	0.11	74	0.00	74	0.00	74	0.02
	60	101	8.89	100	4.50	101	0.71	102	0.05
	80	128	38.26	129	9.40	128	0.06	130	0.08
	100	159	55.77	159	9.20	160	0.08	161	0.11

Table 4. Summary of the results obtained

Algorithm	SCH	GRASP/VND	EA_LGFi	ASH
Number of bins	7243	7241	7239	7269
Average time (s)	7.89	2.20	1.77	0.04
Maximum time (s)	55.77	12.00	27.33	0.11

Our approach is extremely fast, simple to implement and can be an effective approach to solve the bin packing problem or to be used in a bounding scheme on more complex solution methods.

4   Conclusion

We present a heuristic for the rectangular bin packing problem. The heuristic creates, at each iteration, a new sequence of items that will be used to create a cutting plan. The generated sequences retain, in an adaptive manner, characteristics of previous sequences providing intensification and diversification on the explored solution space.

To assess the performance of the proposed heuristic computational experiments have been performed, validating its effectiveness. The next step of this research is to further enhance the results obtained by solving other bin packing problems and to evaluate the effectiveness of this approach to other optimization problems.

Acknowledgement. This project is funded by Portuguese funds through FCT/MCTES (PID-DAC) under the project CIICESI_2017-03.

References

1. Wäscher, G., Haußner, H., Schumann, H.: An improved typology of cutting and packing problems. Eur. J. Oper. Res. **183**, 1109–1130 (2007)
2. Berkey, J.O., Wang, P.Y.: Two-dimensional finite bin-packing algorithms. J. Oper. Res. Soc. **38**, 423 (1987)
3. Martello, S., Vigo, D.: Exact solution of the two-dimensional finite bin packing problem. Manag. Sci. **44**, 388–399 (1998)
4. Lodi, A., Martello, S., Vigo, D.: Heuristic and metaheuristic approaches for a class of two-dimensional bin packing problems. INFORMS J. Comput. **11**, 345–357 (1999)
5. Boschetti, M.A., Mingozzi, A.: The two-dimensional finite bin packing problem. Part I: new lower bounds for the oriented case. Q. J. Belg. Fr. Ital. Oper. Res. Soc. **1**, 27–42 (2003)
6. Boschetti, M.A., Mingozzi, A.: The two-dimensional finite bin packing problem. Part II: new lower and upper bounds. Q. J. Belg. Fr. Ital. Oper. Res. Soc. **1**, 135–147 (2003)
7. Faroe, O., Pisinger, D., Zachariasen, M.: Guided local search for the three-dimensional bin-packing problem. INFORMS J. Comput. **15**, 267–283 (2003)
8. Voudouris, C., Tsang, E.: Guided local search and its application to the traveling salesman problem. Oper. Res. **113**, 469–499 (1999)
9. Monaci, M., Toth, P.: A set-covering-based heuristic approach for bin-packing problems. Informs J. Comput. **18**, 71–85 (2006)
10. Parreño, F., Alvarez-Valdés, R., Oliveira, J.F., Tamarit, J.M.: A hybrid GRASP/VND algorithm for two- and three-dimensional bin packing. Ann. Oper. Res. **179**, 203–220 (2010)
11. Feo, T., Resende, M.G.C.: Greedy randomized adaptive search procedures. J. Glob. Optim. **6**, 109–133 (1995)
12. Hansen, P., Mladenović, N.: Variable neighborhood search. In: Search Methodologies. Introductory Tutorials in Optimization and Decision Support Techniques, pp. 211–238 (2005)
13. Blum, C., Schmid, V.: Solving the 2D bin packing problem by means of a hybrid evolutionary algorithm. Procedia Comput. Sci. **18**, 899–908 (2013)
14. Wong, L., Lee, L.S., Serdang, U.P.M.: Heuristic placement routines for two-dimensional bin packing problem. J. Math. Stat. **5**, 334–341 (2009)
15. Martí, R., Resende, M.G.C., Ribeiro, C.C.: Multi-start methods for combinatorial optimization. Eur. J. Oper. Res. **226**, 1–8 (2013)
16. Lesh, N., Marks, J., McMahon, A., Mitzenmacher, M.: New heuristic and interactive approaches to 2D rectangular strip packing. J. Exp. Algorithmics. **10**, Article no. 1.2 (2005)
17. Lai, K.K., Chan, J.W.M.: Developing a simulated annealing algorithm for the cutting stock problem. Comput. Ind. Eng. **32**, 115–127 (1997)

Hybrid Multi-agent Approach to Solve the Multi-depot Heterogeneous Fleet Vehicle Routing Problem with Time Window (MDHFVRPTW)

Marwa Ben Abdallah and Meriem Ennigrou[(✉)]

SMART Laboratory, Institut Supérieur de Gestion, Université de Tunis,
Tunis, Tunisia
meriem.ennigrou@gmail.com

Abstract. In this article, the multi-depot heterogeneous fleet vehicle routing problem with time window (MDHFVRPTW) is considered. The objective of this work is to minimize the total traveled distance while delivering goods to geographically dispersed customers. In our research we solved the MDHFVRPTW with a multi-agent approach based on the hybridization of three meta-heuristics which are a particle swarm optimization algorithm (PSO), a genetic algorithm (GA) and a memetic algorithm (MA). A mathematical programming model for the problem is presented. In order to show the performance of the proposed approach we tested it on different benchmarks and we compared it with other results obtained from the literature.

Keywords: Multi-agent · Metaheuristic · Vehicle routing problem ·
Optimization · Hybridization · Genetic algorithm ·
Particle swarm optimization · Memetic algorithm

1 Introduction

Vehicle routing problem (VRP) is one of the most known research problems in the areas of combinatorial optimization and operational research. The objective of this work is to minimize the total traveled distance while delivering goods to geographically dispersed customers. Due to NP-hardness of VRP some meta-heuristics are developed to solve it such as genetic algorithms [10,16], ant colony [4,13], particle swarm [9,20], etc. An hybridization between genetic algorithm and particle swarm has been proposed to solve this problem [19]. In this article, the multi-depot heterogeneous fleet vehicle routing problem with time window (MDHFVRPTW) is considered which is the combination of three well-known variants:

© Springer Nature Switzerland AG 2020
A. M. Madureira et al. (Eds.): HIS 2018, AISC 923, pp. 376–386, 2020.
https://doi.org/10.1007/978-3-030-14347-3_37

– the Multi-Depot Vehicle Routing Problem (MDVRP).
– the Heterogeneous Fleet Vehicle Routing Problem (HFVRP).
– the Vehicle Routing Problem with Time Windows (VRPTW).

We solved the MDHFVRPTW with a multi-agent approach based on the hybridization of three meta-heuristics which are a particle swarm optimization algorithm (PSO), a genetic algorithm (GA) and a memetic algorithm (MA).

This paper is organized as follows. The problem is formulated and presented in Sect. 2, including the mathematical model. The proposed approach is presented in Sect. 3, The numerical results are given in Sect. 4. Finally, Sect. 5 concludes the main findings of this research and provides implications for future research.

2   Problem Presentation

2.1   Problem Description

The MDHFVRPTW is represented by a complete graph oriented $G = (N, A)$ where N is the set of nodes consisting of two types of nodes: the customer nodes $NC = 1 \ldots m$ where m is the number of customers and the depot nodes $ND = m + 1 \ldots n$ where n is the total number of nodes. And A is the set of directed edges such that $A = \{(i, j)/i, j \in N \text{ where } N = ND \cup NC$. Such that (i, j) is the edge connecting node i to node j$\}$. A cost, a distance or a travel time are associated to each edge. Each customer has a time service S_i, a request d_i and a time window $[a_i, b_i]$ in which customer i must be delivered, where a_i is the earliest time and b_i is the latest time. In this problem, an heterogeneous fleet is presented $V = 1 \ldots K$ where K is the number of vehicles. Each vehicle k has a capacity Q_k and this capacity cannot be the same for all vehicles. The MDHFVRPTW must satisfy the following constraints:

• Each vehicle starts and ends at the same depot.
• Each customer must be served only once by one vehicle.
• The total capacity assigned to the V_k vehicle must not exceed its capacity Q_k.
• Each customer must be served in his time interval $[a_i, b_i]$.
• Travel time between customers is equal to the distance.

The main objective of this problem is to minimize the total distance.

2.2   Mathematical Model

A mathematical programming formulation for the MDHFVRPTW is presented which is inspired from the mathematical Programming model proposed in [1].

Indices	
n	Total number of nodes (depots and customers)
m	Number of customers
i, j	Nodes indices
k	Vehicle index
K	Number of vehicles
L	Intermediate node index

Parameters	
d_{ij}	Distance between node i and node j
Q_k	Capacity of vehicle k
d_i	Request of customer i
S_i	Service time of customer i
a_i	The earliest time
b_i	The latest time

Decision variables	
Y_{ik}	Service start time of customer i served by vehicle k.
T_{ik}	The time when vehicle k arrives at customer i
$x_{ijk} = \begin{cases} 1 \text{ if vehicle k travels between locations i and j} \\ 0 \text{ else} \end{cases}$	

$$\text{Min } \sum_{i=1}^{n} \sum_{j=1}^{n} d_{ij} \sum_{k=1}^{K} x_{ijk} \tag{1}$$

Subject to:

$$\sum_{i=1}^{m} \sum_{k=1}^{K} x_{ijk} = 1, \qquad \forall 1 \le j \le m \tag{2}$$

$$\sum_{i=1}^{m} \sum_{j=m+1}^{n} x_{ijk} \le 1, \qquad \forall 1 \le k \le K \tag{3}$$

$$\sum_{i=1}^{n} \sum_{L=1}^{n} x_{iLk} = \sum_{j=1}^{n} \sum_{L=1}^{n} x_{Ljk}, \qquad \forall 1 \le k \le K \tag{4}$$

$$\sum_{i=1}^{m} \sum_{j=1}^{m} d_i x_{ijk} \le Q_k, \qquad \forall 1 \le k \le K \tag{5}$$

$$a_i \sum_{i=1}^{m} x_{ijk} \le Y_{ik} \le b_i \sum_{i=1}^{m} x_{ijk}, \qquad \forall 1 \le k \le K, \forall 1 \le j \le n \tag{6}$$

$$Y_{ik} = \begin{cases} d_{ij} \text{ if j is the first customer to be served} \\ \qquad \text{by the vehicle in depot i} \\ T_i + d_{ij} + S_i \text{ otherwise where customer j is} \\ \quad \text{the preceding customer to be served before customer i} \end{cases} \tag{7}$$

$$T_{ik} = \begin{cases} a_i \text{ if } Y_{ik} \le a_i \\ Y_{ik} \text{ else} \end{cases} \tag{8}$$

The objective in Eq. (1) is to minimize the total traveled distance. Equation (2) guarantees that each customer must be served once and only once by a vehicle. Equation (3) assures that if a vehicle is not used, it remains in its depot. Equation (4) assures the conservation of fleet. Equation (5) represents the capacity constraint. Equation (6) represents the time constraint. Equation (7) calculates the time of arrival of vehicle k to customer i. Equation (8) assures that a vehicle must wait customer to be available to be served.

3   Proposed Model

In this section we present the approach used to solve the MDHFVRPTW. This approach is an hybridization between three metaheuristics: particle swarm optimization, genetic algorithm and memetic algorithm performed by a multi-agent system composed of the following agents:

- **PSO Agent:** The agent uses the particle swarm optimization algorithm to build its solution.
- **GA Agent:** the agent uses the genetic algorithm to build its solution.
- **MA Agent:** the agent uses the memetic algorithm to build its solution.

3.1   Common Parameters

Solution Representation. The solution representation is a vector that contains an ordered sequence of customers. To distinguish between the beginning and the end of a route, we separated this sequence by the depot number to which the clients are assigned (Fig. 1).

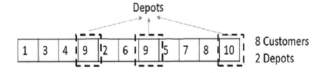

Fig. 1. Solution representation

Fitness Function. In the proposed approach the objective is to minimize the total traveled distance. The distance between customers is calculated using the Euclidean distance formula: $\sqrt{(x_1 - x_2)^2 + (y_1 - y_2)^2}$

Initial Population. In our approach, the initial population used by the agents is randomly created.

3.2 PSO Agent

Swarm particle optimization is a population-based search algorithm. It is inspired by the behavior of migratory birds. It was invented by [8]. The particle swarm optimization algorithm uses the collective behavior of individuals to explore the search space and converge on the optimal solution. At each iteration r, the particle modifies its velocity V_p according to its current position X_p, its best position $pBest_p$ and the best position of the other particles located in its neighborhood gBest. In the next step, the PSO agent converts this position to a feasible solution using a decoder. And in the end it applies a simple local search. The rules for updating the velocity and position of a particle are defined as follows:

$$V_{p,r+1} = \omega V_{p,r} + \alpha * (pBest_p - X_{p,r}) + \beta * (gBest - X_{p,r}) \qquad (9)$$
$$X_{p,r+1} = X_{p,r} + V_{p,r+1} \qquad (10)$$

where ω is the inertia weight; α and β are two random numbers that are uniformly distributed over [0, 1].

Velocity and Position Update. In our approach the PSO agent uses the exchange sequence strategy for the velocity update V_p and the position update X_p of a particle p. This strategy has 3 steps:

- **The exchange operator E (i1, i2):** is an arc representing the indices of the customers that will be permuted in the solution S.
- **The exchange sequence SE = (E1, E2, ..., En):** it is a set of exchange operators applied successively on the vector X_p in order to update the position of the particle p.
- **The basic sequence SEB = S2–S1:** is the exchange sequence which makes it possible to obtain S2 starting from S1 with the minimum number of exchange operators. This step is used to update the velocity V_p of the particle p.

The Decoder. After applying the updates, the PSO agent checks each arc of the Hamiltonian cycle. If the arc satisfies all the constraints of our variant including the constraints of time and capacity, the algorithm continues its course; otherwise, the depot would be inserted between the two nodes and the arc is replaced by two new arcs. The idea of this decoder is inspired by [9].

The Local Search of the PSO. The idea of this local search is inspired by [9]. The local search procedure consists of 3 steps:

- **Step1:** Choose route R which contains the minimum number of customers.
- **Step 2:** For each customer of the chosen route R, the algorithm tries to insert this client in the best position of the best route so that the insertion does not violate the constraints of MDHFVRPTW.
- **Step 3:** If all customers of the route R can be inserted into other routes, the algorithm deletes the R route.

3.3   GA Agent

The genetic algorithm is an evolutionary algorithm. The principle of this algorithm is inspired by the theory of Darwinian human evolution. The idea of this algorithm is to generate new populations by applying the operators of crossover, mutation and selection on some individual of the preceding population. These new populations will form generations.

Selection. The select strategy used in this work is the Tournament Selection Strategy. We select N individuals from the population. We have two options to select an individual from the tournament the first is the selection of the fittest individual in the tournament and the second is the selection of random individual.

Crossover. In our approach the GA agent uses the order crossover (ox) method proposed by [6]. The GA agent randomly selects a sub-string of the two parents. Then the two children will inherit this sub-string in the corresponding position and in the same order. Each parent deletes the customers that are in the other parent's sub-string and passes the sequence of the remaining customers to its child in sequence order. By reaching the end of the chain of child, the algorithm continues its course from the beginning of the chain. At the end of this operator the GA agent applies the same decoder used in the PSO algorithm. Then he chooses the child who generates the best cost (Fig. 2).

Fig. 2. crossover

Mutation. In our approach the GA agent uses the exchange mutation method proposed by [2]. In this operator the algorithm selects an individual. Then it randomly chooses two different positions and reverses them. At the end of this operator the GA agent applies the same decoder used in the PSO algorithm.

3.4   MA Agent

The algorithms belongs to the family of hybrid methods. They represent a hybridization between local search algorithms and genetic algorithms. It was invented by [14]. The general principle of the memetic algorithm is very similar to the genetic algorithm, where a local search operator is added following the crossover operator.

MA Local Research. The idea of this local search is inspired by [17]. This local search is based on the Λ-exchange method. This method uses 3 types of operators.

- **Operator 0–1:** Randomly chooses a customer from a route and tries to insert it in the best place in the other route, while respecting the constraints of time and capacity. If the insertion improves the cost, the customer will be maintained. Otherwise it stays in his route.
- **Operator 1–1:** For each route of these two routes that we have chosen randomly, the algorithm chooses a customer randomly from the first route and inserts it into the second route and the same for the second route.
- **Operator 0–2:** uses the same principle as the 0–1 operator but randomly chooses two customers.

3.5   Dynamic of Agents

In this approach we used Acl message protocol to assure the communication between agents. Initially, each agent receives the totality of knowledge about all customers and begins its local optimization process as described in the previous sections. Every time an agent improves its best current solution it sends it to the other agents which integrate it in their current population. This process continues until reaching the stopping criteria by each agent. The best solution of all agents is then returned as the solution of the problem.

4   Results and Discussion

The Hybrid Multi-agent approach for MDHFVRPTW was implemented using Java Language, and we used a JADE Agent platform as a platform for the agent manipulation. It was tested using an ACER laptop computer with 2.60 GHz Intel Core i7 and 8 GB of RAM. The parameters setting are presented in Tables 1 and 2, respectively. According to our knowledge there is no database presenting the variant MDHFVRPTW in the literature, but to prove the effectiveness of our approach we used Solomon's benchmark [18] and Cordeau's benchmark [5].

Table 1. PSO parameters

Parameters	Values
Number of particles	50
Number of iteration	150
Gama	0.2
Alpha	0.9
Beta	0.5

Table 2. MA and GA parameters

Parameters	Values
Size of the population	150
Probability of crossover	0.8
Probability of mutation	0.2
Stopping criteria	30
Number of generation	30
Number of iteration	300

4.1   Solomon's Benchmark

The authors in [7] modify the C101 instance of Solomon's instances by adding a new depot with coordinate (30, 55) and vehicles with different capacities. In the case of 25 customers they proposed 2 vehicles of 170 units and 1 vehicle of 250 units. For large instances they proposed 3 vehicles of 240 units capacity, 3 vehicles of 200 units capacity and 4 vehicles of 170 units capacity. In this work we adopt these modifications. We consider the same modifications on the other instances of type C1 and RC1 with 25 customers. Tables 3, 4 and 5 represent the results obtained for solomon's modified instances. In Table 3 we compared our results with the results of [7].

Table 3 shows the results obtained for the modified C101 instance with different sizes. Table 3 shows that Our approach has improved the results of the C101 instance for different sizes. The proposed algorithms also improved results with 25 customers compared to [7].

Table 3. The results obtained for the C101 instance

Instance	Customer	PSO	MA	GA	HMA-MDHFVRPTW	Cluster-based approach [7]
C101	25	**162.67**	**164.12**	**181.50**	**162.67**	186.11
	50	320.50	327.99	356.31	**315.29**	-
	100	**784.15**	802.90	824.02	**780.93**	784.90

Table 4. The results obtained for C1 instances

PSO	MA	GA	HMA-MDHFVRPTW	
			Distance	CPU(s)
164.15	165.15	181.11	**162.67**	17.52
165.03	167.40	180.32	**162.42**	12.34
168.23	179.95	189.20	**165.46**	15.17
162.67	163.46	184.57	**162.67**	21.51
162.67	**162.67**	179.41	**162.67**	34.56
162.67	**162.67**	183.58	**162.67**	25.02
167.39	171.08	191.17	**167.06**	22.60
173.62	176.15	195.72	**173.27**	20.87

Table 5. The results obtained for RC1 instances

PSO	MA	GA	HMA-MDHFVRPTW	
			Distance	CPU(s)
381.72	**381.72**	394.12	**381.72**	14.54
319.51	328.58	354.71	**316.07**	22.30
312.25	329.32	358.54	**309.51**	16.61
297.43	305.60	326.49	**294.52**	15.05
389.12	391.45	411.21	**385.07**	27.46
336.37	339.41	378.91	**334.52**	25.19
287.30	292.14	313.57	**282.24**	18.65
282.67	291.40	317.25	**273.47**	11.42

4.2   Cordeau's Benchmark

In this section, we present the results generated by our approach on Cordeau's benchmark. In [12], the authors proposed Multi-Phase Modified Shuffled Frog Leaping Algorithm (MPMSFLA) to solve the MDVRPTW problem. Also in [16], the authors proposed a genetic algorithm to solve the MDVRPTW problem. We compare our results with the results presented in these articles. The authors of these articles compared their results with the tabu search (TS) proposed by [5] and the variable neighborhood search algorithm (VNS) proposed by [15].

The results obtained by the algorithms of our approach are presented in Table 6. From the results presented in Table 6, we notice that the results obtained by the hybrid multi-agent approach are better than the results presented by the PSO, MA and GA algorithms. Our multi-agent approach outperforms the approach of [12] in 5 out of 10 instances. The CPU time provided are presented in Table 7.

Table 6. Results obtained by TS, VNS, MPMSFLA, GA and HMA-MDHFVRPTW for PR instances

Inst	Cust	Dep	TS [5]	VNS [15]	MPM-SFLA [12]	GA [16]	PSO	MA	GA	HMA-MDHF VRPTW
PR01	48	4	1074.12	1074.12	1074.34	1132.12	1083.21	1099.60	1112.25	**1074.12**
PR02	96	4	1762.48	1763.66	1768.23	1825.70	1776.99	1784.99	1802.54	**1765.49**
PR03	144	4	2397.06	2388.73	2378.97	2489.95	2404.49	2423.91	2467.32	2397.06
PR04	192	4	2865.71	2847.56	2858.35	-	2874.98	2892.13	2910.81	2869.86
PR05	240	4	3050.80	3015.27	3021.56	-	3064.55	3077.23	3105.63	3058.28
PR06	288	4	3670.13	3674.6	3685.28	-	3692.41	3704.81	3736.20	3689.17
PR07	72	6	1418.22	1418.22	1418.78	1480.49	1422.60	1430.63	1455.05	**1417.12**
PR08	144	6	2118.50	2103.21	2111.47	2257.94	2112.72	2124.75	2159.18	**2110.89**
PR09	216	6	2760.46	2753.61	2770.61	-	2775.59	2781.37	2809.47	**2768.32**
PR10	288	6	3507.26	3541.01	3555.86	-	3562.43	3570.50	3612.08	3559.18

Table 7. CPU times provided by TS, VNS, MPMSFLA, GA and HMA-MDHFVRPTW for PR instances

Instance	TS [5] (m)	VNS [15] (m)	MPM-SFLA [12] (m)	PSO (m)	MA (m)	GA (m)	HMA-MDHF VRPTW (m)
PR01	28	9.49	0.45	**0.05**	0.13	0.09	0.18
PR02	79	27.63	1.42	**0.21**	0.44	0.32	0.67
PR03	115	75.88	2.96	**0.53**	0.90	0.73	1.53
PR04	144	93.53	4.21	**1.38**	1.87	1.61	3.23
PR05	181	89.36	4.89	**2.14**	2.83	2.60	3.54
PR06	221	96.63	6.33	**3.55**	4.45	3.97	5.94
PR07	53	7.66	1.89	**0.07**	0.17	0.14	0.29
PR08	102	54.92	3.07	**0.38**	0.77	0.64	1.51
PR09	160	69.11	3.85	**1.38**	1.91	1.63	3.36
PR10	227	65.70	6.97	**3.18**	4.05	3.61	5.36

Using the information in Table 7, we note that the PSO algorithm provides the best CPU times compared to other algorithms. The PSO algorithm keeps its performance in different sizes of instances.

5 Conclusion

In this article, we present a new multi-Agent approach based on the hybridization of three meta-heuristics (genetic algorithm, particle swarm and memetic algorithm) to solve Multi-Depot Heterogeneous Fleet Vehicle Routing Problem with Time Window. To evaluate the performance of our approach, we adapted the modifications to Solomon's benchmark and Cordeau's benchmark. Our approach generates good quality solutions in a short computational time. We have already performed other experimentation on the remaining instances in Solomon's benchmark and compared the results obtained by the different algorithms executed separately and those obtained by our multi-agent approach. This comparison demonstrated that the hybridization allowed to improve the results significantly. Moreover, the results presented can be used as base for future comparisons.

References

1. Afshar-Nadjafi, B., Afshar-Nadjafi, A.: Multi-depot time dependent vehicle routing problem with heterogeneous fleet and time windows. Int. J. Oper. Res. **26**, 88–103 (2016)
2. Banzhaf, W.: The molecular traveling salesman. Biol. Cybern. **64**, 7–14 (1990)
3. Belmecheri, F., Prins, C., Yalaoui, F., Amodeo, L.: Particle swarm optimization algorithm for a vehicle routing problem with heterogeneous fleet mixed backhauls and time windows. J. Intell. Manufact. **24**, 775–789 (2013)
4. Benslimane, M.T., Benadada, Y.: Ant colony algorithm for the multi depot vehicle routing problem in large quantities by a heterogeneous fleet of vehicles. Inf. Syst. Oper. Res. **51**, 31–40 (2013)
5. Cordeau, J.F., Gendreau, M., Laporte, G.: A tabu search heuristic for periodic and multi-depot vehicle routing problems. Networks **30**, 105–119 (1997)
6. Davis, L.: Applying adaptive algorithms to epispastics domains. In: Proceedings of the International Joint Conference on Artificial Intelligence, pp. 162–164 (1985)
7. Dondo, R., Cerda, J.: A cluster based optimization approach for the multi depot heterogeneous fleet vehicle routing problem with time windows. Eur. J. Oper. Res. **176**, 1478–1507 (2007)
8. Eberhart, R., Kennedy, J.: Particle swarm optimization, pp. 1942–1948. IEEE (1995)
9. Gong, Y.J., Zhang, J., Liu, O., Huang, R.Z., Chung, H.S.H., Shi, Y.H.: Optimizing the vehicle routing problem with time windows: a discrete particle swarm optimization approach. IEEE Trans. Syst. Man Cybern. **42**, 254–267 (2012)
10. Ivàn, R., Willmer, J., Granada, M.: A metaheuristic algorithm for the multi depot vehicle routing problem with heterogeneous fleet. Int. J. Ind. Eng. Comput. **9**, 461–478 (2018)
11. Kien, J.J., Kim, M.N., Poh, L., Teo, K.M.: Vehicle routing problem with a heterogeneous fleet and time windows. Expert Syst. Appl. **41**, 3748–3760 (2014)

12. Luo, J., Chen, M.R.: Multi-phase modified shuffled frog leaping algorithm with extremal optimization for the MDVRP and the MDVRPTW. Comput. Ind. Eng. **72**, 84–97 (2014)
13. Ma, Y., Han, J., Kang, K., Yan, F.: An improved ACO for the multi-depot vehicle routing problem with time windows. Advances in Intelligent Systems and Computing, vol. 502, pp. 1181–1189 (2016)
14. Moscato, P.: Memetic algorithms: a short introduction. In: Corne, D., Dorigo, M., Glover, F. (eds.) New Ideas in Optimization, pp. 219–234. McGraw-Hill (1999)
15. Polacek, M., Benkner, S., Doerner, F.K., Hartl, R.F.: A cooperative and adaptive variable neighborhood search for the multi depot vehicle routing problem with time windows. Bus. Res. **1**, 207–218 (2008)
16. Ramalingam, A., Vivekanandan, K.: Genetic algorithm based solution model for multi-depot vehicle routing problem with time windows. Int. J. Ad. Res. Comput. Commun. Eng. **3**, 8433–8439 (2014)
17. Shi, Y., Boudouh, T., Grunder, O.: A hybrid genetic algorithm for a home health care routing problem with time window and fuzzy demand. Expert Syst. Appl. **72**, 160–167 (2017)
18. Solomon, M.: Algorithms for the vehicle routing and scheduling problem with time window constraints. Oper. Res. **32**, 254–265 (1987)
19. Xu, S.H., Liu, J.P., Zhang, F.H., Wang, L., Sun, L.J.: A combination of genetic algorithm and particle swarm optimization for vehicle routing problem with time windows. Sensors **15**, 21033–21053 (2015)
20. Yao, B., Yu, B., Hu, P., Gao, J., Zhang, M.: An improved particle swarm optimization for carton heterogeneous vehicle routing problem with a collection depot. Ann. Oper. Res. **242**, 303–320 (2016)

Hybrid System for Simultaneous Job Shop Scheduling and Layout Optimization Based on Multi-agents and Genetic Algorithm

Filipe Alves[1,2], M. Leonilde R. Varela[3(✉)], Ana Maria A. C. Rocha[3],
Ana I. Pereira[1,2], José Barbosa[1], and Paulo Leitão[1]

[1] Research Centre in Digitalization and Intelligent Robotics (CeDRI),
Instituto Politécnico de Bragança, Bragança, Portugal
{filipealves,apereira,jbarbosa,pleitao}@ipb.pt
[2] Algoritmi R&D Centre, University of Minho, Braga, Portugal
[3] Department of Production and Systems, Algoritmi Research Centre,
University of Minho, Braga, Portugal
{leonilde,arocha}@dps.uminho.pt

Abstract. A challenge is emerging in the design of scheduling support systems and facility layout planning, both for manufacturing environments where dynamic adaptation and optimization become increasingly important on the efficiency and productivity. Focusing on the interactions between these two problems, this work combines two paradigms in sequential manner, optimization techniques and multi-agent systems, to better reflect practical manufacturing scenarios. This approach, in addition to significantly improve the quality of the solutions, enables fast reaction to condition changes. In such stochastic and very volatile environments, the manufacturing industries, the fast rescheduling, or planning, are crucial to maintain the system in operation. The proposed architecture was codified in MatLab® and NetLogo and applied to a real-world job shop case study. The experimental results achieved optimized solutions, as well as in the responsiveness to achieve dynamic results for disruptions and simultaneously layout optimization.

Keywords: Multi-agent systems · Meta-heuristics · Scheduling · Optimization

1 Introduction

In order to survive in today's turbulent market environment, which is characterized by dynamic fluctuations of demand patterns across product mix and increasing rates of new products introduction, organizations need to respond agilely in this kind of environment. Since there is no co-operation between the planning and the scheduling functions, production schedule generation processes become inflexible and are restricted by the pre-generated process plan [13].

© Springer Nature Switzerland AG 2020
A. M. Madureira et al. (Eds.): HIS 2018, AISC 923, pp. 387–397, 2020.
https://doi.org/10.1007/978-3-030-14347-3_38

Generally, scheduling problem consist in the identification and optimization of solutions for organizing the operations set execution under certain time constraints as well as capacity constraints in resources [8,14,20]. Thus, in scheduling area there are techniques used that are synthesized with the modern directions referring to applying artificial intelligence methods or optimization algorithms based on meta-heuristics.

Although optimization features are required, the real manufacturing scheduling problems are quite dynamic, considering new orders, delays, failures and unexpected events. Hence, it is hard to handle using only centralized optimization methods [5], since they have a high response time and do not provide autonomous and dynamic behavior. However, Multi-Agent Systems (MAS) [22] in agent-based modeling platforms, offer an alternative way to design, simulate and control systems, with capabilities to adapt to emergence or disruptions through fast local decisions with dynamic behaviors [2,23].

Regarding the literature in this field, there are some issues that present reviews and surveys related to different strategies and techniques of artificial intelligence for job shop scheduling in manufacturing systems, such as the use of meta-heuristics and multi-agent systems [3,6]. In this context some authors address, for example, a flexible job shop scheduling using meta-heuristics based on multi-agents [11,12]. It is also important to note that few studies present the job shop scheduling problem for efficient simulation layouts [17]. On the other hand, some authors integrate in a collaborative way the job shop scheduling with layout planning, that is, presenting a hybrid system capable of optimizing multiple solutions [15,16].

In this sense, this work dealing with a multi-agent system combined with genetic algorithm, proposed to simultaneously perform a job shop scheduling in a flexible layout under a manufacturing dynamic environment. The idea of this system approach, regarding to the existing works, is to consider and enable a cooperation between optimization and agents negotiation by providing dynamic and autonomous simulations that involve disruptions or unexpected events.

The rest of the paper is organized as follows: Sect. 2 describes the system architecture to improve the scheduling solutions in a dynamic and optimized context. Section 3 presents the scheduling and layout tool, with brief descriptions of the Genetic Algorithm and how the layout can be affected and optimized. On the other hand, Sect. 4 describes the agent-based model, developed in NetLogo, for the dynamic decentralized solutions. Section 5 presents the simulation tool applied to a case study, and consequently the discussion of the results. Finally, Sect. 6 rounds up the paper with the conclusions and future work.

2   System Architecture

The information trade-off between the timing of the works arrival, the creation of setups, the assignment of priorities, the testing of disruptions and the obtaining of the solutions for simulation of a real and dynamic job shop environment, will be according to an architecture that encompasses all these parameters in a distributed system. In this context, the system architecture for the manufacturing industry scheduling system integrates off-line and on-line modules as presented in Fig. 1. So, it is possible to deal sequentially with two sub-processes: optimized planning using optimization methods and real-time responsiveness solutions using MAS. The two modules are able to exchange information, balancing the decision-making.

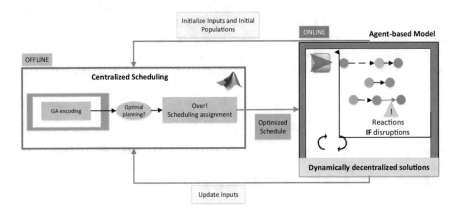

Fig. 1. Developed system architecture.

The module on the left performs the optimized scheduling for the jobs, running off-line, for a given situation under study. The module on the right concerns to the dynamic re-scheduling, in response to disruptions or condition changes, e.g. a broken machine in some situation or a failure in some position. In this module, the re-scheduling is obtained by the interaction of individual entities, each one reasoning about its own schedule. The individual entities (jobs and machines) can share information and have the ability to negotiate in order to overcome disruptions or failures, that is, they can obtain dynamic re-arranging in the layout, with minimal human intervention.

3   Scheduling and Layout Tool

The scheduling activity seeks to make efficient use of its resources to ensure the fast execution of the work so that it can be delivered within the agreed deadlines. Therefore, the scheduling contains varied objectives, evaluated by performance measures, such as the average number of jobs, production profits, minimization of time spent, etc. The present work aims to formulate a dynamic and flexible manufacturing scheduling system, integrated in a layout feature for further process optimization. The possibility of creating different products, requiring different types of processes (jobs) and which in turn require a set of operations, can be performed in certain positions/machines of a particular layout arrangement.

In this context, it is essential to develop efficient decision support methods to solve job shop scheduling, because the operators need to test several scenarios, which makes time requirements crucial.

The off-line module describes the approach to perform the optimized scheduling using Genetic Algorithm (GA). Initially proposed by Holland [9], GA uses a population of individuals to apply genetic procedures: crossover between two different individuals or/and mutation in one individual. The algorithm repeats the crossover and mutation procedures in new populations until the desired diversity of solutions is performed [7]. The optimization method is summarized by the GA presented in Alves et al. [1]. The iterative procedure ends after a maximum number of iterations (NI) or after a maximum number of function evaluations (NFE).

3.1   Layout Optimization

Layout is a feature inherent in any operation. In this way, it is necessary to continuously optimize the layout arrangement, not only for the scheduling of inputs but also to improve the profitability of the outputs (Fig. 2).

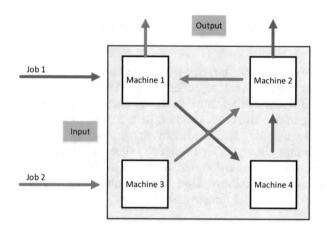

Fig. 2. Facility layout problems and variants.

Through the form and scheduling, appearance and the way materials, information and customers flow through the layout, the fullness of functions will be triggered [18]. The layout optimization problems are quite complex and in general are NP-hard, which requires a high computational complexity. The layout problem focuses on the flexible scheduling, reconfigurable, and agile manufacturing environments where the demand is affected (by machine disruption for example) [18]. With dynamics layout models embedded in optimization models and combined with multi-agent systems, it will be possible to minimize the sum of material handling costs. On the other hand, it will be possible guarantee the best disposition of human and material resources, allowing efficiency, but also dynamic and autonomous re-arranging from the agent-based model.

4   Agent-Based Model

The on-line module considers the use of MAS to implement the dynamic re-scheduling in case of disruption. In addition, it will be possible to interact, simulate and visualize the scheduling performed by the off-line module.

The agent-based model considers three types of agents:

- **Machine agent:** These agents represent the work machines, which will allow the set of manufacturing operations in order to obtain the products. They are immobile, totally passive, and are subject to interruptions or failures.
- **Operation (job) agent:** These agents represent the "jobs" (a job represents a set of operations) that move around to provide the manufacturing of the products according to the scheduling and location of the machines agents. Operation agent only interacts with their own machines.
- **Product agent:** These agents represent the final product to be obtained. After the complete set of operations they require, they send a warning message informing the user of their conclusion.

The Fig. 3 presents the two main categories of global process behaviour: the passive and autonomous behaviours.

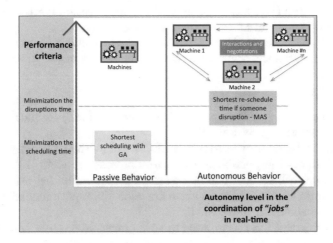

Fig. 3. Dynamic "machines" schedule.

In the passive behavior, the operations follow carefully the planned optimized schedule provided by the off-line module, i.e. using the GA coded in MatLab® [10], without taking into account the disruptions or failures. In the autonomous behavior, the operation agent follows the planned route but is able to dynamically adapt the schedule in case of disruptions through the interaction with other machines, which may be available and with operations in common, to re-arrange the schedule that was previously allocated to the broken machine.

5   Simulation Tool

The described agent-based model was implemented in NetLogo [21], a simulation platform that allows to rapidly instantiate agent-based models to observe the behavior of systems. It provides an intuitive user interface where can be add buttons and control widgets to easily manipulate a model to view different scenarios [4]. The agent-based model developed in NetLogo is connected with MatLab® to allow the exchange of the optimized scheduling solution, through an extension that allows data passing between Netlogo and MatLab® or vice versa, facilitating dynamic integration of these software platforms.

In each simulation, some parameters are used to vary the populations of the agents according to the database. The simulation protocol involved three crucial steps, and it can be applied to many other simulations within the database, thus:

- **Step 1:** The user will have at his disposal the specifications of the problem. Is possible to select the different products to carry out and also to define the quantities of operations (jobs) from the selection already existing in the database. On the other hand, the set of available machines will also be assigned and their coordinates remained unchanged during the experiments.
- **Step 2:** The scheduling is performed by the user. The information will be transmitted and the overall schedule of the optimization method for the jobs planning (output viewer - interface) will be received.
- **Step 3:** It is displayed the jobs schedule sequence and if there is disruption, it can be updated the dynamic reschedule by agents negotiation.

5.1   Case Study

This work presents a Job Shop Scheduling Problem (JSSP) formulated from the Flexible Manufacturing System (FMS).

This data is based on the AIP-PRIMECA cell at the University of Valenciennes which offers the possibility to create different products [19], but in this study a product called "AI" will be created. The operation list also defines the processing times and in this case, there are eight different manufacturing operation types, as shown in the Table 1.

Besides this, the FMS is composed by seven workstations, each one being able to perform a set of operations. The machines are responsible for the completion of manufacturing operations to do the jobs. Some machines are able to complete the same manufacturing operation, while some operations can only be completed on a single machine. Additionally, each machine is continuously available as the system start and each machine can process only one operation at a time. The cell is composed with seven machines and the processing times of the eight manufacturing operations are described on Table 2.

Table 1. Production sequence for each type of job.

	A	I
#1	Plate loading	Plate loading
#2	Axis mounting	Axis mounting
#3	Axis mounting	Axis mounting
#4	Axis mounting	l_comp mounting
#5	r_comp mounting	Screw_comp mounting
#6	L_comp mounting	Inspection
#7	l_comp mounting	Plate loading
#8	Screw_comp mounting	
#9	Inspection	
#10	Plate loading	

Table 2. Processing time of each manufacturing operation (in seconds).

Operation	M1	M2	M3	M4	M5	M6	M7
Plate loading	10						
Plate unloading	10						
Axis mounting		20	20				
r_comp mounting		20		20			
l_comp mounting					20		
L_comp mounting				20		20	
Screw_comp mounting			20	20			
Inspection							10

5.2   Results Discussion

For case study, simulations were carried out on a PC Intel(R) Core(TM) i7 CPU 2.2 GHz with 6.0 GB of RAM. Since the GA is a stochastic method, 10 runs were carried out with random initial points. The values of the control parameters used in GA were adjusted to a suitable experience of the problem, i.e. it was considered a population size ($Ps = 30$) and concerning the probability of the procedures (crossover and mutation), 50% rate was selected. In turn, the function evaluation was fixed at $NFE = 5000$ and the maximum number of iterations as $NI = 100$.

In the present case, one "AI" and two jobs are created. The objective of this round of experiments (i.e., optimized scheduling with passive behavior and disruptions tests with autonomous/dynamic behavior) was to highlight the efficiency of the proposed approach defined by the ability of the latter to minimize the time and automatically optimize disruptions or failures in machines.

The scheduling tool allowed to obtain 100% of successful rate since they found a feasible solution in all runs. In turn, the scheduling is obtained extremely fast,

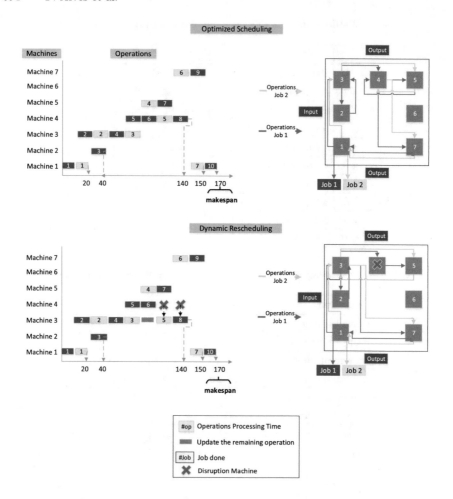

Fig. 4. Optimized solution simulation and dynamic layout rescheduling.

i.e, it only took 25 s. On the other hand, allowed to obtain the solution that characterizes the optimal time of execution of the jobs in question, tested for unprecedented scheduling and another with machine disruptions.

The Fig. 4 presents the optimized solution of the problem under study as well as the subject dynamic rescheduling in tests of machine disruptions.

It is important to note that the scheduling obtained by the GA has always ensured a feasible solution as well as it allows to identify the sets of operations to their respective machines. At the top of the Fig. 4 it is possible to verify optimized scheduling (not subject to disruption testing), which enabled the set of operations to identify their respective machines and allocate themselves properly in the layout for the production of the jobs. Consequently, it is also possible to analyze that the time required to create job 1 and job 2 is respectively 170 and 150 s. After this, using the interface in NetLogo, the same scheduling of the case under study was subjected to a random disruption of a machine after a certain time (in this case machine 4), leaving two operations, one of each job without conclusion. Thus, through the agent-based model, the dynamic and autonomous rescheduling protocol was started to allocate the remaining operations again in the layout. They were allocated to the machine 3 without causing any disturbance and in the same order as they were left. This solution enabled not only the minimal human interaction but also allowed to finish the jobs in the time previously expected, without affecting the end of the remaining operations or priorities imposed.

6   Conclusions and Future Work

This paper addresses a hybrid system for job shop scheduling problems and layout optimization in a dynamic context with numerous constraints, uncertainties and random events. The system architecture has been proposed to solve the two sub-problems presented in a sequential manner. An experimental case study was proposed based on FMS. Thus, through the results presented, it was possible to obtain a fast and optimized scheduling by the off-line module and at the same time interact with the NetLogo platform. In this sense, the same scheduling was subject to random interruption of a machine, which autonomously and with the use of a negotiation protocol between agents, allowed the dynamic rescheduling of the missing operations. In turn, machine 3 allocated the operations and ensured that the final time for completion of the two jobs was not exceeded. This approach not only enables optimized solutions but also ensures minimal human intervention in real-time disruption testing for layout re-arranging.

The main contribution of this paper it was to provide a new way of solving a job shop scheduling problems in simultaneous with layout optimization, using the machines ability to dynamically adapt and design his own schedule in case of external disruptions. Thus, we can improve both flexibility and efficiency in today's competitive manufacturing environments, such as flexible manufacturing systems and just-in-time production, as well as add strategies and dynamics to improve quality and increase profits.

For future work, it is possible to reformulate and increase the complexity of the problem. Another approach to the future could be to use other metaheuristics and implement dynamic scheduling alternatives in case of disruptions according to products and jobs priorities.

Acknowledgements. This work has been supported by FCT – Fundação para a Ciência e Tecnologia under the Project: PEst2015-2020. This work was also supported by COMPETE: POCI-01-0145-FEDER-007043 and FCT - Fundação para a Ciência e Tecnologia within the Project Scope: UID/CEC/00319/2013.

References

1. Alves, F., Pereira, A.I., Fernandes, A., Leitão, P.: Optimization of home care visits schedule by genetic algorithm. In: International Conference on Bioinspired Methods and Their Applications, pp. 1–12. Springer (2018)
2. Barbosa, J., Leitão, P.: Simulation of multi-agent manufacturing systems using agent-based modelling platforms. In: 2011 9th IEEE International Conference on Industrial Informatics (INDIN), pp. 477–482. IEEE (2011)
3. Çaliş, B., Bulkan, S.: A research survey: review of AI solution strategies of job shop scheduling problem. J. Intell. Manufact. **26**(5), 961–973 (2015)
4. Chiacchio, F., Pennisi, M., Russo, G., Motta, S., Pappalardo, F.: Agent-based modeling of the immune system: NetLogo, a promising framework. BioMed Res. Int. **2014** (2014)
5. Eremia, M., Liu, C.C., Edris, A.A.: Advanced Solutions in Power Systems: HVDC, FACTS, and Artificial Intelligence, vol. 52. Wiley, Hoboken (2016)
6. Gen, M., Lin, L., Zhang, H.: Evolutionary techniques for optimization problems in integrated manufacturing system: state-of-the-art-survey. Comput. Ind. Eng. **56**(3), 779–808 (2009)
7. Ghaheri, A., Shoar, S., Naderan, M., Hoseini, S.S.: The applications of genetic algorithms in medicine. Oman Med. J. **30**(6), 406 (2015)
8. Groover, M.P.: Automation, Production Systems, and Computer-Integrated Manufacturing. Prentice Hall Press, Upper Saddle River (2007)
9. Holland, J.H.: Adaptation in Natural and Artificial Systems: An Introductory Analysis with Applications to Biology, Control, and Artificial Intelligence. MIT press, Cambridge (1992)
10. MATLAB: version 8.6.0 (R2015b). The MathWorks Inc., Natick, Massachusetts (2015)
11. Nouiri, M., Bekrar, A., Jemai, A., Niar, S., Ammari, A.C.: An effective and distributed particle swarm optimization algorithm for flexible job-shop scheduling problem. J. Intell. Manufact. **29**(3), 603–615 (2018)
12. Nouri, H.E., Driss, O.B., Ghédira, K.: Simultaneous scheduling of machines and transport robots in flexible job shop environment using hybrid metaheuristics based on clustered holonic multiagent model. Comput. Ind. Eng. **102**, 488–501 (2016)
13. Ouelhadj, D., Petrovic, S.: A survey of dynamic scheduling in manufacturing systems. J. Sched. **12**(4), 417 (2009)
14. Rewers, P., Trojanowska, J., Diakun, J., Rocha, A., Reis, L.P.: A study of priority rules for a levelled production plan. In: Advances in Manufacturing, pp. 111–120. Springer (2018)
15. Ripon, K.S.N., Torresen, J.: Integrated job shop scheduling and layout planning: a hybrid evolutionary method for optimizing multiple objectives. Evol. Syst. **5**(2), 121–132 (2014)
16. Sahin, C., Demirtas, M., Erol, R., Baykasoğlu, A., Kaplanoğlu, V.: A multi-agent based approach to dynamic scheduling with flexible processing capabilities. J. Intell. Manufact. **28**(8), 1827–1845 (2017)
17. Sudo, Y., Matsuda, M.: Agent based manufacturing simulation for efficient assembly operations. Procedia CIRP **7**, 437–442 (2013)
18. Tompkins, J.A., White, J.A., Bozer, Y.A., Tanchoco, J.M.A.: Facilities Planning. Wiley, Hoboken (2010)
19. Trentesaux, D., Pach, C., Bekrar, A., Sallez, Y., Berger, T., Bonte, T., Leitão, P., Barbosa, J.: Benchmarking flexible job-shop scheduling and control systems. Control Eng. Pract. **21**(9), 1204–1225 (2013)

20. Varela, M., Putnik, G., Manupati, V., Rajyalakshmi, G., Trojanowska, J., Machado, J., et al.: Collaborative manufacturing based on cloud, and on other I4.0 oriented principles and technologies: a systematic literature review and reflections. Manag. Prod. Eng. Rev. **9**(3), 90–99 (2018)
21. Wilensky, U.: NetLogo: center for connected learning and computer-based modeling. Northwestern University, Evanston, IL, 1999 (2014)
22. Wooldridge, M.: An Introduction to Multiagent Systems. Wiley, Chichester (2009)
23. Xie, J., Liu, C.C.: Multi-agent systems and their applications. J. Int. Counc. Electr. Eng. **7**(1), 188–197 (2017)

Application of the Simulated Annealing Algorithm to Minimize the *makespan* on the Unrelated Parallel Machine Scheduling Problem with Setup Times

Gabriela Amaral[1(⊠)], Lino Costa[1], Ana Maria A. C. Rocha[1],
Leonilde Varela[1], and Ana Madureira[2]

[1] Algoritmi Research Centre, University of Minho,
4800-058 Guimarães, Portugal
id5731@alunos.uminho.pt,
{lac,arocha,leonilde}@dps.uminho.pt
[2] Interdisciplinary Studies Research Center (ISRC), ISEP/IPP, Porto, Portugal
amd@isep.ipp.pt

Abstract. In this paper, the unrelated parallel machine scheduling problem considering machine-dependent and job sequence-dependent setup times is addressed. This problem involves the scheduling of n jobs on m unrelated machines with setup times in order to minimize the *makespan*. The Simulated Annealing algorithm is used to solve four sets of small scheduling problems, from the literature, on two unrelated machines: the first one has six jobs, the second has seven jobs and the third and fourth has eight and nine jobs, respectively. The results seem promising when compared with other methods referred in literature.

Keywords: Scheduling problem · Setup times · Simulated annealing

1 Introduction

The production scheduling when optimizing configurations, either directly or indirectly, has been an important area for the various types of industries, including plastic, textile, and chemical, as well as some service areas [1–9]. In addition, it has been an area of great importance in research, expressed through a vast set of scientific publications [2–5]. Aydilek et al. [10] address a scheduling problem to minimize order delays, in which the configuration times independently of the order processing times, with the application of algorithms of self-adaptive Differential Evolution and a hybrid and simulated insertion algorithm. The scheduling problem with different approaches to change or setup times, with the objective of reducing complexity in order to minimize the *makespan* or the total production time, with the application of an enhanced version of the ant colony optimization algorithm was studied in [11, 12].

The problem of parallel machine programming has been a growing research area since the early work [13]. In Arnaout et al. [14], it was used an Ant Colony optimization algorithm for minimizing the schedule's *makespan* on unrelated parallel

© Springer Nature Switzerland AG 2020
A. M. Madureira et al. (Eds.): HIS 2018, AISC 923, pp. 398–407, 2020.
https://doi.org/10.1007/978-3-030-14347-3_39

machine with sequence-dependent setup times. For the same problem and objective but considering the setup times, Rabadi et al. [1] introduced a new metaheuristic (MetaRaPS) and its performance was evaluated by comparing its solutions with an existing heuristic for the same problem. The results show that MetaRaPS found all optimal solutions for small problems and outperformed the solutions obtained by the existing heuristics for larger problems. The unrelated parallel machine scheduling problem (PMSP) when minimizing the *makespan* was addressed by Woo et al. [15] where a mixed integer linear programming (MILP) model to find the optimal solution was developed. They proposed a novel rule based on the Genetic Algorithm (GA) with a chromosome representing job assigning sequence to one of the machines and the schedule of jobs completion time rule based on dispatching heuristics during the decoding process of the chromosome. Abreu's work [16] described a hybrid GA for solving the unrelated parallel machine scheduling problem with sequence dependent setup times. A case study on the granite industry is presented and the proposed approach outperformed three traditional dispatch rules presented in the current literature. Gedik et al. [17] studied the non-preemptive unrelated parallel machine scheduling problem with job sequence and machine dependent setup times with the objective of minimizing the *makespan*. This study provided a novel constraint programming (CP) model with two customized branching strategies that uses CP's global constraints, interval decision variables, and domain filtering algorithms. In terms of average solution quality, the computational results indicated that the CP model slightly outperforms all of the state-of-art algorithms in solving small problem instances and was able to prove the optimality of 283 currently best-known solutions. It is also effective in finding good quality feasible solutions for the larger problem instances. Fanjul-Peyro et al. [18] studied the same problem with the same objective function but they modelled the problem by means of two integer linear programming problems. One is based on a model previously proposed in the literature and the other is based on the resemblance to strip packing problems. Since the models were unable to solve medium-sized instances to optimality, they proposed three Metaheuristics strategies for each of these two models. The results show that the Metaheuristics significantly outperform the mathematical models.

Recently, a multi-objective approach about parallel machines was proposed as a new problem, the resource constrained unrelated parallel machine green manufacturing scheduling problem with the criteria of minimizing the *makespan* and the total carbon emission [19]. To solve this problem, a collaborative multi-objective fruit fly optimization algorithm was proposed. The results showed that the multi-objective algorithm was able to obtain more and better non-dominated solutions than other algorithms.

Considering the previous literature review, most of the research addressed the scheduling problem with different objectives and algorithms application in different contexts and industrial environments.

The purpose of this work is motivated by the Rabadi et al. [1] work about the truss manufacturing industry where manufacturing roof trusses requires different setup times depending on the manufacturing sequence and the type of manufacturing apparatus used. Based on data for small instances [1] we propose a Simulated Annealing (SA) algorithm for solving the unrelated parallel machine scheduling problem considering the setup times with the objective of minimizing the *makespan*, C_{max}. In order to assess the performance of this algorithm when solving this kind of problems, a comparison with the solutions obtained by the metaheuristic MetaRaPs in [1] will be made.

Thus, in this study, we intend to evaluate the behaviour of the SA algorithm when solving four types of case studies of unrelated parallel machine scheduling problems with setup times. The first is a collection of 15 instances for two machines and six jobs, the second, 15 instances for two machines and seven jobs, 15 instances for two machines and eight jobs, 15 instances for two machines and nine jobs.

This paper is organized as follows. In Sect. 2, the problem formulation is briefly described, by explaining the main constraints behind the variables and parameters of the problem. In Sect. 3, the results of the numerical experiments are presented and Sect. 4 contains the conclusions of the present study and some ideas for future work.

2   Problem Formulation

This article addresses the unrelated parallel machine scheduling problem considering the scheduling of n jobs that are available at time zero on m unrelated machines (R_m). The objective function is the C_{max}, taking into account machine-dependent and sequence dependent setup time S_{ijk}. When the job's processing times depend on the machines to which they are assigned, and it does not exist any relationship between machine speeds, the machines are unrelated. The setup time S_{ijk} depends on the jobs sequence and the machine because each machine has its own matrix of setup times. The type of problem discussed in this article is classified in the literature as $R_m|S_{ijk}|C_{máx}$ [20]. The basic identical PMSP $P_m\|C_{máx}$ is NP-hard even when $m = 2$ [21–23]. We considered that $R_m|S_{ijk}|C_{máx}$ is also NP-hard since it is a generalization of the former problem [20].

Below we describe the problem used the following assumptions and notations:

- m is the number of parallel machines of the set M;
- n denotes the number of jobs to be scheduled of the set N;
- Each machine can only process one job at a time without interruption;
- For the initial instant that it is at time zero all jobs are available. No restrictions of precedence are imposed among jobs;
- In each machine k, each job i has a processing time p_{ik};
- In each machine k, for processing job j just after job i, there is a setup time s_{ijk}. The setup time is different for each machine.
- The objective is to minimize the *makespan* C_{max}. The term *span* is used to define the completion time of a machine while the term *makespan* is used for the maximum *span* in the solution of the problem.

The nomenclature presented to describe the machines and jobs can be flexible, between i, j, k, as well as the variables describing the sets of machines and jobs (M and N). In this work, we have four case studies involving 2 machines ($m = 2$) and $n = 6$, $n = 7$, $n = 8$ and $n = 9$ jobs.

3   Results and Discussion

In this section, we analyse the behaviour of the SA algorithm when solving unrelated parallel machine scheduling problems considering the setup times and the objective function to be minimized is C_{max}. The instances were taken from [1] and are only considered some small problems and balanced times data belonging to two case studies, each one composed by a collection of 15 instances: the first case for two machines and six jobs, the second for two machines and seven jobs, the third for two machines and eight jobs and the fourth for two machines and nine jobs. The data used is available at Scheduling Research Virtual Centre Homepage [24, 25].

The SA algorithm has undergone some changes to the original algorithm, available at Academic Source Codes and Tutorials [24], developed by S. Mostapha Kalami Heris for parallel machine scheduling problem. The original code of the algorithm was adapted to consider the setup time of the first job, the maximum number of iterations was fixed to 100 iterations within the specific parameters of SA to gain at run time. For each case study and for each instance 20 runs were done, since SA is a stochastic algorithm. The run time, C_{max} and Work Order are outputs of the algorithm SA. For each case study and for each run associated to each instance, two types of graphs can be generated: the work order and the C_{max} along the iterations. The case studies were run in the MatLab® R2017 using a PC i7-4600U CPU @ 2.10 GHz 2.70 GHz.

Table 1 shows the average values of C_{max} (Avg. C_{max}) and run time (Avg Time) obtained by SA when solving each of the 15 instances among the 20 runs.

Table 1. Average solutions of C_{max} and run time

Instance	$m = 2, n = 6$		$m = 2, n = 7$		$m = 2, n = 8$		$m = 2, n = 9$	
	Avg. C_{max}	Avg time	Avg. C_{max}	Avg time	Avg. C_{max}	Avg time	Avg. C_{max}	Avg time
1	390.00	390.30	484.00	486.25	494.00	498.90	627.00	645.00
2	410.00	410.10	527.00	531.00	523.00	523.00	603.00	609.60
3	410.00	411.60	498.00	499.70	508.00	510.30	558.00	565.70
4	380.00	380.70	467.00	470.90	539.00	546.95	618.00	622.60
5	399.00	399.80	495.00	497.00	509.00	512.55	571.00	588.00
6	397.00	397.00	493.00	493.55	502.00	504.75	589.00	596.20
7	394.00	394.20	470.00	470.80	522.00	523.80	611.00	619.40
8	379.00	379.00	516.00	522.05	534.00	536.45	624.00	625.80
9	407.00	407.00	496.00	496.00	518.00	525.15	594.00	613.80
10	394.00	395.50	490.00	493.90	517.00	526.80	616.00	614.10
11	388.00	388.00	489.00	489.00	511.00	525.20	579.00	584.00
12	396.00	401.30	483.00	485.80	509.00	512.10	588.00	598.30
13	384.00	385.10	493.00	493.50	505.00	510.25	599.00	611.50
14	369.00	370.20	488.00	504.40	563.00	565.40	612.00	616.30
15	424.00	424.60	476.00	480.95	507.00	507.00	594.00	608.00

In order to evaluate the performance of SA when solving this set of unrelated parallel machine problems with sequence-dependent setup times, a comparison with another metaheuristic will be done. Thus, Table 2 shows the solutions obtained by MetaRaPs in [1] and the best values of C_{max} (Best C_{max}) obtained by SA when solving each of the 15 instances, among the 20 runs.

Table 2. Best solutions of C_{max} obtained by SA and MetaRaPs

Instance	$m = 2, n = 6$		$m = 2, n = 7$		$m = 2, n = 8$		$m = 2, n = 9$	
	MetaRaPs	Best C_{max}	MetaRaPs	Best C_{max}	MetaRaPs	Best C_{max}	MetaRaPs	Best C_{max}
1	390.00	390.00	484.00	484.00	494.00	494.00	627.00	627.00
2	410.00	410.00	527.00	527.00	523.00	523.00	603.00	603.00
3	410.00	410.00	498.00	498.00	508.00	508.00	558.00	558.00
4	380.00	380.00	467.00	467.00	539.00	539.00	618.00	618.00
5	399.00	399.00	495.00	495.00	509.00	509.00	571.00	571.00
6	397.00	397.00	493.00	493.00	502.00	502.00	589.00	589.00
7	394.00	394.00	470.00	470.00	522.00	522.00	611.00	611.00
8	379.00	379.00	516.00	516.00	534.00	534.00	624.00	624.00
9	407.00	407.00	496.00	496.00	518.00	518.00	594.00	594.00
10	394.00	394.00	490.00	490.00	517.00	517.00	616.00	612.00
11	388.00	388.00	489.00	489.00	511.00	511.00	579.00	579.00
12	396.00	396.00	483.00	483.00	509.00	509.00	588.00	588.00
13	384.00	384.00	493.00	493.00	505.00	505.00	599.00	599.00
14	369.00	369.00	488.00	488.00	563.00	563.00	612.00	610.00
15	424.00	424.00	476.00	476.00	507.00	507.00	594.00	594.00

Analysing Table 2, it can be concluded that the solutions between the two metaheuristics MetaRaPS and SA are very similar. However, it was verified that for the case of two machines and nine works, for the instances 10 and 14, the solutions obtained by SA are better than the ones from MetaRaPS, reaching the optimal known solution.

We noticed that for instance 10 with $m = 2$, $n = 9$, the SA obtained the best solution in 17 of the 20 runs, giving nine different working orders for the same global best solution of $C_{max} = 612$. Figure 1 shows the work sequence that was assigned to the case study for $m = 2$, $n = 9$, instance 10, for the different global solutions. In the figure it is possible to visualize which and how many jobs were allocated to each machine and which machine finishes the execution of the planned tasks, the best achieved C_{max}. The works positioned in the lower part of the figure refer to the machine 1, and those positioned in the upper part, to the machine 2. The empty spaces between the beginning and the end of each job, refer to the setup time allocated between each job.

For instance 14 of the case of two machines and nine jobs, the best solution was obtained in 10 out of the 20 runs, giving in all of them the same working order for a best value of $C_{max} = 610$, as shown in Fig. 2.

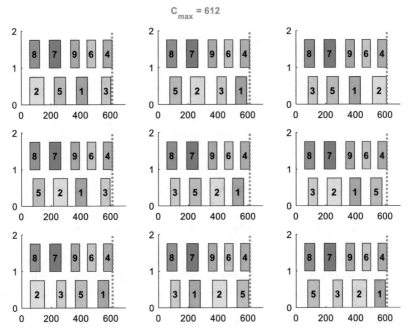

Fig. 1. Work order of instance 10 for $m = 2$ and $n = 9$

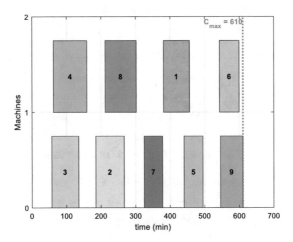

Fig. 2. Work order of instance 14 for $m = 2$ and $n = 9$

Figure 3 shows the value of C_{max} along the iterations for instance 14 with $m = 2$ and $n = 9$. In the first 20 iterations, a noticeable decrease in the C_{max} value is observed. For the remaining iterations, the C_{max} value remains constant for about 60 iterations and a last decrease is observed near iteration 90.

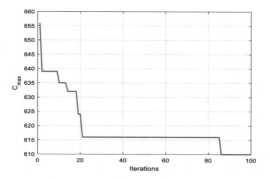

Fig. 3. C_{max} along the iterations of instance 14 for $m = 2$ and $n = 9$

Figure 4 shows the average and best values of C_{max} of the 15 instances, for the case study of two machines ($m = 2$) and six jobs ($n = 6$). For almost all the instances the average of C_{max} values is close to the best values, except for instance 12. Figure 5 plots the average versus best of C_{max} over the 15 instances, for the case study of $m = 2$ and $n = 7$. It presents a large difference between the average and best values for some instances. For the case study of $m = 2$ and $n = 8$, when showing the differences between the average and best values of C_{max}, for the 15 instances, Fig. 6 reveals higher differences for almost all instances. For two machines ($m = 2$) and nine jobs ($n = 9$), in Fig. 7, it is observed that the difference from average and best values of C_{max} has increased for the 15 instances when compared to the other cases.

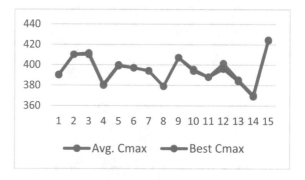

Fig. 4. Average C_{max} vs Best C_{max} for $m = 2, n = 6$

Consistently, Fig. 4, 5, 6 and Fig. 7 show that when the number of jobs increases, the differences between the average and best values of C_{max} also increase. This behaviour can be due to the search space that becomes also larger with the number of jobs. In spite of the growth of the search space, it should be noted that the maximum number of iterations of SA was maintained for all problems. In these figures, we also observe a considerable variability of the average and best values of C_{max} that can be caused by the big difference between the data assigned to each instance.

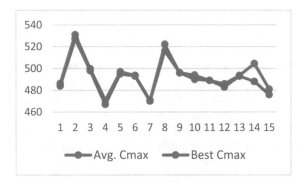

Fig. 5. Average C_{max} vs Best C_{max} for $m = 2, n = 7$

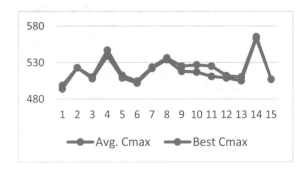

Fig. 6. Average C_{max} vs Best C_{max} for $m = 2, n = 8$

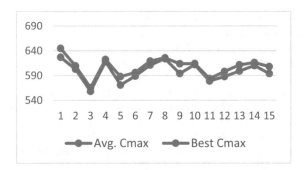

Fig. 7. Average C_{max} vs Best C_{max} for $m = 2, n = 9$

4 Conclusions and Future Work

The area of scheduling problem to optimize setup time has been and remains important for the industries and research. In this work, we intended to study the unrelated parallel machine scheduling problem with the objective of minimizing the C_{max}, considering machine-dependent and job sequence-dependent setup times. This type of problem is classified as $R_m|S_{ijk}|C_{max}$ and the basic identical PMSP $P_m||C_{max}$ is NP-hard even when $m = 2$ [21].

Radabi et al. [1] studied the same problem with the same objective, and introduced a new metaheuristic, the MetaRaPs. In this paper, we used the simulated annealing optimization algorithm for minimizing the C_{max}, of four types of case studies: two machines with a number of jobs varying between six and nine. For each case study, 15 instances were considered and solved by SA algorithm. The results obtained were compared with those obtained by MetaRaPs [1] to evaluate the behaviour of the SA algorithm. The solutions obtained by the two metaheuristics MetaRaPS and SA were very similar. However, it was verified that for the case of two machines and nine works, in two instances, the solutions obtained by SA were better than the ones from MetaRaPS, reaching the optimal known solution. We remark that, for some instances, SA was able to find different working orders for the same global best solution. The performance of SA was also studied when the number of jobs increases. Consistently, it was observed that the differences between the average and best values of C_{max} increase with the number of jobs as the search space becomes larger.

In near future we aim to improve the SA and tune its parameters in order to reduce the time execution. In addition, it is intended to analyse the behaviour the SA through the introduction of more case studies by varying the number of machines and jobs.

Acknowledgement. This project is funded by COMPETE: POCI-01-0145-FEDER-007043 and FCT – Fundação para a Ciência e Tecnologia within the project scope: UID/CEC/00319/2013.

References

1. Rabadi, G., Moraga, R.J., Al-Salem, A.: Heuristics for the unrelated parallel machine scheduling problem with setup times. J. Intell. Manuf. **17**(1), 85–97 (2006)
2. Varela, M.L.R., Aparício, J.N., Silva, S.C.: A web-based application for manufacturing scheduling. In: IASTED International Conference on Intelligent Systems and Control, pp. 400–405 (2003)
3. Reddy, M.S., Ratnam, C., Agrawal, R., Varela, M.L.R., Sharma, I., Manupati, V.K.: Investigation of reconfiguration effect on makespan with social network method for flexible job shop scheduling problem. Comput. Ind. Eng. **110**, 231–241 (2017)
4. Ghadiri Nejad, M., Güden, H., Vizvári, B., Vatankhah Barenji, R.: A mathematical model and simulated annealing algorithm for solving the cyclic scheduling problem of a flexible robotic cell. Adv. Mech. Eng. **10**(1), 1–12 (2018)
5. Hinder, O., Mason, A.J.: A novel integer programing formulation for scheduling with family setup times on a single machine to minimize maximum lateness. Eur. J. Oper. Res. **262**(2), 411–423 (2017)
6. Nikabadi, M., Naderi, R.: A hybrid algorithm for unrelated parallel machines scheduling. Int. J. Ind. Eng. Comput. **7**(4), 681–702 (2016)
7. Rewers, P., Trojanowska, J., Diakun, J., Rocha, A., Reis, L.P.: A study of priority rules for a levelled production plan. In: Hamrol, A., Ciszak, O., Legutko, S., Jurczyk, M. (eds.) Advances in Manufacturing. Lecture Notes in Mechanical Engineering, pp. 111–120. Springer, Cham (2018)

8. Trojanowska, J., Kolinski, A., Galusik, D., Varela, M.L.R., Machado, J.: A methodology of improvement of manufacturing productivity through increasing operational efficiency of the production process. In: Hamrol, A., Ciszak, O., Legutko, S., Jurczyk, M. (eds.) Advances in Manufacturing. Lecture Notes in Mechanical Engineering, pp. 23–32. Springer, Cham (2018)
9. Bülbül, K., Şen, H.: An exact extended formulation for the unrelated parallel machine total weighted completion time problem. J. Sched. **20**(4), 373–389 (2017)
10. Aydilek, A., Aydilek, H., Allahverdi, A.: Minimising maximum tardiness in assembly flowshops with setup times. Int. J. Prod. Res., 1–25 (2017)
11. Zhang, S., Wong, T.N.: Studying the impact of sequence-dependent set-up times in integrated process planning and scheduling with E-ACO heuristic. Int. J. Prod. Res. **54**(16), 4815–4838 (2016)
12. Xu, L., Wang, Q., Huang, S.: Dynamic order acceptance and scheduling problem with sequence-dependent setup time. Int. J. Prod. Res. **53**(19), 5797–5808 (2015)
13. McNaughton, R.: Scheduling with deadlines and loss functions. Manag. Sci. **6**(1), 1–12 (1959)
14. Arnaout, J.P., Musa, R., Rabadi, G.: A two-stage ant colony optimization algorithm to minimize the makespan on unrelated parallel machines—part II: enhancements and experimentations. J. Intell. Manuf. **25**(1), 43–53 (2014)
15. Woo, Y.B., Jung, S., Kim, B.S.: A rule-based genetic algorithm with an improvement heuristic for unrelated parallel machine scheduling problem with time-dependent deterioration and multiple rate-modifying activities. Comput. Ind. Eng. **109**, 179–190 (2017)
16. Abreu, L.R., Prata, B.A.: A hybrid genetic algorithm for solving the unrelated parallel machine scheduling problem with sequence dependent setup times. IEEE Lat. Am. Trans. **16**(6), 1715–1722 (2018)
17. Gedik, R., Kalathia, D., Egilmez, G., Kirac, E.: A constraint programming approach for solving unrelated parallel machine scheduling problem. Comput. Ind. Eng. **121**, 139–149 (2018)
18. Fanjul-Peyro, L., Perea, F., Ruiz, R.: Models and matheuristics for the unrelated parallel machine scheduling problem with additional resources. Eur. J. Oper. Res. **260**(2), 482–493 (2017)
19. Zheng, X.L., Wang, L.: A collaborative multiobjective fruit fly optimization algorithm for the resource constrained unrelated parallel machine green scheduling problem. IEEE Trans. Syst. Man Cybern.: Syst. **48**(5), 790–800 (2018)
20. Pinedo, M.: Scheduling - Theory, Algorithms, and Systems, 3rd edn. Prentice Hall, Upper Saddle River (2008)
21. Karp, R.M.: Reducibility among combinatorial problems. In: Complexity of computer computations, pp. 85–103. Springer, Boston (1972)
22. Garey, M.R., Johnson, D.S.: Computers and intractability: a guide to NP-completeness (1979)
23. Rabadi, G. (ed.): Heuristics, Metaheuristics and Approximate Methods in Planning and Scheduling, vol. 236. Springer, Cham (2016)
24. Academic Source Codes and Tutorials. www.yarpiz.com. Accessed 29 Sept 2018
25. Scheduling Research Virtual Center Homepage. www.SchedulingResearch.com

Ontology-Based Meta-model for Hybrid Collaborative Scheduling

Leonilde Varela[1(✉)], Goran Putnik[1], Vijaya Manupti[2],
Ana Madureira[3], André Santos[1,3], Gabriela Amaral[1],
and Luís Ferreirinha[3]

[1] Algoritmi Research Centre, University of Minho,
4800-058 Guimarães, Portugal
{leonilde,putnikgd}@dps.uminho.pt, abg@isep.ipp.pt,
id5731@alunos.uminho.pt
[2] Mechanical Engineering Department, NIT Warangal, Warangal 506004, India
manupativijay@gmail.com
[3] Interdisciplinary Studies Research Center (ISRC), ISEP/IPP, Porto, Portugal
amd@isep.ipp.pt, l.ferreirinha96@gmail.com

Abstract. In this paper a scheduling meta-model is proposed for supporting hybrid collaboration, regarding machine-machine and human-machine scheduling interactions, based on a scheduling ontology. The utilization of the proposed scheduling ontology-based meta-model is illustrated through an example, which is further analysed, and some main features and advantages of each kind of collaborative interaction are discussed.

Keywords: Collaborative manufacturing management · Scheduling · Ontology · Meta-model · Hybrid collaboration

1 Introduction

In the current fourth industrial revolution (Industry 4.0 or I4.0, for short) [1] collaborative manufacturing and management (Col2M) is becoming even more crucial for Companies' prosperity in general, and more specifically in the context of industrial enterprises. Based on this believe in this paper an ontology-based meta-model for hybrid collaborative scheduling (OMM-HCS) is proposed.

The hybrid collaboration is related in one side to the existence of generated solutions for given manufacturing scheduling problems (MSP), by using automatically selected algorithms through a scheduling knowledge base [2], and which configures a machine-machine collaborative manufacturing and management mechanism (M-MCol2M). Moreover, there is also a human-machine collaborative manufacturing and management mechanism (H-MCol2M), which enables to further dynamically adjust automatically generated solutions previously obtained through the M-MCol2M to enable better fulfilment of specific scheduling requirements occurring in real manufacturing scenarios. Besides, human-human based collaborative practices, which are also possible.

© Springer Nature Switzerland AG 2020
A. M. Madureira et al. (Eds.): HIS 2018, AISC 923, pp. 408–417, 2020.
https://doi.org/10.1007/978-3-030-14347-3_40

In order to properly expose the main ideas underlying this work, the paper is organized as follows: Sect. 2 briefly refers to the collaborative manufacturing and management concept underlying this work. In Sect. 3 the main idea behind hybrid collaborative scheduling is exposed. Section 4 presents the proposed ontology-based meta-model for supporting hybrid collaborative scheduling practices. In Sect. 5 an illustrative example of use of our proposed meta-model is presented. Finally, in Sect. 6 some main conclusions are presented along with planned future work.

2 Collaborative Manufacturing and Management

The Collaborative Manufacturing and Management (Col2M) concept for Industry 4.0 (I4.0) is suggested to integrate a set of sub concepts, hierarchically organized, as it is illustrated in the Fig. 1. On the basis of this proposed Col2M has to be fulfilled the existence of Flexible and Autonomous Manufacturing and Management Resources (F&AM2R), which is related to the existence of corresponding flexible manufacturing resources (FMRs), for instance, in the context of a flexible manufacturing system (FMS) [3], on one side, and on another one these FMRs or FMS have to be autonomous [4]. At a next level appears the Integrated, Networked and Digitalised Manufacturing and Management (I/N&D2M) concept, which implies the integration [5] and digitalisation [6] of manufacturing resources/Companies and their corresponding management functions. Next, is placed the Concurrent Manufacturing and Management (C2M) concept [7], which implies the existence of one or more common goals among the connected manufacturing resources/Companies. Finally, appears the Collaborative Manufacturing and Management [8] concept (Col2M).

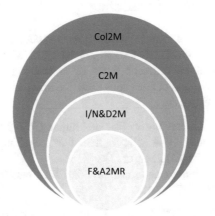

Fig. 1. Venn diagram for the collaborative manufacturing and management concept.

In Fig. 2 the importance of collaboration is expressed in a manufacturing perspective, from the evolution of Computer Integrated Manufacturing (CIM) [9] to the I4.0 manufacturing context (doing axis) and the management practice, from the concurrent perspective up to a complete collaborative level [5]. Through this figure is intended to highlight the perceived growing importance that collaborative manufacturing and management practices are expected to take place in the context of I4.0 manufacturing, and more precisely in the scope of Cyber Physical Systems (CPS) [6].

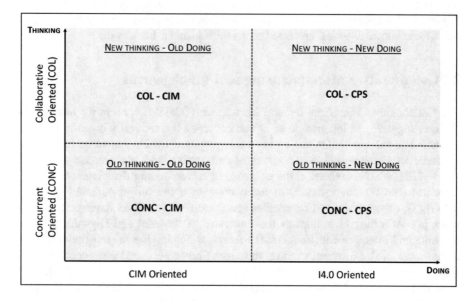

Fig. 2. Collaboration in the context of I4.0

3   Hybrid Collaborative Scheduling

In this paper we propose a ontology-based meta-model for hybrid collaborative scheduling (OMM-HCS), which relies on a proposed Col2M concept, which, in turn, may be considered in three different types of contexts, regarding some kind of underlying kind of flexible and autonomous manufacturing and management resources (F&A2 MR) as being related to:

1. Just human F&A2MR, resulting in Human-Human Col2M (H-H_Col2M);
2. Purely machine oriented F&A2MR or machines, resulting in Machine-Machine Col2M or simply (M-M_Col2M); or
3. A mixed or combined human and machine F&A2MR, resulting in Human-Machine Col2M or simply (H-M_Col2M).

The situation 1. (H-H_Col2M) can, for instance, occur in a context of group decision making [5, 6, 10] among two or more managers from different manufacturing units or Companies, while trying to reach a global solution regarding the division of tasks or production orders among them, based on some underlying negotiation and/or selection mechanism and specified personal contributions through an autonomous and decentralized decision-making process, based on specific decision criteria and one or more common goals to be reached.

Situation 2 (M-M_Col2M) can, for instance, occur in a context of autonomous production control [5], while a set of two or more machines negotiate among them the production order of jobs, which can also be based on some kind of negotiation process, namely through a MAS (Multi-agent System) [11], among other kind of technology, such as based on learning and deep learning [12].

Finally, situation 3 (H-M_Col2M) can, for instance, occur in a manufacturing context with collaborative robots [13] interacting with humans for accomplishing jointly a given task. Moreover, this situation may also occur through a human inter-action with some computer program, for instance while receiving a given manufac-turing scheduling plan and trying to further improve or adjust it.

There is a widened set of publications that has been put forward and also already with a special focus or relation to I4.0 more recently [14], for instance regarding all these three kind of situations (H-H_Col2M, M-M_Col2M, H-M_Col2M) [15], and it is expected that this will continue to happen and even increase further during the next years, and this is one of the main motivations for putting forward our present contri-bution for better clarifying the Col2M concept and its relations with other important ones, considered as fundamental sub concepts of it in the currently and forthcoming context of I4.0, and the underlying Cyber Physical Systems (CPS) reality, on which we are confident that humans will continue to play a crucial role, through H-H_Col2M and H-M_Col2M interactions, as schematized in Fig. 3.

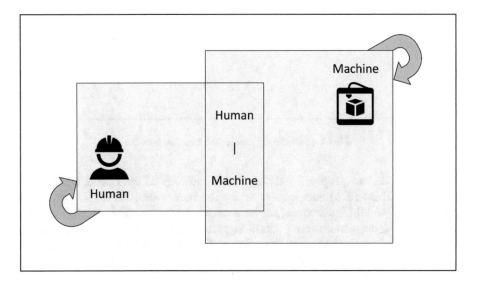

Fig. 3. Hybrid collaborative scheduling

4   Ontology-Based Meta-model for HCS

In this paper an ontology-based meta-model (OMM) is proposed based on a $\alpha|\beta|\gamma$ scheduling ontology [16], illustrated in Fig. 4, for further enabling hybrid collaborative scheduling (HCS), through M-M_Col2M and H-M_Col2M interactions, which are represented through Fig. 3. The scheduling ontology comprises characteristics related to the manufacturing environment, which include the type of production system, being denoted by a set α of problem classification factors and corresponding parameters. Additional characteristics related to the jobs, its operations along with aspects related to the manufacturing resources, such as the jobs' processing machines, are expressed by a set β of problem classification factors and corresponding parameters [16]. Finally, through the γ factor are defined the performance measures that are intended to be optimized through appropriate scheduling algorithms, which are able to be automatically selected and run, based on the proposed scheduling ontology.

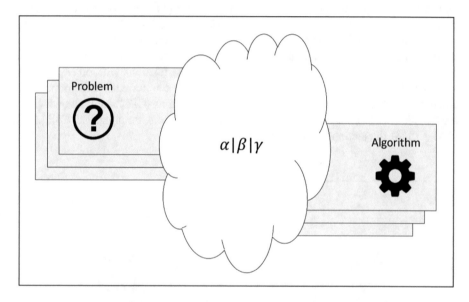

Fig. 4. Scheduling ontology-based meta-model

Furthermore, our proposed ontology-based meta-model for hybrid collaborative scheduling (OMM-HCS) also enables to support human decision-making processes, through H-M_Col2M interactions, for instance when after an automatically generated solution for a given manufacturing scheduling problem is obtained and a manufacturing manager intends to further adapt this solution for better fulfilling some specific production requisites or aims, and an illustrative example about this usage scenario will also be provided next.

5   Illustrative Examples of Use

In this paper a manufacturing scheduling problem regarding a set of m identical multipurpose parallel machines (*PMPM*) [16] for processing a set of n jobs, which have sequence-dependent setup times is considered. The general problem consists on scheduling the set of jobs on the set of identical multi-purpose parallel machines with setup times in order to minimize the maximum completion time of the jobs (C_{max}). This general scheduling problem is classified as $PMPMm|s_{i,j}, d_i|C_{max}$, and the underlying corresponding base problem is $PMPMm||C_{max}$, which is usually a hard problem to be solved [16].

5.1 General Problem Description

Let us consider the following general problem assumptions and notations as described next:

- There are available m identical muti-purpose parallel machines of set M;
- There are available n jobs to be processed of set N;
- Each machine I can just process one job at a time
- The jobs does not allow preemption;
- All n jobs are available for being processed at an initial instant time zero
- No precedence relations does exist among jobs;
- Each job i has a processing time p_i, which is independent of machine k;
- There is a setup time s_{ij} for processing a job j just after job i;
- The objective consists on minimizing the *makespan* (C_{max}) along with the considered machines setup times s_{ij}.

The general problem instance data is presented below, where in Table 1 are expressed the jobs' processing times of each job i (p_i) and the corresponding due dates (d_i), and in Table 2 are presented the jobs' setup times (s_{ij}), which are dependent on the jobs processing order. In this problem instance are considered $n = 8$ jobs, and $m = 3$ identical multi-purpose parallel machines.

Table 1. Jobs' processing times (p_i) and due dates (d_i)

Job	p_i	d_i
1	10	20
2	8	18
3	12	20
4	7	8
5	5	10
6	4	10
7	4	8
8	3	5

Table 2. Jobs' setup times (s_{ij})

Job	1	2	3	4	5	6	7	8
1	–	4	3	0	3	2	1	4
2	2	–	1	2	2	1	4	2
3	3	2	–	5	4	2	2	4
4	5	3	3	–	4	1	0	2
5	3	4	2	2	–	5	2	4
6	0	3	2	1	5	–	2	4
7	4	2	5	1	1	2	–	3
8	5	6	1	2	0	4	3	–

In this work we have considered three application scenarios of the previously presented problem instance, regarding the main general problem described before, for the minimization of the *makespan* (C_{max}). The first problem (P1), belongs to the general problem class $PMPMm|s_{ij}, d_i|C_{max}$, and is automatically solved through a LPT-based scheduling algorithm, in order to minimize the makespan, based on a M-MCol2M described before. The second problem (P2), belongs to the problem class *PMPMm* $|s_{ij}, d_i| C_{max}, \sum s_{ij}$, and is solved through a human-based collaborative approach (H-MCol2M), based on the previous solution automatically generated in Problem 1 (P1), in order not just to minimize the makespan, but also to minimize the sum of the setup times ($\sum s_{ij}$). The third problem (P3), belongs to the problem class *PMPMm* $|s_{ij}, d_i|$ L_{max}, C_{max}, which consists on an extension regarding the first problem (P1), on which the priority performance measure relies on the minimization of both the maximum lateness of jobs, along with the makespan. Therefore, some other approach or algorithm will be needed to solve this variant problem, and the proposed ontology-based model is well suited for accomplishing such a new kind of scheduling request, as it automatically returns the selection of appropriate scheduling problems solving algorithms in a fast and accurate manner, providing not just a variety of alternative algorithms but also the possibility of enabling human adaptations (H-MCol2M), on automatically generated solutions through the scheduling solver (M-MCol2M).

5.2   Problem Results and Discussion

The solutions obtained for the three scheduling problems (P1, P2 and P3) solved are shown below through Figs. 5, 6 and 7, expressed through Gantt charts.

Problem 1 (P1). A problem based on the given problem instance data, under the scope of the scheduling problem class *PMPMm* $|s_{ij}, d_i| C_{max}$.

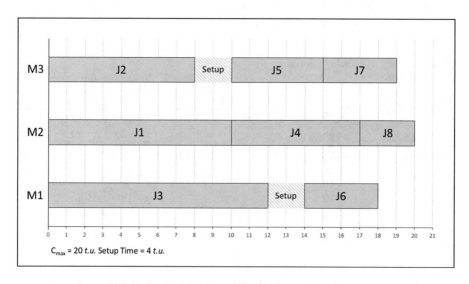

Fig. 5. Gantt chart about solution for problem 1 (P1)

Problem 2 (P2). A problem based on the given problem instance data, under the scope of the problem class $PMPMm|s_{ij}, d_i|C_{max}, \sum s_{ij}$.

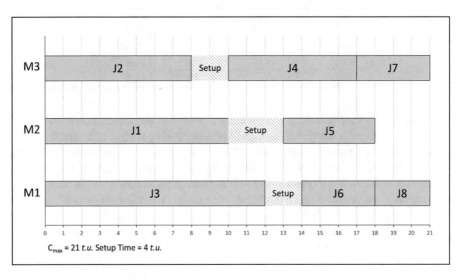

Fig. 6. Gantt chart about solution for problem 2 (P2)

Problem 3 (P3). A problem based on the given problem instance data, under the scope of the problem class $PMPMm|s_{ij}, d_i|L_{max}, C_{max}$.

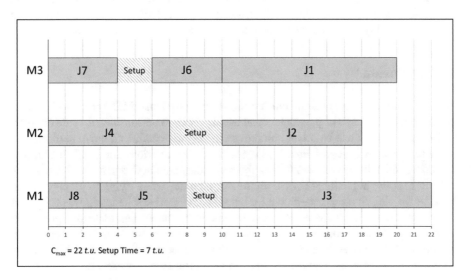

Fig. 7. Gantt chart about solution for problem 3 (P3)

Discussion of Problem Results. According to the summarized results in Table 3 obtained for the three problems (P1, P2 and P3) solved.

Table 3. Summarized problems' results

Problem	C_{max}	$\sum s_{ij}$	L_{max}	L_{mean}	N_L	F_{med}
P1	21	7	16	26	5	15.625
P2	20	4	9	20	5	14.875
P3	22	7	2	−0.875	1	11.5

Based on the three scheduling problems analysed and the different results obtained, which were previously expressed through the corresponding Gantt charts, and the results of the performance measures presented in Table 3, it is possible to easily realise that real-life industrial scheduling problems besides being usually hard to solve, as they are quite demanding and complex in nature, they are also extremely variable, which turns a daily-based scheduling decision making process quite difficult to accomplish in a urgent and precise way in a real-time basis. This happens not just due to the wide range of problem specific conditions or parameters that define each scheduling problem, regarding jobs and production resources characteristics, but also due to the varying demand regarding the performance measures to consider, which can furthermore be conflicting ones, mainly while dealing with multi-objective problems, were usually trade-off situations do occur between different kind of objectives to be optimised. This can be realised through the different solutions and corresponding performance measures' values obtained, for instance regarding the minimisation of the makespan, which reached a minimum value in the human-adapted solution obtained for the problem P2, along with the minimum summation of setup times. Although, if the focus relies on the minimization of the maximum lateness of jobs, then the automatically generated solution reached for problem P3 is considerably better, although the makespan and total setup time become worse.

6   Conclusion

In real industrial systems manufacturing scheduling problems (MSP) are hard in nature due to a widened range of parameters that usually have to be considered while defining each specific problem to be solved. Moreover, usually also different kind of performance measures have to be considered while searching for solutions for MSP in real-live scenarios, which in turn are also frequently conflicting, conduction to trade-off situations difficult to solve in practice. Due to all these difficulties while trying to properly solve MSP, an ontology-based meta-model for hybrid collaborative scheduling was presented in this paper, which intended not just to enable to obtain automatically generated solutions for MSP, through machine-machine interactions, but also to further enable human-machine collaboration to better adjust automatically generated solutions according to real-life adjustment requirements on production schedules.

Acknowledgement. This project is funded by COMPETE: POCI-01-0145-FEDER-007043 and FCT – Fundação para a Ciência e Tecnologia within the project scope: UID/CEC/00319/2013.

References

1. Varela, M.L.R., Putnik, G.D., Manupati, V.K., Rajyalakshmi, G., Machado, J.: Collaborative manufacturing based on cloud, and on other i4.0 oriented principles and technologies: a systematic literature review and reflections. Manage. Prod. Eng. Rev. **9**(3), 90–99 (2018). https://doi.org/10.24425/119538
2. Varela, M.L.R., Aparício, J.N., Silva, S.C.: A Web-based application for manufacturing scheduling. In: IASTED International Conference on Intelligent Systems and Control, pp. 400–405 (2003)
3. Reddy, M.S., Ratnam, C., Agrawal, R., Varela, M.L.R., Sharma, I., Manupati, V.K.: Investigation of reconfiguration effect on makespan with social network method for flexible job shop scheduling problem. Comput. Ind. Eng. **110**, 231–241 (2017)
4. Martins, L., Fernandes, N.O., Varela, M.L.R.: Autonomous production control: a literature review. In: International Conference on Innovation, Engineering and Entrepreneurship, pp. 425–431. Springer (2018)
5. Vieira, G.G., Varela, M.L.R., Putnik, G.D., Machado, J.M., Trojanowska, J.: Integrated platform for real-time control and production and productivity monitoring and analysis. Romanian Rev. Precis. Mech. Opt. Mechatron. **50**, 119–127 (2016)
6. Canadas, N., Machado, J., Soares, F., Barros, C., Varela, L.: Simulation of cyber physical systems behaviour using timed plant models. Mechatronics **54**, 175–185 (2018)
7. Parsaei, H.R., Sullivan, W.G. (eds.): Concurrent Engineering: Contemporary Issues and Modern Design Tools. Springer, Heidelberg (2012)
8. Ming, X.G., Yan, J.Q., Wang, X.H., Li, S.N., Lu, W.F., Peng, Q.J., Ma, Y.S.: Collaborative process planning and manufacturing in product lifecycle management. Comput. Ind. **59**(2–3), 154–166 (2008)
9. Groover, M.P.: Automation, Production Systems, and Computer-Integrated Manufacturing. Prentice Hall Press, Upper Saddle River (2007)
10. Kusi-Sarpong, S., Leonilde Varela, M., Putnik, G., Ávila, P., Agyemang, J.: Supplier evaluation and selection: a fuzzy novel multicriteria group decision-making approach. Int. J. Qual. Res. **12**(2) (2018)
11. Arrais-Castro, A., Varela, M.L.R., Putnik, G.D., Ribeiro, R., Dargam, F.C.: Collaborative negotiation platform using a dynamic multi-criteria decision model. Int. J. Decis. Supp. Syst. Technol. (IJDSST) **7**(1), 1–14 (2015)
12. Pereira, I., Madureira, A., de Moura Oliveira, P.B., Abraham, A.: Tuning meta-heuristics using multi-agent learning in a scheduling system. In: Transactions on Computational Science, vol. XXI, pp. 190–210. Springer (2013)
13. Barattini, P., Morand, C., Robertson, N.M.: A proposed gesture set for the control of industrial collaborative robots. In: 2012 IEEE RO-MAN, pp. 132–137. IEEE, September 2012
14. Thoben, K.D., Wiesner, S., Wuest, T.: Industrie 4.0 and smart manufacturing–a review of research issues and application examples. Int. J. Autom. Technol **11**(1) (2017)
15. Russakovsky, O., Li, L.J., Fei-Fei, L.: Best of both worlds: human-machine collaboration for object annotation. In: Proceedings of the IEEE Conference on Computer Vision and Pattern Recognition, pp. 2121–2131 (2015)
16. Varela, M.L.R., do Carmo Silva, S.: An ontology for a model of manufacturing scheduling problems to be solved on the web. In: Innovation in Manufacturing Networks, pp. 197–204. Springer, Boston (2008)

Decision Support Tool for Dynamic Scheduling

Luís Ferreirinha[1]([✉]), André S. Santos[3], Ana M. Madureira[3],
M. Leonilde R. Varela[2], and João A. Bastos[1]

[1] School of Engineering, Polytechnic of Porto (ISEP/IPP), Porto, Portugal
l.ferreirinha96@gmail.com, jab@isep.ipp.pt
[2] Department of Production and Systems, University of Minho (UM),
Braga, Portugal
leonilde@dps.uminho.pt
[3] Interdisciplinary Studies Research Center (ISRC), School of Engineering,
Polytechnic of Porto (ISEP/IPP), Porto, Portugal
{abg,amd}@isep.ipp.pt

Abstract. Production scheduling in the presence of real-time events is of great importance for the successful implementation of real-world scheduling systems. Most manufacturing systems operate in dynamic environments vulnerable to various stochastic real-time events which continuously forces reconsideration and revision of pre-established schedules. In an uncertain environment, efficient ways to adapt current solutions to unexpected events, are preferable to solutions that soon become obsolete. This reality motivated us to develop a tool that attempts to start filling the gap between scheduling theory and practice. The developed prototype is connected to the MRP software and uses meta heuristics to generate a predictive schedule. Then, whenever disruptions happen, like arrival of new tasks or cancelation of others, the tool starts rescheduling through a dynamic-event module that combines dispatching rules that best fit the performance measures pre-classified by Kano's model. The proposed tool was tested in an in-depth computational study with dynamic task releases and stochastic execution time. The results demonstrate the effectiveness of the model.

Keywords: Dynamic scheduling · Meta heuristics · Hyper heuristics ·
Dispatching rules · Decision support tool · Kano's model

1 Introduction

Scheduling in real-world manufacturing facilities can be extremely difficult. Most manufacturing systems operate in dynamic environments vulnerable to various stochastic real-time events like, machine breakdowns, rush orders, unavailable material, and many others which easily turn preschedules obsolete [1–3]. Those manufacturing systems generate and update production schedules, which are plans that serve as basis for many external activities (e.g., material procurement, preventive maintenance). In an uncertain environment, industries can benefit from better understanding how (re) scheduling strategies affect system's performance, because in this environment the focus of the foreman tends to be on finding efficient and effective ways to adapt current solutions to unexpected events, rather than finding a high-quality schedule that rapidly

© Springer Nature Switzerland AG 2020
A. M. Madureira et al. (Eds.): HIS 2018, AISC 923, pp. 418–427, 2020.
https://doi.org/10.1007/978-3-030-14347-3_41

becomes outdated. In practice, rescheduling is done in a hybrid way, either periodically, i.e., plans for the next period based on current status of the system, or occasionally in response to nonplanned events [1, 4, 5].

When it comes to scheduling systems in dynamic environments, two key elements stand out, the *schedule generation*, which acts as a predictive mechanism and serves as an overall plan for other shop activities, and *schedule monitorization/updating*, which is viewed as the reactive part of the system that attempts to minimize the effect of the disturbances in the performance of the system [4–6].

In view of the above, this paper proposes a decision support tool to dynamic environments that attempts to fill the gap between scheduling theory and practice stated by the author in [7]. Continuing the model proposed in [8] this paper presents a prototype that does not require any interaction with the user besides the definition of the criteria of performance. The definition of the criteria of performance are classified through the degree of satisfaction of the Kano's Model, which classifies the performance measures and balance the interests of the stakeholders. The tool is also connected to the MRP of the company, and when the MRP launches the jobs into the system a meta heuristic is used to generate the schedule. When disruptions happen, the tool begins to reschedule the tasks according to priority rules that best fit the intended performance. To validate the tool, an instance composed of a vast set of tasks with normally distributed stochastic characteristics were executed.

The remaining sections of this paper are organized as follows: Sect. 2 revises production scheduling problems in dynamic environments. Section 3 briefly introduces concepts related to optimization in dynamic environments. The prototype is presented in Sect. 4. In Sect. 5 is presented the computational results. And finally, Sect. 6 presents some conclusions and provides some ideas for future work.

2 Production Scheduling Problems

Production Scheduling (PS) implies the definition of an initial and final moment of the processing of each task and its allocation to the resources, fulfilling certain restrictions that may involve the tasks and/or resources. The purpose of scheduling is to optimize a certain measurement of economic and operational performance. So, a typical production scheduling problem can be seen as, n parts must be processed by using m machines, and each part must be processed in a given order on the respective machine to find the schedule that optimizes certain performance metric [3, 8–10].

In [11] it is mentioned the PS problems can be classified in three levels: 1-*Requirements generation, 2-Processing complexity* and 3-*Scheduling criteria*. The first level refers to *open shop* versus a *closed shop*, i.e., in an *open shop* the production orders are requested by costumers and there is no inventory, while in a *closed shop* the inventory is used to serve the costumers requests, which introduces an inventory replenishment decision. So, in its simplest form, the *open shop* production scheduling is a *sequencing problem*. As for the *closed shop,* PS involve sequencing and also *lot-sizing* decisions associated to the inventory replenishment process. The second level refers to the number of processing steps that are needed for each production task, *One-stage* (one processor or Parallel processors) and Multistage (flow shop or job shop).

The third level refers to the performance measures that the schedule should optimize. Sometimes these three levels are not sufficient to classify the PS problem, whereby the authors in [11] refer two additional levels: 4-*Nature of the requirement specification* and 5-*Scheduling environment*. The fourth level refers to the nature of the parameters, when all the parameters are known and fixed, the problem is classified as *deterministic*, otherwise it is *stochastic*, i.e., when all the parameters are uncertain with specified probability distribution. In the fifth level, when all the tasks are known at the beginning of the schedule the problem is said to be *static*, while when new tasks can arrive unexpectedly, the environment of the problem is *dynamic* [3, 11, 12].

2.1 Dynamic Scheduling

There has been a recent increasing interest in modelling and solving scheduling problems in dynamic environments. In industrial environments, scheduling is an ongoing reactive process where real time events forces reconsideration and revision of pre-established schedules. As mentioned before, in this scheduling environment the tasks are not known at the beginning of the problem which makes the system vulnerable to random, inevitable and unpredictable real-time events that cause a change in the scheduled plans, which makes a previously feasible schedule turn infeasible when it is released to the shop floor. Such unexpected events can occur for a variety of reasons, like related to the resources (e.g., breakdowns, rework, operator illness), or the jobs (e.g., order cancelation, changes in delivery times, late arrivals) [3, 6, 13, 14].

The main goal in solving this problem is no longer to find optimum schedules, because they quickly will become obsolete, but instead be able to efficiently adapt current solutions to the dynamic environment, since near optimal solutions that are easily adaptable, will be preferable to optimum ones. So, when non-planned perturbations occur, it is necessary to find a new schedule with the quality close to the schedule that could have been executed if all of the uncertainty had been revealed *a priori* [8, 15].

Dynamic scheduling has been defined under three categories [4, 6, 14, 16], which are summarized in Table 1.

Table 1. Categories of dynamic scheduling

Completely reactive scheduling	Predictive-reactive scheduling
In completely reactive scheduling, no firm schedule is generated in advance and decisions are made locally in real-time usually by priority dispatching rules	Predictive-reactive scheduling is an iterative process and has two primary steps. The first step generates a production scheduling. The second step updates the schedule in response to a real-time event
Robust pro-active scheduling	
Robust pro-active scheduling is based on predictive schedules that satisfy performance requirements predictably in a dynamic environment	

When it comes to reschedule due to real-time events, two major issues emerge: how and when to react to those events. The first issue lies on the strategy to use to reschedule, which usually leads to schedule repair or complete schedule [4, 5].

Schedule repair refers to adjusting the current schedule, saving CPU times and without compromising the stability of the system. Regarding complete schedule, it literally refers to schedule from scratch. Although this last may be better to maintain the optimal solution, in practice there is usually no time to reschedule [4, 6, 14, 16]. Regarding the second issue, when to reschedule, it basically aims to answer when an event has sufficient impact that a new schedule is necessary. Three policies can be find in the literature [4, 5, 14, 16]: *Continuous rescheduling* or *event driven,* which reschedules each time an event occur, such as an arrival of new tasks, *Periodic rescheduling,* where schedules are generated at regular intervals T and any disruptions between periods of rescheduling are ignored until the next period where it gathers all available information from the shop floor. And last the *hybrid,* which reschedules the system periodically and also when some particular events happen, like machine breakdowns, arrival of urgent jobs, cancelation of jobs, among others [2, 5, 16, 17].

3 Optimization Techniques

3.1 Dispatching Rules

Heuristics are problem specific schedule repair methods, which have the ability to find near-optimum solutions quickly and with little computational effort. Dispatching rules are also heuristics with major importance in completely reactive scheduling and they are used to sort the jobs in the machines queue by some criteria. So, similar to pull mechanisms, like *Kanban cards*, dispatching rules are used to control production.

In [18] an extensive list of dispatching rules is presented and the difficulty of the choice of a dispatching rule arises from the fact that there are $n!$ ways of sequencing n jobs. In the literature, the performance of dispatching rules is usually evaluated experimentally, although in some cases a dispatching rule can be shown to be optimal like the Shortest Processing Time *(SPT)* that minimizes the Mean Flow Time and the Earliest Due Date *(EDD)* that minimizes the Maximum Tardiness, among others [2, 6, 8, 14, 15, 19]. A dynamic scheduling can be viewed as a collection of linked static problems, which implies that methods developed for static scheduling problems become applicable to dynamic ones. Such methods can effectively deal with complex problems and can optimize the quality of the schedules for each static sub-problem. [10]. The authors of [18–20] present a state-of-the-art survey of dispatching rules.

3.2 Meta Heuristics

Scheduling problems are becoming more complex and the exhaustive search for optimal solutions has become impractical, given the computational effort required to find them. In the last decades a new family of approximate algorithms has arisen that dominated the research. These methods are called Meta Heuristics (MHs), a term that was first introduced by Glover [21] in the 1980s. According to [22] meta-heuristics are more advantageous than most heuristics in terms of solution robustness. However, they are more difficult to implement and tune, since they need information about the problem in order to achieve satisfactory results. The author of [23] further states that unlike exact methods, meta heuristics allow dealing with instances of large-scale problems and find satisfactory solutions within a reasonable amount of time.

4 Prototype

The tool was designed with the purpose of not requiring any interaction with the user besides the definition of the performance criteria. The definition of the performance criteria is classified through the degree of satisfaction of the Kano's Model. As an example of how the performance measures can be classified in the tool, see Fig. 1. For the defined objectives in Fig. 1, it is assumed that the user does not want the Maximum Tardiness to surpass a value and, at the same time, wants the Mean Flow Time to be the smallest possible. As Fig. 1 shows, the Maximum Tardiness represents a Must-be attribute in the Kano's model, so if it is not achieved, it results in an extreme dissatisfaction. As for the Mean Flow Time, it represents a One-dimensional, since it corresponds to a degree of satisfaction proportional to the degree of performance of the attribute, i.e., the smaller the better [24, 25].

Fig. 1. Example of performance measures classified through Kano's model

After the definition of the performance criteria, the software itself schedules and reschedules the tasks in the system, over time, even when a real-time event, like the cancelation of a task, changes on due dates and so on, occurs. To generate acceptable solutions in such circumstances the tool is connected to the MRP of the company to generate a predictive schedule using the available information and then, whenever disruptions occur in the system during the execution of the pre-established schedule, rescheduling is performed. Since the prototype is connected to the MRP and most of the time the MRP runs periodically, we are aware that disturbances between two immediate periods of time may happen, which lead us to a hybrid strategy, i.e., rescheduling the system periodically and also when some particular events happen.

When MRP runs, a large number of tasks usually enter the system, so the search for the satisfactory solution is time consuming, and it will hardly remain viable in industrial environment. Thus, the proposed tool generates the predictive schedule by meta heuristics, which can deal with large-scale problems and find satisfactory solutions within a reasonable amount of time. It is worth emphasizing that the ideal would be that all the tasks launched in period T of the MRP would end before the period $T + 1$. However, such a scenario is not likely to occur in real-world environments, so, when the system reaches $T + 1$, the MH will schedule the new tasks that the MRP will launch, as well as the tasks that have not yet been processed. The parameterization of the MHs is done by the design of experiments (DOE) of Taguchi, where the values that are going to compete either are calculated by metrics or are generated randomly within intervals that have been shown to be effective in solving problems of this type in the scheduling community. See for example [23, 26–28] where some metrics and ranges are presented for Simulated annealing (SA). Whenever disruptions happen, the prototype reschedules the tasks with dispatching rules. A brief outline of what has been described here is shown in Fig. 3.

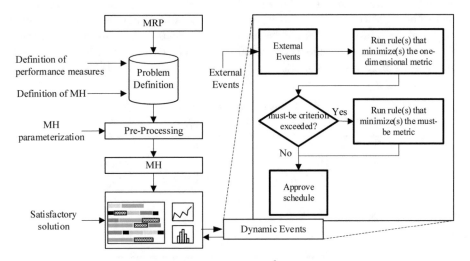

Fig. 2. Scheme of the prototype logic

In Fig. 2 is shown the logic of the prototype. When the MRP runs, the user has to define the performance measures and the MH that is going to generate the predictive schedule, otherwise the tool will maintain the previous choices. After the parameterization a predictive schedule is generated and then is released to the shop floor by a panel where several performance indicators are shown. When disruptions happen, the "Dynamic Events Module" reschedules the tasks. Despite of the prototype presented being a totally autonomous tool, there will be situations where the foreman may want to validate alternative schedules. In this initial phase of the project, such functionality is not operational, but it is in the authors' interest to implement it in the near future.

The main difference of the prototype presented here in relation to other methodologies in the literature, is that it is independent of the machine environment. Much of the literature that analyzes dynamic scheduling presents problem specific methodologies and not a generic model adaptable to several scenarios. For example, in [29] the authors present an efficient hybrid Genetic Algorithm (GA) methodologies where a new KK heuristic + swap and well-known dispatching rules + swap, for minimizing makespan in dynamic job shop problems. In [30] the authors proposed a multi-objective methodology for the FJSP scheduling problem in machine breakdown situations. Another example can be in [31] where the authors proposed a Variable

Neighborhood Search (VNS) algorithm to solve a dynamic Flexible Flow Shop (FFS) problem considering unexpected arrival of new jobs. Although the methodologies presented by the aforementioned authors are efficient for the problems in question, they are hardly immediately applicable to other machine environments.

5   Computational Study

To demonstrate how the prototype operates, 250 jobs with stochastic attributes, shown in Table 2, were created. For this study it was defined that the period between executions of the MRP is 250 time units, the "one-dimensional" criteria is the Mean Flow Time and the "must be" criteria is the Maximum Tardiness, where the maximum value is zero. SA was chosen as the meta heuristic and the obtained results will be compare to the ones achieved if only the rule that best fits the "must be criteria" was used.

Table 2. Distributions used

T	0	250	500	[0–750]
N° of Jobs	50	50	50	100
rj	0	250	500	N(400, 150)
dj	N(300, 25)	N(550, 25)	N(800, 25)	N(300, 25)
pj	N(5, 1)			
wj	N(10, 3)			

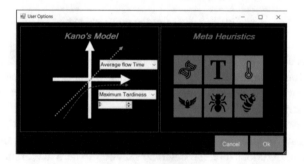

Fig. 3. User options

Depending on the performance criteria selected, the fitness calculation will vary. For the defined test, the fitness calculation is done according to expression 1.

$$F(x) = \alpha.\,\text{Max}\{L\} + \frac{\sum cj}{n}, \text{ where } \alpha = \begin{cases} M, \text{ if } Max(L) > Set\,value \\ 1, otherwise \end{cases} \quad (1)$$

In the expression 1, **M** stands for a major number that forces the meta heuristic to minimize the mean flow time whenever triggered. That **M** must be at least the maximum tardiness of a schedule through the LPT rule. So, whenever the maximum tardiness exceeds the set value, the fitness forces the meta heuristic to find a solution where the Mean Flow Time is minimum from the set of solutions that cannot meet the defined value. Otherwise, it tries to find one that minimizes both. Regarding the "Dynamic Events Module", it will use dispatching rules like SPT and EDD, since these minimize the Mean Flow Time and the Maximum Tardiness, [8].

5.1 Results

The results, Fig. 5, were compared with the results from the system based only on EDD in Fig. 4. Figure 4 compares the Mean Flow Time and the Work in Progress (WIP) between the model and the system based only on EDD. As we can see, the Mean Flow Time in the system based on EDD tends to get a bit higher than the tool. As for

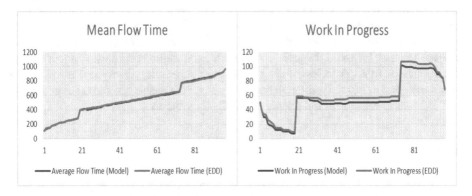

Fig. 4. Obtained results

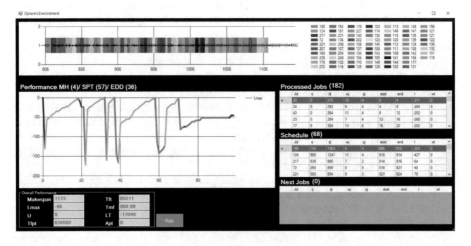

Fig. 5. Prototype results

the work in progress, the system based only on EDD tends to create a bigger WIP. Figure 5 shows the effectiveness of the tool, where the Maximum Tardiness was always less than zero and the Mean Flow Time was minimized whenever possible, reaching 958,89 in the final phase. The SPT rule was used 57 times, gray on the performance chart, the EDD rule was used 36 times, represented in red and the blue stands for the MH. There are 68 tasks in progress and 182 have been processed.

6 Conclusions

This paper intends to demonstrate an approach to the dynamic scheduling production without user intervention. The software itself is connected to the MRP of the company and according to the defined objectives through the Kano's model, whenever disruptions occur it schedules and reschedules through meta heuristics and "Dynamic Events Module", in order to comply with the objectives. The results show the effectiveness of the tool, where the set value of the "must be criteria" was never exceeded, and the "one dimensional" criteria was optimized whenever possible. Regarding future work, it is intended to prove through statistical evidence the operation of the tool, as well as to prepare the tool for more dynamic events such as machine breakdowns, repair times, among others, in order to lessen the gap between scheduling theory and practice. As mentioned previously, it is still in the authors' interest to enable manual intervention by the foreman in relation to the scheduling defined by the tool.

References

1. O'Donovan, R., Uzsoy, R., McKay, K.N.: Predictable scheduling of a single machine with breakdowns and sensitive jobs. Int. J. Prod. Res. 37(18), 4217–4233 (1999)
2. Sabuncuoglu, I., Karabuk, S.: Rescheduling frequency in an FMS with uncertain processing times and unreliable machines. J. Manuf. Syst. 18(4), 268–283 (1999)
3. Madureira, A., Ramos, C., Silva, S.D.C.: Using genetic algorithms for dynamic scheduling. In: 14th Annual Production and Operations Management Society Conference (POMS 2003) (2003)
4. Vieira, G.E., Herrmann, J.W., Lin, E.: Rescheduling manufacturing systems: a framework of strategies, policies, and methods. J. Sched. 6(1), 39–62 (2003)
5. Sabuncuoglu, I., Bayız, M.: Analysis of reactive scheduling problems in a job shop environment. Eur. J. Oper. Res. 126(3), 567–586 (2000)
6. Varela, M.L.R., Ribeiro, R.A.: Distributed manufacturing scheduling based on a dynamic multi-criteria decision model. In: Recent Developments and New Directions in Soft Computing, pp. 81–93. Springer, Cham (2014)
7. Cowling, P., Johansson, M.: Using real time information for effective dynamic scheduling. Eur. J. Oper. Res. 139(2), 230–244 (2002)
8. Ferreirinha, L., Baptista, S., Pereira, A., Santos, A.S., Bastos, J., Madureira, A.M., Varela, M.L.R.: A dynamic selection of dispatching rules based on the Kano model satisfaction scheduling tool. In: International Conference on Innovation, Engineering and Entrepreneurship, pp. 339–346. Springer, Cham (2018)
9. Artiba, A., Elmaghraby, S.E. (ed.).: The Planning and Scheduling of Production Systems: Methodologies and Applications. Springer (1996)

10. Akers Jr., S.B., Friedman, J.: A non-numerical approach to production scheduling problems. J. Oper. Res. Soc. Am. **3**(4), 429–442 (1955)
11. Graves, S.C.: A review of production scheduling. Oper. Res. **29**(4), 646–675 (1981)
12. French, S.: Sequencing and Scheduling: An Introduction to the Mathematics of the Job Shop. Ellis Horwood, Chichester (1982)
13. Goren, S., Sabuncuoglu, I.: Robustness and stability measures for scheduling: single-machine environment. IIE Trans. **40**(1), 66–83 (2008)
14. Ouelhadj, D., Petrovic, S.: A survey of dynamic scheduling in manufacturing systems. J. Sched. **12**(4), 417 (2009)
15. Terekhov, D., Down, D.G., Beck, J.C.: Queueing-theoretic approaches for dynamic scheduling: a survey. Surv. Oper. Res. Manag. Sci. **19**(2), 105–129 (2014)
16. Aytug, H., Lawley, M.A., McKay, K., Mohan, S., Uzsoy, R.: Executing production schedules in the face of uncertainties: a review and some future directions. Eur. J. Oper. Res. **161**(1), 86–110 (2005)
17. Church, L.K., Uzsoy, R.: Analysis of periodic and event-driven rescheduling policies in dynamic shops. Int. J. Comput. Integr. Manuf. **5**(3), 153–163 (1992)
18. Panwalkar, S.S., Iskander, W.: A survey of scheduling rules. Oper. Res. **25**(1), 45–61 (1977)
19. Rajendran, C., Holthaus, O.: A comparative study of dispatching rules in dynamic flowshops and jobshops. Eur. J. Oper. Res. **116**(1), 156–170 (1999)
20. Ramasesh, R.: Dynamic job shop scheduling: a survey of simulation research. Omega **18**(1), 43–57 (1990)
21. Glover, F.: Future paths for integer programing and links to artificial intelligence. Comput. Oper. Res. **13**(5), 533–549 (1986)
22. Xhafa, F., Abraham, A.: Metaheuristics for Scheduling in Industrial and Manufacturing Applications Series. Studies in Computational Intelligence, 1st edn. Springer (2008)
23. Talbi, E.G.: Metaheuristics: From Design to Implementation. Wiley, Chichester (2009)
24. Högström, C., Rosner, M., Gustafsson, A.: How to create attractive and unique custom-er experiences: an application of Kano's theory of attractive quality to recreational tourism. Mark. Intell. Plan. **28**(4), 385–402 (2010)
25. Löfgren, M., Witell, L., Gustafsson, A.: Theory of attractive quality and life cycles of quality attributes. TQM J. **23**(2), 235–246 (2011)
26. Eglese, R.W.: Simulated annealing: a tool for operational research. Eur. J. Oper. Res. **46**(3), 271–281 (1990)
27. Park, M.W., Kim, Y.D.: A systematic procedure for setting parameters in simulated annealing algorithms. Comput. Oper. Res. **25**(3), 207–217 (1998)
28. Kirkpatrick, S., Gelatt, C.D., Vecchi, P.M.: Optimization by Simulated Annealing. Science **220**, 671–680 (1983)
29. Kundakcı, N., Kulak, O.: Hybrid genetic algorithms for minimizing makespan in dynamic job shop scheduling problem. Comput. Ind. Eng. **96**, 31–51 (2016)
30. Ahmadi, E., Zandieh, M., Farrokh, M., Emami, S.M.: A multi objective optimization approach for flexible job shop scheduling problem under random machine breakdown by evolutionary algorithms. Comput. Oper. Res. **73**, 56–66 (2016)
31. Rahmani, D., Ramezanian, R.: A stable reactive approach in dynamic flexible flow shop scheduling with unexpected disruptions: a case study. Comput. Ind. Eng. **98**, 360–372 (2016)

Fuzzy Algorithms for Fractional PID Control Systems

Ramiro S. Barbosa$^{(\boxtimes)}$ and Isabel S. Jesus

GECAD - Knowledge Engineering and Decision Support Research Center,
Department of Electrical Engineering, ISEP/IPP - School of Engineering,
Polytechnic Institute of Porto, Porto, Portugal
{rsb,isj}@isep.ipp.pt

Abstract. This article investigates the use of fuzzy algorithms for fractional PID control systems. The tuning of the fuzzy controllers is based on the prior knowledge of integer or fractional-order control strategy. The suggested fuzzy controllers are fine tuned using a genetic algorithm (GA). The effectiveness and robustness of the proposed methodology is illustrated through its application on the control of a fractional-order plant. The simulation results show the better performance of nonlinear fuzzy algorithms of fractional-order.

Keywords: Fuzzy control · PID controller · Fractional calculus · Fractional PID control · Fuzzy fractional PID control

1 Introduction

In recent years, fractional-order PID (FO-PID) controllers have been a fruitful field of research [1–3]. However, no effective and simple tuning rules still exist for these controllers as those given for the integer PID controllers [4]. It is well known that the FO-PID extends the capabilities of the classical counterpart and, thus, have a wider domain of application, such as in suspension systems, robotics, signal processing, control and diffusion [2,3,5]. On the other hand, the fuzzy logic controllers (FLC) have also been successfully applied in the control of many physical systems, particularly those with uncertainty, unmodelled, disturbed and/or nonlinear dynamics [6,7].

There have been a lot of researches in the application of fuzzy PID control [6–12]. The fuzzy method offer a systematic procedure to design controllers for many kind of systems, that often leads to a better performance than that of the conventional PID controller. It is a methodology of intelligent control that mimics human thinking and reacting by using a multivalent fuzzy logic and elements of artificial intelligence. It has been proved that the use of the fuzzy fractional controllers improved the results for many kind of systems, since it gives additional flexibility to the design [13].

A. M. Madureira et al. (Eds.): HIS 2018, AISC 923, pp. 428–437, 2020.
https://doi.org/10.1007/978-3-030-14347-3_42

In this work, we combine the features of fuzzy controllers with those of fractional controllers of PID-type. The resulting fuzzy fractional PID (FF-PID) controller is investigated in terms of its digital implementation and robustness. The combined advantages of the two controllers results in a better controller with superior robustness and wider domain of application. The tuning methodology of these controllers is based on the prior knowledge of integer or fractional-order control, making the procedure adequate to tune or replace an existent controller in order to improve the system control performance. The main goal is to present a simple methodology for tuning integer as well fractional PID-type controllers, or to replace an existing design with a better one with minor modifications. It is not intended to make an exhaustive comparison between different fuzzy fractional PID structures [13].

The paper is organized as follows. Section 2 presents the basic ideas of continuous and discrete fractional-order PID controllers. Section 3 outlines a tuning methodology for the design of FF-PID controllers. In Sect. 4, we test the proposed fuzzy fractional controllers and assess their applicability and robustness in the control of a fractional-order plant. Finally, Sect. 5 draws the main conclusions.

2 Fractional-Order PID Controllers

The fractional-order controller of PID-type, named $PI^\lambda D^\mu$ controller, may be given as [3]:

$$C\left(s\right) = \frac{U\left(s\right)}{E\left(s\right)} = K_p + \frac{K_i}{s^\lambda} + K_d s^\mu \tag{1}$$

where K_p, K_i and K_d are the proportional, integral and derivative gains, and usually the fractional orders $(\lambda,\ \mu) \in [0,\ 1]$. The time domain equation of the $PI^\lambda D^\mu$ controller is:

$$u\left(t\right) = K_p e\left(t\right) + K_i D^{-\lambda} e\left(t\right) + K_d D^\mu e\left(t\right) \tag{2}$$

where $D^{(*)}$ $(\equiv {}_0 D_t^\alpha)$ denotes the differential operator of integration and differentiation (differintegral) to a fractional-order $\alpha = \{-\lambda,\ \mu\} \in \Re$.

The two most commonly used definitions for the differintegral are the Riemann-Liouville definition and the Grünwald-Letnikov definition. The Riemann-Liouville definition of a fractional derivative is $(\alpha > 0)$:

$$D^\alpha f\left(t\right) = \frac{1}{\Gamma\left(n-\alpha\right)} \frac{d^n}{dt^n} \int_0^t \frac{f\left(\tau\right)}{\left(t-\tau\right)^{\alpha-n+1}} d\tau \tag{3}$$

where $n - 1 < \alpha < n$, n is an integer, $f(t)$ is the applied function, and $\Gamma(x)$ represents the Gamma function of x. For our purpose we use the Grünwald-Letnikov definition, which can be written as $(\alpha \in \Re)$:

$$D^\alpha f\left(t\right) = \lim_{h \to 0} \frac{1}{h^\alpha} \sum_{j=0}^{[t/h]} (-1)^j \binom{\alpha}{j} f\left(t - jh\right) \tag{4a}$$

$$\binom{\alpha}{j} = \frac{\Gamma(\alpha+1)}{\Gamma(j+1)\,\Gamma(\alpha-j+1)} \tag{4b}$$

where h is the time increment and $[v]$ means the integer part of v.

From a control perspective, approach (4) is the most useful and intuitive, particularly for a discrete-time implementation [1,14]. Thus, using (4), a discrete fractional $PI^{\lambda}D^{\mu}$ control equation can be obtained from (2) as ($h \approx T$, T is the sampling period):

$$u(k) = K_p e(k) + K_i D^{-\lambda} e(k) + K_d D^{\mu} e(k) \tag{5}$$

with

$$D^{\alpha} e(k) \approx \frac{1}{T^{\alpha}} \sum_{j=0}^{k} (-1)^j \binom{\alpha}{j} e(k-j) \tag{6}$$

The difference control Eq. (5) is then given by:

$$u(k) = K_p e(k) + \frac{K_i}{T^{-\lambda}} \sum_{j=0}^{k} (-1)^j \binom{-\lambda}{j} e(k-j) + \frac{K_d}{T^{\mu}} \sum_{j=0}^{k} (-1)^j \binom{\mu}{j} e(k-j) \tag{7}$$

Equation (7) shows that the current value of control signal $u(k)$ depends on all previous values of error $e(k)$, making the computation too heavy as time increases and so unsuitable for a practical implementation of these algorithms. For a real implementation of fractional integral and derivative (6) we apply the short memory principle [2], yelding:

$$u(k) = K_p e(k) + \frac{K_i}{T^{-\lambda}} \sum_{j=v}^{k} c_j^{(-\lambda)} e(k-j) + \frac{K_d}{T^{\mu}} \sum_{j=v}^{k} c_j^{(\mu)} e(k-j) \tag{8}$$

where $v = 0$ for $k < L/T$ or $v = k - L/T$ for $k > L/T$; L is the memory length and $c_j^{(\alpha)} = (-1)^j \binom{\alpha}{j}$ are the binomial coefficients which may be calculated recursively as:

$$c_0^{(\alpha)} = 1; \quad c_j^{(\alpha)} = \left(1 - \frac{1+\alpha}{j}\right) c_{j-1}^{(\alpha)}, \quad j = 1, 2, \cdots \tag{9}$$

Note that (8) is given in the form of a FIR filter. Other discrete-time approximations in the form of IIR filters are also possible [1,14,15].

3   Design of Fuzzy Fractional PID Control Systems

In this work, we will apply fuzzy logic control for the design of fuzzy fractional PID controlled systems [6,9], as illustrated in Fig. 1a. The main idea here is to explore the fact that the FLC, under certain conditions, is equivalent to a PID

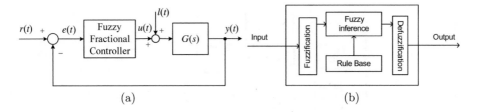

Fig. 1. Fuzzy fractional PID controlled system (a) and fuzzy controller structure (b).

controller and to extrapolate this fact for the case of fractional-order controllers [7,10,16].

The basic form of a fuzzy controller is illustrated in Fig. 1b, consisting of a fuzzy rule base, an inference mechanism, and fuzzification and defuzzification interfaces [9]. In general, the mapping between the inputs and the outputs of a fuzzy system is nonlinear. However, it is possible to construct a rule base with a linear input-output mapping [10,16]. For that, the following conditions must be fulfilled:

- Use triangular input sets that cross at the membership value 0.5
- The rule base must be complete AND combination (cartesian product) of all input families
- Use the algebraic product (*) for the AND connective
- Use output singletons, positioned at the sum of the peak positions of the input sets
- Use sum-accumulation and centre of gravity for singletons (COGS) defuzzification.

It seems reasonable to start with the design of a conventional integer or fractional PID controller and from there to proceed to a fuzzy control design [16]. In this way, the linear fuzzy controller may be used in a design procedure based on integer or fractional PID control, as follows:

1. Build and tune an integer or fractional PID controller
2. Replace it with an equivalent linear fuzzy controller
3. Make the fuzzy controller nonlinear
4. Fine-tune it

With the above procedure, the design of FF-PID controllers will be greatly simplified, particularly if the controller was already implemented and it is desirable to enhance its performance. Moreover, this new type of controllers extends the potentialities of both fuzzy and fractional-order controllers.

3.1   Fuzzy Fractional PID Controller

The fuzzy fractional $PD^{\mu}+I^{\lambda}$ (FF-$PD^{\mu}+I^{\lambda}$) controller adopted in this study combine the fractional-order integral action with a fuzzy PD^{μ}-controller, as illustrated in Fig. 2. The inclusion of an integral action is necessary whenever the

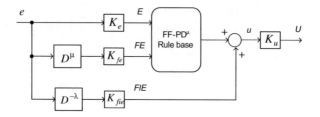

Fig. 2. Fuzzy fractional $PD^{\mu} + I^{\lambda}$ controller.

closed-loop system exhibits a steady-state error. The controller has four tuning gains, K_e, K_{fe} and K_{fie}, corresponding to the inputs and K_u to the output. The following relations are thus verified:

$$E = K_e e, \quad FE = K_{fe} D^{\mu} e, \quad FIE = K_{fie} D^{-\lambda} e, \quad U = K_u u \quad (10)$$

where E, FE and FIE represent respectively the error, fractional change of error, and fractional integral of error. The control signal U is generally a non-linear function of E, FE and FIE:

$$U = \left(f\left(E, FE \right) + FIE \right) K_u = \left(f\left(K_e e\left(k \right) + K_{fe} D^{\mu} e\left(k \right) \right) + K_{fie} D^{-\lambda} e\left(k \right) \right) K_u \quad (11)$$

With a proper choice of design, a linear approximation can be obtained as:

$$f\left(K_e e\left(k \right), \; K_{fe} D^{\mu} e\left(k \right) \right) \approx K_e e\left(k \right) + K_{fe} D^{\mu} e\left(k \right) \quad (12)$$

and

$$U\left(k \right) \approx \left(K_e e\left(k \right) + K_{fe} D^{\mu} e\left(k \right) + K_{fie} D^{-\lambda} e\left(k \right) \right) K_u$$
$$= K_u K_e e\left(k \right) + K_u K_{fe} D^{\mu} e\left(k \right) + K_u K_{fie} D^{-\lambda} e\left(k \right) \quad (13)$$

Comparing (13) with the discrete fractional $PI^{\lambda} D^{\mu}$-controller (5), it yields the relation for the gains of the conventional and fuzzy fractional PID controllers:

$$K_e K_u = K_p, \quad K_{fie} K_u = K_i, \quad K_{fe} K_u = K_d \quad (14)$$

For an equivalent linear $FF\text{-}PD^{\mu}$-controller, the conclusion universe should be the sum of the premise universes and the input-output mapping should be linear. Table 1 lists a linear rule base for the $FF\text{-}PD^{\mu}$ controller composed of four rules. There are only two fuzzy labels (Negative and Positive) used for the fuzzy input variables and three fuzzy labels (Negative, Zero and Positive) for the fuzzy output variable. This rule base should satisfy above mentioned conditions in order to provide a linear mapping. Scaling the input gains by a factor M ($M > 0$) may be necessary to preserve the linearity of the fuzzy controller. However, that should be made without affecting the tuning. This scaling has some advantages, as it will avoid saturation and will provide a simpler design, since the universes ranges of inputs and outputs are normalized to a prescribed interval.

Table 1. Rule base for the FF-PD$^\mu$ controller.

Rule 1	If E is N and FE is N then u is N
Rule 2	If E is N and FE is P then u is Z
Rule 3	If E is P and FE is N then u is Z
Rule 4	If E is P and FE is P then u is P

4 Illustrative Example

Many real dynamical processes are modeled by fractional-order transfer functions [2,5]. Here we consider the fractional-order plant model given in [3]:

$$G\left(s\right) = \frac{1}{0.8s^{2.2} + 0.5s^{0.9} + 1} \tag{15}$$

An integer-order PD controller and a fractional-order PD$^\mu$-controller were designed in [3]:

$$C_{PD}\left(s\right) = K_p + K_d s = 20.5 + 2.7343s \tag{16}$$

$$C_{PD^\mu}\left(s\right) = K_p + K_d s^\mu = 20.5 + 3.7343s^{1.15} \tag{17}$$

Figure 3 shows the unit-step responses of the closed-loop fractional-order system with the conventional PD-controller and with the PD$^\mu$-controller for $G(s)$. The comparison shows that for satisfactory feedback control of the fractional-order system is better to use a fractional-order controller instead of a classical integer-order controller. Note, however, that the control system presents a steady-state error, since no integral action is employed. In all experiments, the simulation parameters are: absolute memory computation of approximation (6), scale factor $M = 0.1$, and sampling period $T = 0.05$ s.

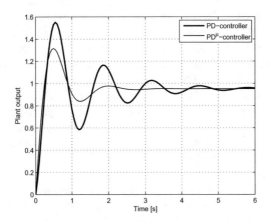

Fig. 3. Unit-step responses of $G(s)$ with the PD and PD$^\mu$ ($\mu = 1.15$) controllers.

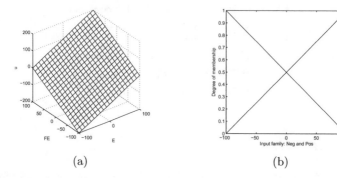

(a)   (b)

Fig. 4. Linear surface (a) with the corresponding input families (b).

Let us design an equivalent linear FF-PD$^\mu$-controller ($K_{fie} = 0$ in Eq. (13)). By configuring the fuzzy inference system (FIS) and selecting three scaling factors, we obtain a FF-PD$^\mu$-controller that reproduces the exact control performance as the fractional PD$^\mu$-controller. We first fix $K_e = 100$, since the error universe is chosen to be percentage of full scale [−100, 100], and the maximum error to a unit step is 1. The values of K_{fe} and K_u are obtained using expressions (14) (with $K_{fie} = 0$). Figure 4 shows the input families and the linear control surface obtained by using the rule base of Table 1 while satisfying the conditions of linear mapping; E is the error, FE is the fractional derivative of error and u is the output of the fuzzy PD$^\mu$ controller. Note that this result represents the step 2 – replace the conventional controller with an equivalent linear fuzzy controller – of the design procedure.

In order to enhance the performance of the control system we now proceed to step 3 of the design – make the fuzzy controller nonlinear. Thus, after verifying that the linear FF-PD$^\mu$-controller is properly designed, we may adjust the FIS settings such as its style, membership functions and rule base to obtain a desired nonlinear control surface. In this work, we choose to change the fuzzy rule base, as illustrated in Table 2. This rule base is commonly used in fuzzy control systems and consists of 49 rules with 7 linguistic terms (NL – Negative large, NM – Negative medium, NS – Negative small, ZR – Zero, PS – Positive small, PM – Positive medium, and PL – Positive large). The membership functions for the premises and consequents of the rules are shown in Fig. 5a. With two inputs and one output the input-output mapping of the fuzzy logic controller is described by the nonlinear surface of Fig. 5b. The fuzzy inference mechanism operates by using the product to combine the conjunctions in the premise of the rules and in the representation of the fuzzy implication. For the defuzzication process we use the centroid method.

Figure 6a shows the unit-step responses of the closed-loop system with $G(s)$ controlled by both the linear and nonlinear FF-PD$^\mu$ ($\mu = 1.15$) controllers. We verify that the nonlinear controller improved the overshoot, settling time, and steady-state error, when compared with the linear fuzzy controller.

Table 2. Fuzzy control rules.

E/FE	NL	NM	NS	ZR	PS	PM	PL
NL	NL	NL	NL	NL	NM	NS	ZR
NM	NL	NL	NL	NM	NS	ZR	PS
NS	NL	NL	NM	NS	ZR	PS	PM
ZR	NL	NM	NS	ZR	PS	PM	PL
PS	NM	NS	ZR	PS	PM	PL	PL
PM	NS	ZR	PS	PM	PL	PL	PL
PL	ZR	PS	PM	PL	PL	PL	PL

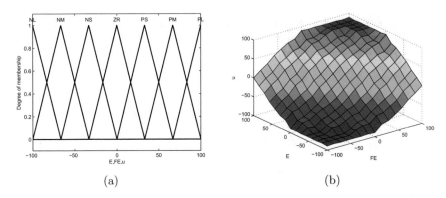

(a)   (b)

Fig. 5. Memberships functions for E, FE and u (a) and nonlinear control surface (b).

For comparison purposes, we also compute the integral of the absolute error (IAE):

$$\text{IAE} = \int_0^t |e(t)|dt \tag{18}$$

For $t = 6$ s we get IAE (linear) $= 0.55$ and IAE (nonlinear) $= 0.4$. The system performance can be further improved. For that, we go to step 4 (and last) of the design procedure – fine tune the controller. The nonlinear fuzzy controller will be adjusted by changing the parameter values of K_e, K_{fe}, and K_u. In this study we make use of a genetic algorithm (GA) to fine tune the gains of the controller [13]. The GA fitness function corresponds to the minimization of the IAE criterion, as defined in (18).

Figure 6a shows the unit-step responses of the closed-loop system with $G(s)$ controlled with the linear and nonlinear PD$^\mu$ ($\mu = 1.15$) controllers, and with the optimal nonlinear FF-PD$^\mu$ ($\mu = 1.15$) controller. The GA parameters are: population size $P = 20$, crossover probability $C = 0.8$, mutation probability $M = 0.05$ and number of generations $N_g = 25$. The interval of the FLC parameters used in the GA optimization are defined around 10% of the nominal parameters obtained with the linear controller. In this case IAE (optimal) $= 0.037$. Clearly,

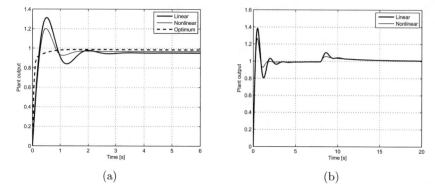

(a)   (b)

Fig. 6. Unit step responses of $G(s)$ with the linear, nonlinear and optimal fuzzy fractional PD$^\mu$ ($\mu = 1.15$) controllers (a) and unit-step and load responses of $G(s)$ with the linear and nonlinear FF-PD$^\mu$+I$^\lambda$ controllers (b).

the optimal nonlinear controller produced better results than other fuzzy controllers, since the transient response (namely the overshoot, rise time, settling time, and steady-state error) and the error IAE are smaller.

Now, let us consider the FF-PD$^\mu$+I$^\lambda$-controller. In order to test the robustness of the fuzzy controller, we introduce a load disturbance of amplitude $l = -2$ after 8 s in system of Fig. 1a. We use the same (K_p, K_d) parameters of the linear FF-PD$^\mu$ ($\mu = 1.15$) controller and tuned the (K_i, λ) for a satisfactory control response. The final tuned parameters are $(K_i, \lambda) = (10, 0.8)$. With $K_e = 100$, and using (14) we obtain K_{fe}, K_u, and K_{fie} of the fuzzy controller.

Figure 6b shows the step and load responses of closed-loop system with FF-PD$^\mu$+I$^\lambda$ controller, $(\mu, \lambda) = (1.15, 0.8)$, for the linear and nonlinear control surfaces. We observe the better response of the fuzzy controller to the reference and disturbance inputs with the nonlinear rule base (Table 2) compared to their linear counterpart (Table 1). If further improved is needed, we can fine tune it using the GA while minimizing the IAE criterion (18), as done in previous examples. Once more, we demonstrate the robustness and effectiveness of these controllers.

5   Conclusions

In this paper were applied fuzzy fractional PID controllers in the control of a fractional plant. It was demonstrated that these controllers are equivalent to the classical fractional PID controllers by using a linear input-output mapping of the rule base of the fuzzy fractional controller. Moreover, by making the controller nonlinear, the performance of the control system proves to be better than its linear counterpart. A methodology for tuning the nonlinear fuzzy fractional PID controllers is also presented. This methodology is simple and effective and can be used to tune or replace an existent fractional or integer PID controller in order to get better performance of the control system.

Acknowledgements. This work is supported by FEDER Funds through the "Programa Operacional Factores de Competitividade - COMPETE" program and by National Funds through FCT "Fundação para a Ciência e a Tecnologia" under the project UID/EEA/00760/2013.

References

1. Barbosa, R.S., Machado, J.A.T., Silva, M.F.: Time domain design of fractional differintegrators using least-squares. Signal Process. **86**, 2567–2581 (2006)
2. Podlubny, I.: Fractional Differential Equations. Academic Press, San Diego (1999)
3. Podlubny, I.: Fractional-order systems and $PI^\lambda D^\mu$-controllers. IEEE Trans. Autom. Control **44**, 208–214 (1999)
4. Åström, K.J., Hägglund, T.: PID Controllers: Theory, Design, and Tuning. Instrument Society of America, San Diego (1995)
5. Oldham, K.B., Spanier, J.: The Fractional Calculus. Academic Press, New York (1974)
6. Lee, C.C.: Fuzzy logic in control systems: fuzzy logic controller - Part I & II. IEEE Trans. Syst. Man Cybern. **20**, 404–435 (1990)
7. Li, H.-H., Gatland, H.B.: Conventional fuzzy control and its enhancement. IEEE Trans. Syst. Man Cybern.-Part B: Cybern. **26**, 791–797 (1996)
8. Mann, G.K.I., Hu, B.-G., Gosine, R.G.: Analysis of direct action fuzzy PID controller structure. IEEE Trans. Syst. Man Cybern.-Part B: Cybern. **29**, 371–388 (1999)
9. Passino, K.M., Yurkovich, S.: Fuzzy Control. Addison-Wesley, Menlo Park (1998)
10. Mizumoto, M.: Realization of PID controls by fuzzy control methods. J. Fuzzy Sets Syst. **70**, 171–182 (1995)
11. Carvajal, J., Chen, G., Ogmen, H.: Fuzzy PID controller: design performance evaluation, and stability analysis. Inf. Sci. **123**, 249–270 (2000)
12. Eker, I., Torun, Y.: Fuzzy logic control to be conventional methods. J. Energy Convers. Manage. **47**, 377–394 (2006)
13. Jesus, I.S., Barbosa, R.S.: Genetic optimization of fuzzy fractional PD+I controllers. ISA Trans. **57**, 220–230 (2015)
14. Machado, J.A.T.: Analysis and design of fractional-order digital control systems. SAMS J. Syst. Anal. Model. Simul. **27**, 107–122 (1997)
15. Chen, Y.Q., Vinagre, B., Podlubny, I.: Continued fraction expansion to discretize fractional order derivatives-an expository review. Nonlinear Dyn. **38**, 155–170 (2004)
16. Jantzen, J.: Foundations of Fuzzy Control. Wiley and Sons, Chichester (2007)

A Semantic Web Architecture for Competency-Based Lifelong Learning Support Systems

Kalthoum Rezgui[1,2(✉)] and Hédia Mhiri[1]

[1] SMART Lab, Institut Supérieur de Gestion de Tunis, Université de Tunis,
Tunis, Tunisia
kalthoum.rezgui@gmail.com
[2] Institut Supérieur des Arts du Multimédia de Manouba, Université de Manouba,
Manouba, Tunisia

Abstract. In this paper, we present a semantic Web architecture of a competency-based lifelong learning support system which falls within the scope of lifelong learning and attempts to deal with the issue of competency tracking, management, and development in learning networks. In particular, the proposed system aims to track learners' competencies in learning networks and to provide them with different competency assessment procedures, such as learner positioning, competency gap analysis, and competency profile matching. In addition, it integrates a personalized learning path generator module that enables learners fill the gap between available and expected competencies. This system is currently under construction and will include semantic Web services capabilities for supporting automatic matchmaking tasks.

1 Introduction

During the last ten years, lifelong learning [1] has become an important topic of academic and policy debates on promoting new partnerships and curricula aiming at promoting the acquisition and continuous development of new competencies. According to [6], both competency development and lifelong learning provide a good answer to market requirements. Today, learners have to learn during their entire lifetimes in order to be well prepared to answer to market requirements and future jobs that do not yet exist or still unknown [6]. Indeed, professional success depends on the constant renewal of existing competencies and acquirement of new ones. The concept of lifelong learning can be considered as a continuous education process which covers any form of learning (i.e., formal, non-formal and informal) undertaken throughout the life and leading to fostering of continuous development and improvement of knowledge, skills and competencies. In this context, learning networks emerged as alternative models that merge pedagogical, organizational, and technological perspectives to support the provision of a broad variety of learning opportunities [7,8]. This paper focuses on the description of a semantic Web architecture of a competency-based

© Springer Nature Switzerland AG 2020
A. M. Madureira et al. (Eds.): HIS 2018, AISC 923, pp. 438–448, 2020.
https://doi.org/10.1007/978-3-030-14347-3_43

lifelong support system. The functional architecture of this system consists of two repositories and seven main modules which provide lifelong learners with different competency assessment procedures. The core component of this architecture is a competency ontology which enables modeling competencies, competency profiles, and associated evidences [10]. Besides, it integrates a learning path generator module which responsible for generating the most suitable learning path for each learner and that helps him develop target competencies. The remainder of this paper is structured as follows: First, the functional architecture and its corresponding modules of the proposed system are described in detail in Sect. 2. A proof-of-concept of the semantic competency profile matching algorithm is demonstrated in Sect. 3. Finally, conclusions and future work are presented in Sect. 4.

2   System Architecture and Modules

In this section, we present the functional architecture of the proposed system as well as its main modules (see Fig. 1).

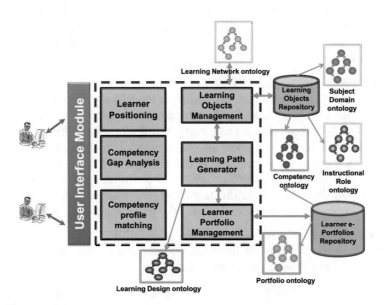

Fig. 1. Functional architecture of the system

User Interface (UI) Module
The UI module manages interactions between the lifelong learner and the system.

Algorithm 1. Learner Positioning

1: **Input:** ePort /*A collection of a lifelong learner's evidences*/
        compModel /*A set of competency profiles associated with a learning network in a specific domain*/
2: **Output:** learnerACP /*The actual competency profile of the lifelong learner in the learning network*/
3: **Initialization:** $Evid_{eport} \leftarrow \emptyset$
4: **Begin**
5:   Extract_Evidences(ePort, $Evid_{eport}$)
6:   **for each** $CP_i \in$ compModel **do**
7:     **for each** $C_j \in CP_i$ **do**
8:       **if** $EvidenceSource(C_j) \subseteq Evid_{eport}$ **then**
9:         learnerACP $\leftarrow$ learnerACP $\cup \{C_j\}$
10:        **end if**
11:     **end for**
12:   **end for**
13: **End**

Learner ePortfolio Management (LePM) Module

This module is responsible for handling requests for accessing and updating the repository of learners' e-Portfolios, such as creating new competency/evidence records and updating proficiency levels after a successful completion of all learning activities that have as learning objectives these proficiency levels. This repository stores learners' e-Portfolios in accordance with an e-Portfolio ontology. For more details on this ontology, we refer the reader to [12].

Learner Positioning (LP) Module

The LP module is involved upon enrollment of the lifelong learner in the learning network with a set of artifacts collected in his e-Portfolio and which evidences the mastery of a set of competencies previously acquired from different learning networks. It enables interpreting these evidences in order to estimate the proficiency levels of the learner for one or several competencies of the competency model associated with the current learning network. The process of *Learner Positioning* is described by Algorithm 1: In this algorithm, the *Extract_Evidences* procedure enables to extract a set of evidences from the lifelong learner's e-Portfolio.

Learning Objects Management (LOsM) Module

This module is responsible for handling the learning objects repository (LOR) and supports the following functionalities:

- Manage (i.e., add, edit, and delete) learning objects and the corresponding metadata. Instances of metadata are created based on the IEEE LOM RDF Binding [9] application profile that we have developed. Details of the proposed application profile are provided in [11].
- Search the LO repository for learning objects of different types, instructional roles, and competencies.

Competency Profile Matching (CPMatch) Module

The CPMatch module provides different matching procedures that enable comparing the lifelong learner's actual profile with the requirements of a required competency profile. In particular, two algorithms of competency profile matching have been employed: a simple *exact matching* algorithm and a *semantic matching* algorithm that involves different types of similarity measures.

[A.] Exact matching

The exact matching algorithm proceeds as follows: Each required competency from the required competency profile is researched with exactitude in the lifelong learner's acquired competency profile. In addition, a weight is assigned to each required competency which assesses its importance in the calculation of the matching coefficient [3,4].

[B.] Semantic matching

In the semantic matching algorithm, each required competency from the required competency profile is compared with each competency in the lifelong learner's profile taking into account its skill and knowledge parts, its proficiency level, and its context. In addition, the best match is justified by a weight factor which reflects the importance of the required competency in the computation of the matching coefficient [3,4,13]. Accordingly, the formula proposed for computing the similarity of a learner's competency profile and a required competency profile is as follows:

$$sim(rcp, acp) = \frac{(\sum_{i \in I} weight(C^i_{rcp}) * Max[sim(C^i_{rcp}, C^j_{acp}) * sim_{Level}(C^i_{rcp}, C^j_{acp}) * sim_{Context}(C^i_{rcp}, C^j_{acp}) | j \in J])}{\sum_{i \in I} weight(C^i_{rcp})} \tag{1}$$

where:

- $weight(C^i_{rcp})$ is the weight of the required competency C^i_{rcp}
- $sim(C^i_{rcp}, C^j_{acp}) = sim_{skill}(skill(C^i_{rcp}), skill(C^j_{acp})) * sim_{knowledge}(knowledge$ $(C^i_{rcp}), knowledge(C^j_{acp}))$
- sim_{skill} computes the similarity between two generic skills $s1$ and $s2$ by applying the formula 2 proposed by [3]: α is the number of generic skills (it is equal to 10).
- $sim_{knowledge}$ computes the similarity between two concepts $c1$ and $c2$ from a domain ontology by applying the formula 3.
- sim_{Level} computes the similarity between two proficiency levels $pl1$ and $pl2$ using formula 2 with $\alpha = 4$ (the number of proficiency levels).
- $sim_{Context}$ computes the similarity between two contexts $ctx1$ and $ctx2$ using formula 2 with $\alpha = 5$ (the number of contexts).

$$sim(a_1, a_2) = \begin{array}{ll} 1 - 1/\alpha(a_1 - a_2) & If a_1 - a_2 \geq 0 \\ 1 & If a_1 - a_2 \prec 0 \end{array} \tag{2}$$

The algorithm for semantic matching is outlined in what follows:

Algorithm 2. Semantic Competency Profile Matching

1: **Input:** RCP /*A required competency profile*/
        learnerACP /*A competency profile acquired by the lifelong learner*/
2: **Output:** Cmatch /*A matching coefficient*/
3: **Initialization:** SumWeight ← 0 /*Accumulated weight of required competencies*/
        CalcWeight ← 0 /*Accumulated weight of acquired competencies*/
        simList ← ∅ /*A list of competency similarity values*/
4: **Begin**
5: **for each** rc ∈ RCP **do**
6:    i ← 0
7:    SumWeight ← SumWeight + rc.weight
8:    **for each** ac ∈ LearnerACP **do**
9:       Identify_Skill(ac,rc,skillac,skillrc)
10:      Identify_Knowledge(ac,rc,knowledgeac,knowledgerc)
11:      Match(knowledgerc,knowledgeac,$sim_{knowledge}$)
12:      $simList_i := sim_{skill}(skillrc, skillac) * sim_{knowledge} * sim_{Level}(rc, ac) * sim_{Context}(rc, ac)$
13:      i ← i+1
14:    **end for**
15:    CalcWeight ← CalcWeight + (rc.weight) * Max($simList_i$)
16: **end for**
17: Cmatch ← (CalcWeight / SumWeight) * 100
18: **End**

- The *Identify_Skill* procedure enables to identify the skill part of a required competency in *skillrc* and a learner's competency in *skillac*.
- The *Identify_Knowledge* procedure enables to identify the knowledge part of a required competency in *knowledgerc* and a learner's competency in *knowledgeac*.
- The *Match* function is used to calculate the similarity between two knowledge elements *knowledgerc* and *knowledgeac*.

In the *Match* function, we employed a similarity measure which relies on ideas from [2,3,14] to calculate the similarity between two given concepts in a concept hierarchy. The measure used (see Eq. 3) determines the similarity between two concepts $c1$ and $c2$ by computing the distance $dist(c1, c2)$ between them, as represented by their respective positions in the underlying concept hierarchy:

$$Sim(c1, c2) = 1 - dist(c1, c2) \qquad (3)$$

However, since the distance between upper level concepts is greater than lower level concepts, a *milestone* value is assigned to every node in the concept hierarchy, which is calculated as follows:

$$milestone(n) = \frac{1/2}{k^{l(n)}} \qquad (4)$$

Where k is a factor larger than 1 that indicates the rate at which the milestone value decreases throughout the hierarchy (k is set to 2 in our case as used in Corese[1]) and $l(n)$ is the depth of the node n in the hierarchy.

---

[1] http://wwwsop.inria.fr/edelweiss/software/corese/.

Given that the path between two concepts $c1$ and $c2$ passes through the closest common parent, the distance between concepts is, then, measured by their milestones and the milestone of their closest parent $ccp(c1, c2)$ as follows:

$$dist(c1, c2) = dist(c1, ccp) + dist(c2, ccp)$$
$$dist(c, ccp) = milestone(ccp) - milestone(c) \tag{5}$$

The Competency Gap Analysis (CGA) Module

The CGA module allows the calculation of a competency gap between the target competency profile and the learner's actual competency profile by adjusting the target profile based on competencies already mastered by the learner. In fact, for each target competency, the learner's competency profile is fetched in order to find out if this competency is already mastered.

The process of competency gap analysis is described by the *CGA* algorithm: the inputs of this algorithm are the competencies targeted by a lifelong learner and the competencies that he/she actually has. The output of this algorithm is an adjusted target competency profile, i.e., the set of competencies that are new to the learner. This profile describes the target learning objectives of the learning path to be generated by the Learning Path Generator module.

Algorithm 3. CGA

1: **Input:** TCP /*A target competency profile*/
           learnerACP /*A set of competencies acquired by the lifelong learner*/
2: **Output:** Target /*A set of competencies that are new to the learner*/
3: **Initialization:** $Target \leftarrow \emptyset$
4: **Begin**
5: **for each** $Competency_i \in TCP$ **do**
6:     $ProficiencyLevel_j \leftarrow$ findCompetency($Competency_i, learnerACP$)
7:     **if** $ProficiencyLevel_j < Competency_i.hasProficiencyLevel$ **then**
8:         add(Target,$Competency_i$)
9:     **end if**
10: **end for**
11: **End**

The *findCompetency* function searches a target competency in the set competencies acquired by a lifelong learner and returns its corresponding proficiency level if it exists, 0 otherwise.

The Learning Path Generator (LPG) Module

This module is responsible for the generation of personalized learning paths that help lifelong learners in developing target competencies. The first step in the LPG process consists in finding out which learning objects could fulfill the

---

Algorithm 4. LOSelection

1: **Input:** TCP /*A target competency profile*/
            LOR /*A collection of learning objects and associated metadata*/
2: **Output:** Selected /*A selection of learning objects*/
3: **Begin**
4: **for each** $Competency_i \in TCP$ **do**
5:     selected ← searchLO (LOR, $Competency_i$, $Competency_i.hasProficiencyLevel$)
6:     **if** ($Competency_i.requires$ OR $Competency_i.iscomposedOf \neq$ null) **then**
7:         **for each** $competency_j \in (competency_i.requires \cup competency_i.iscomposedof)$ **do**
8:             Selected   ←   Selected   ∪   LOSelection   (LOR, $Competency_j$, $Competency_i.hasProficiencyLevel$)
9:         **end for**
10:     **end if**
11: **end for**
12: **End**

---

need for competencies calculated by the CGA algorithm. This step is described by Algorithm 4: The inputs of the LOSelection algorithm are a target competency profile and a repository of learning objects and metadata provided by learning object creators during their upload to the repository collection. Each competency in this target profile is used to identify relevant learning objects. In particular, the search function is based on the *lom-cls:educationalObjective* metadata element of learning objects stored in the repository. In addition, the *requires* and *isComposedOf* relationships between competencies are used to trigger the search for learning objects that might potentially be relevant. The output of this algorithm is a set of learning objects from the collection that fill the gap between the learner's needs and the learning objectives.

The second step of the learning path generation process is to filter the set of retrieved learning objects based on the lifelong learner's learning style. In our system, we chose to implement the Felder-Silverman learning style model (FSLSM) [5] since it enables describing the learning style of a learner with more details. In particualr, the FSLSM learning style model defines four dimensions related to how people process information. Each dimension has two possible values: *Active/Reflective, Sensing/Intuitive, Visual/Verbal, Sequential/Global*. In this work, the cognitive style of the learner has been deducted from the questionnaire filled out by the lifelong learner upon registration with the system. Since we have adopted the Felder-Silverman learning style model, we used their questionnaire proposed by [5], known as *"Index of Learning Styles Questionnaire"*, to determine the learning style of the lifelong learner.

The filtering step is described by the *FilterSelectedLO* algorithm. This algorithm takes as inputs the set of learning objects discovered by the previous step and the learning style of the lifelong learner. The output of the algorithm is a set of filtered learning objects. In particular, the system queries the lifelong learner's e-Portfolio for the value of the learning style category. For example, if the learner belongs to the category of visual learners (i.e., who prefer to see what they learn through graphs, charts and images), then learning objects that are annotated with the metadata elements *md:type = Simulation or Illustration or Demonstration* and *dc:format = Graphics or Multimedia or Movies* are added to the set of filtered learning objects.

The third step of the learning path generation process is to assemble selected learning objects in a learning path (unit of learning) compliant with the IMS LD model.

Algorithm 5. FilterSelectedLO

1: **Input:** Selected /*A selection of learning objects*/
   LearningStyle /*The preferred learning style of the lifelong learner*/
2: **Output:** filteredSelection
3: **Initialization:** filteredSelection ← ∅
4: **Begin**
5: **if** (LearningStyle.value = Visual) **then**
6:     **for each** LO_i ∈ Selected **do**
7:         **if** ((LO_i.metadata.type = Simulation | Illustration | Demonstration) AND (LO_i.metadata.format = Video | Multimedia |Graphics)) **then**
8:             filteredSelection := filteredSelection ∪ LO_i
9:         **end if**
10:     **end for**
11: **else if** (LearningStyle.value = Sequential) **then**
12:     **for each** LO_i ∈ Selected **do**
13:         **if** ((LO_i.metadata.type = Exercise | Question | AnswerToQuestion | Self-assessment) AND (LO_i.metadata.format = Text | Audio)) **then**
14:             filteredSelection ← filteredSelection ∪ LO_i
15:         **end if**
16:     **end for**
17: **else if** (LearningStyle.value = Active) **then**
18:     **for each** LO_i ∈ Selected **do**
19:         **if** ((LO_i.metadata.type = Exercise | Simulation | Question | AnswerToQuestion | Experiment | Case | Self-assessment) AND (LO_i.metadata.format = Text) AND (LO_i.metadata.interactivityType = Active)) **then**
20:             filteredSelection ← filteredSelection ∪ LO_i
21:         **end if**
22:     **end for**
23: **else if** (LearningStyle.value = Sensitive) **then**
24:     **for each** LO_i ∈ Selected **do**
25:         **if** ((LO_i.metadata.type = Case | Fact | Procedure | Example | Experiment) AND (LO_i.metadata.format = Video | Multimedia | Graphics | Audio) AND (LO_i.metadata.semanticDensity = Medium | High | Very high)) **then**
26:             filteredSelection ← filteredSelection ∪ LO_i
27:         **end if**
28:     **end for**
29: **end if**
30: **End**

---

3   Proof of Concept

To better illustrate the results of the proposed semantic matching algorithm, we chose a required competency profile (RCP) and two learners' profiles (Learner-ACP1 and LearnerACP2) (see Fig. 2). A snippet of an OOP (Object-Oriented

Programming) teaching domain ontology with the milestones values for the corresponding ontology levels is depicted in Fig. 3. Table 1 below summarizes the results of matching the required RCP with LearnerACP1 and LearnerACP2. In Table 2, we present a detailed matching result of comparing the required competency profile with the learner1's competency profile. Table 3 shows a detailed matching result of comparing the required competency profile with the learner2's competency profile.

Required Competency Profile
Weight Competency Proficiency Level Context
60 Describe the basic features of OOP Expert Higher Education
40 Use an integrated development environment (IDE) Advanced Higher Education
Learner1's Competency Profile
Competency Proficiency Level Context
Use socket Intermediate Higher Education
Use an integrated development environment (IDE) Expert Higher Education
Learner2's Competency Profile
Competency Proficiency Level Context
Identify control structures Expert Higher Education
Use an objcet database management system (ODBMS) Beginner Higher Education

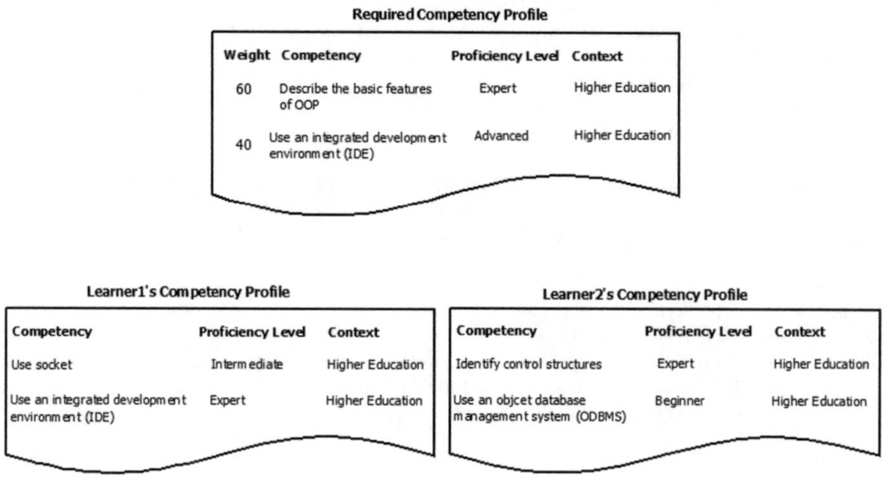

Fig. 2. Three sample competency profiles

m(2)=1/4=0.25
m(3)=1/8=0.125
m(4)=1/16=0.0625
m(5)=1/32=0.0313
OOPElement
OOPConcept
OOTools
OOPBasicFeatures
OOPAdvancedFeatures
IDE
ODBMS
Class
ControlStructures
Object
Socket
Thread

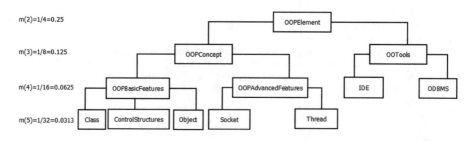

Fig. 3. OOP domain ontology segment with corresponding milestone values

Table 1. Matching coefficients of RCP with LearnerACP1 and LearnerACP2

Matching coefficient	Final	Related to required C1	Related to required C2
		Describe the basic features of OOP (Expert, Higher education)	Use an integrated development environment (IDE) (Advanced, Higher education)
LearnerACP1	77.5%	62.5%	100%
LearnerACP2	57.5%	90%	43.75%

Table 2. A detailed matching result of RCP with LearnerACP1

Required competencies	Learner1'competencies	Similarity
Describe the basic features of OOP (Expert, Higher education)	Use socket (Intermediate, Higher education)	42.19%
	Use an integrated development environment (IDE) (Expert, Higher education)	**62.5%**
Use an integrated development environment (IDE) (Advanced, Higher education)	Use socket (Intermediate, Higher education)	44.53%
	Use an integrated development environment (IDE) (Expert, Higher education)	**100%**

Table 3. A detailed matching result of RCP with LearnerACP2

Required competencies	Learner2'competencies	Similarity
Describe the basic features of OOP (Expert, Higher education)	Identify control structures (Expert, Higher education)	**90%**
	Use an object database management system (ODBMS) (Beginner, Higher education)	15.62%
Use an integrated development environment (IDE) (Advanced, Higher education)	Identify control structures (Expert, Higher education)	41.56%
	Use an object database management system (ODBMS) (Beginner, Higher education)	**43.75%**

4 Conclusion

In this paper, we proposed the design of a competency-based lifelong learning support system that aims to track competency development in learning networks and to provide lifelong learners with different competency assessment procedures. The functional architecture of the proposed system, its constituent modules and the underlying algorithms have been identified and specified. Further research will focus on the use of a blockchain-based smart contracts approach to competency assessment and validation.

References

1. Beer, S.: Lifelong learning: debates and discourses. Paper submitted to Niace's lifelong learning inquiry (2007). http://www.niace.org.uk/lifelonglearninginquiry/docs/ConceptualisingLLL.pdf
2. Billig, A., Sandkuhl, K.: Match-making based on semantic nets: the XML-based approach of BaSeWeb. In: Proceedings of the 1st Workshop on XML-Technologien für das Semantic Web, pp. 39–51. Springer, London (2002)
3. Bizer, C., Heese, R., Mochol, M., Oldakowski, R., Tolksdorf, R., Eckstein, R.: The impact of semantic web technologies on job recruitment processes. In: International Conference Workshop on Computer Science (WI 2005), Bamberg, Germany (2005)
4. Boucetta, Z., Boufaïda, Z., Yahiaoui, L.: Appariement sémantique des documents à base d'ontologie pour le E-Recrutement. In: Colloque sur l'Optimisation et les Systèmes d'Information (COSI 2008), pp. 1–12, June 2008
5. Felder, R., Silverman, L.: Learning and teaching styles in engineering education. Eng. Educ. **78**(7), 674–681 (1988)
6. Hoffmann, H.: Framework of actions on the lifelong development of competencies and qualifications (2012). http://erc-online.eu/wp-content/uploads/2014/04/2012-00426-E.pdf
7. Kew, V.: The TENCompetence personal competence manager (2007). http://ceur-ws.org/Vol-280/p08.pdf
8. Koper, R., Rusman, E., Sloep, P.: Effective learning networks. J. Lifelong Learn. Eur. 18–27 (2005)
9. Nilson, M.: IEEE learning object metadata RDF binding (2002). http://kmr.nada.kth.se/static/ims/md-lomrdf.html
10. Rezgui, K., Mhiri, H., Ghédira, K.: An ontology-based approach to competency modeling and management in learning networks. In: Proceedings of the 8th International KES Conference on Agents and Multi-agent Systems - Technologies and Applications (KES AMSTA 2014), pp. 257–266 (2014)
11. Rezgui, K., Mhiri, H., Ghédira, K.: An ontology-based multi-level semantic representation model for learning objects annotation. In: Proceedings of the 14th International Conference on Computer Systems and Applications (AICCSA), pp. 1391–1398 (2017)
12. Rezgui, K., Mhiri, H., Ghédira, K.: Towards a common and semantic representation of e-portfolios. Data Technol. Appl. **52**(4), 520–538 (2018)
13. Sicilia, M.: Ontology-Based Competency Management: Infrastructures for the Knowledge Intensive Learning Organization, pp. 302–324. IGI Global, Hershey, April 2005
14. Zhong, J., Zhu, H., Li, J., Yu, Y.: Conceptual graph matching for semantic search. In: Proceedings of the 10th International Conference on Conceptual Structures: Integration and Interfaces, pp. 92–106. Springer, London (2002)

Modified and Hybridized Monarch Butterfly Algorithms for Multi-Objective Optimization

Ivana Strumberger[1], Eva Tuba[1], Nebojsa Bacanin[1], Marko Beko[2,3], and Milan Tuba[1(✉)]

[1] Singidunum University, Belgrade, Serbia
tuba@ieee.org
[2] COPELABS, Universidade Lusófona, Lisbon, Portugal
[3] CTS/UNINOVA, Lisbon, Portugal

Abstract. This paper presents two improved versions of the monarch butterfly optimization algorithm adopted for solving multi-objective optimization problems. Monarch butterfly optimization is a relatively new swarm intelligence metaheuristic that proved to be robust and efficient method when dealing with NP hard problems. However, in the original monarch butterfly approach some deficiencies were noticed and we addressed these deficiencies by developing one modified, and one hybridized version of the original monarch butterfly algorithm. In the experimental section of this paper we show comparative analysis between the original, and improved versions of monarch butterfly algorithm. According to experimental results, hybridized monarch butterfly approach outperformed all other metaheuristics included in comparative analysis.

Keywords: Monarch butterfly optimization · Swarm intelligence · NP hardness · Multi-objective · Metaheuristics

1 Introduction

Optimization is one of the most widely used research domains due to the fact that almost any real life and practical problem can be modeled as an optimization problem. Many continuous, as well as combinatorial, optimization problems exist that include search for global optimum of highly irregular and non-linear functions with huge number of local optima and possibly complicated constraints. Many of these problems belong to the group of NP hard tasks. For such problems, it is relatively easy to code the optimization algorithm, but the algorithm cannot produce satisfying results within a reasonable amount of computational

This research is supported by Ministry of Education, Science and Technological Development of Republic of Serbia, Grant No. III-44006.

A. M. Madureira et al. (Eds.): HIS 2018, AISC 923, pp. 449–458, 2020.
https://doi.org/10.1007/978-3-030-14347-3_44

time. Well-known example of such problems is the traveling salesman problem (TSP).

Unconstrained (or bound constrained) continuous optimization problem can be defined as D-dimensional minimization or maximization problem:

$$\min \left(or \max\right) f(x), \quad x = (x_1, x_2, x_3, ..., x_D) \in S, \tag{1}$$

where x is a real vector with $D \geq 1$ components and $S \in R^D$ is an D-dimensional hyper-rectangular search space constrained by lower and upper bounds.

The nonlinear constrained optimization problem in the continuous space can be formulated as in Eq. (1), but in this case $x \in F \subseteq S$. S is again D-dimensional hyper-rectangular space and $F \subseteq S$ is the feasible region defined by the set of m linear or non-linear constraints:

$$g_j(x) \leq 0, \quad for \quad j \in [1, q] \tag{2}$$
$$h_j(x) = 0, \quad for \quad j \in [q + 1, m]$$

where q is the number of inequality constraints, and $m - q$ is the number of equality constraints.

An unconstrained k-objective continuous optimization problem can be expressed as:

$$\min \left(or \max\right) F(x) = (f_1(x), f_2(x), ..., f_m(x))^T$$
$$\text{subject to } x \in \Omega, \tag{3}$$

where Ω is the decision variable space, x is a decision vector, $F : \Omega \to R^m$ consists of m real-valued objective functions, and R^m is the objective space. The goal is to calculate the points that yield the optimum values for all m objective function simultaneously.

Since standard, deterministic algorithms cannot be applied to NP hard problems, metaheuristics can be used to generate satisfying, and in some cases optimal solutions. One of the most prominent categories of the metaheuristics algorithms is the group of swarm intelligence metaheuristics. Swarm intelligence simulates a group of organisms in nature, and rely on four self-organization principles: positive feedback, negative feedback, multiple interactions and fluctuations [1]. Some of the recent swarm intelligence algorithms are fireworks algorithm (FWA) [2] and elephant herding optimization (EHO) [3]. After the initial implementation, many modified and upgraded versions of the FWA were devised [4–7]. EHO was also applied on support vector machine parameters tuning [8], robot path planning [9], multilevel image thresholding [10] and static drone placement [11].

In this paper we show modified and hybridized versions of the monarch butterfly optimization (MBO) algorithm. MBO is a novel swarm intelligence metaheuristic that was firstly proposed by Wang and Deb in 2015 [12]. MBO models the migration behavior of monarch butterfly insect. Original MBO was tested on thirty-eight global optimization benchmarks and proved to be robust optimization metaheuristics that outperforms other state-of-the-art algorithms [12].

We tested our approaches on four standard multi-objective benchmarks and performed comparative analysis with other state-of-the-art algorithms tested on the same problem sets.

2   Original Monarch Butterfly Optimization Algorithm

The search process of intelligent algorithms models two properties (attributes) of monarch butterflies: Lévy flights, that are performed by some butterflies during migration and the process of laying eggs by female butterflies for generating offspring [13]. These two characteristics were the main source of inspiration for the devision of MBO metaheuristic.

The search process of the MBO metaheuristic is conducted by two operators: migration operator and butterfly adjusting operator.

For simplification purposes, the migration process is idealized as in [12]: the whole population of monarch butterflies stay in Land 1 for 5 months and stay in Land 2 for 7 months. In this paper, instead of Land 1 and Land 2, terms subpopulation 1 and subpopulation 2, respectively will be used. The number of monarch butterflies in subpopulation 1 and subpopulation 2 can be calculated by using Eqs. (4) and (5), respectively [12]:

$$ceil(p \cdot NP) \cdot NP_1 \tag{4}$$
$$(NP - NP_1) \cdot NP_2, \tag{5}$$

where function $ceil(a)$ is the function that rounds argument a to the nearest integer greater than or equal to a, NP represents the total size of the population (number of butterflies), NP_1 and NP_2 are number of monarch butterflies in subpopulation 1 and subpopulation 2, respectively, and p denotes ratio of monarch butterflies in the subpopulation 1. The process of MBO's migration can be mathematically formulated as [12]:

$$x_{i,k}^{t+1} = x_{r_1,k}^t, \tag{6}$$

where $x_{i,k}^{t+1}$ is the k-th element (parameter) of the x_i individual at generation (iteration) $t+1$, and $x_{r_1,k}^t$ is the k-th parameter of the individual x_{r_1} at current generation t. Individual r_1 is randomly selected from subpopulation 1.

When the expression $r \leq p$ holds, k-th parameter of the newly created butterfly is calculated by using Eq. (6). Ratio r is calculated as $r = rand \cdot peri$, where $peri$ indicates the migration period, and $rand$ is uniformly distributed number between 0 and 1. In the original MBO implementation, the value of $peri$ was set to 1.2 [12]. The same value for $peri$ is used in our experiments.

In the opposite case, if the expression $r \geq p$ is satisfied, the parameter k of the newly generated monarch butterfly is calculated using the following equation:

$$x_{i,k}^{t+1} = x_{r_2,k}^t, \tag{7}$$

where $x_{r_2,k}^t$ represents the k-the parameter of the randomly chosen butterfly r_2 from subpopulation 2 in the current generation t.

By performing analysis of Eqs. (6)–(7), it can be concluded that the MBO algorithm balances the direction of migration operator by adjusting the value of ratio p. If the ration p is relatively large, more parameters from monarch butterflies in subpopulation 1 will be selected when generating offspring. In the other case, if the value of p is relatively small, the influence of individuals from the subpopulation 2 on the offspring is greater. In our MBO implementation, p is set to 5/12, as in [12].

As stated above, second mechanism that guides the monarch butterfly's search process is butterfly adjusting operator. This operator performs as follows: for all parameters in individual j, if the pseudo-random number $rand$ is smaller than or equal to p, the new solution is generated by the following equation [12]:

$$x_{j,k}^{t+1} = x_{best,k}^{t},$$
(8)

where $x_{j,k}^{t+1}$ is the k-th parameter of the new solution j, and $x_{best,k}^{t}$ is the k-th parameter of current best solution in the whole population.

Contrarily, if the uniformly distributed number $rand$ is greater than p, the new solution is created by employing the following expression [12]:

$$x_{j,k}^{t+1} = x_{r_3,k}^{t},$$
(9)

where $x_{r_3,k}^{t}$ indicates the k-th parameter of randomly selected solution r_3 from Subpopulation 2, and $r_3 \in \{0, 1, 2, ..., NP_2\}$.

In the final case, if the condition $rand \geq BAR$ holds, the k-th parameter of the offspring solution is generated by using following expression [12]:

$$x_{j,k}^{t+1} = x_{j,k}^{t} + \alpha \times (dx_k - 0.5),$$
(10)

where BAR denotes butterfly adjusting rate, and dx is the walk step of the monarch butterfly j that can be calculated by using Lévy flights.

The parameter α in Eq. (10), denotes the weighting factor, given as [12]:

$$\alpha = S_{max}/t^2,$$
(11)

where S_{max} denotes the maximum walk step that individual butterfly can move in one step. In the presented equations, the parameter α controls the balance between exploitation (intensification) and exploration (diversification).

3   Modified Monarch Butterfly Optimization Algorithm

In the original MBO [12] implementation, the search process in the late iterations of algorithm's execution can be improved. We came to this conclusion by conducting empirical experiments.

At late phases of MBO's execution, the rational assumption is that the algorithm has converged to the right part of the search space, and that more search power should be used for exploitation around the current best solutions in the population. However, in the original MBO, search power is unnecessary used on exploration of the search space.

To tackle the stated issue, and to enhance the MBO search process in late iterations of algorithm's execution (for the case of multi-objective optimization problems last 15% of iterations), we introduce hybrid solutions, that are created by applying the crossover operator on the current best solutions in subpopulation 1 and subpopulation 2. Crossover operator was adopted from GA. Besides the adoption of the crossover operator, we also introduced additional pseudo-random number ϕ that controls the creation of hybrid solutions. Modified MBO that uses hybrid solutions in the late 15% of algorithm's execution was simply named modified MBO (MMBO). If the condition $\phi > 0.5$ is satisfied, the parameter k of the hybrid solution x_h for iteration $t + 1$ is created as:

$$x_{h,k}^{t+1} = x_{ws1,k}^t \tag{12}$$

On the other hand, if the condition $\phi \leq 0.5$ holds, the parameter k of the hybrid solution x_h in iteration $t + 1$ is generated as follows:

$$x_{h,k}^{t+1} = x_{ws2,k}^t \tag{13}$$

In the Eqs. (12) and (13), $x_{ws1,k}$ and $x_{ws2,k}$ are k-th parameters of the worst solutions in subpopulation 1 and subpopulation 2, respectively.

In this way, in the last 15% of iterations, we replace worst solutions from subpopulation 1 and subpopulation 2 with hybrid 1 and hybrid 2, respectively. According to the presented equations and pseudo-codes, the MMBO algorithm can be summarized as Algorithm 1.

Algorithm 1. Pseudo-code of the MMBO algorithm

Initialization.
Fitness evaluation. Evaluate each monarch butterfly against the objective function and calculate fitness.
while $t < MaxGen$ **do**
    Sort all individuals in the population according to its fitness.
    Divide the whole population into subpopulation 1 and subpopulation 2.
    for $i = 1$ **to** NP_1 (all butterflies in the subpopulation 1) **do**
        Generate new individuals in subpopulation 1 by using migration operator
    end for
    for $j = 1$ **to** NP_2 (all butterflies in the subpopulation 2) **do**
        Generate new individuals in subpopulation 2 by using adjusting operator
    end for
    if $t \geq (MaxGen \cdot 0.85)$ **then**
        Evaluate subpopulation 1 and subpopulation 2 according to the newly updated positions.
        Generate two hybrid solutions by using Eq. (12) and Eq.(13).
        Replace worst solutions in subpopulation 1 and subpopulation 2 with hybrid 1 and hybrid 2, respectively.
    end if
    Merge newly generated subpopulation 1 and subpopulation 2 into one whole population.
    Evaluate the population according to the newly updated positions.
    Increase the iteration counter t by one.
end while
return Best individual in the whole population

---

4   Hybridized Monarch Butterfly Optimization Algorithm

FA was inspired by the biochemical and social characteristics of the fireflies. Basic idea behind this approach is that each firefly in the population moves towards the position of the brighter firefly. First implementation of the FA was proposed for unconstrained optimization [14]. Many successful implementations of the FA can be found in the literature survey [15, 16].

In the case when algorithm by accident, in early iterations, hits the right part of the search space, this search process around the current best yields good results. Unfortunately, in most algorithm's runs, the MBO does not hit the right part of the search space in early iterations, which leads to the worse mean values with high dispersion. To overcome this deficiency, in early iterations of algorithm's execution, we incorporated FA's search equation in the original MBO, that replaces search process conducted by using Eq. (8)

In the FA metaheuristics, the movement of a firefly i (process of exploration and exploitation) towards the brighter, and thus more attractive firefly j, for each solution's parameter, is determined by [17]:

$$x_i^{t+1} = x_i^t + \beta_0 r^{-\gamma r_{i,j}^2}(x_j^t - x_i^t) + \alpha(rand - 0.5), \tag{14}$$

where $t + 1$ is the next iterations, t is the current iteration, β_0 is attractiveness at $r = 0$, α is randomization parameter, $rand$ is random number uniformly distributed between 0 and 1, and $r_{i,j}$ is distance between fireflies i and j.

The distance between fireflies is calculated using Cartesian distance form [14]:

$$r_{i,j} = ||x_i - x_j|| = \sqrt{\sum_{k=1}^{D}(x_{i,k} - x_{j,k})^2}, \tag{15}$$

where D is the number of problem parameters. For most problems, $\beta_0 = 0$ and $\alpha \in [0, 1]$ are adequate settings. Parameter α for FA search process is being gradually decreased from its initial value:

$$\alpha(t) = (1 - (1 - ((10^{-4}/9)^{1/nGen}))) * \alpha(t - 1), \tag{16}$$

where t is the current iteration.

By conducting experimental tests for multi-objective benchmarks, we found that the algorithm obtains optimum results when FA's search equation (Eq. 14) replaces MBO butterfly migration equation (Eq. (8)) in the first 50% of iterations. Pseudo-code of the MBO-FS metaheuristics is shown in Algorithm 2

5   Experimental Results, Setup and Discussion

For tackling multi-objective optimization problems, one approach is to combine all objectives into a single objective with techniques such as weighted coefficients. Here, we extended our implemented algorithms (MBO, MMBO and MBO-FS) to produce Pareto front directly.

Algorithm 2. Pseudo-code of the MBO-FS algorithm

Initialization. Set the iterations counter $t = 1$; generate the population P of NP monarch butterfly individuals randomly; set the maximum generation number $MaxGen$, monarch butterfly number in land 1 NP_1 and in land 2 $NP2$, max step length S_{max}, butterfly adjusting rate BAR, migration period $peri$ and the migration ratio p.

Fitness evaluation. Evaluate each monarch butterfly against the objective function and calculate fitness.

while $t < MaxGen$ **do**
    Sort all individuals in the population according to its fitness.
    Divide the whole population into subpopulation 1 and subpopulation 2.
    for $i = 1$ **to** NP_1 (all butterflies in the subpopulation 1) **do**
        Generate new individuals in subpopulation 1 by using migration operator
    end for
    for $j = 1$ **to** NP_2 (all butterflies in the subpopulation 2) **do**
        if $t < MaxGen \cdot 0.5$ **then**
            Generate new individuals in subpopulation 2 by using adjusted MBO-FS operator
        else
            Generate new individuals in subpopulation 2 by using adjusting operator
        end if
    end for
    Merge newly generated subpopulation 1 and subpopulation 2 into one whole population.
    Evaluate the population according to the newly updated positions.
    Increase the iteration counter t by one.
end while
return Best individual in the whole population

Each iteration starts with the evaluation of objective function for all solutions, and each pair of solutions is being compared. After that a random weight is generated with the sum equal to 1, so that the combined best solution g_*^t can be found. The non-dominated solutions are passed to the next iteration.

Obtained current best solution g_*^t which minimizes a combined objective via the weighted sum, enables the algorithm to perform search process more efficiently [18].

General parameters for all three metaheuristics (MBO, MMBO and MBO-FS) were set as follows: $S_{max} = 1.0$, butterfly adjusting rate (BAR) is set to $5/12$, migration period $(peri)$ is adjusted to 1.2, and the migration ratio r is set to $5/12$. Maximum number of generations (iterations) in one algorithm's execution $(MaxGen)$ is set to 200. Population size parameters were set as: total population number $NP = 50$, subpopulation 1 $NP_1 = 21$ and subpopulation 2 $NP_2 = 29$. With 200 generations and 50 individuals, total number of function evaluations is 10.000. To make the comparative analysis more realistic, we used the same number of function evaluations as in [19]. As already stated in Sect. 3, when the condition $t \geq MaxGen * 0.85$ is satisfied, then we replaced worst solutions in subpopulation 1 and subpopulation 2 with newly created hybrid solutions using Eqs. (12) and (13). Parameters that are specific to MBO-FS metaheuristics were set as follows: initial value of randomization parameter α to 0.5, attractiveness at $r = 0$, β_0 to 0.2, and light absorption coefficient γ to 1.0.

We tested our approaches on four well-known multi-objective optimization benchmarks $ZDT1$, $ZDT2$, $ZDT3$ and $ZDT6$. Two performance metrics were taken into consideration: convergence metric (γ) [20], and diversity metric (Δ) [20].

Table 1. Comparative analysis between MBO, MMBO and MBO-FA

Algorithm	MBO	MMBO	MBO-FS
Function	*ZDT1*		
Convergence metric (γ)			
Average	0.0185	0.0092	**0.0069**
Median	0.0173	0.0081	**0.0050**
Best	0.0063	0.0049	**0.0038**
Worst	1.0642	**0.0165**	0.0176
Std	0.0053	0.0004	**0.0003**
Diversity metric (Δ)			
Average	0.6439	0.6405	**0.6379**
Median	0.6593	**0.6385**	0.6462
Best	0.6335	0.6229	**0.6113**
Worst	0.7032	0.6978	**0.6920**
Std	**0.0462**	0.0009	**0.0005**
Function	*ZDT2*		
Convergence metric (γ)			
Average	0.6998	0.6711	**0.6486**
Median	0.0029	**0.0019**	0.0031
Best	0.0021	**0.0007**	0.0011
Worst	0.0059	0.0063	**0.0045**
Std	**0.0665**	0.9930	0.9890
Diversity metric (Δ)			
Average	0.6678	0.6539	**0.6489**
Median	0.7681	0.7005	**0.6449**
Best	0.7637	0.5698	**0.5290**
Worst	0.7385	0.7329	**0.7313**
Std	0.0531	0.0452	**0.0432**
Function	*ZDT3*		
Convergence metric (γ)			
Average	**0.0049**	0.0063	0.0051
Median	0.0042	0.0066	**0.0035**
Best	0.0097	0.0043	**0.0029**
Worst	**0.0045**	0.01051	0.0049
Std	0.0005	0.0013	**0.0001**
Diversity metric (Δ)			
Average	0.6736	0.6473	**0.6397**
Median	0.6379	**0.6309**	0.6312
Best	0.6001	0.5923	**0.5896**
Worst	0.7022	0.6983	**0.6635**
Std	0.0422	0.0371	**0.0315**
Function	*ZDT6*		
Convergence metric (γ)			
Average	0.0005	0.0069	**0.0001**
Median	**0.0007**	0.0022	0.0066
Best	0.0009	0.0038	**0.0003**
Worst	0.0062	**0.0038**	0.0039
Std	0.0049	0.0402	**0.0032**
Diversity metric (Δ)			
Average	0.6573	0.6489	**0.6481**
Median	0.6392	**0.6301**	0.6312
Best	0.5009	0.4053	**0.3989**
Worst	**0.7956**	0.9549	0.7918
Std	0.1049	0.1003	**0.0823**

To evaluate to robustness and efficiency of proposed metaheuristics for multi-objective benchmarks, we first conducted comparative analysis between MBO, MMBO and MBO-FS. Obtained results and comparative analysis for $ZDT1$, $ZDT2$, $ZDT3$ and $ZDT6$ benchmarks are shown in Table 1. Best results for each tested function and for each category are marked bold.

From Table 1 it is clear that MBO-FS outperforms original MBO and MMBO metaheuristics. MBO-FS enhances original MBO approach by introducing FA's search equation in early cycles of algorithm's execution by modifying butterfly adjusting operator mechanism. In this way, the convergence speed, as well as average results of the original MBO were significantly improved.

The MMBO metaheuristics tackles with deficiency of the original MBO algorithm in late iterations of algorithm's execution by enhancing the search process around the current best solution in the population. Both algorithms, the MMBO and the MBO-FS show improvements over the original MBO, but the improvement is greater in the case of MBO-FS.

6   Conclusion

In this paper, two improved version of the original MBO algorithm were presented. The modified version, MMBO, enhances exploitation in the late cycles of algorithm's execution, while hybridized version, MBO-FS, improves exploration in the early execution cycles. Algorithms were adapted for solving multi-objective optimization problems. In the experimental section of this paper, a comparative analysis between original MBO, MMBO and MBO-FS was presented. Experimental results show that the MBO-FS obtains on average better results than all other approaches included in comparative analysis.

Acknowledgements. This research is supported by the Ministry of Education, Science and Technological Development of Republic of Serbia, Grant No. III-44006.

References

1. Bonabeau, E., Dorigo, M., Theraulaz, G.: Swarm Intelligence: From Natural to Artificial Systems. Oxford University Press, Oxford (1999)
2. Tan, Y., Zhu, Y.: Fireworks algorithm for optimization. In: Advances in Swarm Intelligence, LNCS, vol. 6145, pp. 355–364 (2010)
3. Wang, G.-G., Deb, S., Coelho, L.D.S.: Elephant herding optimization. In: Proceedings of the 2015 3rd International Symposium on Computational and Business Intelligence (ISCBI), pp. 1–5, December 2015
4. Tuba, E., Tuba, M., Simian, D., Jovanovic, R.: JPEG quantization table optimization by guided fireworks algorithm, vol. 10256, pp. 294–307. Springer International Publishing, Cham (2017)
5. Bacanin, N., Tuba, M.: Fireworks algorithm applied to constrained portfolio optimization problem. In: Proceedings of the 2015 IEEE Congress on Evolutionary Computation (CEC), pp. 1242–1249, May 2015

6. Tuba, E., Tuba, M., Beko, M.: Node localization in ad hoc wireless sensor networks using fireworks algorithm. In: Proceedings of the 5th International Conference on Multimedia Computing and Systems (ICMCS), pp. 223–229, September 2016

7. Tuba, E., Tuba, M., Dolicanin, E.: Adjusted fireworks algorithm applied to retinal image registration. Stud. Inform. Control **26**(1), 33–42 (2017)

8. Tuba, E., Stanimirovic, Z.: Elephant herding optimization algorithm for support vector machine parameters tuning. In: Proceedings of the 2017 International Conference on Electronics, Computers and Artificial Intelligence (ECAI), pp. 1–5, June 2017

9. Alihodzic, A., Tuba, E., Capor-Hrosik, R., Dolicanin, E., Tuba, M.: Unmanned aerial vehicle path planning problem by adjusted elephant herding optimization. In: 25th Telecommunication Forum (TELFOR), pp. 1–4. IEEE (2017)

10. Tuba, E., Alihodzic, A., Tuba, M.: Multilevel image thresholding using elephant herding optimization algorithm. In: Proceedings of 14th International Conference on the Engineering of Modern Electric Systems (EMES), pp. 240–243, June 2017

11. Strumberger, I., Bacanin, N., Beko, M., Tomic, S., Tuba, M.: Static drone placement by elephant herding optimization algorithm. In: Proceedings of the 24th Telecommunications Forum (TELFOR), November 2017

12. Wang, G.-G., Deb, S., Cui, Z.: Monarch butterfly optimization. Neural Comput. Appl. 1–20 (2015)

13. Breed, G.A., Severns, P.M., Edwards, A.M.: Apparent power-law distributions in animal movements can arise from intraspecific interactions. J. Roy. Soc. Interface 12 (2015)

14. Yang, X.-S.: Firefly algorithms for multimodal optimization. In: Stochastic Algorithms: Foundations and Applications, LNCS, vol. 5792, pp. 169–178 (2009)

15. Bacanin, N., Tuba, M.: Firefly algorithm for cardinality constrained mean-variance portfolio optimization problem with entropy diversity constraint. Sci. World J. **2014**, 16 (2014). Special issue Computational Intelligence and Metaheuristic Algorithms with Applications, Article ID 721521

16. Tuba, M., Bacanin, N.: Improved seeker optimization algorithm hybridized with firefly algorithm for constrained optimization problems. Neurocomputing **143**, 197–207 (2014)

17. Yang, X.-S.: Firefly algorithm, stochastic test functions and design optimisation. Int. J. Bio-Inspired Comput. **2**(2), 78–84 (2010)

18. Yang, X.-S.: Multiobjective firefly algorithm for continuous optimization. Eng. Comput. **29**, 175–184 (2012)

19. Ma, L., Hu, K., Zhu, Y., Chen, H.: Cooperative artificial bee colony algorithm for multi-objective RFID network planning. J. Netw. Comput. Appl. **42**, 143–162 (2014)

20. Deb, K.: Running performance metrics for evolutionary multi-objective optimization. In: Proceedings of the 4th Asia-Pacific Conference on Simulated Evolution and Learning (SEAL 2002), pp. 13–20 (2002)

Usage of Textual and Visual Analysis to Automatically Detect Cyberbullying in Online Social Networks

Carlos Silva[(✉)], Ricardo Santos[(✉)] [iD], and Ricardo Barbosa[(✉)] [iD]

CIICESI - Center for Research and Innovation in Business Sciences
and Information Systems, School of Management and Technology,
Polytechnic Institute of Porto, Felgueiras, Portugal
{8120333, rjs, rmb}@estg.ipp.pt

Abstract. The Internet is more and more present in everyone life, crossing all age groups, being possible to access to it from anywhere, through the computer, smartphone or tablet. With this increase in presence, some social hazards begin to manifest in the digital world, like bullying. The digitalization of bullying, or cyberbullying, is a very common practice among young people nowadays, becoming much more present due to their constant online activity, especially through online social networks. This work presents a model proposal that combines the analysis of textual and visual content published across online social networks, with the goal to see it implemented in a system who can be autonomous and intelligent to identify and classify the interactions between the users in this kind of online applications. The proposed model also aims to become a tool that can aid the identification of bullies and victims, through the identification of sociodemographic and cognitive characteristics prominent in each role, contributing to the prevention of cyberbullying situations through the identification of possible bullies and/or victims.

Keywords: Online social networks · Bullying · Textual analysis · Image analysis

1 Introduction

Being connected to the Internet is more frequent than ever in modern society. The Internet is a great asset because it allows us to be connected to any part of the world in a fraction of a second, enabling us to access any information in real time, and it can override other methods that would take us much longer to achieve certain goals. However, among those benefits, it also has its negative part and each user is constantly susceptible to some problems such as virus, phishing, or social hazards.

Present in the generality of homes, there are a lot of devices connected to the Internet and, due to simple user interfaces, are easier to use even by children that initiate their contact to the digital world earlier and earlier. Even without knowing to read or write, infants start accessing the Internet, essentially to play games or see some videos for their age but, by growing, they begin to join popular online social networks. Also, as this is a much common process, children keep accessing these technologies in

© Springer Nature Switzerland AG 2020
A. M. Madureira et al. (Eds.): HIS 2018, AISC 923, pp. 459–469, 2020.
https://doi.org/10.1007/978-3-030-14347-3_45

an uncontrolled way by their parents or tutors which may lead them to use it improperly, visualizing content that they should not have access to so early and making them a target for social hazards. The usage of online social networks is also correlated to a set of problems that can be faced while using this kind of applications such as identity theft, lack of privacy, or be the target of insult or verbal abuse situations, a process known as bullying.

Bullying is a complex social dynamic, motivated essentially by differences of the domain, social capital or culture [1]. The desire for dominance, acquisition and maintenance of social capital are the main factors of motivation for the initiation and prolongation of the practice of bullying. For example, the difficulty by the victims of acquiring some good, may impede them from getting a better social, which may lead to disregard from other people. The aggressors, also known as bullies, keep stalking their victims trying to humiliate them and making them feel worse, which leads to negative effects on such people, like anger and depression, that in some cases can lead to tragical endings such as suicides.

Traditional bullying has a faster negative impact on the victim as he or she is contacting with the bully at that moment, face to face and without a chance to escape from the insult. Despite not having a fast effect, cyberbullying is more constant than the traditional form of bullying, as the bully can select his target from a diverse amount of options or can even use multiple ways to reach is objective. Cyberbullying can be described as when the internet, mobile phones or other devices are used to send text or images that can hurt, humiliate or embarrass other people, and it is the most constant version of the traditional bullying [2].

These bullying attacks are more personal if we compare them with some others that we can face by using the Internet. Since the young people are frequently publishing in social networks, creates an opportunity to someone feel anger and start to rip against another user, so this kind of applications are allowing this practice of bullying to be more frequent and constant than ever, and it is a situation that must be kept in mind in order to better teach and educate all those involved, and also to prepare them to face such situations.

The present work aims to propose a model to combat cyberbullying situations manifested in online social networks using two types of classification techniques: textual and visual. Due to the nature of these platforms, most of the activity is composed of textual or visual content (or a combination of both), making them the most common type of content shared across them. In addition to an individual approach to textual or visual content, the proposed work also contemplates the combination of both in order to attend a common practice where photographs contain textual references that are posteriorly digitally included.

This work is structured as follows: Sect. 2 presents the main problems that can be faced while using the Internet and includes an introduction to the topic of cyberbullying; Sect. 3 is intended to explain the main characteristics of a cyberbullying case scenario, presenting textual and visual approaches of analysis. In Sect. 4 we describe our proposal, intended to mitigate this type of problem in online social networks, by combining the analyses of textual and visual characteristics of a publication, aiming to identify the bullies, their victims, and bullying content. This work ends with a conclusion and a reference to future work.

2   The Downside of the Internet

We feel a need to relate to the world daily, either by exchange some emails with our customers or partners, to shopping, check the news, share some moments with our friends, or just to pass the time. The Internet has ceased to be a technology seen only for those who were in search of leisure and fun and already occupies a determining place in the functioning of society.

The importance of its usage has grown over the years, the problems that eventually occur become even more worrying and as it has more and more information available every day. The constant danger associated with Internet usage is not restricted by demographic groups. Children start using the Internet at increasingly younger ages, so it is important to their parents to be alert and to know the risks that their kids may be subject to and to teach them to be more careful. They spend twice as long online as their parents think they do and start using the Internet at the average age of three. The two average hours they spent online are primarily to play games and to watch videos, but when they grow up, the interest in online social networks begins to emerge [3]. By joining the school, they start to need to use the web for research to the accomplishment of some academic works and as they want to communicate with their schoolmates, the need of online social interactions increases.

Recently, with the entrance of the General Data Protection Regulation in European Union [4], organisations are taking more attention to these situations trying to increase security and user protection, especially to prevent privacy breaches and data loss. Some additional actions focusing on user security would be necessary to help avoid facing these situations, even if there are still programs that work as advisors to alert people for some of these cases that are spread all over the Internet. Resort to automatic mechanisms to combat these problems would be advantageous in order to find these threats with elevated metrics of accuracy and precision.

2.1   Cyberbullying

Cyberbullying is always related to traditional bullying, is just a different way to execute this practice recurring to technological devices, which mostly allow the connection to the Internet. This type of technology-based practice mainly manifested through web applications or telephone messages, consists of constant acts of psychological violence perpetrated on one individual by another or by a group of individuals who in most cases know each other personally. These acts do not directly provide physical contact, unless those involved cross in the day-to-day, and tend to be seen almost like a constant chase that is made by the aggressor to the victim, to try to diminish and make her feel worse psychologically. Furthermore, these situations are more frequent among young people, because of the time they spend online and due to their lack of maturity. This practice of violence across online social platforms is becoming more constant, harder to identify, and more conducive to humiliation with a higher number of people reached.

McClowry et al. [5] divide bullying into two types: direct, entails blatant attacks on a targeted young person; indirect, involves communication with others about the targeted individual (spreading harmful rumours). They also refer out that bullying can be physical, verbal or relational (denying friendship) and may involve property damage.

Bullying begins with the aggressor feeling of superiority over other individuals when he or she notices that they are more susceptible or vulnerable to verbal or moral aggression that causes them distress and pain, especially when it occurs in a school environment. In most cases, the aggressors were also victims of these practices in the past, having feelings of anger in themselves and start attacking other individuals to feel better. Aggressive teens have authoritarian personalities combined with a strong need to control or dominate, as well as being victims of abuse at home and releasing all their anger over the weaker classmates as revenge [6]. Often, these attacks are linked to sensitive topics such as race and culture, sexuality, intelligence, physical appearance, and aspects that people cannot change about themselves [2].

Cyberbullying is always carried out in an intentional and repeated way, and through the presence of young people in online social networks, it becomes even more repercussive since many of the other users can see the situation and the impact on the victim increases. However, although most of these users have passive behaviour whenever they encounter content of this kind, some people try to intervene in the discussion, nevertheless, this is not always solved in the best way [7].

Some work has already been done in order to combat this kind of threat, however, an ideal solution is still to be presented. Huang et al. [8] used a Twitter corpus to identify social and textual features to create a composite model to automatically detect cyberbullying. They built graphs related to the social network and derived a set of features, to see the context of "me", "my friends" and the relationship between them.

Zhao et al. [9] present a mechanism for the automatic detection of cyberbullying in social networks. First, they define a list of bad words. Then, based on word embeddings, they extend these linguistic resources, setting different weights to each feature based on the similarity between the word embeddings to form a vector representation using word2vec. The insulting seeds list contains 350 words that indicate insult or negative emotions ("nigga", "fuck", etc.). Using word embeddings, they verify the similarity between words through the weights assigned to each one of them. When the final representation of each Internet message is obtained, the linear classifier SVM is used to detect the existence of cyberbullying.

Lightbody et al. [10] assure that combining sentiment analysis with image processing techniques is considered an appropriate platform for categorization of textual and visual connotations of content. With this, the authors intend to show that it is not only through text that the attack can be made, since the offensive text can be presented as an image, or else, the offence can be tried by editing a photo. They consider that the most relevant pictures will be those that may contain nudity, evidence of editing and analysis of text within the image. The existence of text related to the image helps to determine the risk of content negativity and the associated category.

Contrasting to traditional bullying, it may seem easier for an end to a cyberbullying situation, trying to just block the means used by the attacker or simply to avoid accessing the same types of applications. However, the threat by the attacker to divulge certain private information or intimate photos can arrest the victim and making he or she be like a marionette, which can do the will of the aggressor even if they are acts that the victim does not intend to do. Some work has been done, but an optimal solution is yet to be presented.

3 Text and Image as Tools for Bullying Identification

For the identification of bullying situations in online social networks, is necessary to understand the main types of content that can be available in this type of applications. Analysing the main social networks, it is identified that the content available is essentially composed of: text, which can be used as the main element of a publication, as a description of other content or as a comment to a main element; image, which may relate to a photography or to a graphic art, and may be accompanied by a textual description; video, which, like an image, can present different categories of content and can be also accompanied by a textual description. Regardless of the type of content published, it is practically transversal to all online social platforms the possibility of comment on any post, essentially by text.

A text that is considered bullying has certain characteristics, namely the fact that it always has an abusive, insulting, or threatening tone, and can generate negative repercussions on the people to whom they are headed to (such as depression, anxiety, violence or even suicide). Usually, these texts are composed of rude words and insults to increase the negative emotional load and the idea of hate that is intended to convey. Just by checking the context of the sentence it is possible to see if these words are enough to generate the emotions described above. As such, referral to another person may be essential to classify the phrase as it may be directed to someone, resulting in a need to find the presence of personal or user account names.

It is not only through the presence of a nominal reference that a threat or a directed insult is identified. The usage of certain expressions, known as n-gram words, is quite common and is used together with some keywords (like insults), increases the probability that the text corresponds to a bullying situation. Some examples of these expressions are: "you are", "he is", "they are", that refers to someone, so the adjective that will follow these expressions will be decisive for the final verdict, which was explored by Dinakar et al. [2]. If words of negative content follow it, the text is more likely to be classified as bullying. On the other hand, if the following words convey positive content, the phrase is likely to be classified as non-bullying.

To decipher a text as being bullying or not bullying, it is extremely essential to know a list of insults, either words or expressions, to look for them whenever a new text is analysed. However, these linguistic resources may appear to be camouflaged in certain cases, for different reasons, either by the user own willingness to write in this way, as a slang form or by a way of trying to overcome a system that has a simple function to reduce the presence of insults. It is common in some applications, whenever words from their insults list appear, they are replaced in part by another character, usually asterisks. Thus, these hieroglyphs - words that use symbols to replace casual characters - must be included in the list of insults to look for in the text. Some examples of this hieroglyphs are "b1tch", "5hit" or "@ss", as Huang et al. referred to in their work [9].

The presence of emojis may be an additional indicator for this type of analysis. Very common in the analysis of sentiment of the phrases, these resources may allow identifying an increase of emotions in the phrase. For example, if a phrase contains many symbols of this kind that translate into joy or sympathy, it is more likely that the purpose of its use is to convey some positive emotional charge and vice-versa.

When the text presented in a publication is accompanied by multimedia content, like images or videos, other aspects must be considered. Posting another person photography on social networks can, right away, draw attention. The photography alone may already be enough element to in-break the person who appears there, as the person may feel that he or she does not look well in the way it appears in it, or it may present an action that the person performed that he did not intend to show publicly. If the image is accompanied by a textual description that is considered bullying, it is very likely that it is being used to denigrate someone who is present in the image. In addition, the comments that appear in these publications that contain images may also translate into insults, referring to the presence of a person or the acts that the person is performing.

Some of these images may contain text written on it due to graphical editions [10], which is becoming a common practice in online social networks where you just can publish some update if you upload a photograph, as the case of Instagram. Also, there are a lot of images shared across social networks, known as "memes", that have a purpose to be a joke, and have written text within. However, those "memes" can be adapted to be another way of a cyberbullying attack. That text could be the way the bully found to attack its target, even if the victim is not present in the image. In some cases, other components may be added to the photo to make it look different from the reality, trying to embarrass someone.

Regarding videos, due to their nature of being a sequence of individual frames, the scenario is the same as images. The only difference that can exist in this type of content in relation to the images, is the possible existence of sound elements in the video, and the insult or reference to someone can be made by this way.

Knowing now the main characteristics of the text that constitutes a threat and that relates to a bullying situation, as well as the different types of content that can arise in the social networks, where this study is intended to focus, let us to construct a model that allows to identify and mitigate these situations, trying to design a workflow for the different scenarios that may arise and try to decipher who is involved and which is the victim and which is the aggressor.

4   Proposed Model

Recognising the main characteristics of textual and visual content that can be identified as a bullying situation, we propose a model capable to identify different types of content published in online social networks, aiming towards the identification of bullying situations, as well as those involved. Additionally, this model allows the identification of determinant characteristics that can aid the identification of bullies or victims. This identification relies on the history of bullying occurrences recorded by the model, through the identification of sociodemographic or even cognitive aspects that are a common factor among them.

The present model contemplates the textual analysis, being this the main form of insult, and the visual analysis to the content of images and videos that can be published, something that had not yet been done in previous related works, which tends to just perform textual or visual analysis independently. This model is desired to integrate an intelligent system that can adapt to different situations and that is able to learn autonomously over time, based on its experience, increasing its knowledge base as more analyses are carried out, to be more effective on each execution. Thanks to the vast techniques and technologies of machine learning, the implementation of this model is assured, and several options will allow that it can be constructed to be fast, effective, and reliable as possible.

An online social network content has at least a text or a photo, in the beginning, having afterwards some comments and reactions added to it. So, we should start to look at the existence of some new content and identify its type to proceed with the classification. After extracting the text from the description, comments and from an optical character recognition task over the present images, it should be classified between the classes bullying or not bullying. Then, we should keep analysing the images, when they are available, to look for the human presence, trying to identify some possible victim in a picture or video. If the insult is present in the description or comments, by checking who wrote that and who publish the content, we can easily identify which person is the victim and which one is the aggressor, so we can implement some measures for them to try to make them away from this situation in the future.

An oversimplification of this model application can be explained as follows: (1) an online social publication is set as input; (2) the model analyses it and classify its content between the classes bullying and non-bullying; (3) return the output of this classification.

4.1   Main Use Cases

As previously mentioned, content considered essential for detecting the practice of cyberbullying is arranged on the form of text or image, as such, these are a vital part in the definition of the main use cases for this model.

Whenever a post or comment is made, it is possible that they are being used to attack, offend, harass or threaten someone, so we are facing a potential bullying situation if it is included in one of the following cases:

Generic Scenarios:

- The published content is a text. It has characteristics of bullying and aims to attack, offend, harass or threaten someone;
- The published content is a photo or video. No person is identified in the images. The present description, if any, contains bullying characteristics;
- In any of the previous points, if there are comments from other users, and the content of these is composed of characteristics of bullying.

Scenarios in Which the Victim is the One Who Made the Publication:

– The published content is a photo or video, and the person that posted is identified in the images. There are comments from other users, and their content is made up of bullying features.

Scenarios Where the Harasser is the One Who Made the Publication:

– Published content is a photo or video, a person is identified in the images, but this is not the same as the one who wrote the post. The present description, if any, is made up of characteristics of bullying;
– Published content is a photo or video, and it has, at least, text displayed on it, and it is made up of characteristics of bullying.

To have a system capable of responding to the situations that these cases refer to, it will be necessary to analyse and classify text, recognize text inside images, detect human presence and perform facial recognition tasks. There are several libraries, many of them open-source, that allow the implementation of this type of mechanisms in an environment that is intelligent and capable of learning autonomously.

The analyses that are made during execution can be separated into distinct modules to facilitate any changes that are intended to be made, either for adding new functionalities or for adapting the model to other problems. For its proper functioning, it is a requirement to collect different texts that are classified for each one of the categories, so the system has a starting point to make its next comparisons. In addition, a prior configuration of the system will be necessary to introduce new photos or to indicate where they can be collected, so the system can be trained to recognize the face of each person. It is also important to remember that some pictures containing textual elements need to be collected so the system can be trained in different ways to recognize and translate that content to plain text.

4.2 Model Workflow

Considering the main use cases presented in the previous point, and always focusing on social networks, where there is an identification account for each user, the main workflow of the model should be as follows:

1. **Verify the existence of a new publication:** when a new publication appears, it should be analysed and classified;
2. **Check content type:** it is necessary to verify if the new publication contains text for analysis to classify it as being susceptible to bullying or not, and to complement the analysis verifying if there is any multimedia content;
3. **Classify found text:** either a description or a comment, it must be analysed and classified to know if this text corresponds to a bullying situation;
4. **Recognize text inside images:** if some text is detected inside the image content, it must be recognized to be analysed the same way as the description and comments;

5. **Search for human presence in multimedia content:** If there is any multimedia content attached to the text, such as an image or a video, objects must be detected to find people;
6. **Recognize the person present in the content:** if the human presence is detected in the multimedia content, it is necessary to perform a facial recognition task, in order to verify if the present person refers to the person who made the publication or not;
7. **Generate output:** after the analysis and classification of the publication, the output must be generated with the probability that it belongs to each of the classes, and then other actions that are ideal for the treatment of this information can be applied;

The final classification will be generated based on a probabilistic value for each of the classes and a situation should be considered as bullying if that value corresponds to a minimum percentage of 80%, giving a margin to the system for some error in the analysis process. However, this value should not be a constant limit for all situations, considering that it is intended to verify the recidivism of the person to whom the analysed content belongs, in the practice of bullying. If this person already has a history with some classifications such as bullying, it is also more likely to confirm the new occurrence of this practice, so the value to consider in these cases should start at a lower value. Similarly, if a person has never been involved in such a situation, and this interaction is classified as such, there is a possibility of being a false positive, for example, the phrase in question may correspond to irony or can just be a joke.

Figure 1 allows us to follow the steps of the workflow, which allows us to identify when it is a situation that should be classified as bullying as well as to decipher (when possible) who may be the victim and who may be the aggressor by what is presented in the analysed publication. We always should start with content type check, in order to follow the right tasks. If the content type only corresponded to text, it will only be needed to analyse it and classify based on itself and in the classifications history of the person who wrote it. Otherwise, if there is multimedia content, we will deepen our analysis. The image and video content will follow the same workflow, intending that the videos will be converted into images, frame by frame, as we will not perform sound recognition tasks. The first thing to do in these cases is looking to a description or comments presence. If there is any textual content, either in description, comments or in image optical character recognition, we advance with its textual analysis. If some of this is classified as bullying, we proceed with image analysis, to try to identify the victim and the bully. First, we look for human presence in the content, and if someone is found, we advance to facial recognition tasks. For facial recognition, the same way the text classifier was already trained, it needs to be trained with faces of the people that would be in this real-time analysis. If we can recognize a person who is not who published the content, and as the previously classified texts were bullying, we can consider that the victim is in the image and the bully is the one who shared it, unless the person identified wrote some bullying comment in that publication, and then the roles changes. If we identify a person in the images and it is the same who published it, and it has some comment from another person which was classified as bullying, the victim is again in the image and the bully wrote an inappropriate comment. For other cases, where there is no text classified as bullying, the final classification should indicate not bullying.

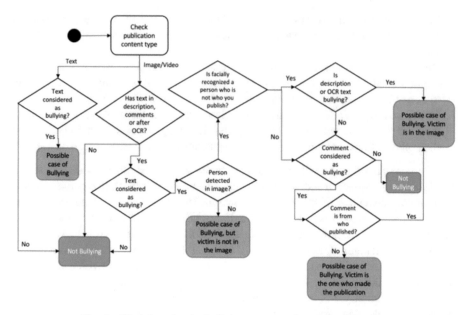

Fig. 1. Workflow for the bullying analysis classification process.

After identifying those involved in the situation, knowing specifically who has had a certain role in the case, some measures can be applied, from performing user account locks, to sending alerts to the tutors, so that they can follow closer these young people. In the future, we may look for patterns in the data trying to identify which sociodemographic and cognitive characteristics a victim or an aggressor traditionally present, in order to predict which types and profiles of people can easily go to attack or be attacked other online.

5 Conclusion and Future Work

Through the constant use of social networks by the younger population, there is a risk that they may incur in bullying practice, trying to threaten somebody else, and that person may feel constantly inferior or persecuted, or even perform tragic actions. After identifying this problem, it was decided to propose a model for the detection of cases of cyberbullying in social networks, trying to connect a strand of analysis of multimedia content to textual analysis, something that until the moment has not been thoroughly investigated by the scientific community. This addition aims to increase the effectiveness of a system to implement this model, intended to be intelligent and autonomous, to learn over time and adapt to different scenarios. The combination of these techniques will certainly increase the effectiveness of the classification of bullying cases that may appear in publications available in various social networks, offering an opportunity to follow some user more closely, and to take actions such as account blocks or the generation of alerts for those responsible. Content that sensitizes the non-practice of these actions can be presented in the feed if it is noticed that the user is involved in these situations. In the future, the data collected in each classification can be used to try to trace the personality of the people who normally advance with these acts and the people who are victims.

This model also has the advantage of being easily adaptable to other common problems in social networks, for example, to detect the presence of fake news or clickbait, being only necessary to make changes in the textual base that the system trying to find in the posts, and in the types of objects that are wanted to be found in the images that joins these texts. As future work and in order to improve the proposed model, is important to include sound analysis, allowing the inclusion of video content as another source for the classification process.

References

1. Evans, C., Smokowski, P.: Theoretical Explanations for Bullying in School: How Ecological Processes Propagate Perpetration and Victimization. University of Kansas, USA. University of North Carolina, USA (2016)
2. Dinakar, K., Reichart, R., Lieberman, H.: Modeling the Detection of Textual Cyberbullying. MIT Media Lab, Massachusetts Institute of Technology, Cambridge, USA (2011)
3. The Telegraph – Children using internet from age of three, study finds. https://www.telegraph.co.uk/technology/internet/10029180/Children-using-internet-from-age-of-three-study-finds.html. Accessed 27 Sept 2018
4. European Union – General Data Protection Regulation. https://eugdpr.org/. Accessed 01 Oct 2018
5. McClowry, R., Miller, M., Mills, G.: Theoretical Explanations for Bullying in School: What Family Physicians Can do to Combat Bullying. Department of Family and Community Medicine, Thomas Jefferson University, Philadelphia, USA (2017)
6. Farrington, D., Baldry, A.: Bullies and delinquents: personal characteristics and parental styles. J. Community Appl. Soc. Psychol. 10(1), 17–31 (2000)
7. Seixas, S., Fernandes, L., Morais, T.: Bullying e cyberbullying em idade escolar. J. Child Adolesc. Psychol. 7, 1–2 (2016)
8. Huang, K., Singh, V., Atrey, P.: Cyber bullying detection using social and textual analysis. ACM, New York (2014)
9. Zhao, R., Zhou, A., Mao, K.: Automatic Detection of Cyberbullying on Social Networks based on Bullying Features. School of Electrical and Electronic Engineering, Nanyang Technological University, Singapore (2016)
10. Lightbody, G., Bond, R., Mulvenna, M., Bi, Y., Mulligan, M.: Investigation into the Automated Detection of Image based Cyberbullying on Social Media Platforms. School of Computing and Mathematics, University of Ulster, Northern Ireland. Carnbane Business Centre, Newry, Northern Ireland (2014)

A Proposal for Avoiding Compensatory Effects While Using ELECTRE TRI with Multiple Evaluators

Helder Gomes Costa[1](✉)(iD), Livia Dias de Oliveira Nepomuceno[2], and Valdecy Pereira[1]

[1] Universidade Federal Fluminense, Niterói, Brazil
heldergc@id.uff.br
[2] Centro Federal de Educao Tecnolgica Celso Suckov da Fonseca, Rio de Janeiro, Brazil

Abstract. This paper describes ELECTRE TRI ME as a proposal for multicriteria modelling of group decisions. It avoids the common inconsistency of inputing compensatory data into outranking models. It also allows evaluators to have their own sets of criteria, scales and weights. To describe the Proposal works, an hypothetical multicriteria-group decision problem was modelled. The results shown and highlight the inconsistency of adopting compensatory mechanisms of aggregating preferences while modelling multicriteria group decision problems.

1 Introduction

ELECTRE TRI is a method reported in [1] and [2] to deal with sorting or classification problems regarding multiple criteria or viewpoints. As the other members of the ELECTRE family of multicriteria decision methods, it takes into account the non-compensatory outranking algorithm proposed in [3] and was designed to work with only one decision unity, which implies that each alternative should receive only one evaluation in each criterion. In other words, ELECTRE TRI was designed to receive evaluations from only one decision unit (or evaluation unit), in each criterion.

On the other hand, there are several situations in which evaluations come from more than one evaluator, as in a group decision situation. In this context, when using ELECTRE in multiple evaluators situations it is common to reduce the problem to a single lone evaluator, by applying a consensus dialogue-based technique or by pre-processing the data by using a technique similar to weighted mean in order to reach a number that means the "consensus" from a voting system - as it occurs in [4] and in Nepouceno et al. [5], among others. In general, this second approach is more commonly used when the number of evaluators is high - which is a barrier for using dialogue-based techniques and the alternative is to use voting systems.

© Springer Nature Switzerland AG 2020
A. M. Madureira et al. (Eds.): HIS 2018, AISC 923, pp. 470–480, 2020.
https://doi.org/10.1007/978-3-030-14347-3_46

A problem that appears in the second approach is the following: weighted mean is a compensatory technique in such way it is in conflict against the non-compensatory principles that are in the core of outranking methods.

Facing the issue described above, this paper presents a modeling that applies ELECTRE TRI ME to an hypothetical and aleatory sorting problem in order. The goal here is to highlight the differences between ELECTRE TRI ME and the commonly used method of adopting weighted mean as an input to ELECTRE TRI.

2 Background on ELECTRE TRI

As reported in [6–9], decision making in a complex environment involves consideration of multiple criteria. The seminal reference of ELECTRE family of muticriteria methods, the ELECTRE I, proposed in [3] as ELECTRE, is based on non-compensatory outranking principles. As reported in [10] and also in [11], to well-establish the core difference between a compensatory and a non-compensatory approach, one should consider an analogy with a volleyball match. So, taking into account a volleyball game in which team A wins team B by 25 to 5 in the first set; but loses all the three following sets to team B by 25 to 20. In this situation, one could follow one of the procedures bellow to identify the winner of the match:

(a) A compensatory procedure: use the sum of the points gained by the team in each set. In this case, A would be the winner by 85×80 points.
(b) A non-compensatory procedures: use the number of game sets that each team wins. In this case, team B would be the winner by 3×1 sets. This procedure, which is actually adopted in volleyball matches, could be classified as an outranking approach.

The main principle of outranking is very similar to the second procedure described above, if one considers each game set as a criterion.

[1] and [2] describe3d ELECTRE TRI as an outranking-based method that focuses in solving sorting problems under a multiple criteria perspective. The sorting problem consists in assigning an alternative into a ranked category, as shown in Fig. 1.

As one can see in Fig. 2, the categories in the set $\underline{C}$ are delimited by a set of profiles or borders $\underline{B}$. The categories are ranked from worst (C_1) to best (C_{h+1}). Observe that a generic profile b_h is both the superior limit of C_h, and the inferior limit of C_{h+1}.

In other words, ELECTRE TRI is designed to approach sorting problems, in which alternatives of a set $\underline{A} = (a_1, a_2, a_3, ..., a_m)$ are allocated into a set of categories $\underline{C} = (C_1, C_2, C_3, ..., C_{h+1})$, considering the performance of the elements in $\underline{A}$ under the set or family of criteria $\underline{F} = (k_1, k_2, k_3, ..., k_j, k_{j+1}, ..., k_n)$. The classes in $\underline{C}$ are delimited by a set $\underline{B} = (b_1, b_2, ..., b_h)$ of profiles. The matrix of grades or performances is $G = G(A) = ((g_1(a_1), ..., g_n(a_1)), (g_1(a_2), .., g_n(a_2)), ..., (g_1(a_m), .., g_n(a_m)))$.

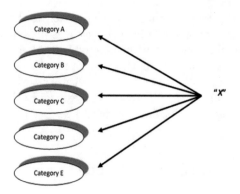

Fig. 1. The sorting problem

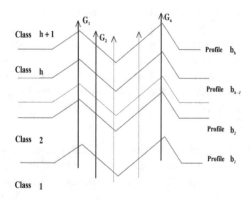

Fig. 2. Class of the ELECTRE TRI

3   The ELECTRE TRI ME

The following assumptions were made while using ELECTRE TRI ME (the Multiple Evaluators (ME) version of ELECTRE TRI):

(a) Alternatives

$\underline{A} = (a_1, a_2, ..., a_i, ..., a_m)$ is a set composed by m alternatives.

(b) Evaluators

$\underline{E} = (e_1, e_2, ..., e_j, ..., e_n)$ is a set of n evaluators.

(c) Criteria

$\underline{F}_{e1} = (k_{1e1}, k_{2e1}, ..., k_{ve1})$ is the subset composed by the v criteria adopted by the evaluator e_1.

$\underline{F}_{e2} = (k_{1e2}, k_{2e3}, ..., k_{xe2})$ is the subset composed by the x criteria adopted by the evaluator e_2.

...

$\underline{F}_{ej} = (k_{1ej}, k_{2ej}, ..., k_{yej})$ is the subset composed by the y criteria adopted by the evaluator e_j.

$$\cdots$$

$\underline{F}_{en} = (k_{1en}, k_{2en}, ..., k_{zen})$ is the subset composed by the z criteria adopted by the evaluator e_n.

Observe that in the ELECTRE TRI ME, each one evaluator should have his own set of criteria.

(d) Weights of the criteria

$\underline{W}_{e1} = (w_{1e1}, w_{2e1}, ..., w_{ve1})$ is the subset composed by the weights of the v criteria under the perspective of the evaluator e_1. So that, in this vector w_{2e1} means the weight of criterion 2, under the opinion of evaluator e_1.

$\underline{W}_{e2} = (w_{1e2}, w_{2e2}, ..., w_{xe2})$ is the subset composed by the weights of the x criteria under the perspective of the evaluator e_1. So that, in this vector w_{1e2} means the weight of criterion 1 under the opinion of evaluator e_2.

$$\cdots$$

$\underline{W}_{en} = (w_{1en}, w_{2en}, ..., w_{zen})$ is the subset composed by the weights of the z criteria under the perspective of the evaluator e_n. So that, in this vector, w_{2en} means the weight of criterion 2, under the opinion of evaluator e_n.

(e) Performance of the alternatives

$\underline{G}_{e1}(a_1) = (g_{1e1}(a_1), g_{2e1}(a_1), ..., g_{ve1}(a_1))$ is the subset composed by the performance of alternative a_1 under the perspective of the evaluator e_1 and the v criteria set $\underline{F}_{e1}$ adopted by this evaluator. So that, in this vector, $g_{2e1}(a_1)$ means the performance of alternative a_1 under criterion 2, under the viewpoint of evaluator e_1.

$\underline{G}_{e1}(a_i) = (g_{1e1}(a_i), g_{2en1}(a_i), ..., g_{ze1}(a_i))$ is the subset composed by the performance of alternative a_i under the perspective of the evaluator e_1 and his own set of x criteria set $\underline{F}_{e1}$. So that, in this vector, $g_{2e1}(a_i)$ means the performance of alternative a_i under criterion 2, under the viewpoint of evaluator e_1.

$$\cdots$$

$\underline{G}_{ej}(a_i) = (g_{1ej}(a_i), g_{2enj}(a_i), ..., g_{zej}(a_i))$ is the subset composed by the performance of alternative a_i under the perspective of the evaluator e_j and his criteria set $\underline{F}_{ej}$. So that, in this vector, $g_{2ej}(a_i)$ means the performance of alternative a_i under criterion 2, under the viewpoint of evaluator e_j.

$$\cdots$$

$\underline{G}_{en}(a_m) = (g_{1en}(a_m), g_{2en}(a_m), ..., g_{zen}(a_m))$ is the subset composed by the performance of alternative a_m under the perspective of the evaluator e_n and his criteria set $\underline{F}_{e1[n}$. So that, in this vector, $g_{2en}(a_m)$ means the performance of alternative a_m under criterion 2, under the viewpoint of evaluator e_n.

(f) Categories

$\underline{C} = (C_1, C_2, ..., C_{h+1})$ is the set of $h + 1$ categories, which are ranked from the worst (C_1) to the best (C_{h+1}). All the evaluator should have the same number or categories, in such away h has the same value for all the evaluators.

(g) Profiles description

$\underline{B}_{kej} = (b_1, b_2, \cdots, b_i, \cdots, b_h)$ is a set composed by h profiles or boundaries that delimit the categories for a criterion k, under the view point of an evaluator e_j. As the categories are adjacent, b_i is both the superior limit of C_i, and the inferior limit of C_{i+1}.

This proposal differs from the previous ELECTRE-based approaches in two ways:

(a) It allows each evaluator to have its own set of criteria, and is its own scales for evaluation of criteria weights and alternatives performances.
(b) It makes use of a non-compensatory approach to deal with the individual preferences of the evaluators.

To do this, it makes a unique and simple assumption: it assumes the whole criteria set $\underline{F}$ as the union of every subset of criteria from each evaluator. So that:

$$\underline{F} = \underline{F}_{e1} \ U \ \underline{F}_{e2} \ U \ldots \ U \ \underline{F}_{ej} \ U \ldots U \ \underline{F}_{en-1} \ U \ \underline{F}_{en} \tag{1}$$

where

$\underline{F}_{e1}$ is the subset of criteria from evaluator e_1, $\underline{F}e2$ is the subset of criteria that the evaluator e_2 takes into account, and so on.

As a consequence of this assumption, it follows that:

$$\underline{W} = \underline{W}_{e1} \ U \ \underline{W}_{e2} \ U\ldots \ U \ \underline{W}_{ej} \ U \ldots \ U \ \underline{W}_{en-1} \ U \ \underline{W}_{en} \tag{2}$$

$$\underline{B} = \underline{B}_{e1} \ U \ \underline{B}_{e2} \ U\ldots \ U \ \underline{B}_{ej} \ U\ldots U \ \underline{B}_{en-1} \ U \ \underline{B}_{en} \tag{3}$$

$$\underline{G}(a_j) = \underline{G}_{e1}(a_j) \ U \ \underline{G}_{e2}(a_j) \ U \cdots \ U\underline{G}_{en-1}(a_j) \ U \ \underline{G}_{en}(a_j) \tag{4}$$

Where a_j is a generic alternative in the set of alternatives $\underline{A}$.

As one can see, the Eq. 4 above refers to the performance of the alternative a_j alone.

So, if we have a situation in which there are 4 evaluators E_1, E_2, E_3 and E_4 having, respectively, 2, 1, 3 and 2 criteria for evaluating 4 alternatives, the resulting matrix of performance $\underline{G}$ should look as follows:

$$\underline{G} = \begin{bmatrix} & E_1 & & E_2 & E_3 & & & E_4 & \\ & c1_{e1} & c2_{e1} & c1_{e2} & c1_{e3} & c2_3 & c3_{e3} & c1_{e4} & c2_{e4} \\ a_1 & 6 & 8 & 8 & 3 & 5 & 9 & 5 & 3 \\ a_2 & 5 & 7 & 4 & 2 & 3 & 8 & 6 & 9 \\ a_3 & 2 & 8 & 3 & 6 & 59 & 6 & 4 & 6 \\ a_4 & 6 & 5 & 6 & 8 & 4 & 4 & 9 & 1 \end{bmatrix}$$

- c_{1e1}, c_{2e1} are the criteria adopted by E_1, c_{1e2} is the one taken into account by E_2, c_{1e3}, c_{2e3} and c_{3e3}, are those adopted by E_3, and c_{1e4}, c_{2e4} are the criteria that E_4 chose to take into account.

- $\underline{A} = (a_1, a_2, a_3, a_4)$ is the set of alternatives or objects to be evaluated. So, in this sample, the value 7 corresponds to $g_{c2_{e1}}(a_2)$ (the performance of alternative a_2 on criterion $c_{2_{e1}}$, that is evaluated by the evaluator E_2).

After these assumptions, ELECTRE TRI ME performs in the same way that ELECTRE TRI does, using the same procedures to sort the alternatives.

4 Hypothetical Example

In this section an hypothetical example is given in order to highlight some features of the proposal and how it differs from the usual approaches that use ELECTRE TRI for modelling situations with multiple evaluators.

4.1 Data of the Example

Suppose we have a situation with six evaluators ($\underline{E} = (E_1, E_2, E_3, E_4, E_5, E_6)$ that have to classify an alternative $a\varepsilon\underline{A}$ into a set of 5 categories ($\underline{C} = (VeryPositive, Positive, Neutral, Negative, VeryNegative)$).

Without loss of generality, let us suppose that the evaluators uses the same number of criteria and decided to adopt the scales that appear in Table 1, for evaluating the importance or weight of the criteria and the performance of the alternative a_1.

Table 1. Scales for evaluating performance of alternatives and criteria importance

Performance		Weigth	
Verbal	Numeric	Verbal	Numeric
Very positive	2.0	Very high importance	4
Positive	1.0	High importance	3
Neutral	0.0	Medium importance	2
Negative	−1.0	Low importance	1
Very negative	−2.0	Not important	0

According to the the values in Table 1, the values of the boundaries of the categories are $\underline{B} = (-1.5, -0.5, 0.5, 1.5))$, as one can see in Table 2, that shows the categories and their lower and upper boundaries.

Suppose that, each evaluator used a set $\underline{F}$ composed by three criteria ($\underline{F} = (c_{E1}, c_{E2}, c_{E3})$) and that based on the scales shown in Table 1, the evaluators estimated the criteria's weights and the Perfomance of a_1 as shown in Table 3. Notice that, for simplicidty of this example and without loss of generality, the weights were all assumed as 3.

As a whole, based on the data of the example, we have:

Table 2. Scales for evaluating performance of alternatives and criteria e and their-boudaries

Class	Code	Lower boundary	Upper boundary
Very positive	A	1.5	$+\alpha$
Positive	B	0.5	1.5
Neutral	C	0.0	0.5
Negative	D	-0.5	0.0
Very negative	E	$-\alpha$	-0.5

Table 3. Weight of criteria and performance of alternative a"

	Performance of a_1			Weights		
	c_{E1}	c_{E2}	c_{E3}	c_{E1}	c_{E2}	c_{E3}
E_1	2	2	2	3	3	3
E_2	2	2	2	3	3	3
E_3	1	1	1	3	3	3
E_4	2	2	2	3	3	3
E_5	2	2	2	3	3	3
E_6	1	1	1	3	3	3

(a) Alternatives: $\underline{A} = (a_1)$ is a unitary set composed by one alternative.
(b) Evaluators: $\underline{E} = (E_1, E_2, E_3, E_4, E_5, E_6)$ is a set of 6 evaluators.
(c) Criteria:
 - $\underline{F}_{E1} = (c_{1E1}, c_{2E1}, c_{3E1})$ is the subset composed by the criteria adopted by the evaluator E_1.
 - $\underline{F}_{E2} = (c_{1E2}, c_{2E3}, c_{3E2})$ is the subset composed by the x criteria adopted by the evaluator E_2.
 - $\underline{F}_{E2} = (c_{1E3}, c_{2E3}, c_{3E3})$ is the subset composed by the y criteria adopted by the evaluator E_3.
 - $\underline{F}_{E4} = (c_{1E4}, c_{2E4}, c_{3E4})$ is the subset composed by the criteria adopted by the evaluator E_4.
 - $\underline{F}_{E5} = (c_{1E5}, c_{2E5}, c_{3E5})$ is the subset composed by the x criteria adopted by the evaluator E_5.
 - $\underline{F}_{E6} = (c_{1E6}, c_{2E6}, k_{2E6})$ is the subset composed by the y criteria adopted by the evaluator E_6.
 - So, $\underline{F} = \underline{F}_{E1} U \underline{F}_{E2} U \underline{F}_{E3} U \underline{F}_{E4} U \underline{F}_{E5} U \underline{F}_{E6}$ Or $\underline{F} = (c_{1E1}, c_{2E1}, c_{3E1}, c_{1E2}, c_{2E3}, c_{3E2}, c_{1E3}, c_{2E3}, c_{3E3}, c_{1E4}, c_{2E4}, c_{3E4}, c_{1E5}, c_{2E5}, c_{3E5}, c_{1E6}, k_{2E6}, k_{3E6})$
(d) Weights:
 - $\underline{W}_{E1} = (w_{1E1}, w_{2E1}, w_{3E1}) = (3, 3, 3)$
 - $\underline{W}_{E2} = (w_{1E2}, w_{2E2}, w_{3E2})$
 - $\underline{W}_{E3} = (w_{1E3}, w_{2E3}, ..., w_{3E3})$

- $\underline{W}_{E4} = (w_{1E4}, w_{2E4}, w_{3E4}) = (3, 3, 3)$
- $\underline{W}_{E5} = (w_{1E5}, w_{2E4}, w_{3E5})$
- $\underline{W}_{E6} = (w_{1E6}, w_{2E6}, ..., w_{3E6})$
- So, $\underline{W} = \underline{W}_{E1} U \underline{W}_{E2} U \underline{W}_{E3} U \underline{W}_{E4} U \underline{W}_{E5} U \underline{W}_{E6} = (3, 3, 3, 3, 3, 3, 3, 3, 3,$ $3, 3, 3, 3, 3, 3, 3, 3, 3)$

(e) Performance of the alternative a_1:

$\underline{G}_{E1}(a_1) = (g_{1E1}(a_1), g_{2E1}(a_1), g_{3E1}(a_1)) = (2, 2, 2)$ $\underline{G}_{E2}(a_1) = (g_{1E2}(a_1), g_{2E2}(a_1), g_{3E2}(a_1)) = (2, 2, 2)$ $\underline{G}_{E3}(a_1) = (g_{1E3}(a_1), g_{2E3}(a_1),$ $g_{3E3}(a_1)) = (1, 1, 1$ $\underline{G}_{E4}(a_1) = (g_{1E4}(a_1), g_{2E4}(a_1), g_{3E4}(a_1)) = (2, 2,$ $2)$ $\underline{G}_{E5}(a_1) = (g_{1E5}(a_1), g_{2E5}(a_1), g_{3E5}(a_1)) = (2, 2, 2)$ $\underline{G}_{E6}(a_1) = (g_{1E6}(a_1), g_{2E6}(a_1), g_{3E6}(a_1)) = (1, 1, 1)$

(f) Categories:

$\underline{C} = (Very\ positive, Positive, Neutral, Negative, Very\ negative)$ Or $\underline{C} = (A, B, C, D, E)$ is the set of 5 categories, which are ranked from the worst $(Very negative)$ to the best $(Very positive)$. The center of these categories are, respectively 2.0, 1.0, 0.0, −1.0 and −2.0, as it appears in Table 2.

(g) Profiles: $\underline{B} = (b_1, b_2, b_3, b_4) = (-1.5, 0.5, 0.0, 0.5, 1.5)$, as it appears in Table 2.

4.2   Applying ELECTRE TRI ME

From this point, ELECTRE TRI ME performs as the seminal ELECTRE TRI outranking procedure. So when applying the equation for calculating the credibility degree, it results in the values shown in Table 4.

Table 4. Credibility degree from applying ELECTRE TRI ME

	A	B	C	D	E
σ	0.67	1.00	1.00	1.00	1.00

4.3   Applying ELECTRE TRI

Although ELECTRE TRI was proposed for modelling decision situations with only one decision unit or evaluator, it has been applied in situations with group decisions. On the majority of these cases, the weighted mean of the perceptions collected from the data are applied. The consequences of applying the mean to the data of the hypothetical example are:

- Three criteria $\underline{F} = (C_1, C_2, C_3)$
- All the criteria having the same weight $\underline{W} = (3, 3, 3))$
- The alternative a_1 having performance $\underline{G}(a_1) = (1.67, 1.67, 1.67)$

So, when applying the equation for calculating the credibility degree, it results in the values that appears in Table 5.

Applying the same reasoning using the mode of the evaluations as the input data, it results in:

Table 5. Credibility degree from applying ELECTRE TRI and mean as input

	A	B	C	D	E
σ	1.00	1.00	1.00	1.00	1.00

- Three criteria $\underline{F} = (C_1, C_2, C_3)$
- All the criteria having the same weight $\underline{W} = (3, 3, 3))$
- The alternative a_1 having performance $\underline{G}(a_1) = (2.00, 2.00, 2.00)$

As a consequence, the credibility degrees that comes from applying the statistical mode as an input are those shown in Table 6.

Table 6. Credibility degree from applying ELECTRE TRI and mode as input

	A	B	C	D	E
σ	1.00	1.00	1.00	1.00	1.00

4.4   Comparing the Results

Table 7 summarizes the results from adopting mean, mode, and ME values as input to ELECTRE TRI. It can be noted that by using both the mean or the mode as the input data, the credibility degree would appear as if the set of evaluators agree with a total or maximum credibility that a_1 should be classified into category *Very positive*, which is a fake conclusion based on the data shown in Table 7.

Table 7. Sumary of credibility degree

Categories	Credibility degree		
	Mean	Mode	ME
A	1.00	1.00	0.67
B	1.00	1.00	1.00
C	1.00	1.00	1.00
D	1.00	1.00	1.00
E	1.00	1.00	1.00

In other words, once the credibility degree of alternative a_1 is at least *Very positive* (once $\sigma(a_1, A) = 1.00$)), it may induce a misled conclusion that all the evaluators agree that a_1 has a performance *Very positive* in all criteria - which is an error, once the evaluator E_3 evaluated a_1 as *Positive* and not as *Very positive*. This occurs because the compensatory effects of the weighted mean and mode algorithms.

On the other hand, as one can see in the fourth column of Table 7, by inserting the ME principle into ELECTRE TRI the compensatory effect is avoided, and the result is more closer to the fact that not all the evaluators consider3ed a_1 as *Very positive*.

5   Conclusion

This paper analysed the inconsistency of using compensatory inputs in multicriteria group decision problematic. It proposes a easy way to avoid such inconsistences by introducing a singele adaptation in ELECTRE TRI sorting method. More specifically, it proposes a hybridism between ELECTRE TRI and ME principles to approach sorting problems (ELECTRE TRI ME). The proposal, named as ELECTRE TRI ME, also allows each evaluator to has his own criteria, scales and profiles.

In order to show how it works, ELECTER TRI ME was applied to a hypothetical situation and the results were compared to those that come from using the mean and the mode as input in ELECTRE TRI.

As focus for further researches, we suggest to apply the ELECTRE TRI ME, for developing solutions for real evaluation problems, inserting ME principles into MCDA techniques.

Acknowledgments. This study was financed in part by the Coordenação de Aperfeiçoamento de Pessoal de Nivel Superior - Brasil (CAPES)- Finance Code 001, and also by the Conselho Nacional de Desenvolvimento Cientfico e Tecnologico - Brasil (CNPq) grant: 312228/2015-5.

References

1. Mousseau, V., Slowinski, R., Zielniewicz, P.: A user-oriented implementation of the ELECTRE-TRI method integrating preference elicitation support. Comput. Oper. Res. **27**, 757–777 (2000)
2. Mousseau, V., Slowinski, R.: Assignment examples. J. Glob. Optim. **12**, 157–174 (1998)
3. Roy, B.: Classement et choix en presence de points de vue multiples (la methode ELECTRE). La Revue d'Informatique et de Rech. Oprationelle (RIRO) **8**, 57–75 (1968)
4. SantAnna, A.P., Costa, H.G., Nepomuceno, L.D.O., Pereira, V.: A probabilistic approach applied to the classification of courses by multiple evaluators. Pesquisaesqui. Operacional **36**(3), 469–485 (2016)
5. Nepomuceno, L.D.O., Costa, H.G.: Analyzing perceptions about the influence of a master course over the professional skills of its alumni: a multicriteria approach. Pesquisaesqui. Operacional **35**(1), 187–211 (2015)
6. Arrow, K.J.: Social Choice and Individual Values. Wiley, London (1963)
7. Saaty, T.L.: The Analytic Hierarquic Process, 1st edn. RWS Publications, Pittsburg (1980)

8. Zeleny, M.: Multiple Criteria Decision Making, 1st edn. McGraw-Hill, New York (1982)
9. Roy, B., Boyssou, D.: Mthodologie Multicritre d'Aide la Dcision, 1st edn. Economica, Paris (1985)
10. Costa, H. G.: An multicriteria approach to evaluate consumer satisfaction: a contribution to marketing. In: Proceedings of the VIII International Conference on Decision Support Systems (ISDSS 2005), Porto ALegre, RS, Brasil (2005)
11. Costa, H.G., Mansur, A.F.U., Freitas, A.L.P., De Carvalho, R.A.: ELECTRE TRI aplicado a avaliao da satisfao de consumidores. [ELECTRE TRI applied to costumers satisfaction evaluation]. Production 17, 230–245 (2007)

A Hybrid Variable Neighborhood Tabu Search for the Long-Term Car Pooling Problem

Imen Mlayah[1(✉)], Imen Boudali[2(✉)], and Moncef Tagina[1(✉)]

[1] COSMOS Lab, ENSI, University of Manouba, Manouba, Tunisia
imene.mlayah@gmail.com, moncef.tagina@gmail.com
[2] ENIT, University of Tunis ElManar, Tunis, Tunisia
imen.boudali@gmail.com

Abstract. The car pooling is an environmental and economical way to travel in and outside urban areas since it consists in sharing a trip on a private car of a driver with other passengers. Given its benefits in reducing traffic congestion, fuel consumption and carbon emissions, it has aroused the interest of some researchers. In this paper, our interest is focused on the Long-term Car Pooling Problem which consists in defining a set of car pools that are served by each user in turn on different days. Our contribution for this NP-complete problem is to propose a new hybrid metaheuristic method for its solving: Hybrid Variable neighborhood Tabu Search algorithm. A comparative study of the hybrid algorithm with the existing approaches was conducted on the base of several test problems inspired from the Vehicle Routing Problem instances.

1 Introduction

The expanding use of private cars has caused traffic congestion, noise, increasing fuel consumption and loss of time. Public transport systems can greatly reduce traffic congestion effects. However, they are limited to urban areas and cannot provide as much flexibility and comfort as private transportation. Carpooling is a flexible and efficient way to travel as sharing journeys reduces carbon emissions, air pollution and traffic congestion and the need for parking spaces. Given the environmental and sustainable impact of this alternative, the carpooling problem has recently received a growing interest from the community of operational research and artificial intelligence. Two forms of carpooling problem are distinguished: Daily Car Pooling Problem - DCPP and Long Term Car Pooling Problem - LCPP. In the first variant, a number of users declare their availability as a server for picking or bringing back other persons or clients in one day. The clients are assigned to servers by specifying the routes to be driven with respect of car capacity and time windows constraints. The DCPP is commonly considered as a particular case of Dial-a-Ride Problem - DARP [1]. Therefore, the successful approaches for solving the DARP in literature were effectively

© Springer Nature Switzerland AG 2020
A. M. Madureira et al. (Eds.): HIS 2018, AISC 923, pp. 481–490, 2020.
https://doi.org/10.1007/978-3-030-14347-3_47

adapted for the DCPP [2]. In LCPP, each user is acting as a server and a customer for a long term period. The problem consists in defining a set of car pools where each user on different days in turn picks up the remaining pool members. The main objective of this variant is to minimize the number of implied vehicles as well as the total travelled distance by all the users when acting as servers with respect of time windows and car capacity constraints. Different approaches have been proposed in literature for solving the NP-complete LCPP [3]. We find some heuristic methods such as the Saving Functions Based Algorithm [4], the Simulation Based algorithm [5] and Multi-Matching System [6]. In the first method, the authors mainly focused on modelling the problem while the solving phase is performed by a simple matching of different users with the support of the car pool model. This approach provided good performances in saving the travel distances for real applications. When assessing benchmarking problems, the method generates good results for instances with clustered and distributed users. The Simulation Based algorithm [5] uses a divide-and-conquer approach. K-means clustering algorithm [7] is used in the divide stage. It allows classifying objects based on attributes into a number of groups. In Multi-Matching System [6], the authors define several constraints based on geographical distances, ideal gap of departure time and arrival time between each couple of users. The model also considers some additional preferences from users such as smoking habit and gender in order to facilitate the grouping among users. Besides heuristics, some metaheuristics has been proposed for solving LCPP. In [8], the authors proposed an application of an ANTS metaheuristic to Long-term Car Pooling Problem where ants construct complete problem solutions, with the objective of maximizing the size of user pools and minimizing the cost of the travelled paths by the drivers. They presented two different mathematical formulations of the problem, and they used them to derive lower bounds to the cost of an optimal solution of the problem. The local search procedure used in this algorithm is the variable neighborhood local search. Guo et al. [9] have presented a Clustering Ant Colony algorithm - CAC for LCPP. In CAC algorithm, the ant is vested the ability of clustering during its tour and memorizes its clustering experience to direct the search of the future ants in order to provide an effective and efficient method for LTCPP. Thus, the classic Ant Colony Algorithm - ACO algorithm has been transformed into a clustering method. Guo et al. have also presented a multi-agent based self-adaptive genetic algorithm to solve long-term car pooling problem [10]. The system is a combination of multi-agent system and genetic paradigm, which is guided by a hyper-heuristic dynamically adapted by a collective learning process. A Guided Genetic Algorithm has been presented by Guo et al. [11] to solve LCPP. To improve the classic GA, they have introduced a new concept called the constitution information, which is considered as the preference of one client willing to be pooled in the same car pool with another. In guided genetic algorithm - GGA, the constitution information of the better individuals will always be memorized and updated. Then this information will be used for guiding the genetic operators, in order to produce more feasible offspring solutions. With this mechanism, the time required for repairing the infeasible solutions is significantly decreased.

2   Mathematical Formulation of the LCPP

In this section, we provide the mathematical formulation of the Long term Car Pooling Problem. So, we will define the employed notation. Then, we will introduce the decision variables, constraints and the objective function. According to graph perspective, the problem can be stated as follows [3]:

- $G = (V, A)$: a directed graph, where V is the set of nodes, and A is the set of arcs.
- $V = \{0, ..., n\}$ with 0 the node associated to the destination or workplace and the node i corresponds to location of customer i, $0 \leq i \leq n$.
- $A = (i, j)|i, j \in V$: the set of directed weighted arcs (i, j) with a positive travel cost c_{ij} and a travel time t_{ij}.

Each user of the long term car pooling must specifies some parameters:

- T: the maximal extra driving time the user is willing to accept for picking up colleagues, in addition to the time needed to drive directly from his home to workplace.
- e: the acceptable time for leaving home
- r: the acceptable time for arriving at work
- Q: the capacity of user's car.

Note that pools are assumed stable for a period of time and will not change frequently. So, the number of members in a pool must be equal or less than the capacity of the smallest implied car. The LCPP is a multi-objective problem, aiming to minimize the number of vehicles that are travelling to or from the workplace and the total travel cost of all users. However, it is possible to combine the two objectives in a single objective function by using a penalty concept. The LCPP then can be formulated as an integrated program as follows [3]. Let k represents a pool of users and $|k|$ its size. Each user of pool k, on different days, will use his car to pick up the other members and then go to the workplace or destination. Thus, the driver has to find a Hamiltonian path originating from his node and then passing through the other member nodes. Notice that each node must be crossed exactly once and that the path must terminate at the workplace. We denote by:

- $ham(i, k)$: the shortest path starting from $i \in k$ and ending at 0 by connecting all j in pool k.
- Q_k: the minimal capacity of all cars in pool k
- c_{i0}: the cost for a user i driving directly to the workplace from its location.
- p_i: the penalty associated to user i when travelling alone

So, the cost of pool k could be defined by the following statement:

$$cost(k) = \begin{cases} \sum_{i \in k} \frac{cost(ham(i,k))}{|k|} & if \quad |k| > 1, \\ \sum_{i \in k} c_{i0} + p_i & otherwise. \end{cases} \quad (1)$$

Since a solution to the LCPP includes a set of pools K, the total cost is then stated as the sum of $cost(k)$:

$$cost(K) = \sum_{k \in K} cost(k) \quad (2)$$

The problem formulation assumes the following decision variables:

- $x_{ij}^{hk} = \begin{cases} 1 & \text{if arc (i,j) is travelled by a server } h \text{ of a pool } k \\ 0, & \text{otherwise.} \end{cases}$
- $y_{ik} = \begin{cases} 1 & \text{if user } i \text{ is in pool } k \\ 0, & \text{otherwise.} \end{cases}$
- $\xi_{ij} = \begin{cases} 1 & \text{if user } i \text{ is not pooled with any other user} \\ 0, & \text{otherwise.} \end{cases}$

We denote by:

- s_i^h: The pick-up time of user i by server h;
- f^h: The arrival time of server h at the workplace;
- c_{ij}: The travel cost between users i and j;
- t_{ij}: The travel time between users i and j;
- Q_k: The minimum car capacity of pool k;
- T_i: The extra driving time specified by user i;
- e_i: The acceptable time for leaving home of user i;
- r_i: The acceptable time for arriving at work of user i;
- p_i: The penalty for user i when he travels alone;
- K: Index set of all pools;
- U: Index set of all users;
- A: Index set of all arcs.

The objective function can be formulated by Eq. (3):

$$f_{LCPP} = min(\sum_{k \in K} \frac{\sum_{h \in U} \sum_{(i,j) \in A} cost_{ij} x_{ij}^{hk}}{\sum_{i \in U} y_{ik}} + \sum_{i \in U} p_i \xi_i) \quad (3)$$

Subject to:

$$\sum_{j \in U/\{h\}} x_{ij}^{hk} = y_{ik} \qquad i, h \in U, k \in K \quad (4)$$

$$\sum_{j \in U} x_{ji}^{hk} = y_{ik} \qquad i, h \in U, k \in K \quad (5)$$

$$\sum_{j \in U} x_{ij}^{hk} = \sum_{j \in U} x_{ji}^{hk} \qquad i, h \in U, k \in K \quad (6)$$

$$\sum_{k \in K} y_{ik} + \xi_i = 1 \qquad i \in U \tag{7}$$

$$\sum_{(i,j) \in A} x_{ij}^{hk} \leq Q_h \qquad h \in U, k \in K \tag{8}$$

$$\sum_{(i,j) \in A} x_{ij}^{hk} t_{ij} \leq T_h \qquad h \in U, k \in K \tag{9}$$

$$S_i^h \geq e_i \qquad i, h \in U \tag{10}$$

$$S_j^h - S_i^h \geq t_{ij} - M(1 - \sum_{k \in K} x_{ij}^{hk}) \qquad (i,j) \in A, h \in U \tag{11}$$

$$F_i^h \geq S_i^h + t_{i0} - M(1 - \sum_{k \in K} x_{i0}^{hk}) \qquad i, h \in U \tag{12}$$

$$F_i^h \leq r_i + M(1 - \sum_{k \in K} \sum_{j \in U} x_{ij}^{hk}) \qquad i, h \in U \tag{13}$$

Equations (4) and (5) assume that a user i must belongs to only one pool and that if there is a path originated in h going from i to j or j to i Eq. (6) is a continuity constraint. Equation (7) assumes that each user has to be assigned to a pool or to be penalized, while (8) and (9) are car capacity and maximal driving time constraints, respectively. Equations (10) and (11), where M is a big constant, collectively set feasible pick-up times, while (12) and (13) set minimum and maximum values of feasible arrival times, respectively.

3   The Proposed Hybrid VNTS Method

Our proposed approach to deal with the LCPP is an hybridization of two meta-heuristics: the Variable Neighborhood Search - VNS and Tabu Search- TS. The VNS is a powerful trajectory-based metaheuristic that follows more than one trajectory [12]. Its basic idea is to apply a systematic change of neighborhoods within a local search. Here, several neighborhood structures are used instead of a single one, as it is generally the case in many local search implementations. Furthermore, the systematic change of neighborhood is applied during both a descent phase and an exploration phase, allowing to get out of local optima. VNS has been recently applied to the green vehicle routing problem by Affi et al. [13]. In this section, we present the different stages of our method by specifying the integrated heuristics in designing the neighborhoods. The overall algorithm is detailed in Algorithm 1. Three main stages are defined in this approach: initialization phase, shaking phase and local search phase.

3.1   Initialization Phase

The method used to generate the initial solution is based on the sweep approach. Sweep heuristic is the construction heuristic method for the well-known Vehicle Routing Problem. The customers are sorted according to increasing order of

---

Algorithm 1. Hybrid Variable Neighborhood Tabu Search Algorithm

---

Define the set of neighborhood $N_k(k = 1, ..., n)$;
Generate(s_0); /* Generate initial solution */
$s = s_0$;
repeat
  $k = 1$;
  repeat
    Generate solution s' at random from $N_k(s)$; /* Shaking */
    $s'' = $ **TabuSearch** (s'); /* Apply Tabu search to obtain local optimum s'' */
    if $Q(s'') \leq Q(s)$ **then**
      $s = s''$; /* Move */
      $k = 1$; /* Start the next search in the first neighborhood of solution s */
    else
      $k = k + 1$
    end if
  until $k = n$
until stopping criteria is met

---

polar angle. So, to construct the initial solution, a customer is randomly selected and put in a car pool to start the sweep process. The other customers are allocated to the current car pool according to their sorting order, until a constraint violation occurs. This method gives a feasible solution with reasonable quality. The main advantage of sweep approach is that near and far customers are mixed in the same route. This makes the solution more balanced, as there are no extremely good routes and extremely bad routes.

3.2 Shaking Phase

Once the initialization phase is achieved, the optimization process is launched with a shaking phase. The objective of this phase is to disturb the solution so as to provide a good starting point for the local search. The set of neighborhood structures used for shaking is the core of the VNS. The main difficulty at this stage is to find a balance between effectiveness and the chance to get out of local optima. So, we define a set of neighborhood structures $N_k(k = 1, ..., k_{max})$ from which we randomly choose one neighborhood k at each iteration of the process.

3.3 Local Search Phase

The obtained solution from the shaking phase is submitted to the local search phase for optimization. We apply a tabu search method as a local search method to improve the solution. The Tabu Search metaheuristic is a local search algorithm that was proposed by Glover [14]. This metaheuristic is based on neighborhood search which iteratively change the current solution to another one until a stopping criteria is met. In order to avoid visiting of already explored solutions, a systematic use of memory is performed. The tabu list records recently explored solutions. Thus, the search process is diversified to unexplored region of the search space. In Algorithm 2, we give a brief description of the employed Tabu Search. During the iterative scheme of TS, a move to better unexplored neighbouring solution is performed until a stopping criteria is met such as the maximal iteration number.

Algorithm 2. Tabu Search Algorithm

Given a solution s and a neighborhood structure y
Initialise tabu list TL_y of t size
$Best = s$,$counter = 0$, $t = 10$;
while ($couter \leq maxIters$) **do**
   $s' =$ Best in neighborhood y of s;
   $s = s'$
   if $c(s') \leq c(Best)$ **then**
      $Best = s$;
   end if
   update tabu list();
end while
return Best

---

4   The Design of Neighborhoods

A set of neighborhood structures is defined in order to search for the local optimal. The *Variable neighborhood Search* procedure consists of a main loop considering each neighborhood in turns. A local optimum is obtained from the use of each neighborhood, after that the next neighborhood is considered. The procedure ends when no improvement in the current solution is recorded.

- **Swap neighborhood.** It consists in swapping any user i with any user j who can pick up and deliver each of user i's car pool members within his/her maximum driving time. The two selected users i and j are deleted from their original clusters and inserted into the each other's cluster. If the overall cost decreases, then accept the new solution.
- **Gravity center exchange neighborhood.** The operator applies the ejection chain idea. The first step is to sort the car pools in a list as follows: A car pool i is randomly selected to be the first element of the list then another car pool j is added to the list such as its gravity center is the closest to the one of car pool i. The procedure is repeated until all car pools are added to the list. Then, the operator consists in selecting any car pool on the list as the start point and proceeding with the following procedures. Suppose the kth pool of the list is selected, the user who is the farthest from the gravity center of the pool is moved to the $(k + 1)^{th}$ pool of the list. Then, if the $(k + 1)^{th}$ pool violates the car capacity constraint, the same procedure will be applied to it. The chain is cyclic, so that the car capacity constraint is always satisfied. If

the resulting cost of the new solution is lower than the cost of the current one, then we accept the new solution.

- **Divide neighborhood.** The operator consists in dividing any car pool into two non-empty car pools with all possible combinations. If the cost of the two new pools decreases the overall solution cost, then we accept the new solution.
- **Merge neighborhood.** The operator merges two non-full car pools with respect to the car capacity and time constraints. Any non-full car pool i and car pool j, which are able to satisfy the car capacity and time constraints after merging, are combined together by the operator. If the resulting cost is lower than the sum of the costs of Pool 1 and Pool 2, then we accept the new solution.

5   Computational Results

This section reports about the computational results obtained by applying HVNTS to different sets of test problems. HVNTS was implemented in JAVA language, under Eclipse, and all results were obtained running the code on a Windows operating system with Intel Core i2 T6570 2.10 GHz CPU and 3 GB RAM. The LCPP instances were originally derived from VRP hard instances presented in the literature and adapted as DCPP instances [2]. The same data files are used with just considering the elements specific to LCPP. They are classified into the following classes: C, R and M. The first class C (Cluster) involves problems that are clustered into regions contrary to problems of class R (Random) which are randomly dispersed. In our work, we present the experimental study that we carried out in order to assess the effectiveness and the performance of the proposed HVNTS algorithm. We have compared our algorithm with two other algorithms. The first is the Variable neighborhood Search developed in [15] and the second is the Tabu Search algorithm that we develop to deal with the long term Car Pooling problem. We performed this comparative study in order to assess the performance of our hybrid approach and to show its effectiveness in comparison with the basic algorithms of VNS and TS. The experiment consists in performing 30 simulation runs for each problem instance. In the following two Tables 1 and 2 of experimental results, the first column 'Instance' shows the name of the instance, while the column 'Size' shows the size of the problem, the column 'Best' corresponds to the value of the best found solution, the column 'Avg' gives the average solution value of the 30 independent runs and the column 'Max' denotes the value of the worst solution found by the corresponding algorithm. In Tables 1 and 2, we illustrate the obtained results for 6 instances problems in the case of 50 and 100 customers, respectively. For each problem we consider the best cost that we obtained after a number of tests and we compare it with the best solution of VNS algorithm existing in literature [15] and Tabu Search algorithm that we have implemented. Table 1 shows the results of the R and C set instances with the size of 100 clients. The HVNTS outperforms the VNS approach on 3 instances considering the best solution quality of 30 runs.

Table 2 shows the results of the R and C set instances with the size of 50 clients. The HVNTS outperforms the TS approach on all the instances considering the best solution quality of 30 runs.

Table 1. Results of set C instances and R instances for 100 clients

Instance	Size	TS			HVNTS			VNS[15]	
		Best	Avg	Max	Best	Avg	Max	Best	Avg
C101	100	1722.4	1803.2	1913.6	**1577.9**	1777.4	1899	1644.9	1684.6
C102	100	1813.3	1860.8	1928.8	**1717.1**	1906.8	2080.4	1729.2	1753.8
C103	100	1949.7	2036.8	2087.5	1893	2051.1	2259	1545.9	1563.6
R101	100	2149.8	2201.8	2390.8	**2108.1**	2313.4	2633.5	2211.2	2286.6
R102	100	2192.4	2134.4	2301.3	2160.7	2270.5	2552.2	1856.7	1898.7
R103	100	2354.4	2341.5	2524	2348.4	2508.2	2697.3	2288.3	2379.8

Table 2. Results of set C instances and R instances for 50 clients

Instance	Size	TS			HVNTS		
		Best	Avg	Max	Best	Avg	Max
C101	50	749.8	786.1	870.3	**736**	790.2	849.2
C102	50	613.2	692.4	769.7	**575.5**	695.6	852
C103	50	921.6	997.7	1063.2	**920.2**	997.8	1061.1
R101	50	1083	1211.1	1423	**1008.7**	1201.2	1404.9
R102	50	1164.8	1239.5	1339.2	**1133.1**	1220.3	1320
R103	50	1295	1384.1	1454.6	**1285.1**	1401.2	1500.8

6   Conclusions

In this paper, a hybrid Variable neighborhood Search Tabu Search approach is proposed to deal with the Long-term Car Pooling problem. An initialization of the search process is performed by using the Variable neighborhood Search method with a sweep approach. We propose four neighborhood structures: *Swap neighborhood, Gravity center exchange neighborhood, Divide neighborhood* and *Merge neighborhood*. Then, the optimization process is ensured by Tabu Search. The method is afterwards assessed on the based of instances that were inspired from benchmarking problems of the Vehicle Routing Problem. In order to show the efficiency of our method, we compare it to the basic VNS and the classical Tabu Search. The hybrid method gives a promising and competing results in comparison with the existing trajectory-based metaheuristics for solving LCPP with medium and small size instances. As future works, we intend to include other heuristics for guiding the solution construction such as preference mechanisms. In addition, it will be useful to extend our proposed approach to study larger scale benchmarks.

References

1. Molenbruch, Y., Braekers, K., Caris, A.: Typology and literature review for dial-a-ride problems. Ann. Oper. Res. **259**, 295–325 (2017)
2. Baldacci, R., Maniezzo, V., Mingozzi, A.: An exact method for the car pooling problem based on lagrangean column generation. Oper. Res. **52**, 422–439 (2004)
3. Varrentrapp, K., Maniezzo, V., Stützle, T.: The long term car pooling problem on the soundness of the problem formulation and proof of NP - completeness (2002)
4. Ferrari, E., Manzini, R., Pareschi, A., Persona, A., Regattieri, A.: The car pooling problem: heuristic algorithms based on savings functions. J. Adv. Transp. **37**, 243–272 (2003)
5. Correia, G., Viegas, J.M.: A structured simulation-based methodology for carpooling viability assessment. Transp. Res. Board **5662**, (2008)
6. Yan, S., Chen, C.Y., Lin, Y.F.: A model with a heuristic algorithm for solving the long-term many-to-many car pooling problem. IEEE Trans. Intell. Transp. Syst. **12**, 1362–1373 (2011)
7. Macqueen, J.: Some methods for classification and analysis of multivariate observations. In: Proceedings of the Fifth Berkeley Symposium on Mathematical Statistics and Probability, Volume 1: Statistics, pp. 281–297. University of California Press (1967)
8. Maniezzo, V., Carbonaro, A., Hildmann, H.: 15 an ANTS heuristic for the long - term car pooling problem. Stud. Fuzziness Soft Comput. **141**, 411–430 (2004)
9. Guo, Y., Goncalves, G.: A clustering ant colony algorithm for the long-term car pooling problem. In: Icsi11.Eisti.Fr, pp. 1–10 (2011)
10. Guo, Y., Goncalves, G., Hsu, T.: A multi-agent based self-adaptive genetic algorithm for the long-term car pooling problem. J. Math. Model. Algorithms **12**, 45–66 (2013)
11. Guo, Y., Goncalves, G., Hsu, T.: A guided genetic algorithm for solving the long-term car pooling problem. In: IEEE SSCI 2011 - Symposium Series on Computational Intelligence - CIPLS 2011: 2011 IEEE Workshop on Computational Intelligence in Production and Logistics Systems, pp. 60–66 (2011)
12. Mladenović, N., Hansen, P.: Variable neighborhood search. Comput. Oper. Res. **24**, 1097–1100 (1997)
13. Affi, M., Derbel, H., Jarboui, B.: Variable neighborhood search algorithm for the green vehicle routing problem. Int. J. Ind. Eng. Comput. **9**, 195–204 (2018)
14. Glover, F.: Heuristics for integer programming using surrogate constraints. Decis. Sci. **8**, 156–166 (1977)
15. Guo, Y.: Metaheuristics for solving large size long-term car pooling problem and an extension. Ph.D. thesis, Université Lille Nord de France (2012)

Early Diagnose of Autism Spectrum Disorder Using Machine Learning Based on Simple Upper Limb Movements

Mohammad Wedyan[1] ⬡, Adel Al-Jumaily[1(✉)] ⬡,
and Alessandro Crippa[2] ⬡

[1] Faculty of Engineering and Information Technology,
University of Technology Sydney, Sydney, NSW, Australia
Adel.Al-Jumaily@uts.edu.au
[2] IRCCS Eugenio Medea, Scientific Institute, Bosisio Parini, Lecco, Italy

Abstract. The importance of early diagnosis of autism that leads to early intervention such thing shall increase the results of treating it. The Autism Spectrum Disorder (ASD) affects the children activities and caused difficulties in interaction, impairments in communication, delayed speech, and weak eye contact. These activities used as the base for ASD diagnosis decision. Children move their upper limb before some of the other activities. Moving upper limb can be based for ASD diagnosis decision for autistic children. Such paper examines diagnosing the ASD that depends on motioning the children's upper-limb aged between two and four years based on executing specific procedures and machine learning. The approach that such study utilized is both (LDA) Linear Discriminant Analysis in order to elicit the features and (SVM) Support Vector Machines for classifying thirty children such study selected fifteen autistic children out of fifteen non-autistic children by testing the collected data that are collected from doing an easy task. The results of such study have accomplished an optimal sortation accuracy of 100% and the average accuracy of 93.8%. Such outcomes provide more proof of simple brachium motioning that might be utilized in sorting poor performance of autistic children precisely.

Keywords: Autistic children · Autism · LDA · SVM · Early autism detection

1 Introduction

Autism disease refers to complex body's neurotic disturbance, that considers as a portion of series of disturbance that is commonly recognized as autism spectrum disorders that manifest in human's life [18, 24–26], such disorders ranged from simple condition to more complicated condition.

The diagnosis of ASD at the beginning of children's life increases the chance for better treatment outcomes. Delayed speech, difficulty in communicating, difficulty in creating social relationships with others are the main symptoms that assist experts to detect autism depending on observation only [26]. The trial of searching on the early indications didn't inhibit any attempt to accomplish the early diagnosing of autism. The

A. M. Madureira et al. (Eds.): HIS 2018, AISC 923, pp. 491–500, 2020.
https://doi.org/10.1007/978-3-030-14347-3_48

study of child motion like walking and brachium considers one of the autism marks [16]. Such determined mechanics marks are valuable since it reflects the early marks of the autistic child even before the infants utter any word. Moreover, it can be measured [6, 16].

2   Related Work

There are a few research that investigated the relationship between the upper limb movements and for children with autism. But they reached an important conclusion, which is there a little upper limb motioning might be utilized to precisely categorize autistic children who suffer from low-functioning.

In [6] Fisher discriminant ratio was used to select the features from data then SVM was also used to classify preschool-elderly participants with autism have been in comparison with 15 typically developing participants who had been matched by using mental age the result show classification accuracy of 84.9%.

Wedyan et al. reported their studies in [24–26] upper limb is valuable in enhancing the medical practice of ASD diagnosing. Consequently, such a thing might enhance the diagnosis with the computer assistance. They used LDA as eliciting features, and SVM as a classification method. Furthermore, such outcomes provide a view on practicable motor autism marks.

In this study [16] classifying approach SVM to upper limb movements was used. To recognize special features in "reach-and-throw" motion. Ten autistic kids with pre-school age kids and ten non-autistic kids doing the same tasks were examined. The SVM approach shows that be able to differentiate the two collections: the result was achieved 100% and 92.5% with a soft margin algorithm and with an additional conservative one, respectively. These outcomes have acquired with a "radial basis function kernel," and they suggested that a "non-linear analysis" is perhaps wanted.

3   Methods

This part of the paper describes steps for the achievement of this study. These steps are collect motions measurements, describe the task, features elicitation, classification, and evaluation. The study concludes that the two children groups executed the task in different performance. Such classification was implemented by utilizing SVM. The overall amount of mechanics data was (i.e., 17 criteria) LDA was implemented to elicit features from raw mechanics information. After that the extracted features (EF) were divided into two categories, the first category was a testing group that contains only one section of the original EF, while the training group contains the remaining EF groups that are executed by utilizing machine learning modes, SVM.

3.1   Participants

Such study compared fifteen autistic children aged between two and four with fifteen non-autistic children (TG). IRCCS Eugenio Medea- a scientific firm that is specialized for research by utilizing "Griffiths Mental Development Scales"- estimated their both IQ and their intellectual age [9]. The participants with low-functioning as a standard clinical showed a little result on the measurement of Griffiths regarding the children - ranged between one and two years- academic weakness [3]. Indeed, all children have an ordinary or corrected-to-ordinary vision.

The children in the ASD class were recruited at IRCCS Eugenio Medea - Scientific Institute for research over an eighteen month. Every child in the medical center was earlier diagnosed as stated of the standard that reported in the "Diagnostic and Statistical Manual of Mental Disorders-IV-TR" [1] with the assistant of the practitioner expertise in children neuropsychiatry by applying ASD.

3.2   Collect Motions Measurements

The approach that was used for recording the motion information is an optoelectronic approach. 3D motion information was recorded via 8 infrared movement cameras that are positioned in the four places in front of the children, tiny passive markers were placed on the ulnar and radial portions of the children's campus and to the back of the hand particularly on the fourth and fifth of hands bones. Moreover, around four markers were placed on the box border under the target area and two on the ball. The software was utilized for calculating the motion apportionment and parameters approximation [6].

3.3   Task

The kids sat on a variable height seat that was adjusted to the body of sharers in front of a desk. The supervisor sat on the desk fronting portion. Moreover, the parents have existed. The overall estimations that started with the sharers' hands remaining at a group position that far from the ball around 20 cm. The mission of such study concentrated on holding the small rubber ball which was placed for assistance. The children recognized the movement before dropping such ball inside the opened box. Such box was inserted within a clear square shaped box its size was sufficiently large in a manner that doesn't require soft movements. It has been made ten trials for each child, such trials were distributed as five consecutive trials on the left portion and five consecutive trials on the other side. The order of a group of experiments was equally distributed among children, the tester carried out the first mission because it clarifies the mission's demand (i.e. holding the ball and dropping it on the box). If the following marks did not appear, such as oral marks, trial training, and the portion that differs from person to person that has been delivered to children before recording to assert the child's realization of the trial. The children were permitted to take a break for resting during the experiment. The experiment's task was both simple and interesting enough to guarantee to obtain the motivation and commitments for the whole groups of children.

3.4   Feature Extraction

In this paper, LDA was utilized in such paper to elicit features. LDA is a feature extraction method is usually applied to raw information. The manner's significance is to remove the excessive and unrelated specifications. Therefore, such classification of the modern examples will be more precise [11]. Moreover, when the group of features enhances the computational value, it will also increase to settle such matter it is vital to search for a manner to decrease the group of features. Features eliciting manner generates modern variables as features for reducing the dimensions of the determined raw information or features. The merits of specification eliciting are represented as follows higher distinguishing force and control that considers more suitable when it is not subjected to supervision. Furthermore, the lack of explained information and transformation might cost a lot [10].

LDA considers one of the well-known methods in extracting the features that might be utilized for both supervised and unsupervised learning [28]. The merits of LDA consider as a resolve that might be obtained by resolving a generalized "eigenvalue" technique. Thus, LDA permits for a rapid and massive transformation of information samples [7]. Their procedures are used for classification and decreasing the dimensions of information [2]. Such study utilized LDA in order to generate features by decreasing the dimensions of information.

Such study utilized LDA in order to generate specifications by decreasing the dimensions of information. The method of LDA process is represented in exploring "the projection hyperplane" that minimizes the class variations and maximizes the gap among the dual collections [17]. Such hyperplane can be utilized for various goals, such as categorization, dimension decreasing, and for exploring the significance of determined features [28].

LDA equation is [4]

$$DS = W_1C_1 + W_2C_2 + \ldots + W_nC_n \tag{1}$$

Where DS = discriminant score, W = discriminant weights, and C = independent variables.

The previous study [6] employed Fisher's discriminant ratio FDR, FDR is a successful feature selection technique [23] which selected seven features from the 17 features the selected features were, the overall period, the triangle carpus angle, the total of units' motion, the time of peak dragging, peak rev, time of peak speed, and peak speed.

By maximizing FDR ratio, the ratio is between class variance to within class variance for the training group we can obtain LDA. Using LDA for the two classes, the data will be transformed to one-dimensional subspace, and for multiclass (C) then the reduction dimensionality is $(C - 1)$ [14].

3.5   Support Vector Machine Model (SVM)

It considers a learning pattern that designates the stamp to things. Therefore, it is known as learning by practicing instances. It considers a supervised learning that employs for both regression and classification problem [5, 15]. SVM work principle relies on analyzing a statistically training set and finds suitable function from a group of function; the best function is that minimizes a determined risk (experiential risk). The risk is contingent on the complication of the group of functions chosen and on the example set (training set) [19]. It is a well-known classification process that utilized for obtaining the great learning precision. Thus, such paper utilizes SVM as a categorized process the information concerning the SVM manner is thoroughly explained in [21]. This work applied the SVM with a linear kernel and selected the SVM because; it is usually the most common manner to execute the classification of complex pattern classification tools.

3.6 Classification Algorithm Evaluation

In our study, we deal with binary classification data which is the most common classification task and means that the object is to be categorized in only 1 class out of 2 incoherent classes [20]. Therefore, the are many different classification evaluations and they are considered fundamental in measuring the quality of classification and learning algorithm [8]. In this paper, the cross-validation approach was used to confirm the progress of the classification performance. By and large, CV approaches depending on dividing the whole information source into binary integral sub-groups: the first group (learning group) and the second group (testing group). However, the first group consists of a series of features linked to stamp; it is utilized to execute the learning the algorithm. The second group is not linked to a stamp and utilized to execute the validity of the algorithm, by applying various segments of the information, various times of CV might be implemented.

One of the specified instances of CV is Leave-one-out cross-validation, and it is usually applied to test the validation in research because it is a neutral assessment of the probability of flaw. LOO strategy relies on the principle of breaking the whole into N sub-collections, One of them is utilized for experimenting, while the remaining group of (N − 1) is utilized for learning of the algorithm. Thus, In each round, N − 1 sub-collections is used learning group and the remaining item is utilized for testing and so on. Therefore, it is applicable that the group of articles is applied the number N [6, 12].

In binary classification data the Sensitivity (Recall), Specificity, and Accuracy are three of the most measurements that used to inspect the correctness of a classifier, which relies on four counts. Correctly categorized and from to the (correct positives) class, correctly classified and not belong to the class (correct negatives), incorrectly classified as well as belong to the class (false positives), and incorrectly classified and not belong to the class (false negatives) [20]. These measurements were computed to measure the accomplishment of the classifier.

Sensitivity = true positives/(correct positives + wrong negatives)

Specificity = true negatives/(wrong positives + true negatives)

Accuracy = (correct   positives + correct   negatives)/(correct   positives + correct negatives + wrong positives + wrong negatives)

In this study, defines as the value of a correctly classified kind in the (ASD) class, particularity defines as the value of correctly categorized kids in the (TG) class, Lastly, the precision of classification that means the average of correctly classified children in both ASD AND TG groups [29].

The whole machine learning steps (extraction and classification) were performed on the Matlab program (version R2017a). In specific, Such paper has precisely utilized specifications of the devices of Matlab 2017 to perform the categorization calculation.

3.7   Evaluation

In this paper, as mentioned before, Sensitivity, Specificity, and Accuracy were used as the main criterias to assess the performance of categorization. These measurements of all rounds were computed by Utilizing CV five-folds. For every fold, the selection of each considers like the checking group, while the remaining folds are used as a learning group. The medium measurement of 5 allocations is applied for every round. Moreover, for obtaining the optimal empirical results the statistical analyses was utilized. As mentioned above, such experiments were established on data collection.

3.8   Results

Such paper applied SVM code [13, 22, 27] that divided the group features randomly into dual classes of both learning and testing group. The SVM calculation was frequently repeated five rounds (N = 5), the first four times by picking various six children as a test group and other twenty-four children as an educating group. The SVM permitted the successful children classification by calculating the overall mean categorization and achieved the following results; average sensitivity is 91.0%, average specificity is 98.1%, and mean accuracy is 93.8%.

However, such an outcome is consistent with various studies. For instance, who has used the study that called a proof of concept in order to define if a little motioning of a hand might be utilized precisely in categorizing the autistic kids who suffer from low-functioning. The percentage of such achieved to eighty-four point nine percent. While the purpose of Wedyan et al. in [26] is to define if a little upper limb motor motion could be utilized in categorizing the children with riskiness for autism infants as well as comparing them with the lowest riskiness for autism. And their research results reached 71.9% accuracy, but results of this study exceed that previous studies results; the accuracy reached 93.8%. Table 1 and Figs. 1, 2, and 3 show the enhancement that was achieved after applied LDA instead of Fisher discriminant ratio (FDR) used the same classification measures Accuracy, Sensitivity, and Specificity, respectively.

Table 1. Compares the results of the proposed work with those of the related studies.

	The algorithms were applied	Mean accuracy	Mean sensitivity	Mean Specificity
[6]	FDR	84.9%	82.2%	89.1%
[26]	LDA	78.5%	66.7%	72.5%
[24]	LDA	75.0%	76.47%	73.33
Proposed work	LDA	93.8%	91.0%	98.1%

The column charts in Figs. 1, 2, and 3. Show the enhancement of Accuracy, Sensitivity, and Specificity after using LDA instead of FDR. The maximum enhancement was in the mean Specificity, which is equal to 9%. Then the mean Accuracy, which reached 8.9%. Finally, the mean Sensitivity enhancement was 8.8% compared to FDR algorithm. Overall, it is clear that LDA algorithm has a clear favorite.

4 Discussion and Future Work

The importance of early diagnosis of autism that leads to early medication such thing shall increase the results of treating. It such kids motion firstly their brachium after that they move to other activities, the diagnosis of ASD depends on such actions. Such current work explored the diagnosing of the ASD that depends on kids, aged between 2–4, little upper-limb motioning. The usage of LDA is for eliciting features from the gathered sub-motions that consider the more distinguishing matter for the ASD comparing with TG. For searching significant group differences that depend on kinematic information. Generating features was applied by utilizing LDA techniques that are utilized for various goals, such as information classification and decreasing data dimensionality, such research used LDA for both generating features and decreasing information dimensionality.

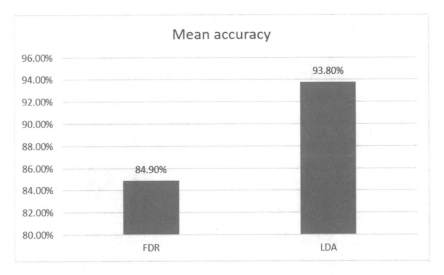

Fig. 1. Shows the mean accuracy enhancement after applied LDA comparing to FDR

In this study, SVM was applied as a classification algorithm. SVM considers a learning algorithm that determines labels for data. Therefore, it is known as learning by practicing instances. The SVM allowed categorizing kids successfully by calculating the overall mean classification and reached the following results; average sensitivity is 91.0%, average specificity is 98.1%, and mean accuracy is 93.8%.

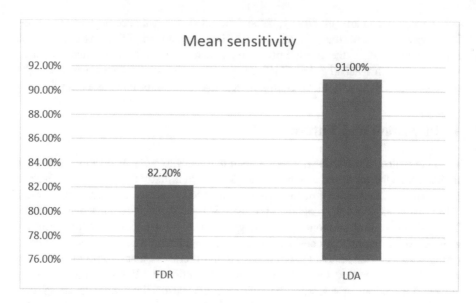

Fig. 2. Shows the mean sensitivity enhancement after applied LDA comparing to FDR.

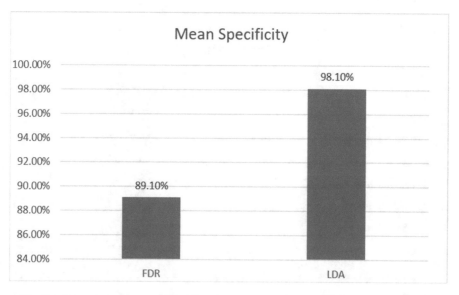

Fig. 3. Shows the mean specificity enhancement after applied LDA comparing to FDR.

Using LDA in such study represents outcomes that surpass the results of the previous study. Therefore, a little brachium motioning might be utilized to precisely categorize autistic children who suffer from low-functioning. The considerable predictive value of LDA eliciting features is somehow valuable in enhancing the medical practice of ASD diagnosing. Consequently, such a thing might enhance the diagnosis with the computer assistance. Furthermore, such outcomes provide a view on practicable motor autism marks. That might be beneficial in determining the clearly defined section of children. Thus, decreasing the medical incoherence within the huge range of behavioral phenotype.

The upcoming work will require prompting a much greater model to manifest bigger possibly enjoyable variations. For instances, six variations in both TG and ASD sharers to define significant inferences. Also, the dire need to examine the manner of sharers' reaction towards remedies in case if we are trying to improve a better remedy for sharers with ASD.

Acknowledgments. Researchers acknowledge for Scientific Institute IRCCS "Eugenio Medea in Italy" for permitting the authors to access this data set.

References

1. American Psychiatric Association: DSM-IV-TR: diagnostic and statistical manual of mental disorders, text revision, vol. 75. American Psychiatric Association, Washington, DC (2000)
2. Balakrishnama, S., Ganapathiraju, A.: Linear discriminant analysis-a brief tutorial, vol. 18. Institute for Signal and information Processing (1998)
3. Barnett, A., Guzzetta, A., Mercuri, E., et al.: Can the Griffiths scales predict neuromotor and perceptual-motor impairment in term infants with neonatal encephalopathy? Arch. Dis. Child. **89**, 637–643 (2004)
4. Bramhandkar, A.J.: Discriminant analysis, applications in finance. J. Appl. Bus. Res. **5**, 37–41 (2011)
5. Burges, C.J.: A tutorial on support vector machines for pattern recognition. Data Min. Knowl. Discov. **2**, 121–167 (1998)
6. Crippa, A., Salvatore, C., Perego, P., et al.: Use of machine learning to identify children with autism and their motor abnormalities. J. Autism Dev. Disord. **45**, 2146–2156 (2015)
7. Dwinnell, W., Sevis, D.: LDA: linear discriminant analysis, vol. 29673. Matlab Central File Exchange (2010)
8. Ferri, C., Hernández-Orallo, J., Modroiu, R.: An experimental comparison of performance measures for classification. Pattern Recogn. Lett. **30**, 27–38 (2009)
9. Griffith, R.: The Ability of Young Children. A Study in Mental Measurement. University of London Press, London (1970)
10. Guyon, I., Elisseeff, A.: An introduction to feature extraction. In: Feature Extraction, pp. 1–25. Springer (2006)
11. Hira, Z.M., Gillies, D.F.: A review of feature selection and feature extraction methods applied on microarray data. Adv. Bioinform. **2015**, 1–13 (2015)
12. Kohavi, R.: A study of cross-validation and bootstrap for accuracy estimation and model selection. In: IJCAI, pp. 1137–1145 (1995)
13. Lauer, F., Guermeur, Y.: MSVMpack: a multi-class support vector machine package. J. Mach. Learn. Res. **12**, 2293–2296 (2011)
14. Perner, P.: Machine Learning and Data Mining in Pattern Recognition. Proceedings of the 10th International Conference, MLDM 2014, St. Petersburg, Russia, 21–24 July 2014. Springer (2014)
15. Noble, W.S.: What is a support vector machine? Nat. Biotechnol. **24**, 1565–1567 (2006)

16. Perego, P., Forti, S., Crippa, A., et al.: Reach and throw movement analysis with support vector machines in early diagnosis of autism. In: 2009 Annual International Conference of the IEEE Engineering in Medicine and Biology Society, EMBC 2009, pp. 2555–2558. IEEE (2009)

17. Prince, S.J., Elder, J.H.: Probabilistic linear discriminant analysis for inferences about identity. In: 2007 IEEE 11th International Conference on Computer Vision, ICCV 2007, pp. 1–8. IEEE (2007)

18. Rojas, E.M., Ramirez, M.R., Moreno, H.B.R., et al.: Autism disorder neurological treatment support through the use of information technology. In: Innovation in Medicine and Healthcare 2016, pp. 123–128. Springer (2016)

19. Shmilovici, A.: Support vector machines. In: Data Mining and Knowledge Discovery Handbook, pp. 231–247. Springer (2009)

20. Sokolova, M., Lapalme, G.: A systematic analysis of performance measures for classification tasks. Inf. Process. Manag. **45**, 427–437 (2009)

21. Tong, S., Koller, D.: Support vector machine active learning with applications to text classification. J. Mach. Learn. Res. **2**, 45–66 (2001)

22. Vedaldi, A.A.: MATLAB wrapper of SVMstruct (2011)

23. Wang, S., Li, D., Wei, Y., et al.: A feature selection method based on fisher's discriminant ratio for text sentiment classification. In: International Conference on Web Information Systems and Mining, pp. 88–97. Springer (2009)

24. Wedyan, M., Al-Jumaily, A.: Early diagnosis autism based on upper limb motor coordination in high risk subjects for autism. In: 2016 IEEE International Symposium on Robotics and Intelligent Sensors (IRIS), pp. 13–18. IEEE (2016)

25. Wedyan, M., Al-Jumaily, A.: An investigation of upper limb motor task based discriminate for high risk autism. In: 2017 12th International Conference on Intelligent Systems and Knowledge Engineering (ISKE), pp. 1–6. IEEE (2017)

26. Wedyan, M., Al-Jumaily, A.: Upper limb motor coordination based early diagnosis in high risk subjects for Autism. In: 2016 IEEE Symposium Series on Computational Intelligence (SSCI), pp. 1–8 (2016)

27. Weston, J., Watkins, C.: Multi-class support vector machines. Citeseer (1998)

28. Xanthopoulos, P., Pardalos, P.M., Trafalis, T.B.: Linear discriminant analysis. In: Robust Data Mining, pp. 27–33. Springer (2013)

29. Zhu, W., Zeng, N., Wang, N.: Sensitivity, specificity, accuracy, associated confidence interval and ROC analysis with practical SAS® implementations. In: NESUG Proceedings: Health Care and Life Sciences, Baltimore, Maryland, pp. 1–9 (2010)

Orientation Sensitive Fuzzy C Means Based Fast Level Set Evolution for Segmentation of Histopathological Images to Detect Skin Cancer

Ammara Masood[1] and Adel Al-Jumaily[2(✉)] ⓘ

[1] University of New South Wales, Sydney, Australia
[2] Faculty of Engineering and Information Technology,
University of Technology Sydney, Sydney, NSW, Australia
Adel.Al-Jumaily@uts.edu.au

Abstract. Malignant Melanoma in one of the most deadly skin cancer. In latest cancer diagnosis, pathologists examine biopsies for analysing cell morphology and tissue distribution for making diagnostic assessments. However, this process is quite subjective and has considerable variability. Automated computational diagnostic tools based on quantitative measures can help the diagnosis done by pathologists. The first and foremost step in automated histopathological image analysis is to properly segment the tissue structures such as nest, irregular distribution and melanocytic cells which indicate some disorder or potential cancer. This paper presents a novel technique for automatic segmentation of histopathological skin images, without user intervention. It is based on an innovative approach for utilizing the concepts of clustering and level set evolution. Firstly, the image is pre-processed to enhance the differentiating structural details. Then a novel orientation sensitive Fuzzy C mean clustering is used to generate the initial coarse segmentation and to calculate the controlling parameters for level set evolution. Later, refined fast level set based algorithm is used to finalize the segmentation process. Experimental analysis on a database of 150 histopathological images display the accuracy of the proposed method for detecting the melanocytic areas of the image, with true detection rate of 87.66% and Dice similarity coefficient of 0.88, when segmentation results are compared with images marked by expert pathologists.

Keywords: Histopathological images · Skin cancer · Diagnosis · Segmentation

1 Introduction

Malignant melanoma is the deadliest forms of cancer. The number of melanoma patients showed rapid increase in Europe, America, and Australia over the last 20 plus years. In US, death rate is 1 every 57 min [1]. Cancer treatment costs more than $3.8 billion (7.2%) of health system costs.

Pathologists use biopsy samples removed from patients and examine them based on their expertise. Development of computational tools for automated diagnosis that

© Springer Nature Switzerland AG 2020
A. M. Madureira et al. (Eds.): HIS 2018, AISC 923, pp. 501–510, 2020.
https://doi.org/10.1007/978-3-030-14347-3_49

operate on quantitative measures can improve diagnosis accuracy. Automatic grading of pathological images has been investigated in various fields, including brain tumor, breast cancer, skin cancer and oral sub-mucous fibrosis detection [2–5].

Challenges faced during segmentation of histopathological skin images include elimination of noise, touching and overlapping cells and inhomogeneous interior of a nucleus that adds to the difficulty of the overall process. Numerous techniques are present in literature for tackling the problem of image segmentation [6–8]. These include various thresholding, clustering, region growing, and edge detecting methods. In this paper, a new method is proposed based on the concepts of orientation sensitive fuzzy C-means clustering and level set evolution to segment the cancerous area in the histopathological images of skin. The method is tested on 150 samples and segmentation results are compared with group of state of the art methods used in literature for the segmentation of histopathological images.

The paper is structured as follows: Sect. 2 gives details of the pre-processing stage and the segmentation method. Section 3 presents experimental results and comparisons. Finally, conclusion is in Sect. 4.

2 Materials and Methods

The clinical database included 150 histopathological images of biopsies of skin cancer patients. The images were obtained from Sydney Melanoma Diagnostic Centre, Royal Prince Alfred Hospital. The images cover most frequent cases of invasive melanoma. For the slides for the images, staining is H and E stain, the magnification is times 20 with 200 and 400, the original section is 8 mm $\times$ 4 mm. The theoretical/mathematical details are explained in the following section.

2.1 Pre-processing

Pre-processing stage is required to reduce the noise that can affect segmentation. Noise arises mainly from the staining process which makes focal area identification difficult, especially, at cellular level. Segmentation results highly depend on good preprocessing.

Proposed Pre-processing method is shown in Fig. 1. Initially, RGB image is converted to Lab colour space. The Luminance component (L channel) is subjected to low pass filtering using median filter. This helps in correction of grey level shading and reducing noise. After median filtering intensity values are mapped from 0 to 255 for contrast enhancement. The L-channel is then combined with chrominance channels (a-Channel and b-channel) thus converting the image back to RGB colour space. Next, RGB image is converted to grayscale image by forming weighted sum of R, G & B components.

Afterwards, anisotropic diffusion is used for smoothing the homogenous parts of the images while sharpening the edges and enhancing important details. Coarser resolution results for larger values of t while lower values of t gives fine resolution. This is represented by solution of the anisotropic diffusion equation given in (1). Where Δ denotes Laplacian operator, ∇ represents the gradient and c(x, y, t) is the diffusion coefficient which controls the rate of diffusion. After the anisotropic diffusion stage, the output is an enhanced image that can improve the segmentation results.

$$\frac{\partial I}{\partial t} = div(c(x, y, t)\nabla I) = \nabla c.\nabla I + c(x, y, t)\Delta I \tag{1}$$

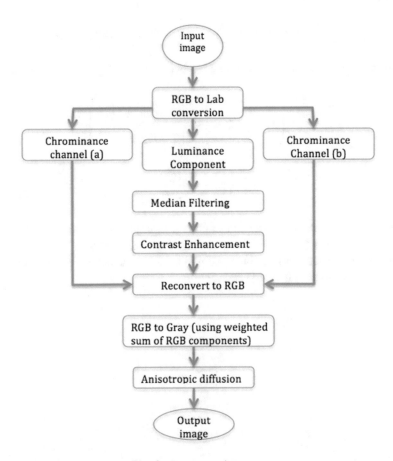

Fig. 1. Pre-processing stage

2.2 Segmentation

The segmentation algorithm proposed here is described in the flowchart in Fig. 2. Initially, Orientation Sensitive Fuzzy C Means clustering is applied on the pixels of the pre-processed image to classify them into clusters. It results in a rough image for the area under consideration. The rough image is then used as initial level set function for the Level Set Evolution method and to estimate its controlling parameters. The steps of our segmentation algorithm are discussed below in detail.

The aim of Fuzzy C mean Clustering (FCM) algorithm is to get an optimal fuzzy c-partition of the image by iteratively evolving fuzzy partition matrix $U = [u_{ij}]$ and computing cluster centres. The fuzzy c-partition of data set is fuzzy partition matrix $U = [u_{ij}]$ with $i = 1, 2, 3....C$ and $j = 1, 2, ...N$, where u_{ij} denotes membership value for ith pixel to jth cluster. The membership functions need to satisfy following conditions (2).

$$\sum_{j=1}^{C} u_{ij} = 1 \, \forall i; \, 0 < \sum_{i=1}^{N} u_{ij} < N \, \forall i; \, 0 \leq u_{ij} \leq 1 \, \forall j, i \tag{2}$$

Given the number of clusters, following two recurrent Eqs. (3, 4) are used.

$$U_{ij} = \left(1 + \sum_{\substack{k=1 \\ k \neq j}}^{C} \left(\frac{\|x_i - \mu_j\|^2}{\|x_j - \mu_j\|^2}\right)^{1/m-1}\right)^{-1} \tag{3}$$

$$\mu_j = \frac{\sum_{i=1}^{N} \mu_{ij}^m \cdot x_i}{\sum_{i=1}^{N} \mu_{ij}^m} \tag{4}$$

where U_{ij} is fuzzy membership of xi to class j, μ_j is jth class centre, C is number of classes taken here as 3, $m \in [1, \infty)$ is a weighting exponent that defines fuzziness of the membership values and can be a real number >1. Calculations reveal that the best choice of m lies in the interval [1.5, 2.5].

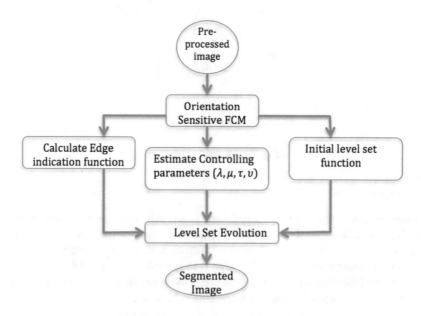

Fig. 2. Proposed segmentation method

Orientation Sensitive Fuzzy C Mean Clustering
FCM clustering algorithm can be used to compute the cluster centers precisely [9]. However, for the case of segmentation of the histopathological images the distance between clusters depends not only on the cluster centers, but also on their orientation, thus the fuzzy membership matrix is updated using (5) and the FCM algorithm becomes sensitive for the orientation. We have referred to this modified FCM algorithm as Orientation Sensitive Fuzzy C Means (OS-FCM).

$$
U_{ij} = \left(1 + \sum_{\substack{k=1 \\ k \neq j}}^{c} \left(\frac{(x_i - \mu_j)^T A_j (x_i - \mu_j)}{(x_i - \mu_k)^T A_j (x_i - \mu_k)} \right)^{1/m-1} \right)^{-1}
\tag{5}
$$

where the matrix A_j which is used for computing the distance. $A_j = V_j^T L_j V_j$ where L_j is a diagonal matrix with inverse of the eigenvalues of co-variance matrix C_j & matrix V_j is composed of corresponding eigenvectors.

OS-FCM algorithm works on iterative optimizing of objective function for weighted similarity measure between pixels of image and that in c-cluster centres. A local extreme of objective function gives optimal clustering for input data. The objective function is represented by (6)

$$
Q = \sum_{i=1}^{C} \sum_{j=1}^{N} (u_{ij})^m \|x_j - \mu_i\|^2
\tag{6}
$$

For optimization, the algorithm tries to minimize objective function Q by updating the cluster centres and membership functions in iterative way. After optimisation pixels in the pre-processed image are labelled to generate the initial coarse segmented image. The results of fuzzy clustering regularize Level set evolution algorithm provided in next section. It will result in a refined segmentation of the cancerous regions present in the image.

Fast Level Set Evolution
The proposed level set methods are established on the bases of partial differential equations (PDEs) and dynamic implicit interfaces. In traditional Level Set (LS) formulation, the curve which is denoted by V is represented implicitly with a Lipschitz function $\emptyset$, by V = {(x, y)| $\emptyset$ (x, y) = 0), and the evolution of curve is taken as the zero level curve at time t for the function $\emptyset$ (t, x, y). The equation of the evolving curve can be represented in terms of level set function $\emptyset$ and is provided below (7).

$$
\frac{\partial \emptyset}{\partial t} + F|\nabla \emptyset| = 0, \quad \emptyset(0, x, y) = \emptyset_0(x, y)
\tag{7}
$$

It is referred to as levels set equation, where F = div $(\nabla \emptyset(x, y)/|\nabla \emptyset(x, y)|)$ is speed function to provide the effecting forces which includes both the internal force from the geometry of the interface and the external force due to the image gradient/artificial momentums.

For making sure that the process of level set evolution stops as close to the optimal solution as possible, the advancing force will to be regularized with the help of an edge indication function denoted here as r. The edge indication function r is given by (8) with $I^*_{OS\_FCM}$ referring for initial segmented image obtained through proposed Orientation Sensitive Fuzzy C Mean clustering.

$$r = 1/\left(1 + \left|\nabla I^*_{OS\_FCM}\right|^2\right) \tag{8}$$

A geometric active contour model is based on level set evolution algorithm [10] and the motion of mean curvature is given by Eqs. (9, 10).

$$\frac{\partial \emptyset}{\partial t} = r|\nabla\emptyset|\left(div\left(\frac{\nabla\emptyset}{|\nabla\emptyset|}\right) + v\right) \tag{9}$$

$$\emptyset(0, x, y) = \emptyset_0(x, y) \, in \, R^2 \tag{10}$$

where $div\left(\frac{\nabla\emptyset}{|\nabla\emptyset|}\right)$ provides the approximation of the mean curvature and v is customable balloon force which is responsible for pushing curve towards the cancerous regions of the image. For the final step of segmentation, the full iteration for levels set evolution is given by (11).

$$\emptyset^{j+1} = \emptyset^j + \tau\left[\xi\left(r, \emptyset^j\right) + \vartheta\xi\left(\emptyset^j\right)\right] \tag{11}$$

where weighting coefficient ϑ of the penalty term $\xi(\emptyset)$ is calculated by taking the ratio of area of on pixels in the image ($I^*_{OS\_FCM}$) to its perimeter pixels, time step τ is taken as $0.2/\vartheta$ so that ($\tau \times \vartheta$) remains always smaller than 0.25 which is required to ensure stable evolution and $\xi(r, \emptyset) = \lambda\delta_\varepsilon(\emptyset)div\left(r\frac{\nabla\emptyset}{|\nabla\emptyset|}\right) + rv\delta_\varepsilon(\emptyset)$ is the term indicating the measurement to attract $\emptyset$ towards the varying boundary. The penalty term is given as $\xi(\emptyset) = \left(\nabla^2\emptyset - \frac{\nabla\emptyset}{|\nabla\emptyset|}\right)$ which forces $\emptyset$ to automatically approach the genuine signed distance function. λ is coefficient of contour length used for smooth control and its value is taken here as $0.1/\vartheta$, which is the best value found through experimental analysis. If the evolution process needs to be accelerated, the value of λ can be increased but that can lead to over-smoothened contours which can be very critical in the case of histopathological images due to the presence of minor details, that cannot be ignored. Thus, such images need more care, as over smoothened images may lose significant details about the boundary of cells, which is important for accurate cancer diagnosis.

3   Results

Segmentation results are presented in Fig. 3 for sample. Results show that our method is able to identify the affected tissue areas efficiently. In Fig. 3(a–d) the segmentation was quite accurate. The method was able to segment most of the malignant cell regions, however, due to the complexity of the images, we noted that in some case like Fig. 3(e) and (f) the segmentation method resulted in under-segmentation and over-segmentation respectively. We are continuing to work on improving the segmentation method to deal with more complex images, however, the results we got so far are convincing to prove the success of proposed method. The automated detection of the affected region of interest can be very helpful during the radiotherapy or other treatment stages.

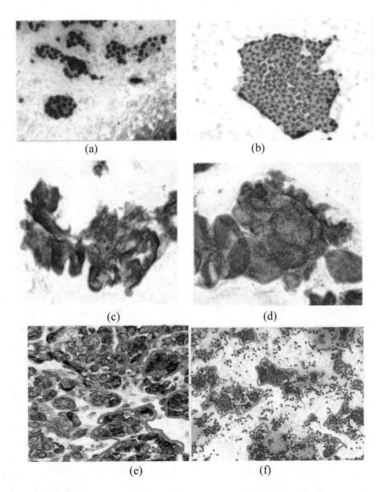

(a)   (b)

(c)   (d)

(e)   (f)

Fig. 3.   Segmentation results of the proposed method

3.1   Comparative Analysis

For evaluating the efficiency of proposed method, its performance is compared with some of the well-known segmentation methods over the same collection of histopathological skin images. The metric used for measurements is based on pixel-by-pixel comparison of pixels enclosed in the segmented result (SR) obtained through the automated methods and the Ground Truth result (GT) marked by the experts. The evaluation parameters used in this research are provided in Table 1.

Table 1. Evaluation parameters

True detection rate (TDR)	$TDR(SR, GT) = \dfrac{\#(SR \cap GT)}{\#(GT)}$	Higher TDR better segmentation method
False positive error (FPE)	$FPE(SR, GT) = \dfrac{\#(SR \cap G\overline{T})}{\#(GT)}$	Lower FPE better segmentation method
False negative error (FNE)	$FNE(SR, GT) = 1 - \dfrac{\#(SR \cap GT)}{\#(GT)}$	Lower FN better segmentation method
Dice similarity coefficient (DSC)	$DSC = \dfrac{2\,TP}{((FP+TP)+(TP+FN))}$ Where true positive (TP), false positive (FP), false negative (FN)	A value of 0 for DSC indicates no overlap between the automated segmented output and actual ground truth; a value of 1 indicates perfect agreement

Table 2 shows the comparative results of the proposed method with 6 methods (codes developed based on respective optimised algorithms) used for the segmentation of histopathological images. First method used is watershed algorithm [10] where the local minima of the gradient of the image are chosen as markers. It has been noticed that it is useful in detecting boundary lines between touching cells but sometimes resulted in over segmentation as found in some other medical imaging studies referred in. To overcome the over segmentation issue a second step involving region merging is used which help in reducing the false positive error to 19.08. However, it is still much higher compared to the proposed method. Second method used in comparative analysis is the gradient flow vector (GVF) [11]. This method was unable to provide good segmentation for images with very small contrast between cancerous and non-cancerous regions. The third method is expectation maximization-based level set which is one of the region-based contour detection methods [12]. This method showed good true detection rate, but false positive rate was high. Similar to GVF, expectation maximization based level set (EM-LS) also showed relatively poor performance for images with low contrast. We also compared the results with some of the popular segmentation methods used in literature for skin images, including FCM Clustering, Discriminative clustering, Region Growing, Spatial Level set and Thresholding [5]. However, none of the methods performed very well as compared to the proposed methods. We optimized all comparative methods to the best possible level based on available literature. It is evident from the results that the proposed method has shown reasonably better performance compared to other methods.

Table 2. Comparative analysis of proposed segmentation method

Method	TDR (%)	FPE (%)	FNE (%)	DSC
Watershed algorithm	82.01	19.08	17.99	0.82
EM-LS	83.42	22.2	16.58	0.81
GVF	75.23	16.21	24.77	0.79
FCM clustering	77.05	22.54	22.95	0.77
Region growing	79.21	21.07	20.79	0.79
Thresholding	76.92	21.42	23.08	0.78
Spatial level set	77.82	19.35	21.06	0.75
Discriminative clustering	77.03	21.34	20.92	0.78
Proposed method	87.66	11.66	12.34	0.88

4   Conclusion

An efficient segmentation method is presented for segmentation of histopathological images of skin cancer. The method combines the benefits of clustering and evolutionary algorithm like level set for providing fine segmentation of cancerous areas. The enhanced fuzzy C mean clustering with orientation sensitivity helps in better approximating the cancerous area. In addition, OS-FCM estimates controlling parameters for level set evolution. This helps the level set evolution to stabilize to the genuine boundary earlier, suppresses boundary leakage and alleviates the need for manual intervention. All these improvements enhanced the segmentation results. Performance comparison has also been carried out with some of the popular methods used so far for histopathological image segmentation and it has been shown that the results of the proposed algorithm seem promising.

References

1. American Cancer Society: Cancer Facts & Figures (2014). http://www.cancer.org/acs/groups/content/@research/documents/webcontent/acspc-042151.pdf. Accessed 5 Jan 2015
2. Sharma, D., Srivastava, S.: Automatically detection of skin cancer by classification of neural network. Int. J. Eng. Tech. Res. **4**(1), 15–18 (2016)
3. Li, B., Zhao, Y., et al.: Melanoma segmentation and classification in clinical images using deep learning. In: 10th International Conference on Machine Learning and Computing, pp. 252–256 (2018)
4. Masood, A., Al-Jumaily, A., Anam, K.: Self-supervised learning model for skin cancer diagnosis. In: 7th Annual International IEEE EMBS Conference on Neural Engineering, pp. 1012–1015. IEEE (2015)

5. Babu, M., Madasu, V.K., Hanmandlu, M., Vasikarla, S.: Histo-pathological image analysis using OS-FCM and level sets. In: IEEE 39th Applied Imagery Pattern Recognition Workshop (AIPR), pp. 1–8 (2010)
6. Silveira, M., et al.: Comparison of segmentation methods for melanoma diagnosis in dermoscopy images. IEEE J. Sele. Top. Sig. Process. **3**(1), 35–45 (2009)
7. Masood, A., Al-Jumaily, A.: Computer aided diagnostic support system for skin cancer: a review of techniques and algorithms. Int. J. Biomed. Imaging **2013**, 1–22 (2013)
8. Hoshyar, A., Al-Jumaily, A., Sulaiman, A.R.: Review on automatic early skin cancer detection. In: International Conference on Computer Science and Service System, pp. 4036–4039 (2011)
9. Masood, A., Al-Jumaily, A.: Fuzzy C mean thresholding based level set for automated segmentation of skin lesions. J. Sig. Inf. Process. **4**(3), 66–71 (2013)
10. Grau, V., Mewes, A.U.J., Alcaniz, M., Kikinis, R., Warfield, S.K.: Improved watershed transform for medical image segmentation using prior information. IEEE Trans. Med. Imaging **23**(4), 447–458 (2004)
11. Mahmoud, M.K.A., Al-Jumaily, A.: Segmentation of skin cancer images based on gradient vector flow (GVF) snake. In: Proceedings of International Conference on Mechatronics and Automation (ICMA), pp. 216–220 (2011)
12. Fatakdawala, H.J., et al.: Expectation maximization-driven geodesic active contour with overlap resolution: application to lymphocyte segmentation on breast cancer histopathology. IEEE Trans. Biomed. Eng. **57**(7), 1676–1689 (2010)

Electrogastrogram Based Medical Applications an Overview and Processing Frame Work

Ahmed Al Taee and Adel Al-Jumaily$^{(\boxtimes)}$ (iD)

University of Technology Sydney, 15 Broadway, Sydney, Australia
Ahmed.A.Ahmed@students.uts.edu.au,
Adel.Al-Jumaily@uts.edu.au

Abstract. Surface Electrogastrogram (sEGG) examination is a noninvasive method based on recording the electrical signals of stomach muscles that can be used for investigation of a stomach wave propagation. sEGG records by placing electrodes on the abdominal skin. The recording of gastric dysrhythmias is an important tool for the clinician when patients have symptoms that suggest gastric dysfunction such as unexplained nausea, bloating, postprandial fullness, and early satiety. On the other hand, these upper gastrointestinal (GI) symptoms are nonspecific, and diseases or disorders of other organ such as esophagus, gallbladder, small bowel, colon, and non-GI diseases also can be considered. The progress of the EGG based application is very slow as EGG still suffers from several limitations. The aim of this study is to review the Electrogastrogram detection, analyzing methods and its application to enter the clinical world with recommendations for future study.

Keywords: EGG · Electrogastrography · Electrogastrogram ·
Gastric slow waves · Medical application · Detection

1 Introduction

A Biomedical signal is a collective of electrical signals acquired from any organ, as the nervous system always controls the muscle activity (contraction/relaxation). Thus, this signal measures electrical currents generated in muscles during contraction/extraction. This Biomedical signal is describable in terms of its frequency, amplitude and phase and it is a function of time [1].

Biomedical instrumentation is widely used to collect the Bio-potential signals such as: Electrocardiogram (ECG) where use to collect the signal from heart; Electroencephalogram (EEG) to collect the signal from outside the brain, Electrocorticogram (ECoG) to collect the signal from the surface of the brain, Electromyogram (EMG) to collect the signal from the skeleton muscle, Electroneurogram (ENG) to collect the signal from the neurons in the central nervous system (brain, spinal cord…etc.), Electroretinogram (ERG) and Electro-oculogram (EOG) to collect the signal from the surface of the cornea and Electrogastrogram (EGG) to collect the signal from the stomach [2].

This paper is studying the Electrogastrogram (EGG) signal, which is a method of recording gastric electrical activity through non-invasive electrodes placed on the

---

The original version of this chapter was revised: Author's name has been corrected from 'Ahmad A. Al-Tae' to 'Ahmed Al Taee'. The correction to this chapter is available at https://doi.org/10.1007/978-3-030-14347-3_57

abdominal skin. The EGG provides information about the gastric myoelectric frequency and the amplitude of the EGG signal in the case of normal and abnormal frequency ranges.

In 1921, Walter Alvarez recorded the first human EGG signal [3]. In this experiment, two electrodes were placed on the abdominal surface of a woman and connected them to a sensitive string galvanometer. A sinusoid signal with a frequency of 3 cycles per minute (0–05 Hz) was then recorded. As he stated in his paper "the abdominal wall was so thin that her gastric peristalsis was easily visible" [3]. Later in the mid of 1970s Electrogastrography gained renewed interest. The first team who applied the spectral analysis technique to extract the information about the frequency of the EGG signal and the time variations of the frequency was Stevens and Worrall [4]. After that in 1980, it was believed that the EGG provides not only information about the frequency of the contractions of the smooth muscles of the stomach but also information about the degree of the contractile activity [5]. Smout and his colleagues were the first team who pointed out that the amplitude of the EGG increases when contractions occur. During the last two decades a fast increase in the development of electrocardiography has taken place, however EGG still suffers from several limitations such as; difficulties in recording and analyzing. Because EGG signal consists of low frequency components, it is usually associated with some interferences caused by other organs placed near the stomach. Also, the lack of databases, which could allow for both comparison and research purposes. In addition, the EGG signal is very vulnerable to motion artefacts which overlap with that of gastric myoelectrical activity [6]. Accordingly, an efficient solution for automated detection, analysis and suppression of motion artefacts in the EGG is required.

The increased knowledge about the relationship among gastric dysrhythmias (delayed gastric emptying, nausea…etc.) and appropriate application of non-invasive EGG can provide more information and insight in understanding the mechanisms that regulate stomach motility and diagnose disease.

This paper attempts to report the state of the art of the pathophysiological background of gastric electrical activity, the recording and processing methodology of the EGG and the possible clinical applications.

2   EGG Signal Properties

The basis for emptying of solids from the stomach is gastric peristaltic contractions. These events begin in the mid to high corps region, develop into a ring around the stomach, and spread down the length of the stomach to the pylorus. The pressure wave resulting from gastric peristalsis pushes the contents of the stomach toward the pyloric sphincter, but a nearly simultaneous contraction of the ring of muscle in the pyloric canal and the terminal antrum ultimately forces much of the food in a retrograde direction, toward the body of the stomach. Sheer forces that develop as a result of this forceful retropulsion cause mechanical disruption of solid particles [9].

In pathophysiology, Gastric myoelectrical activity consists of two components: slow waves or Electrical Control Activity (ECA) and spike potentials or Electrical Response Activity (ERA) (Chen and McCallum) [7]. Both the pattern and timing of gastric slow wave propagation are fundamental components of digestive function. Gastric slow waves are autonomously generated and propagated by specialized pacemaker cells called the Interstitial Cells of Cajal (ICC), were initially identified by Cajal [8]. There are different types of ICC with different functions. Myenteric Interstitial cells of Cajal [ICC-MY] serve as a pacemaker which creates the bioelectrical slow wave potential that leads to contraction of the smooth muscle. Slow waves are spread out from one cell to another on longitudinal layers causing electronic currents in the circular layer. It originates in the proximal stomach and propagates distally toward the pylorus as shown in Fig. 1. The slow wave is ubiquitous and occurs regularly whether the stomach contracts or not. This slow wave determines the maximum frequency, propagation velocity and propagation direction of gastric contractions and it is almost sinusoidal and typically identified by its slow frequency and low amplitude (between 100 and 500 µV), with frequency in human around 3 cycles per minute (cpm). However, a spike potential or (ERA), is superimposed on the gastric slow wave, and a strong lumen occluded contraction occurs. Therefore, stomach muscular contraction is produced by ERA, which can only appear at the top of depolarization of the slow wave [9].

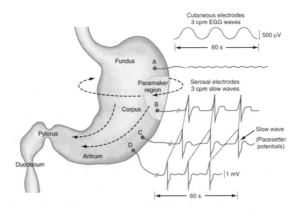

Fig. 1. Gastric pacesetter potentials or slow waves originate from the pacemaker area on the greater curve. Pacesetter potentials travel in a circumferential and aboral direction at a rate of approximately 3 cycles per minute (cpm). The cutaneously recorded electrogastrogram shows 3-cpm wave pattern. The fundus has no rhythmic electrical activity [10].

Abnormalities in the regularity of the gastric slow wave are associated with chronic digestive diseases such as nausea and vomiting, anorexia, gastroparesis, dyspepsia, and gastroesophageal reflux disease. Disturbances of the gastric dysrhythmias can occur in different patterns that include an increase in the dominant frequency of the myoelectrical activity of the stomach from 3 cpm to 4–9 cpm regular activity is defined as tachygastria; a decrease in the dominant frequency of the myoelectrical activity of the stomach from 3 cpm to 1–2 cpm regular activity is defined as bradygastria. To sense the combined signal from many muscle fibers an electrode may be placed on the skin over the muscle.

3   Procedure to Measure

The procedure for a gastric measurement study consists of many variables, including meal composition, patient positioning, instrumentation, frequency of data acquisition, study length and data analysis. EGG activities measured by several electrodes attached on the abdomen. Usually, EGG has to be recorded in a quiet room to minimize interferences of electrical signals that might be detected by the electrodes. Patients should be advised to remain still and quiet whenever possible. In addition, the patient is asked to fast overnight or for at least 4 h prior to the study. Diabetic patients should be studied in the morning, 20–30 min after their normal insulin dosage. A wide variety of drugs can affect gastric emptying and therefore most medication should be discontinued prior to the test evaluation. Smoking could delay gastric emptying and should be avoided.

The patient should be placed in a comfortable position ranging from supine to 45° inclination. The position should be maintained duration the testing time. Typically, the EGG signals are collected for relatively long time (120 to 150 min). There is no optimal length for an EGG examination, however the best practice is divided into three phases [11] as below:

Pre-prandial	Meal	Postprandial
20–40 min	5–15 min	30–150 min

- The first phase (pre-prandial) or no contractions, usually takes no longer than 40 min and it should be before a standardized meal (tested person should be fasting).
- The second phase or intermittent contractions usually takes between 5 and 15 min, during which time the person under test eats a standardized meal. The test meal consumption which often takes the form of light liquid meal (400 ml). (standard depends on the health care center performing the test).
- The third phase (postprandial) or regular rhythmic contractions, usually takes 30–150 min after the meal.

All EGG measuring equipment, should be plugged into a medical grade isolation transformer. Also, to ensure unidirectional current flow from the skin to the amplifiers all the EGG, recording equipment must be individually isolated electrically. Some institutions use a signal generator to calibrate the EGG signal against a test sinusoidal wave of known frequency. The abdominal skin where the electrodes are to be positioned should be thoroughly cleaned to ensure that the impedance between the pair of electrodes is below 10 kΩ. To do so, it is advise that any hair on the abdominal surface is shaved and abraded the skin. Then apply a thin layer of electrode gel for 1 min to penetrate into the skin. Before placing the electrode, the excessive gel must be completely wiped away. Pre-gelled adhesive Ag/AgCl electrocardiographic electrodes are recommended, because of their performance to acquire a cutaneous signal during the period of the study [12].

4   Electrode Locations

The channel number and position are varying depends on the investigators' preferences. Thus, different configurations of EGG electrodes have been provided with different strengths and limitations. Some maps use cutaneous reference points as landmarks for stomach shape and position while alternative setups take into account the actual position and shape of the stomach evaluated by means of imaging diagnostics (X-ray or ultrasound).

The best configurations used by Chen et al. [7]. Where the surface EGG signals are captured by six disposable electrodes: four signals electrodes A1 to A4, reference electrode R and ground electrode U placed on the anterior abdominal wall overlying the patient's stomach as shown in Fig. 2. The main electrode (A3) is located 2 cm above the mid-point between the xiphoid process and the umbilicus. Two more electrodes (electrodes A2 and A1) were located on an upper 45° angle, with an additional electrode (electrode A4) located 4 cm to the right of the central electrode. The common reference electrode was placed at the cross point of two lines, one horizontal-connecting electrode A1 and the other vertical-connecting electrode A3. The ground electrode was placed on the left costal margin.

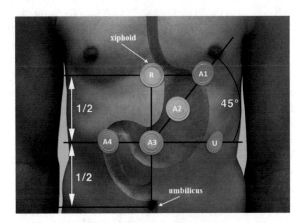

Fig. 2. Electrode location in a human model

5   Data Analysis

Previous studies of EGG signal detection used Fourier transform (or Fast Fourier Transform, FFT) to extract the frequency information from the gastric electrical activity. The main drawback of this method is that Fourier analysis is suitable for stationary signals and EGG is a non-stationary and non-deterministic signal therefore, this method is not suitable with EGG analysis [11]. To extract the information from EGG it is required to get not only information about the frequency of the gastric signal but also information about the time variation of the frequency, therefore, running spectral analysis is required (RSA) [13]. RSA method was is based on FFT and it is calculated as follows: for a given data set of the EGG, a window with a length of W samples is applied, an FFT with the same length is calculated and a sample spectrum is obtained for the first block of data. The sample spectrum at the next time step is obtained in the same way by shifting the window S (where S ≤ W) samples forward. This method is easy to implement. However, its drawback is then it is not suitable of tracking fast frequency changes of the gastric signal. In addition, EGG applications need several minutes' data for each RSA to yield reasonable frequency resolution. However, dysrhythmic events may occur within 1–2 min [14]. As a result, gastric dysrhythmia of brief duration may not be represented well using the RSA method [15, 16].

To avoid that (Chen et al.) [17], developed a modern running spectral analysis technique based on an Adaptive Autoregressive Moving Average (ARMA) model, which gives more satisfactory results, especially in detecting dysrhythmias with short duration in the EGG. It yields higher frequency resolution and more precise information about the frequency variations of the gastric electrical activity. It is especially powerful in detecting dysrhythmic events of the gastric electrical activity with short durations. Adaptive ARMA method provides more satisfactory results than the FFT, higher and narrower peaks indicating distinct frequency components and more precise information about the appearance and disappearance of the tachygastrial event [17].

Later Wavelet Transform (WT) is designed to address the problem of nonstationary signals. The WT can be categorized into continuous (CWT) and discrete wavelet transform (DWT). DWT is often used because a considerable effort and a vast amount of data is necessary to calculate a continuous wavelet of a signal. DWT, has the ability to compute the power spectrum not only at a particular time interval but at a particular time instant thereby resulting in the instantaneous determination of the rhythmic variation. Therefore, the DWT analysis method enables to determine whether the EGG recording was normal or abnormal. The expression of DWT is [18].

$$U = \frac{1}{M} \sum_{n=0}^{M-1} W^2(n)$$

Where W(n) is the window function of length M.

Recently the Neural Network also has been used by many researchers due to flexibility, high performance and suitability for real time application. Many pattern recognition systems in biomedical signal processing use Neural network. Haddab and Laghrouche et al. [2] used backpropagation neural networks to detect and eliminate motion artifacts for the recorded EGG signal. Where three layers neural network have been used; Input, hidden and output layer. Hidden layers used tan-Sigmoid function and hard-limit functions used in the output layer [18].

6 EGG Medical Application

When the patients have a motility disorder which normally appears as vomiting, nausea or any signs indicate that the stomach not emptying the stomach normally, then Electrogastrography (EEG) is required to diagnose the case. So, the EGG can be used to diagnose the pathophysiology of the diseases which relate to stomach slow wave or dysrhythmia. The possible clinical results from EGG test; (a) normal, (b) tachygastria, (c) bradygastria, (d) the rhythm pattern is not specifying, (e) lack of signal power increase at postprandial and (f) lack of a single dominant frequency (DF).

Many methods used to classify the EGG signal, like using Support Vector Machines (SVM), Neural Network and Genetic Algorithms [19–21]. However, more accurate results have been found using the Self Organizing Map (SOM) Classifier [22], to classify the EGG signal. The overall performance of the classifier can be enhanced by the tuning of the classifying method. So, for SOM Classifier to get better results a fine tuning of the map parameters such as; the map size, neighborhood definition, lattice, the rate of the learning, learning algorithm and normalization the input data should be considered.

Until now EGG tests alone cannot specify the disease, however it may give indications and evidence of stomach motor dysfunctions and with the help of other tests like manometry test and gastric empty test, then disease can be diagnosed. This section summarizes the diseases and disorders which EGG test could help to diagnose:

Cyclic Vomiting Syndrome (CVS): This disease appears in all ages and it appears as episodes of severe vomiting. Episodes can last for hours or days and alternate with symptom-free periods. The EGG test in children shows the CVS have tachygastria only postprandially and abnormal EGG signal when higher tachygastria activity occur [23].

Central Nervous System (CNS) disorders in children: Children with this disease suffer from vomiting and gastroesophageal reflux. 62% of children with this disorder had gastric dysrhythmia and 32% of them had gastroesophageal reflux and gastric dysrhythmias. EGG can be used for diagnosing as gastric dysrhythmia is a function of disorders symptoms [24].

Chronic Idiopathic Intestinal Pseudo-Obstruction (CIIP) appears mainly in children and in adults with diabetes or scleroderma. This intestinal neural syndrome is caused by severe impairment in the ability of the intestines to push food through. Symptoms of this disease are nausea, abdominal pain, severe distension, dysphagia, vomiting, diarrhea and constipation. EGG has been used to diagnose CIIP in children. As patients with this disorder have tachygastria, it can be highly suggestive of a neuropathic dysmotility [25].

Irritable Bowel Syndrome (IBS) is a common, long-term disorder of the GI that affects the large intestine (bowel or colon). The symptoms of IBS can be abdominal pain and cramps, abdominal bloating, diarrhea and/or constipation. If dyspepsia is present, then EGG can diagnose IBS. EGG tests show that both the temporal and spatial regularity of gastric slow waves have a negative relation with GI, which suggests that the uncoupling may be responsible for the delayed GE in patients with functional dyspepsia [2].

Helicobacter pylori (Hp) is a microaerophilic bacterium usually found in stomach. It is the main cause of gastritis, peptic ulcer disease and gastric cancer. EGG tests show that tachygastria and antral hypomotility is increased in patients with Hp. The test for patients with Hp shows significant dysrhythmia for (ECA) activities. Also, the power of slow wave signal is too high with this patient. [26].

Gastroparesis disease is a condition that affects the normal spontaneous movement of the muscles (motility) in the stomach. Gastroparesis is a condition in which the stomach cannot empty itself of food in a normal fashion. The frequency of EGG in these patients is more abnormal than other patients with diabetic gastroparesis [26].

Pregnant women often suffer from nausea and vomiting. EGG can be used to find the pathophysiology of this symptoms. EEG tests show that women with nausea have gastric dysrhythmias which not exist with women without nausea [25]. The test shows that the dysrhythmias reduce, with liquid food and food containing a predominant protein, more than food containing carbohydrate, fat meals and solid meals [27–30].

Diabetic gastropathy is a number of neuromuscular dysfunctions of the stomach, including abnormalities of gastric contractility. Patients often suffer from nausea, early satiety, intermittent vomiting after meals and vague epigastric discomfort. This disorder appears in both type 1 and type 2 diabetes. EGG diagnosis of gastric dysrhythmias provides new insights into gastric neuromuscular abnormalities and guides therapies to improve upper GI symptoms in patients with diabetes mellitus [31, 32].

7   Conclusion

Biomedical signals are controlled by the nervous system and are dependent on the physiological and anatomical properties of muscles. The EGG is a technique used to collect the data from the stomach. EGG is a complicated signal, collected by using a non-invasive procedure to provide information on a slow wave rhythm. Gastric dys-rhythmias are found in many disorders and diseases in which nausea and vomiting are prominent symptoms. Therefore, EGG has been used to investigate in a wide range of disorders associated with gastric function. The understanding of the basic stomach motility pathophysiology may have extended EGG applications and make it as a new domain for exploring stomach motor dysfunction.

Until now, there has been no standard recommendation for recording position, period and test meal throughout the whole world. The progress in EGG application has been very slow. Generally speaking, research on the EGG is still on the experimental level due to many factors like poor signal quality. This paper gives an overview for how EGG signal can be analyzed, and the main diseases can be related to the gastric dysarthria. In future studies if researchers of EGG can use an advanced analysis, then it will provide more information in terms of slow wave coupling, power and phase shifting. Consequently, a properly indicated EGG measurement may extend our knowledge of slow wave stomach dysmotility.

References

1. Riezzo, G., Russo, F., Indrio, F.: Electrogastrography in adults and children: the strength, pitfalls, and clinical significance of the cutaneous recording of the gastric electrical activity. BioMed Res. Int. **2013**(11), 282757 (2013)
2. Haddab, S., Laghrouche, M.: Microcontroller-based system for electrogastrography monitoring through wireless transmission. Meas. Sci. Rev. **9**(5), 122–126 (2009)
3. Alvarez, W.C.: The electrogastrogram and what it shows. JAMA **78**, 1116–1119 (1922)
4. Stevens, J.K., Worrall, N.: External recording of gastric activity: the electrogastrogram. Physiol. Psychol. **2**, 175–180 (1974)
5. Smout, A.J.P.M.: Myoelectric activity of the stomach: gastroelectromyography and electrogastrography. Ph.D. Dissertation, Erasmus Universiteit Rotterdam, Delft University Press (1980)
6. Chen, J., Schirmer, B.D., McCallum, R.W.: Serosal and cutaneous recordings of gastric myoelectrical activity in patients with gastroparesis. Am. J. Physiol. **266**, G90–G98 (1994)
7. Chen, J.: New interpretation of the amplitude increase in postprandial electrogastrograms. Am. J. Physiol. **98**, A335 (1990)
8. Lee, H.T., Hennig, G.W., Fleming, N.W., et al.: Septal interstitial cells of cajal conduct pacemaker activity to excite muscle bundles in human jejunum. Gastroenterology **133**(3), 907–917 (2007)
9. Smout, A.J.P.M., Jebbink, H.J.A., Samsom, M.: Acquisition and analysis of electrogastrographic data—the Dutch experience. In: Electrogastrography: Principles and Applications, Raven, New York, pp. 3–30 (1994)
10. Koch, K.L.: Electrogastrography. In: Schuster, M., Crowel, M., Koch, K.L. (eds.) Atlas of Gastrointestinal Motility, pp. 185–201. BC Decker, Ontario (2002)
11. Mintchev, M.P., Kingma, Y.J., Bowes, K.L.: Accuracy of cutaneous recordings of gastric electrical activity. Gastroenterology **104**(5), 1273–1280 (1993)
12. Patterson, M., Rintala, R., Lloyd, D., Abernethy, L., Houghton, D., Williams, J.: Validation of electrode placement in neonatal electrogastrography. Dig. Dis. Sci. **46**(10), 2245–2249 (2001)
13. Koch, K.L., Stern, R.M.: Handbook of Electrogastrography. Oxford University Press, New York (2004)
14. Chen, J.D.Z., Zou, X., Lin, X., Ouyang, S., Liang, J.: Detection of gastric slowwave propagation from the cutaneous electrogastrogram. Am. J. Physiol. **277**(2), G424–G430 (1999)
15. van der Schee, E.T., Grashius, J.L.: Running spectral analysis as an aid in the representation and interpretation of electrogastrographic signals. Med. Biol. Eng. Comput. **25**, 57–62 (1987)
16. Pfister, C.J., Hamilton, J.W., Nagel, N., Bass, P., Webster, J.G., Tompkins, W.J.: Use of spectral analysis in the detection of frequency differences in the electrogastrograms of normal and diabetic subjects. IEEE Trans. Biomed. Eng. **35**, 935–941 (1988)
17. Chen, J.: Adaptive filtering and its application in adaptive echo cancellation and in biomedical signal processing. Ph.D. thesis, Department of Electrical Engineering, Katholieke Universiteit Leuven, Belgium (1989)
18. Price, C.N., Westwick, D.T., Mintchev, M.P.: Analysis of canine model of gastric electrical uncoupling using recurrence quantification analysis. Dig. Dis. Sci. **50**, 885–892 (2005)
19. Świerczyński, Z., Zagańczyk, A.: Application of neural network and genetic algorithms in computer aided gastric diagnostic system. BBE **25**(1), 49–58 (2005)

20. Świerczyński, Z., Mazur, J.: Application of SVM in computer aided gastric diagnostic system. BBE **24**(4), 19–30 (2004)
21. Świerczyński, Z., Guszkowski, T.: Application of SOM in classification of EGG signals. Metrol. Measur. Syst. **13**(3), 279–285 (2006)
22. Proceedings of the Biocybernetyka i inżynieria biomedyczna. XIII Krajowa Konferencja Naukowa, pp. 231–236 (2003). (in Polish)
23. Leahy, A., Besherdas, K., Clayman, C., Mason, I., Epstein, O.: Abnormalities of the electrogastrogram in functional gastrointestinal disorders. Am. J. Gastroenterol. **94**, 1023–1028 (1999)
24. Kara, S., Dirgenali, F., Okkesim, S.: Estimation of wavelet and short-time Fourier transform sonograms of normal and diabetic subjects' electrogastrogram. Comput. Biol. Med. **36**, 1289–1302 (2006)
25. Levanon, D., Zhang, M., Chen, J.D.Z.: Efficiency and efficacy of the electrogastrogram. Dig. Dis. Sci. **43**(5), 1023–1030 (1998)
26. Cucchiara, S., Riezzo, G., Minella, R., Pezzolla, F., Giorgio, I., Auricchio, S.: Electrogastrography in non-ulcer dyspepsia. Arch. Dis. Child. **67**(5), 613–617 (1992)
27. Ravelli, A.M., Ledermann, S.E., Bisset, W.M., Trompeter, R.S., Barratt, T.M., Milla, P.J.: Foregut motor function in chronic renal failure. Arch. Dis. Child. **67**(11), 1343–1347 (1992)
28. Jednak, M.A., Shadigian, E.M., Kim, M.S., et al.: Protein meals reduce nausea and gastric slow wave dysrhythmic activity in first trimester pregnancy. Am. J. Physiol. **277**(4), G855–G861 (1999)
29. Leahy, A., Besherdas, K., Dayman, C., Mason, I., Epstein, O.: Abnormalities of the electrogastrogram in functional gastrointestinal disorders. Am. J. Gastroenterol. **94**(4), 1023–1028 (1999)
30. Koch, K.L., Stern, R.M., Vasey, M., Botti, J.J., Creasy, G.W., Dwyer, A.: Gastric dysrhythmias and nausea of pregnancy. Dig. Dis. Sci. **35**(8), 961–968 (1990)
31. Walsh, J.W., Hasler, W.L., Nugent, C.E., Owyang, C.: Progesterone and estrogen are potential mediators of gastric slow-wave dysrhythmias in nausea of pregnancy. Am. J. Physiol. **270**(3), G506–G514 (1996)
32. Thor, P., Lorens, K., Tabor, S., Herman, R., Konturek, J.W., Konturek, S.J.: Dysfunction in gastric myoelectric and motor activity in Helicobacter pylori positive gastritis patients with non-ulcer dyspepsia. J. Physiol. Pharmacol. **47**(3), 469–476 (1996)

A Novel MAC Scheme for Reliable Safety Messages Dissemination in Vehicular Networks

Muhammad Alam[1(✉)], João Rufino[2], Kok-Hoe Wong[1], and Joaquim Ferreira[2,3]

[1] Department of Computer Science and Software Engineering,
Xi'an Jiaotong-Liverpool University, Suzhou, China
{m.alam,kh.wong}@xjtlu.edu.cn
[2] Instituto de Telecomunicações, Universidade de Aveiro, Aveiro, Portugal
{joao.rufino,jjcf}@ua.pt
[3] ESTGA, Universidade de Aveiro, Águeda, Portugal

Abstract. Safety messages dissemination in vehicular networks requires real-time and deterministic medium access. The existing standard, IEEE 802.11p, for vehicular networks is based on Carrier Sense Multiple Access (CSMA) which is not a deterministic mechanism as collision may occur to access the medium that can cause unbounded delays. Therefore, in this paper, we presented a new deterministic Medium Access Control (MAC) scheme for vehicular communications that ensure reliable safety messages dissemination in vehicular networks. In the proposed scheme, each Road Side Unit (RSU) and On-Board Units (OBU) is assigned a time slot in a fix medium access time (elementary cycle). Each RSU controls time access using the knowledge of the connected vehicles. Two realistic scenarios, Vehicle-to-Infrastructure (V2I)/Infrastructure-to-Vehicle (I2V) and Vehicle-to-Vehicle (V2V), are presented to test the proposed scheme. The proposed scheme is simulated and the presented results validate its effectiveness in terms of higher safety packets success rate and low communications delays.

Keywords: Vehicular networks · Safety messages · Gateways · Medium access control

1 Introduction

Over the past couple of decades, transportation systems began to receive widespread attention from the scientific community and emerged towards Intelligent Transportation Systems (ITS). Effective vehicular connectivity techniques can significantly enhance transportation efficiency, reduce traffic incidents and improve overall safety, alleviating the impact of congestion; devising the so-called Intelligent Transportation Systems (ITS) experience [4]. Furthermore, during the past decades the volume and density of vehicles increased significantly, especially the road traffic; leading to a dramatic increase in the number

© Springer Nature Switzerland AG 2020
A. M. Madureira et al. (Eds.): HIS 2018, AISC 923, pp. 521–529, 2020.
https://doi.org/10.1007/978-3-030-14347-3_51

of accidents and congestion, with negative impacts on economy, environment and quality of peoples lives. In particular, according to the World Health Organization (WHO), road traffic injuries are estimated to be the leading cause of death for young people (aged 15 to 29) and the ninth cause of death worldwide in 2015 [2]. The enabling communication technologies are intended to realize the frameworks that will spur an array of applications and use cases in the domain of road safety, traffic efficiency, and drivers assistance. Although these applications allow dissemination and gathering of useful information among vehicles and between transportation infrastructure and vehicles in pursuance of assisting drivers to travel safely and comfortably;

Traffic safety applications, such as hazard location warnings and collision warnings, rely heavily on the timely delivery of safety-critical data. Most of these applications demand strictly bounded timing response and are highly dependent on the performance of the underlying wireless vehicular communication technology. In most cases, these systems are required to have dependable timeliness requirements since data communication must be conducted within predefined temporal bounds along with fulfilling other requirements such as reliability, security etc. This is mainly because the unfulfillment of these requirements may compromise the expected behaviour of the system and cause economic losses or endanger human lives. In addition, the broadcast nature of wireless communications in an open environment makes it more vulnerable to un-wanted external entities compared to the wired communications. Therefore, the consideration of the real-time aspects in the implementation of ITS services offers great potential to improve the level of safety, efficiency and comfort on our roads. In addition, European ITS considered the deterministic medium access methods as one of the criteria for the evaluation of possible ITS communication technologies.

In addition to the fact that applications for traffic safety rely heavily on the support of wireless vehicular communications, the existing standards such IEEE 802.11p [1] do not provide deterministic real-time support and the other enabling technologies such LTE-V from 3GPP and traffic sensors are either in their definition stage or not fully deployed and tested for dependable real-time communication. Therefore, various MAC protocols have been proposed in the literature to guarantee the safety messages and ensure deterministic wireless vehicular communications. For instance, in [3] the authors have proposed Vehicular Flexible Time Triggered (V-FTT) protocol that provide a deterministic access to wireless medium. The V-FTT protocol divides the medium access time into elementary cycles of 100 ms each. These elementary cycles are composed of a collision-free period and contention-based period. The Collision-free period is reserved for the RSUs and OBU to communication whereas contention-based period is composed of Free Period that is used for no-VFTT complaint vehicles to communicate. The work presented in paper is based on V-FTT protocol. Since MAC protocols are considered very important for deterministic wireless communications a number protocols and schemes are presented in the literature. In [5], the authors have presented a regulated and unregulated contention period based MAC. However, the analysis presented in the work is not enough to validate the deterministic vehicular communication. Other related works are presented in [6–9].

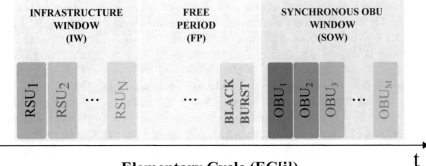

Fig. 1. V-FTT protocol overview

The rest of this paper is organized as follows: Sect. 2 presents the overview of the proposed MAC scheme. Section 3 provides the detailed scenarios tested for the proposed scheme. The details of simulations and results are presented in Sect. 4. The paper is concluded in Sect. 5.

2   Proposed MAC Scheme

Our proposed mechanism is based on the real-time MAC protocol called Vehicular-Flexible Time Triggered (V-FTT) proposed in [3]. Like V-FTT, the medium time is divided into 100 ms frames called Elementary Cycles (ECs). Each EC is further divided into time slots and reserved for RSUs, OBUs and free period (FP). As shown in Fig. 1, there are predefined windows for I2V and for V2I/V2V real-time communication. The FP between these windows allows other vehicles (non V-FTT aware) to communicate making the protocol interoperable with other solutions. In V-FTT each RSU sends the scheduling information and warning messages to the connected OBUs by broadcasting these and other messages. The vehicles operate in coordinated manner communicating in the reserved time slot.

One of the main downfalls of V-FTT is strongly related with the reallocation of used time-slots. In highly congested networks, where the number of vehicles is superior to the time-slots available, RSUs decide which vehicles can communicate in the next EC. Moreover, in such scenarios the FP is highly reduced or even removed from the EC, blocking the operation of other protocols. Following V-FTT, time slots are arbitrarily allocated to OBUs, without prioritization, which makes the protocol potential unsuitable for such scenarios. One of the most blatant problems concerns the delivery of CAMs, a periodic message that describes the vehicles current status.

In order to face this handicap, we employ the Earliest Deadline First (EDF) concept. RSUs use the information provided by the surrounding environment to perform an educated schedule of the next EC. Using the validity of the information, the data age, speed and status of vehicles, RSUs know the real-time traffic demands in the network and schedule traffic accordingly. In specific cases, OBUs can be assigned to multiple time slots in current and future superframes/elementary cycles, e.g., priority vehicle or correlated road warning messages. Creating a solution to the delivery of frequent and highly deadline-dependent data traffic needed for safety applications. Using this solution RSUs can optimize the medium utilization, but most importantly ascertain timeliness on the delivery of vital messages. Furthermore, by enjoying from a overall perception of its surroundings they can reduce or even remove the FP without jeopardizing other systems. Adapting the next EC to the current traffic demands.

3   Scenarios

To evaluate the overall performance of the proposed MAC scheme we considered two types of scenarios that put to test the reliability of the proposed solution. The two scenarios are depicted in Figs. 2 and 3, the first is infrastructure-based and the latter multi-hop. Both are depicted in the next subsections.

3.1   V2I/I2V Communication

The first approach uses the RSU as a focal point, centralizing all communications and information of vehicles in the transmission range. Therefore, each RSU must have an available communication slot for these vehicles. In addition, RSUs can exchange the information of the associated vehicles with each other and therefore, reserve free slots for incoming vehicles; highly reducing the contest for free time slots. Thus, each vehicle is always connected by having a reserved slot to communicate as shown in the Fig. 2. In addition, the information of the each RSU is shared with the nearby static gateways (SGWs).

3.2   V2V/V2I Communication

Infrastructure based vehicular networks are very reliable compared to the V2V communications however they are also very expensive due to the costs of deployment of extra hardware. Since vehicles can operate as any node in traditional wireless networks, acting as the source, destination or router of the information, in places where it is difficult to install extra RSUs or gateways, V2V communication plays a vital role. In this scenario, we aim to analyze the behavior of nodes in a V2V wireless communication in terms of reliability of the safety messages dissemination. As depicted in Fig. 3 two RSUs were placed near the roadside connecting set of highly mobile vehicles. Since the coverage of the RSU is limited and the connected vehicles are moving in high speed, the wireless medium is not stable, seriously affect the packet transmission success rate. Therefore, we adopted the concept of mobile gateways (MGWs) where vehicles having benefiting from a quality connection with the RSU are selected as information routers. The vehicles that are not in the RSU coverage area or suffering from weak signal

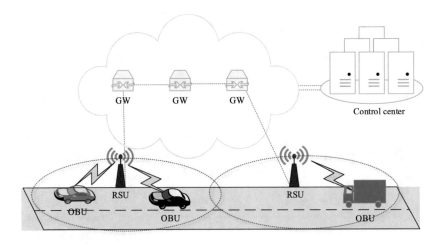

Fig. 2. V2I/I2V communication scenario

connect with the nearby MGW and start a V2V communication, scheduled in the contention based medium slots. For testing purposes only one hope V2V communication was considered for safety messages dissemination, but the same concept can be applied to multi-hop communication.

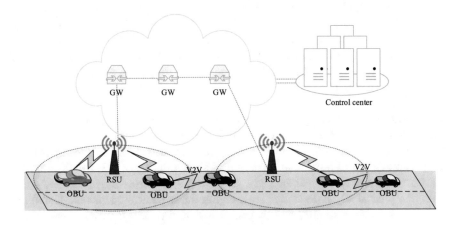

Fig. 3. V2V/V2I communication scenario

4   Simulations and Results

The proposed scheme along with the native IEEE 802.11p is simulated using SUMO [10] and NS3 [11] simulators. Concretely, a freeway mobility model representing the highway scenario was used, where the behaviour of 100 vehicles was simulated during 100s. In simulations, the transmission range of the each

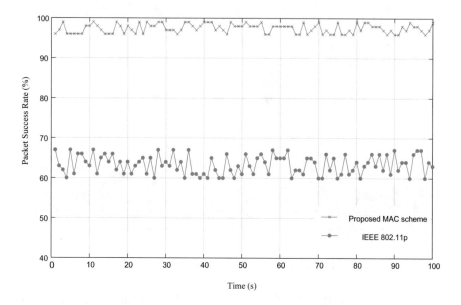

Fig. 4. Percentage of Packet success rate in V2I/I2V communication scenario.

RSU is kept 500 m. Figure 4 depicts the percentage of packet success rate (PSR) comparison of the both MAC mechanisms. Our proposed scheme gives a much higher PSR compared to the original IEEE 802.11p. This is mainly because our proposed mechanism guarantees the slot availability and reservation for safety messages, allowing a contention free access to the medium by OBUs. Contrary to this, the native IEEE 802.11p MAC constantly drops the packets due to congestion that causes the contention to access the medium and results in higher packet loss.

When it comes to analyze the performance of real-time or deterministic communication one of the most important points to consider is the delay of communication. Therefore, we performed a series of simulations test to check the communication delays of both MAC schemes as shown in Fig. 5. It is worth to notice that our proposed MAC scheme has a very low delay. The average recorded delay is about 27 ms to 30 ms. In some cases, higher delay values (58 ms and 45 ms) are recorded and this is mainly due to packet lost and re-transmission. Overall, we recorded a very contact delay values for constant vehicle density (100 vehicles) which is very much a realistic scenario. On the other hand, the native IEEE 802.11p MAC suffers from higher delays and in most cases the values are not acceptable for safety messages dissemination. The main reason for higher delays are the lack of prior knowledge about the scheduling that causes contention and vehicles have to go in a back off mode. However, in more than 50% times the native MAC also fulfills the requirement of the real-time packets dissemination in terms of delays. The highest recorded delay is 136 ms and 134 ms where the vehicles have to wait for longer in back off period to start the transmission. The

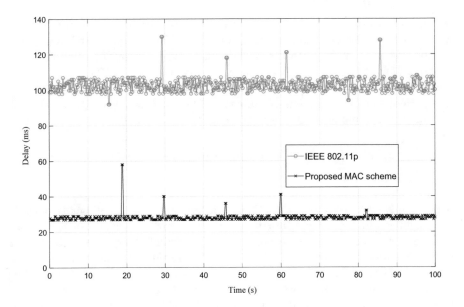

Fig. 5. Delay comparison in V2I/I2V communication scenario.

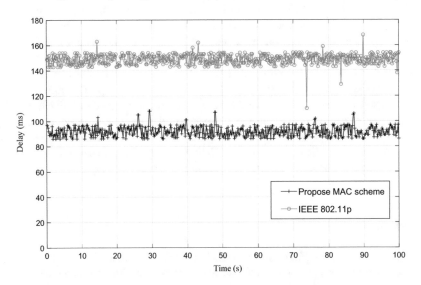

Fig. 6. Delay comparison in V2V communication scenario.

system offers higher delays in case of V2V based communication. This is important to measure as there are number of cases where the vehicles cannot find any nearby RSU and they need to send or receive real-time messages. Therefore, in these tests we analyzed the delays of both V2V communication schemes. However, we have restricted the test to one hop neighbors only. Again, our proposed scheme offers less delays as the OBUs are communicating in the free periods while the connection of the relay or mobile GW OBU is guaranteed in the contention free window. The average recorded delay for one hop V2V and then V2I is about 87 ms to 98 ms. In some case the delays jump up to 108 ms and 109 ms but this is mainly because of the contention based access to the medium in the free period. On the other hand, the native MAC offers much higher delays in case of V2V communication and in some tests it reached 168 ms as can be seen Fig. 6.

5   Conclusions

Strict real-time behaviour and safety guarantees are typically difficult to attain in vehicular adhoc networks, but they are even harder to attain in high speed mobility scenarios, where the response time of distributed algorithms may not be compatible with the dynamics of the system. In addition, in some operational scenarios, the IEEE 802.11p MAC may no longer be deterministic, possibly leading to unsafe situations. Therefore, in this paper we have presented a MAC scheme based on previously proposed protocols called V-FTT. We have introduced the earliest deadline first (EDF) concept in which each RSU can use the status of nearby vehicles in real-time to schedule their communication slots in the next elementary cycle. Thus ensuring the real-time operation of the medium and at the same time releases the slot once the communication of the done on a particular time slot. The proposed scheme is test for V2I and V2I/V2V based communication and the results are compared with the native IEEE 802.11p in terms of packet success rate and delays. For V2I/I2V scenario, the percentage of PSR of our proposed mechanism is 96 to 99 compared to the 58 to 68 of IEEE 802.11p. Similarly, our proposed scheme offers very small delays compared to the original standard. For V2V/V2I scenario, the average delay of our proposed scheme 98 ms compared to 158 ms record for IEEE 802.11p. The presented results show that our proposed scheme guarantees a reliable safety messages dissemination in vehicular networks.

References

1. IEEE Standard for Information technology. Telecommunications and information exchange between systems Local and metropolitan area networks. Specific requirements Part 11: Wireless LAN Medium Access Control (MAC) and Physical Layer (PHY) Specifications, IEEE Std 802.11-2012 (Revision of IEEE Std 802.11-2007), p. 12793, March 2012
2. European Commission. Statistics accidents data (2015). http://ec.europa.eu/transport/road_safety/specialist/statistics/index_en.htm
3. Meireles, T., Fonseca, J., Ferreira, J.: The Case For Wireless Vehicular Communications Supported By Roadside Infrastructure, in Intelligent Transportation Systems Technologies and Applications. Wiley, Hoboken (2014)

4. Alam, M., Ferreira, J., Fonseca, J.: Introduction to intelligent transportation systems. Studies in Systems, Decision and Control, pp. 1–17 (2016). https://doi.org/10.1007978-3-319-28183-4_1
5. Bohm, A., Jonsson, M.: Real-time communication support for cooperative, infrastructure-based traffic safety applications. Int. J. Veh. Technol. **2011** (2011)
6. Mak, T.K., Laberteaux, K.P., Sengupta, R.: A multi-channel VANET providing concurrent safety and commercial services. In: Proceedings of the 2nd ACM International Workshop on Vehicular Ad Hoc Networks, ser. VANET 2005, pp. 1–9. ACM, New York (2005). https://doi.org/10.1145/1080754.1080756
7. Bansal, G., Kenney, J., Rohrs, C.: LIMERIC: a linear adaptive message rate algorithm for DSRC congestion control. IEEE Trans. Veh. Technol. **62**(9), 4182–4197 (2013)
8. Torrent-Moreno, M., Mittag, J., Santi, P., Hartenstein, H.: Vehicle to vehicle communication: fair transmit power control for safety-critical information. IEEE Trans. Veh. Technol. **58**(7), 3684–3703 (2009)
9. Rufino, J., Alam, M., Almeida, J., Ferreira, J.: Software defined P2P architecture for reliable vehicular communications. Perv. Mob. Comput. **42**, 411–425 (2017). https://doi.org/10.1016/j.pmcj.2017.06.014
10. SUMO: URL: 8. http://sumo-sim.org/
11. Network simulator 3. http://www.nsnam.org/

DSS-Based Ontology Alignment in Solid Reference System Configuration

Alexandre Gouveia[1]([✉]), Paulo Maio[1], Nuno Silva[1], and Rui Lopes[2]

[1] School of Engineering, Polytechnic of Porto, 4249-015 Porto, Portugal
{aas,pam,nps}@isep.ipp.pt
[2] DigitalWind, Lda., 3810-498 Aveiro, Portugal
rlopes@digitalwind.pt

Abstract. uebe.Q is a managing software for solid referential information systems, such as ISO 9000 (for quality) and ISO 1400 (for environment). This is a long-term developed software, encompassing extensive and solid business logic with a long and successful record of deployments. A recent business model change imposed that the evolution and configuration of the software, shifts from the company (and especially the development team) to consultants and other business partners, along with the fact that different systems and respective data/information need to be integrated with minimal intervention of the development team. The so far acceptable rigidity, fragility, immobility and opacity of the software became a problem. Especially, the system was prepared to deal with a specific database respecting a specific schema and code-defined semantics. This paper describes the approach taken to overcome the problems derived form the previous architecture, by adopting (i) ontologies for the specification of business concepts and (ii) an information-integration Decision Support System (DSS) for mapping the domain specific ontologies to the database schemas.

Keywords: Solid referential information system · Ontology ·
Alignment · DSS

1 Introduction

uebe.Q[1] is an information managing software for solid referential information systems, such as ISO 9000 (for quality) and ISO 1400 (for environment). This is a long-term developed software, encompassing extensive and solid business logic with a long and successful record adoption by companies in different business areas and countries. Currently and for the last decade, every time uebe.Q software is deployed for/on a company, an extensive and specific configuration process takes places involving DigitalWind's ISO 9000 and ISO 1400 experts, software development and operation teams. So, typically, DigitalWind's collaborators were playing the roles of the developer, the seller and the business consultant altogether.

---

[1] http://www.uebeq.com.

© Springer Nature Switzerland AG 2020
A. M. Madureira et al. (Eds.): HIS 2018, AISC 923, pp. 530–539, 2020.
https://doi.org/10.1007/978-3-030-14347-3_52

A recent business model change determined that the configuration and adaptation tasks of the software have to shift from the DigitalWind's development and operation teams to third-party consulting and other business partners. Furthermore, it is imperative to provide the system with the ability to consume/generate data from/to different data sources, independent of their data model (e.g. relational, document), schema and semantics.

However, uebe.Q configuration and deployment requires strong business and technological expertise, thus constraining the adoption of the new business model. In addition, because configuration often requires specific refinement of the software, its rigidity, fragility, opacity and immobility [1] is emphasized, hence compelling its re-engineering.

Consequently, the goal of the dySMS project is to facilitate the software configuration by business experts and consultants while improving the configurability and adaptability of the software. For that, it is prescribed an approach combining:

- ontologies, that allows the configuration of the business concept according to the enterprise's business, department or individual employee;
- rules, that allows the specification of business rules/constraints upon the ontological concepts, which goes beyond the expressivity of the ontological constraints.
- DSL (Domain-Specific Language) [2,3], developed with the purpose to facilitate the business experts/consultants to configure the system with minimal or no technical skills.
- An information-integration Decision Support System (DSS) [4] that helps configuring the system to deal with different sources of information in a seamless and transparent way, representing the information according to the configured domain ontologies.

This is, therefore, a project aiming to solve a re-engineering problem combining software engineering and knowledge engineering best practices and respective available technology.

Therefore, this is not a new Rapid Application Development approach [5], but a high-level declarative configuration of a new system that exploits data from different legacy systems and respective data sources. To the best of our knowledge, nothing has been done as proposed in this paper and with such amplitude of adoption.

The rest of the paper describes the information systems software context (Sect. 2), a short introduction to the concepts and problems addressed by the configurability dimension of the system (Sect. 3), and to DSS for ontology mediation (Sect. 4). In Sect. 5, the ontology alignment DSS is presented. Conclusions and future work are described in Sect. 6.

2   Context

The legacy software (left-hand side of Fig. 1) adopts a client-server architectural style, whose backend (i.e. the server part) evolved into a large monolithic software information system, developed in a myriad of frameworks (e.g. ASP, ASP.Net) and programming languages (e.g. C#, VB.Net). The backend is responsible for the data access (from/to relational database), business logic and server-side generation of the Rich Internet Application client-side UI pages (in HTML and JS).

While the business logic captured by the backend is a relevant asset, it suffers from rigidity, fragility, opacity and immobility [1], frustrating a conservative and traditional evolution. In that sense, the system has been re-engineered, exploiting the functionalities (i.e. business logic) of the legacy system, as well as the incremental redefinition of the existing functionalities to the new version (i.e. at configuration time). The added components (at the right-hand side of Fig. 1) include the dySMS Core, the Single Page Application (SPA) frontend and the DSL-based configurator.

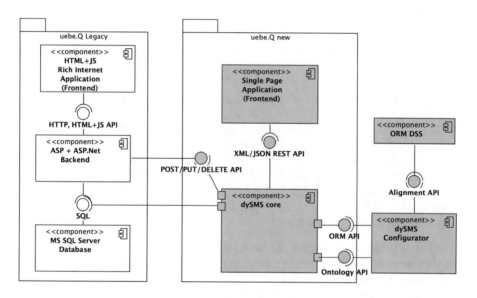

Fig. 1. Component-based diagram of the new uebe.Q and configurator.

The dySMS Core component reads data directly from the legacy database, but the original business logic might be needed to update, delete or create data.

The DSL-based Configurator allows the business consultant to define the domain-specific ontology (model) and respective (business) rules. The DSL has been developed in the JetBrains Meta Programming System[2] which provides a projectional editor that drives the configurator tasks according to the DSL-captured structure and semantics.

The semantics defined through the ontology and rules follows a hierarchical policy, in which the semantics defined at a hierarchy level binds to the semantics defined in top hierarchy levels, promoting organizational reasoning consistency.

---

[2] https://www.jetbrains.com/mps/.

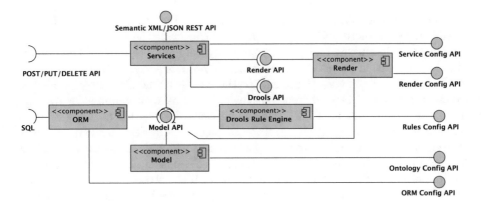

Fig. 2. Detailed component-based diagram of the new uebe.Q.

Figure 2 depicts the inner architecture of the dySMS Core (backend), including the identification of the parts/responsibilities and the details of the configuration interfaces. The system has the following components:

- Model is the set of classes and respective instances that are managed by the system. The classes, associations and constraints correspond to the configured ontology(ies). Ontologies are conceptual models that formally describe the structure and semantics of a specific domain/business [6]. In order to keep up-to-date the model regarding the configured ontology(ies), a reflection-based approach generates on-the-fly the configuration-time-defined classes, allowing the system to operate with whatever classes structure is required.
- ORM, which is responsible for the data access of the relational databases and the instantiation of the configured ontologies from the data sources.
- Drools[3] is a rule engine that reasons upon the model/ontology using a Closed World Assumption (CWA). Rules execute upon all sort of managed data, including business data and user authentication data, which allows both monotonic and non-monotonic business calculations, validations and authorization control.
- Render is responsible for the projection and serialization of the data (instances and attributes) according to the configured authorization, projection and serialization rules.
- Services is the set of semantic REST web services that are responsible for the availability of the new functionalities. The component delegates the responsibilities in the other components as suggested in the diagram depicted in Fig. 2.

3   Ontology Mediation

The system configurability is a key part of the system, as it provides the system's ability to accommodate different domain/business, information and services.

---

[3] https://www.drools.org.

While ontologies are often used for information/knowledge sharing, they are used here as a structure and as an artifact for capturing the semantics across the organization hierarchy, and as a dynamically configurable model. The domain ontology definition across the organization is not addressed in this paper.

Ontology or schema alignment is the process whereby correspondences between entities of two different ontologies/schemas with common or overlapping domains are established [7] and is particularly relevant in many areas of application of ontologies [8–10]. Automatic alignment systems make use of automatic matching algorithms (ontology matchers) which evaluate the similarities between pairs of source and target ontologies/schemas' entities, exploring different dimensions of ontologies, namely the lexical, the structural (relations, hierarchy) and the semantics (defined at the language level and domain constraints) [7] dimensions.

Ontology/schema alignment have different applications and complexity, ranging from Level 0 (the simplest) to Level 2 (the more expressive and complex) [7]. A simple correspondence establishes a binary relation (e.g. equivalence ($\equiv$), subsumption ($\leq$) between a single entity of the source ontology and a single entity of the target ontology. For example, in Fig. 3, the simple correspondence ($O_1 : Book, O_2 : Volume, =, 1.0$) expresses the semantic equivalence between the entity $Book$ of the ontology O_1 and the entity $Volume$ of the ontology O_2, with a degree of confidence of 1.0. A complex correspondence establishes a non-binary relation (e.g. concatenation, split, arithmetic operation) between sets of ontology entities. In Fig. 3, the complex correspondence ($O_1 : name, \{O_2 : title, O_2 : subtitle\}, Split, 0.9$) expresses that the instance values of the entities $title$ and $subtitle$ should be the result of the split of the instance values of the entity $name$. In this correspondence, the degree of confidence is 0.9.

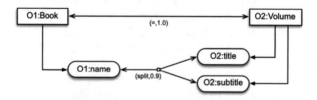

Fig. 3. Simple and complex correspondences.

Information-integration alignments are those that provide a univocal and functional alignment for adoption in Ontology Mediation tasks (e.g. data transformation, integration and migration) [7,9,11]. I.e. its adoption allows generating target repository data that adheres to the structure and semantics of the target repository, from the source repository data. Level 0–2 alignment are required depending on the ontology/schemas characteristics.

Analysis of state of the art automatically-generated alignments shows that ambiguous alignments are quite common and prevent direct application of these alignments in Ontology Mediation tasks. In fact, most of the matching algorithms generate incomplete, incorrect and mutually contradictory alignments, preventing their application in scenarios demanding high quality and completeness. The results obtained with the automatic alignment systems are in fact below the requirement for ontology mediation, requiring the user/expert intervention [12,13], by correcting and completing the automatic alignments into data integration suitable alignments. This is a time-consuming, error-prone and often incomplete process.

4   DSS and Ontology Alignment

Although the term DSS has many connotations, [14] defined three major characteristics (i) DSS are designed specifically to facilitate decision processes; (ii) DSS should support rather than automate decision-making and (iii) DSS should be able to respond quickly to the changing needs of decision-makers. Power [15] differentiates five types of DSS according to the mode of assistance:

- Data-driven, which emphasizes analysis of large amounts of data. It supports decision making by analyzing given time-series of internal company and external data and returning new information gained by those analysis;
- Model-driven, which emphasizes access to and manipulation of a statistical, financial, optimization, or simulation model. It uses data and parameters provided by users to assist decision-makers in analyzing a situation;
- Knowledge-driven, which provides specialized problem-solving expertise stored as facts, rules, procedures, or in similar structures. The expertise consists of knowledge about a particular domain;
- Document-driven, which manages, retrieves, and manipulates unstructured information in a variety of electronic formats;
- Communication-driven, which uses network and communications technologies to facilitate collaboration and communication. It supports more than one person working on a shared task.

The importance of user involvement in ontology matching is gaining more and more attention. An evidence of this importance is the observation in [16] that the automatic matching is only a first step towards a final alignment and a validation by a user/expert is needed. Another evidence was the introduction in the OAEI 2013 campaign [17] of the Interactive Matching track, where user validation was simulated using an oracle. The number of user interventions and the time spent were also measured in addition to the influence of the validation on the quality of the alignments. According to [9], there are at least three areas in which users may be involved:

- by providing input to the system (before matching), such as an initial alignment, which may be used as anchor by the matching system, and obviously the ontologies to be matched;

- by configuring and tuning the system, which includes strategy and parameter selection, such as matcher weights, thresholds for filtering the results and aggregation parameters; and
- by providing feedback to matchers, validating the decisions, during or after the automatic matching, in order for them to adapt and improve the results.

Some studies have already been made on how to involve users in ontology matching, including [9]:

- enhancing the creation of candidate correspondences through query logs;
- designing of time matcher interaction;
- designing of alignments manually or by checking and correcting them, e.g. through learning;
- graphical visualization of alignments based on cognitive studies;
- providing an environment for manually design complex alignments through the use of a connected perspective that emphasize the connections between relevant entities of the ontologies being matched;
- explicitly specifying structural transformations using a visual language.

More recently, Dragisic and colleagues [18] identify the issues of the most common kind of user involvement, i.e. user validation, by reviewing existing systems and literature related to ontology matching. The authors focus on three main issues and provide a categorization that shows how different state-of-the-art systems deal with these issues. The relevant aspects highlighted are:

- The influence of user expertise;
- The exploration of user input by the ontology matching systems services;
- The impact of user interfaces in supporting interaction and involvement.

5   DSS and DSL Ontology Alignment Process

To address the ontology/schema configuration problem, two complementary approaches are combined:

- DSL-based Configurator. The designed DSL is a high-level, simple and yet powerful ontology mapping language that follows state of the art best practices [7,9,19,20]. Despite facilitating the correct and complete specification of the alignment, along with its transformation to different requirements (e.g. ontology and ORM specification), this manual tool remains very time consuming;
- Information-integration Decision Support System. The alignment between the source database schema and the business ontology(ies) defined in the DSL-based configurator is the first approach using the ontology mapping information integration DSS [21].

The information integration alignment Decision Support System (DSS) assists the business consultant in specifying the mapping between source and target ontologies/schemas, balancing the precision and recall of the system with the user's participation in the process, i.e. minimizing the user participation, and maximizing the precision and recall. The DSS has three conceptual operation modes (cf. Fig. 4):

- a completely automatic operation, in which all the decisions are taken either by automatic matchers [7] or randomly (in case no information is available for decision);
- a completely human-dependent operation, in which all the decisions are taken by the human expert/consultant;
- a user-supported operation, in which the system takes some decisions (rule-defined heuristics), and the human takes some other decisions, such that the expert's decisions will constrain the decisions afterwards in an automatic or heuristic phase.

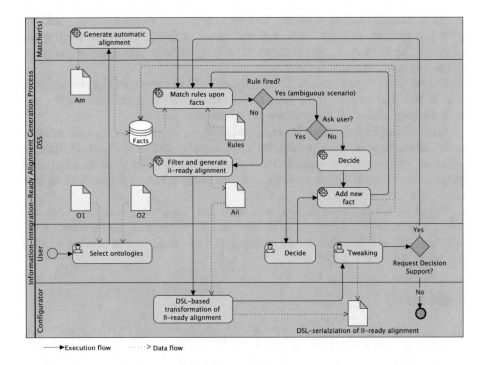

Fig. 4. Combination of DSS and DSL information integration alignment process.

This process is implemented in a DSS whose core is (i) the Drools rule-engine and (ii) the rules that represent a set of heuristics designed according to the formal and multi-dimensional analysis of the ontologies and alignment, yielding a strong formal rational to the system. In particular, one of the major contributions of this tools is the system's ability to search for alignments with longer contextualization paths (i.e. n > 1-step paths), which is useful in complex heterogeneous alignment scenarios, such as those observed in uebe.Q applications.

When no more rules are found to fire, i.e. when no more ambiguous scenarios (scenarios-to-resolve) are found, the filtering process prepares an information-integration-ready alignment.

The resulting alignment is then forward to the DSL component (where further tweaking can occur) for generating the ORM mappings.

The resulting alignment is then read by the dySMS DSL-based configurator, which provides the business consultants with an extra level of alignment specification/tweaking. Once terminated, the alignment is pushed to the ontology and ORM configuration interfaces for dynamically configuration of the dySMS Core.

6   Summary

Several parts of the uebe.Q original database is now managed by dySMS system through different company-specific, role-specific and user-specific ontologies. The system is able to deal with different structures of the same semantics, relying solely in the DSS and DSL configuration (i.e. without any specific programming/development).

Experiments in the context of uebe.Q and dySMS show that the combination of DSL and DSS reduces the configuration time while improving the quality and completeness of the schema/ontology alignments.

Despite the limited experiments conducted so far, the results are evidence of the advantages of the combination of automatic and user-driven systems in complex, subjective, time-consuming and error-prone tasks such as generating information integration-ready alignment for solid referential information systems.

Based on the achieved results, we argue that the proposed architecture and adopted technology represent a valid approach for re-engineering business legacy software with configurability and extensibility in mind, in the context of solid referential information systems.

References

1. Martin, R.C.: Designing Object Oriented Applications using UML, 2nd edn. Prentice Hall, Upper Saddle River (2000)
2. van Deursen, A., Klint, P., Visser, J.: Domain-specific languages: an annotated bibliography. SIGPLAN Not. **35**, 26–36 (2000)
3. Kattenstroth, H.: DSMLs for enterprise architecture management: review of selected approaches. In: Proceedings of the 2012 Workshop on Domain-Specific Modeling (DSM 2012), pp. 39–44. ACM, New York (2012)
4. Zarat, P., Liu, S.: A new trend for knowledge-based decision support systems design. Int. J. Inf. Decis. Sci. **8**, 305–324 (2016)
5. Schmidt, R.F.: Software Engineering: Architecture-Driven Software Development. Morgan Kaufmann, Burlington (2013)

6. Fensel, D.: Ontologies: Silver Bullet for Knowledge Management and Electronic Commerce. Springer, Heidelberg (2001)
7. Euzenat, J., Shvaiko, P.: Ontology Matching, 2nd edn. Springer, Heidelberg (2013)
8. Otero-Cerdeira, L., Rodrguez-Martnez, F.J., Gmez-Rodrguez, A.: Ontology matching: a literature review. Expert Syst. Appl. **42**(2), 949–971 (2015)
9. Shvaiko, P., Euzenat, J.: Ontology matching: state of the art and future challenges. IEEE Trans. Knowl. Data Eng. **25**(1), 158–176 (2013)
10. Staab, S., Studer, R.: Handbook on Ontologies, 2nd edn. Springer, Berlin (2009)
11. de Bruijn, J., Ehrig, M., Feier, C., Martns-Recuerda, F., Scharffe, F., Weiten, M.: Ontology mediation, merging, and aligning. Semant. Web Technol. (2006)
12. Lambrix, P., Kaliyaperumal, R.: A session-based ontology alignment approach enabling user involvement. Semant. Web J. **8**(2), 225–251 (2017)
13. Balasubramani, B.S., Taheri, A., Cruz, I.F.: User involvement in ontology matching using an online active learning approach. In: 10th International Workshop on Ontology Matching (2015)
14. Alter, S.L.: Decision Support Systems: Current Practice and Continuing Challenge. AddisonWesley Publishing Co., Reading (1980)
15. Power, D.J.: Decision Support Systems: Concepts and Resources for Managers. Quorum Books, Westport (2002)
16. Euzenat, J., Meilicke, C., Shvaiko, P., Stuckenschmidt, H., Trojahn dos Santos, C.: Ontology alignment evaluation initiative: six years of experience. In: Spaccapietra, S. (ed.) Journal on Data Semantics XV, vol. 6720, pp. 158–192. Springer, Heidelberg (2011)
17. Paulheim, H., Hertling, S., Ritze, D.: Towards evaluating interactive ontology matching tools. In: Cimiano, P., Corcho, O., Presutti, V., Hollink, L., Rudolph, S. (eds.) The Semantic Web: Semantics and Big Data, vol. 7882, pp. 31–45. Springer, Heidelberg (2013)
18. Dragisic, Z., Ivanova, V., Lambrix, P., Faria, D., Jimnez-Ruiz, E., Pesquita, C.: User validation in ontology alignment. In: Proceedings of the 15th International Semantic Web Conference (ISWC), pp. 200–217. Springer, Kobe (2016)
19. David, J., Euzenat, J., Scharffe, F., Trojahn dos Santos, C.: The Alignment API 4.0. Semantic Web – Interoperability, Usability, Applicability, no. 2 (2011)
20. Bauer, C., King, G., Gregory, G.: Java Persistence with Hibernate. Manning (2015)
21. Gouveia, A., Silva, N., Martins, P.: A rule-based DSS for transforming automatically-generated alignments into information integration alignments. 19th International Conference on Information Integration and Web-Based Applications and Services, Austria (2017)

Clustering of PP Nanocomposites Flow Curves Under Different Extrusion Conditions

Fátima De Almeida, Eliana Costa e Silva[✉], and Aldina Correia

CIICESI, ESTG / P.PORTO - Center for Research and Innovation in Business Sciences and Information Systems, School of Management and Technology/Polytechnic of Porto, Margaride, 4610-156 Felgueiras, Portugal
{mff,eos,aic}@estg.ipp.pt

Abstract. For assessing the structural features of organoclay C15A dispersions in PP/PP-g-MA melts under different processing conditions along the screws of Twin Screw Extruder, in this work four different clustering algorithms are considered. The best algorithm and number of clusters is selected using three internal validation measures: connectedness, Dunn's index and silhouette width. The results show that hierarchical clustering is the algorithm that yield better results and two is the best number of clusters. Two clusters are well identified $L/D = 9.5$ and 32.

Keywords: Hierarchical clustering · K-means · Fuzzy clustering · Internal validation · Rheology · Twin screw extruder · Processing conditions

1 Introduction

Engineering of polymers is not an easy exercise with evolving technology, it often involves complex concepts and processes. The preparation and characterization of nanocomposites with different polymer matrices shows its considerable scientific and technological potential [1–4]. Advanced nanocomposites stem from the use of slight fractions of nanomaterials with high levels of dispersion to generate remarkable property enhancement as well as favourable cost-effectiveness, including better mechanical, thermal, electrical, and barrier properties [2–6]. Nanomaterials have garnered interest due to their structural diversity and excellent properties, nanostructured materials and their composites have been extensively studied also for energy storage/conversion related applications [3]. The choice of the organic modifier is considered a key parameter for the preparation of nanocomposites. Exfoliation/dispersion cannot develop properly, even in the presence of a compatibilizer, if the organic modifier is not adapted to the chemical nature of the matrix, whatever the processing conditions and the specific mechanical energy provided during mixing [7].

Continuous melting is more suitable to the optimization of nanocomposites materials, since it is easier to control the thermo-mechanical history imparted

A. M. Madureira et al. (Eds.): HIS 2018, AISC 923, pp. 540–550, 2020.
https://doi.org/10.1007/978-3-030-14347-3_53

on the samples [4, 7–10]. For continuous melting, the extrusion process is more easily up scalable to an industrial level and thus offers the best possibility form future development of nanocomposites materials. The main largest extrusion process is Twin Screw Extruder (TSE), mainly used for the preparation and process of polymeric materials, through compounding and reactive extrusion, in which the residence times are significantly shorter [7, 11]. It is recognized that the extruder process conditions are important variables that must be optimized to effect a high degree of delamination and dispersion [8, 10, 12]. However, inconclusive observations have been made, both the relationship between thermomechanical stresses applied during processing and dispersion levels remain to be fully understood. Numerical modeling can be a very efficient tool to overcome these difficulties. However, it remains a real challenge, as it necessitates to couple flow simulation in complex geometry, reaction kinetics and evolutionary rheological behavior [11]. Moreover, the information needed is sometimes difficult to obtain with the required accuracy. Melting experiments of polyolens on modular co-rotating TSE presented melting initiation mechanisms and melting propagation mechanisms [13, 14]. Several mathematical models have been developed to predict how melting will occur under different prescribed conditions [13–15]. Also, a multiple regression analysis, where the platelet count is determined by transmission electron microscopy (TEM). A model including up to quadratic terms in mean residence time, $\bar{t}$, and variance, σ^2, but with no cross terms [9]. A particle analysis by light microscopy was performed using the software Analysis (Olympus, Japan). A regression equation between sample thickness and area ratio, allowed to calculated the normalized dispersion index value and used to study filler dispersion and distribution [6]. However, due to the limited number of observations in the dataset, the number of fitting parameters must be kept small in order to be meaningful.

Simulations of flow conditions along the co-rotating TSE were performed using Ludovic® software, which enables the calculation of the main local flow variables like shear rate, filling ratio, melt temperature, pressure, residence time and dissipated energy, from melting zone to die exit [11, 12]. These simulations were carried out by rheological data (flow curves) to establish relationships between nanocomposites structure and processing parameters along the screw profile, where local physical values cannot be precisely measured (or measured at all) during processing.

A multi-objective optimisation method for composite design in real manufacturing industry for promoting sustainability is used in [1, 16, 17]. The interrelationships between multiple objectives (requirements) were analyzed and classified by hierarchical clustering analysis [1]. The linear regression model obtained was compared to other linear regression models which were obtained by other mathematical techniques, and it has been proved to be the best model which has the smallest gap between predicted values and experimental results. With regard to the PP nanocomposites, previously in [18] the focus was on the interaction of the experimental parameters and properties measure results of PP compatibilized nanocomposites. The results obtained using multivariate analysis of variance (MANOVA) allowed the generalization of observed relations between the processing conditions and the properties studied.

The present study aims at a comprehensive assessment of the different PP nanocomposites properties. The rheological properties of polymer–based nanocomposites have deserved an intensive scientific interest, particulary the oscillatory or dynamic flow [9,12,19,20]. Dynamic behavior of the nancomposites, under different processing conditions, are highly dependent on elastic properties of polymeric materials. Oscillatory flow results of PP/organoclay nanocomposites are analyzed considering the processing conditions effect of nanostructure on organoclay C15A particles, since the presence of a nanostructure in the amorphous polymer matrix alters considerably the viscoelastic behavior in the terminal zone. For that, in this work, several clustering techniques are used to aggregate similar experimental results obtained under different processing conditions.

2   Experimental Setup

Maleated polypropylene (PP-g-MA) was used as a compatibilizer to allow clay dispersion. For continuous melting, the PP nanocomposites were prepared at fixed composition (80/15/5; w/w/w, wt.%) of PP/PP-g-MA/C15A in a modified co-rotating TSE, a Leistritz LSM 30.34. Mixing element is the combination of diversified Kneading Disks (KD) with left-handed/right-handed configuration and/or the adjustable thickness [15]. The 29D long screws, are built through mixing zones separated by feeding elements [18]. Two different screw configurations, with the same numbers and position of conveying and mixing elements, with KD staggered positively (90°) and negatively (−30°) were applied to vary the flow and shear conditions. The KD positively and negatively staggered are considered as the low screw intensity (P1) and high (P2), respectively. Relatively to the last one, it is believed that a high shear screw profile (P2) is adequate to ensure the necessary dispersion levels. Different screw speed (N), temperatures (T) and outputs (Q) were chosen.

The processing temperatures selected (T1, T2 and T3, respectively, 170, 180 and 190 °C) for compounding were set along the barrel and die. The screw speed (N1 = 150 rpm or N2 = 300 rpm) and feed rate (Q1, Q2 or Q3, respectively, 2, 4 or 6 kg.h^{-1}) were also set. Melting of PP nanocomposites in a modular TSE was carried out as a function of: (i) screw profile, (ii) screw speed, (iii) barrel temperature and (iv) feed rate. Samples were collected along the extruder and at die exit (Fig. 1) and immediately frozen in liquid nitrogen, prior to characterization.

Rheological measurements were performed using a parallel plate rheometer, referenced as a Paar Physica MCR 300. The experiments were carried out in small amplitude oscillatory shear. The samples were characterized in the linear domain through frequency sweep measurements, at a temperature of 180 °C and a constant strain of 5%, from 0.01 to 100 rad.s^{-1}. This type of characterization allows to highlight the presence of a percolated network formed by the nanofillers [19,20]. Indeed, for the range of shear rates encountered in extrusion

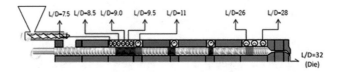

Fig. 1. Sampling devices and location along the TSE

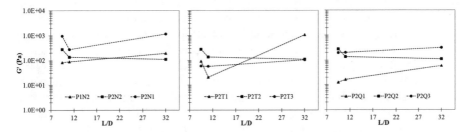

Fig. 2. Effect of P and N, T and Q on the evolution of G' along the extruder

$(10$–100 s$^{-1})$, there is quite no difference in viscosity between matrix and nanocomposite [10]. Thus, rheological data (flow curves) of the PP/PP-g-MA nanocomposites were considered at low frequency (up to 7.2 rad.s^{-1}) to test the samples.

For PP based nanocomposites whatever the processing condition and position along the extruder, an increase of the complex viscosity is observed, which is characteristic of a percolated network formed by the clay platelets [19,20]. This clearly indicates that the clay tactoids were exfoliated during mixing, at least partially [18]. In the tested conditions, C15A layers are easily dispersed in the PP matrix, compatibilized with PP-g-MA, even though P2N1 and P2T1 provides better results. Morphological evolution along the extrusion profile, independently of processing conditions, revelead that microscale dispersion primarily happens in the melting zone, whereas discontinuous exfoliation is, generally, observed along the mixing zones, up to the die exit (Fig. 2). The results indicate that exfoliation is mainly issued from clay tactoids and small aggregates, emphasizing the crucial role of primary microscale dispersion on the final structure and properties of the nanocomposites [3,7–9]. This result showed that not only screw profile and processing conditions are important processing parameters but the also evolution of dispersion along the screw axis. Actually, flow along the restrictive screw elements generates the stresses and residence time levels required for a good development of both intercalation and exfoliation. Still, local influences of stress and residence time may cause significant viscous dissipation [10–12] which, in turn, could decrease matrix viscosity and facilitate its draining out from the C15A clay galleries, total or partially, and the evolution will either develop much slower, remaining basically constant, or reversion may be observed, depending on the selected processing conditions. To identify the local differences a new methodology is used and presented next.

3　Methodology

Classification and clustering analysis are commonly used in industrial applications. While classification attempts to assign new subjects into existing groups based on observed characteristics, clustering divides subjects into different clusters so that the within-cluster variations are small compared with the variation between clusters. The goal of this unsupervised learning tecnique is to assign each subject into one of the clusters based on some similarity measure [22–24]. There are many methods available in literature for clustering analysis.

In this work four clustering algorithms are considered, namely: **(i)** Hierarchical Agglomerative Clustering (HAC) algorithm, that initially places each observation in its own cluster and then the clusters are successively joined together by their "closeness", determined by a dissimilarity matrix. The number of clusters can be estimated by applying cuts at a chosen height in the dendrogram; **(ii)** iterative method K-means which minimizes the within-class sum of squares for a given number of clusters. It starts with an initial guess of the clusters center and placing each observation in the cluster to which it is closest to. Next the cluster centers are updated until the cluster centers no longer move; **(iii)** Partitioning Around Medoids (PAM), that although similar to K-means, is consider more robust since it admits the use of other dissimilarities besides the Euclidean distance; **(iv)** Fanny performs fuzzy clustering, where each observation may present partial membership to each cluster, indicated by a vector of probabilities.

In order to assess the cluster quality, different measures may be used [22]. Internal validation measures (IVM) consider only intrinsic information in the data. In fact, only the dataset and the clustering partition as input are used. Three very often used IVM are: **(i)** connectedness, **(ii)** compactness, and **(iii)** separation of the cluster partitions. Connectedness indicates the extension to which observations are placed in the same cluster as their nearest neighbors. Compactness evaluates cluster homogeneity, by looking at the intra-cluster variance. Separation quantifies the degree of separation between clusters, typically by measuring the distance between clusters centroids. Since compactness increases with the number of clusters, but separation decreases, popular methods combine the two measures into a single score. Dunn's index and silhouette width are examples of non-linear combinations of the compactness and separation [25]. The silhouette width lies in $[-1, 1]$, while Dunn's index and silhouette width vary from zero to ∞, and both should be maximized. In contrast, the connectivity takes a value between zero and ∞ and should be minimized.

4　Results and Discussion

In this section we present the results concerning G' for low frequencies, more precisely frequencies up to $7.2\,\mathrm{rad.s^{-1}}$. The samples were collected at three locations along the TSE: $\mathrm{L/D} = 9.5$, 11 and 32 (see Fig. 1). Figure 3 shows the evolution of G' as the frequency increases. For $\mathrm{L/D} = 9.5$ Fig. 3 suggest that G' curves may be separated into two groups. The curves relative to P2N1 present a very similar behavior and appear to distinguishing from all the others. While for $L/D = 11$ the curves all seem very similar. Finally, for $L/D = 32$ the G' curves seem divided into two groups, the first includes P2N1 and P2T1 experiment curves.

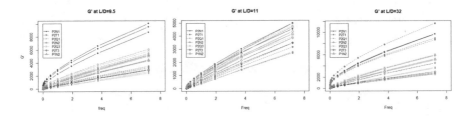

Fig. 3. G' at $L/D = 9.5$ (left), 11 (center) and 32 (right)

For assessing if the groups of curves depicted in Fig. 3 are in fact significant, four clustering algorithms were consider, namely: HAC[1], K-means, PAM and fanny, briefly distinguished in Sect. 3. Previously to the analysis, a z-score standardization was applied to the variables. The most adequate clustering algorithm and number of clusters was selected from the four algorithms and three validation measures, described in Sect. 3. The results were obtained using the clValid [21] package from R and are depicted in Figs. 4, 5 and 6.

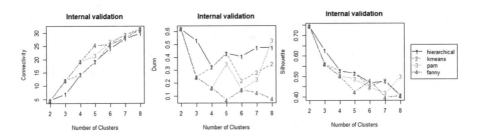

Fig. 4. Internal validation for G' at $L/D = 9.5$

For $L/D = 9.5$, we can observed that for all algorithms the connectivity is minimized for two clusters, also Dunn's index and silhoutte width are maximized for two clusters (Fig. 4). Considering $L/D = 11$, two is the number of clusters that minimizes the connectivity, for any of the algorithms (Fig. 5). The number of clusters that maximizes the Dunn's index is five for HAC, K-means and PAM algorithms, and four for the fanny algorithms. The silhoutte width is maximized for two clusters. Finally, for $L/D = 32$, we can observed that for all algorithms the connectivity is minimized for two clusters, also the Dunn's index and the silhoutte width are maximized for two clusters.

---

[1] With Euclidean distance and Ward method.

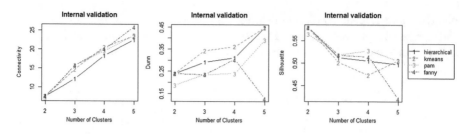

Fig. 5. Internal validation for G' at L/D = 11

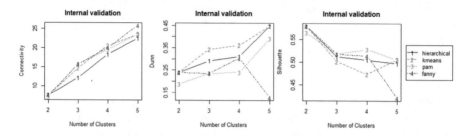

Fig. 6. Internal validation for G' at L/D = 32

Table 1 summarizes the internal validation results. For most cases Dunn's index and silhouette width the best solution was obtained using the HAC methods for all locations along the TSE. Concerning with connectivity PAM is the best algorithm for L/D = 11 and L/D = 32, while HAC is the best for L/D = 9.5. The number of clusters (#C) is two for all cases except for L/D = 11 with Dunn's index (Table 1). Taking this into consideration, two clusters and HAC method are considered. The two clusters are also suggested by the screeplots in Figs. 7, 8 and 9.

Table 1. Optimal scores

TSE location IVM	L/D = 9.5			L/D = 11			L/D = 32		
	Score	Method	#C	Score	Method	#C	Score	Method	#C
Connectivity	4.5786	HAC	2	7.1226	PAM	2	4.2282	HAC	2
Dunn's index	0.6219	HAC	2	0.4478	HAC	5	1.2887	HAC	2
Silhouette	0.7446	HAC	2	0.5784	HAC	2	0.8192	HAC	2

Concerning the TSE location L/D = 9.5 (Fig. 7), the first cluster consists of the replications of the experiment P2N1, that corresponds to the processing conditions with the highest shear screw profile (P2) and the lower screw speed (N1). Therefore the different replications under these processing conditions present similar behavior, with high dispersion. The second cluster includes the flow curves for all the other processing conditions, indicating the different conditions do not yield significant differences in the nanolayers dispersion.

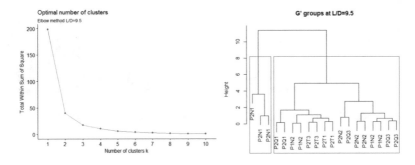

Fig. 7. Scree plot and dendrogram for G' curves at $L/D = 9.5$

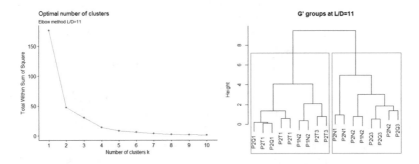

Fig. 8. Scree plot and dendrogram for G' curves at $L/D = 11$

The cut height of the clusters for the TSE location $L/D = 11$ (Fig. 8) is smaller than the one for $L/D = 9.5$ (Fig. 8), which is in line with the flow curves behavior observe in Fig. 3. In fact, at $L/D = 11$ the flow curves present quite similar behavior for different processing conditions. From the two clusters in Fig. 8 it is not possible to asses that the processing conditions favor, for this particular TSE location, the material dispersion. At die $(L/D = 32)$, the two clusters indicate that high screw profiles (P2) with low temperatures (T1) or high screw profiles (P2) with low velocities (N1) favor nanolayers dispersion. In fact, all the replications under conditions P2N1 and P2T1 present very similar flow patterns (Fig. 9). For the other processing conditions the dispersion pattern is quite different from the one presented for P2N1 and P2T1.

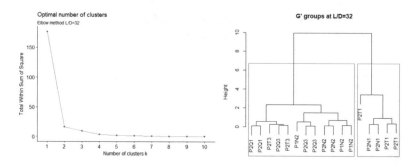

Fig. 9. Scree plot and dendrogram for G' curves at L/D = 32

5 Conclusion and Future Work

In this paper the structural features of organoclay C15A dispersions in PP/PP-g-MA, under different processing conditions were considered. The rheological results (G' curves) at frequencies up to $7.2 \, \text{rad.s}^{-1}$, of the resulting PP nanocomposite samples are investigated in detail along the screws (L/D = 9.5, 11 and 32). For this purpose four clustering techniques and three internal validation measures are used to aggregate similar experimental results obtained under different processing conditions.

The results show that hierarchical clustering yields the best results and two is the best number of clusters. Two clusters are evident for L/D = 9.5. The first cluster contains P2N1 experiments, corresponding to highest shear screw profile and lower screw speed, where higher dispersion is obtained. The other cluster contains all the other experiments. Therefore the different processing conditions have no significant effect for the second cluster. These evidence is not so clear for L/D = 11, with G' values very similar between experiments. For this particular TSE location, it is not possible to asses that processing conditions favor material dispersion. At die exit the two clusters are well divided and high screw profiles with low temperatures or high screw profiles with low speeds favor dispersion.

In future we intend to study if other material properties are grouped according for the same processing conditions. The same clustering techniques can be used and one may compare if the membership is similar with the one obtained for G'. We intend also to investigate other material measures such as TEM.

References

1. Gu, F., Hall, P., Miles, N.J.: Development of composites based on recycled polypropylene for injection moulding automobile parts using hierarchical clustering analysis and principal component estimate. J. Clean. Prod. **137**, 632–643 (2016)
2. Mohan, N., Senthil, P., Vinodh, S., Jayanth, N.: A review on composite materials and process parameters optimisation for the fused deposition modelling process. Virtual. Phys. Prototyp. **12**(1), 47–59 (2017)

3. Mochane, M.J., Mokhena, T.C., Mokhothu, T.H., Mtibe, A., Sadiku, E.R., Ray, S.S.: The importance of nanostructured materials for energy storage/conversion. In: Handbook of Nanomaterials for Industrial Applications, pp. 768–792 (2018)
4. Gonçalves, J., Lima, P., Krause, B., Potschke, P., Lafont, U., Gomes, J.R., Abreu, C.S., Paiva, M.C., Covas, J.A.: Electrically conductive polyetheretherketone nanocomposite filaments: from production to fused deposition modeling. Polymers 10(8), 925 (2018)
5. Dong, Y., Pramanik, A., Liu, D.I., Umer, R.: Manufacturing, characterisation and properties of advanced nanocomposites. J. Compos. Sci. 2, 46 (2018)
6. Abdulkhani, A., Hosseinzadeh, J., Ashori, A., Dadashi, S., Takzare, Z.: Preparation and characterization of modified cellulose nanofibers reinforced polylactic acid nanocomposite. Polym. Test. 35, 73–79 (2014)
7. Normand, G., Mija, A., Pagnotta, S., Peuvrel-Disdier, E., Vergnes, B.: Preparation of polypropylene nanocomposites by melt-mixing: comparison between three organoclays. J. Appl. Polym. Sci. 134(28), 45053 (2017)
8. Cho, S., Hong, J.S., Lee, S.J., Ahn, K.H., Covas, J.A., Maia, J.M.: Morphology and rheology of polypropylene/polystyrene/clay nanocomposites in batch and continuous melt mixing processes. Macromol. Mater. Eng. 296(3–4), 341–348 (2011)
9. Dennis, H., Hunter, D.L., Chang, D., Kim, S., White, J.L., Cho, J.W., Paul, D.R.: Effect of melt processing conditions on the extent of exfoliation in organoclay-based nanocomposites. Polymer 42(23), 9513–9522 (2001)
10. Domenech, T., Peuvrel-Disdier, E., Vergnes, B.: The importance of specific mechanical energy during twin screw extrusion of organoclay based polypropylene nanocomposites. Compos. Sci. Technol. 75, 7–14 (2013)
11. Vergnes, B., Berzin, F.: Modeling of reactive systems in twin-screw extrusion: challenges and applications. C. R. Chim. 9(11–12), 1409–1418 (2006)
12. Lertwimolnun, W., Vergnes, B.: Effect of processing conditions on the formation of polypropylene/organoclay nanocomposites in a twin screw extruder. Polym. Eng. Sci. 46(3), 314–323 (2006)
13. Jung, H., White, J.L.: Investigation of melting phenomena in modular co-rotating twin screw extrusion. Int. Polym. Proc. 18(2), 127–132 (2003)
14. Jung, H., White, J.L.: Modeling and simulation of the mechanisms of melting in a modular co-rotating twin screw extruder. Int. Polym. Proc. 23(3), 242–251 (2008)
15. Malik, M., Kalyon, D.M., Golba Jr., J.C.: Simulation of co-rotating twin screw extrusion process subject to pressure-dependent wall slip at barrel and screw surfaces: 3D FEM Analysis for combinations of forward-and reverse-conveying screw elements. Int. Polym. Proc. 29(1), 51–62 (2014)
16. Gurrala, P.K., Regalla, S.P.: Multi-objective optimisation of strength and volumetric shrinkage of FDM parts. Virtual Phys. Prototyp. 9(2), 127–138 (2014)
17. Villmow, T., Potschke, P., Pegel, S., Häussler, L., Kretzschmar, B.: Influence of twin-screw extrusion conditions on the dispersion of multi-walled carbon nanotubes in a poly (lactic acid) matrix. Polymer 49(16), 3500–3509 (2008)
18. De Almeida, M.F., Correia, A., Costa e Silva, E.: Layered clays in PP polymer dispersion: the effect of the processing conditions. J. Appl. Stat. 45(3), 558–567 (2018)
19. Cassagnau, P.: Melt rheology of organoclay and fumed silica nanocomposites. Polymer 49(9), 2183–2196 (2008)
20. Domenech, T., Zouari, R., Vergnes, B., Peuvrel-Disdier, E.: Formation of fractal-like structure in organoclay-based polypropylene nanocomposites. Macromolecules 47(10), 3417–3427 (2014)

21. Brock, G., Pihur, V., Datta, S., Datta, S.: clValid, an R package for cluster validation. J. Stat. Softw. (2011)
22. Kaufman, L., Rousseeuw, P.J.: Finding Groups in Data: An Introduction to Cluster Analysis, vol. 344. Wiley, Hoboken (2009)
23. Ledolter, J.: Data Mining and Business Analytics with R. Wiley, Hoboken (2013)
24. Zhao, Y.: R and Data Mining: Examples and Case Studies. Academic Press, Cambridge (2012)
25. de Amorim, R.C., Hennig, C.: Recovering the number of clusters in data sets with noise features using feature rescaling factors. Inf. Sci. **324**, 126–145 (2015)

Building a Decision Support System to Handle Teams in Disaster Situations - A Preliminary Approach

Isabel L. Nunes[1,2], Gabriel Calhamonas[1], Mário Marques[3],
and M. Filomena Teodoro[3,4](✉) (iD)

[1] Faculty of Sciences and Technology, Universidade Nova de Lisboa,
2829-516 Caparica, Portugal
[2] UNIDEMI, Department of Mechanical and Industrial Engineering,
Faculty of Sciences and Technology, Universidade Nova de Lisboa,
2829-516 Caparica, Portugal
[3] CINAV, Naval Academy, Base Naval de Lisboa, Alfeite, 1910-001 Almada, Portugal
maria.alves.teodoro@marinha.pt
[4] CEMAT, Instituto Superior Técnico, Lisbon University,
Av. Rovisco Pais, 1, 1048-001 Lisbon, Portugal

Abstract. The objective of the present work is to build and implement a decision support system with the ability to prioritize certain teams for specific incidents taking into account the importance of each team that acts in case of emergency, the sequence of tasks that should perform all possible orders to be given. With such purpose, a collaborative estimation or forecasting technique that combines independent analysis with the maximum use of feedback was applied to build consensus among experts who interact anonymously (Delphi forecasting). The topic under discussion is distributed (in a series of rounds) between the participating experts who comment on it and modify the opinion(s) until a certain degree of mutual consensus is reached. In our case, the collection and summary of knowledge of a group of experts from a given area was done through various phases of questionnaires, accompanied byan organized feedback. Simultaneously, it was necessary to make a survey about the existing emergency teams, as well as of all possible tasks in case of emergency. With this purpose were held meetings with experienced individuals in emergency cases, some documents about some exercises performed were consulted, also direct observation of a simulation of a catastrophe exercise in which humanitarian aid is required allowed to collect and analyze some important data. The decision support system is not completed but we hope to have given an important contribution to its construction.

Keywords: Decision support system · Expert · Delphi method · Questionnaire · Catastrophe

A. M. Madureira et al. (Eds.): HIS 2018, AISC 923, pp. 551–559, 2020.
https://doi.org/10.1007/978-3-030-14347-3_54

1 Introduction

The development of technology and the information management has been a constant in modern times. The construction of decision support systems (DSS) were no exception to the rule. With the emergence of the necessity to get optimal decisions in short time, these systems allowed to simplify the decision chain, to improve the performance of tasks execution allowing a reduction of expected costs.

DSS appear in a wide array of different areas of real life such as industrial enginneering [1], transportation systems [2], teaching, medicine, nursing [3], etc.

In the context of naval operations, the project THEMIS, promoted by portuguese Navy, aims to build a DSS to handle with disaster relief operations, where responders receive information on the incidents, tasks to execute, navigation orientation to incidents and advice how to perform the tasks. Under naval context, with a parallel approach to the same problem, applying an emergent technique, the augmented reality, the authors of [4–6] described two applications built for supporting different types of naval operations. Also an empirical study about user experience in naval context can be found in [7].

We are particularly interested to build and implement a DSS with the ability to prioritize certain teams for specific incidents taking into account the importance of each team that acts in case of emergency, the sequence of tasks that should perform all possible orders to be given.

To project and implement such system we have considered the facilities and high qualified staff of Portuguese Navy and used the Delphi method, a method that is exceptionally useful where the judgments of individuals are needed to "address a lack of agreement or incomplete state of knowledge ... the Delphi is particularly valued for its ability to structure and organize group communication" [3].

Data collection is already complete. Although the analysis of results is still taking place partially using the data not yet complete, we have the evidence of promising results.

The outline of this work consists in four sections. In Sect. 2 we present some details about the methodology implemented in the present work. The empirical application is done in Sect. 3, where are described the procedures to get the data and how the experiment was implemented. In last Section, we do some conclusions and final remarks.

2 Methodology

In order to be able to analyze all the possible approaches to the problem in question, this work began with a brief review of the existing literature on the Delphi method and alternatives.

Notice that, since middle 20th century, we can find numerous references in very distinct areas where the Delphi method was applied. For example, in industrial engineering, namely the construction of an index in automotive supply chain

[8,9], transportation [2], in paper pulp production [10], education [11], natural sciences [3], etc. In [12] we can achieve a numerous description of Delphi method applications.

It was chosen to apply the Delphi method, a structured process for collecting and summarizing knowledge of a group of experts from a given area through several phases of questionnaires, with organized feedback [13]. The Delphi method has the advantage of avoiding the problems associated with evaluation techniques based on traditional group opinions, such as Focus Groups, that may create problems of bias in responses due to the presence of opinion leaders [14].

The scope of these questionnaires was defined through meetings with a Captain-Lieutenant and Rear Admiral belonging to the Portuguese Navy. The scope consists in the establishment of the importance of each team that acts in case of emergency should perform a certain task and which all possible tasks can be given, providing to the DSS the ability to prioritize each team for certain incidents and specific tasks.

Firstly, it was necessary to survey about the existing emergency teams as well as all possible tasks(orders) in case of emergency. With this purpose, were scheduled meetings with navy officers with large experience in emergency cases, some documents/memos were collected and analyzed, some exercises were performed and all details registered, and also by direct observation of a simulation - an exercise of a catastrophe in which humanitarian aid was required. Subsequently, the data collected was organized and validated by a Lieutenant Captain and Rear Admiral, with large experience in catastrophe situations.

In order to proceed with the questionnaires, several platforms were analyzed and tested if they could support the questionnaire form according to the characteristics of Delphi method. The `SurveyMonkey.com` website was rejected because the free version has a limit for questions, the `GoogleForms` platform was rejected due to the impossibility of submitting the questionnaire data in future questionnaires, the Armstrong website `.wharton.upenn.edu` developed by J. Scott Armstrong in partnership with the International Institute of Forecasters was rejected because of the inability of respondents to review their given answers in completed questionnaires. It was then chosen to perform the questionnaires in `Excelformat`, due to the ease of sending (e-mail), ease of data processing and in order to always maintain the same platform throughout the several rounds of questionnaires to be performed.

3   Empirical Application

In this section, we will describe shortly the three main rounds of Delphi method approach.

3.1   Round I

To proceed with Delphi method, in a preliminary stage to first round of questionnaires, the potential group of experts were inquired about some individual characteristics of their profile: age, gender, professional rank, training class, type of experience in response to disasters (real versus train exercise) and the total ship boarding time (less than 1 year, 1–3 years, 3–5 years, more than 5 years).

With the objective of identifying the degree of priority that which each team should carry out each task, the same group also performed a questionnaire classified in a Likert scale of importance from 1 (Not Important) to 6 (Extremely Important) all possible tasks to be carried out during a humanitarian disaster relief operation for each existing team that can provide its service in the concerned operation (consult all tasks in Table 1 and all possible teams in Table 2).

For example, consider a medical team. A nurse is perfectly capable of performing the task "Handling and distributing food for the injured", but due to hers specific skills and abilities it is more important to assign a nurse to the task "Give 1st aid". Thus, it is expected that the medical team has assigned the task "Give first aid" with a rating closer to 6, while for the task "Handling and distributing food for the injured" is expected a rating closer to 1.

Taking into consideration the similarities of the available teams, they were reorganized into different brigades in order to facilitate the response:

- Reconnaissance Brigade - Constituted by the Reconnaissance team;
- Search and Rescue Brigade (SAR) - Consisting of the Search and Rescue (SAR) team, SAR-URB (Urban Search and Rescue team) and SAR-EST;
- Technical Brigade - Consisted of the Firefighting team, Water and Sanitation team, Electricity and Mechanical team;
- Logistics Brigade - Made up of the Supply and Food team;
- Medical Brigade - Constituted by the medical team.

The questionnaire was organized considering 52 tasks and 11 teams with a total of 572 questions classified on a Likert scale of 1 to 6. Each expert should answer to all questions, indicate the degree of confidence of the given answer and how much experience had with each team previously.

In the first round of questionnaires, 12 experts were considered, all males with at least 5 years on board, all aged between 35 and 54 years, with positions of Captain-lieutenant, Captain-of-sea-and-war and Captain-of-frigate (the most common). Between all experts only 25% have real past experience of disaster response. The remaining 75% experts are training experienced.

To identify the tasks that reached consensus is necessary to determine the Inter-Quartile Range (IQR), that is, the difference between the 1st Quartile and the 3rd Quartile. IQR represents 50% of the observations closest to the median. When IQR is less than 1 it means that more than 50% of the responses obtained are within 1 point of the Likert scale [11].

To process the questionnaires data and classify each one as consensual/no consensual, was necessary to compute a weighted mean (WM), where the weights are based on the experience of each expert responded to with each team. These weights evaluate the individual time of service. Notice that the opinion of a more experienced expert has a more significant weight in the weighted mean of the importance of a certain team and vice-verse. See Table 3.

Table 1. Collected tasks

Identify incidents	Repair electrical power system
Screen survivors/homeless and injured	Repair communications system
Provide 1st aid	Repair of lighting system [point of interest]
Census individuals	Repair mechanical energy production system
Identify location for [point of interest]	Repair power distribution system
Transport equipment to install [point of interest]	Recover basic sanitation
Mount [point of interest]	Create safety perimeter
Carrying severe injuries	Ensuring perimeter safety
Stabilize serious injury	Impose order and safety
Carry minor injured	Carry out order and safeguard of goodies and property
Rescue imprisoned victim	Carry out flight operations to transport material
Rescue victim from altitude	Perform flight operations for medical evacuations
Rescue victim in collapsed structure	Transport material [type] from [local] to [local]
Rescue isolated victim by land	Shift escort
Rescue single victim by air	Convey distribute food for the wounded
Rescue an isolated victim via water	Convey and Distribute food
Stabilize structures	Dead transport to morgue
Clear paths	Support the funeral ceremony
Build support structures for rescue	Status report
Fire fighting	Evacuate equipment to [point of interest]
Fighting floods	Evacuating population to [point of interest]
Carrying out shoring	Diving for minor repairs
Diving drinking water	Diving for rescue
Restoration of water supply	Evacuation of animals
Control of water leaks	Distribution of animal feed
Repair electric pumping system	Burying dead animal

For the tasks with an IQR greater than 1, is assumed that the experts have not reached a consensus on how important a certain team is to perform a specif task. In this way it will be considered in the next round of questionnaires.

Table 2. Available teams.

Reconnaissance	Water and sanitation technique brigade
Search and rescue - SAR	Mechanical engineering brigade
Search and rescue urban - SAR URB	Technical brigade of electricity
Search and rescue structures - SAR EST	Supply
Brigade firefighting technique	Food
Medical	

Table 3. Considered weights as function of level of experience of each expert.

Level of experience	Weigth
1	0.1
2	0.5
3	0.9
4	1.1
5	1.5
6	1.9

When a task has an IQR less or equal than one, is necessary to analyze its distribution of frequencies. Can be seen that the answer distribution per task is, come times, non consensual. In this case, it is proposed a new criterion: there is no consensus on all questions when there are 2 or more non-adjacent levels of importance that account for at least 20% of the sum of weighted levels of experience.

Applying the proposed rule, the first round of questionnaires can be summarized that between the 572 questions, there was no consensus on 290 questions (50.7%), from which 282 have an IQR higher than 1; between the tasks with IQR less or equal to one, only 8 did not meet the requirements of the proposed methodology. These total of 290 questions are included in next round of questionnaires.

3.2   Round II

In the second round of Delphi questionnaires, a personalized questionnaire was sent to all experts involved in the previous round, 11 responses were obtained, 91.6% of the total number of experts who responded to the previous round. The questionnaire counts with the 290 questions that did not reach the necessary consensus previously, and in order to try to achieve it, the weighted response frequency of each question was calculated, as well as the weighted average and the previous expert response.

The respondents were asked to review the data provided and revise their previous response, changing it if necessary, using the same scale, and justifying their choice. The results obtained, from the 290 questions, 265 (91.4%) reached a consensus. Then, the 3rd round is constituted by 25 questions, all of them identified as non-consensual due an IQR higher than 1 and none with an IQR less or equal to 1.

3.3   Round III

In the third and final round, from the 572 questions initially considered, only 25 did not reach consensus during the first two rounds.

A simultaneous interview with two respondents was scheduled with the objective of defining the final result of the tasks in which there was no consensus in the previous round and validating all the answers obtained previously, in order to identify any possible errors.

The experts chosen to participate in last round were two Commanders with the Captain-of-the-Frigate position that are used to coordinate and develop training and training actions, such as the DISTEX exercise, a simulation of a natural catastrophe. During the interview the tasks in disagreement were presented, as well as the justifications that were given by the experts to base their classification in the previous round, in order that the interviewees reached a consensus for the final value of importance to be assigned to each assignment.

The Fig. 1 display the number of tasks in which each team is the most indicated, and, in principle, the team called to respond to a certain order. It can

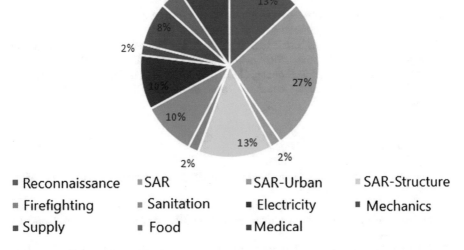

Number of tasks in which each team is the most indicated

Fig. 1. Distribution of tasks per team. Percentage of tasks in which each team is the most indicated.

be observed that 42% of the tasks are carried out by a SAR brigade team, 13% by the Reconnaissance team, 24% by the Technical brigade teams, 12% by the Logistics brigade and 10% by the Medical team.

4 Final Results and Conclusions

Attending that we are particularly interested to build and implement a DSS with the ability to prioritize certain teams for specific incidents, taking into account the importance of each team that acts in case of emergency, in the present work we have contributed positively for such objective.

We have performed the Delphi method, that consisted in 3 rounds of completed questionnaires. After the consensus on all issues, could be computed and evaluated an index allowing to associate a number (color) scale which evaluate the order of teams that must be called to perform a certain task - if the team is green (red) means that was more (less) suggested to carry out such order.

We also could identify which of tasks is the most indicated for each team, and, in principle, the team is called to respond to a certain order of priority. It can be observed that majority of the tasks are carried out by a SAR brigade team, followed by the Reconnaissance team, the Technical brigade teams, the Logistics brigade and the Medical team respectively.

The details about the analysis of results per round will be available in an extended version of the this article. Also, the tasks associated with priority to each team, will be detailed. Some suggestions for future work can be proposed. For example, to consider a more detailed weight for experts experience using hierarchical classification, where we can classify the experience of each expert evaluating the similarity between the individuals in the group of proposed experts. Another option is to experiment another dispersion measure, a nonparametric one for example, and compare with the number of consensus per round presented in the present work.

Acknowledgements. This work was supported by Portuguese funds through the *Center for Computational and Stochastic Mathematics* (CEMAT), *The Portuguese Foundation for Science and Technology* (FCT), University of Lisbon, Portugal, project UID/Multi-/04621/2013, and *Center of Naval Research* (CINAV), Naval Academy, Portuguese Navy, Portugal.

References

1. Reuven, L., Dongchui, H.: Choosing a technological forecasting method. Ind. Manage. **37**(1), 14–22 (1995)
2. Duuvarci, Y., Selvi, Ö., Günaydin, H.M., GüR, G.: Impacts of transportation projects on urban trends in İzmir, pp. 1175–1201, Digest 2008 (Teknik Dergi 19, Special Issue), December 2008
3. Powell, C.: The Delphi technique: myths and realities. Methodol. Issues Nurs. Res. **41**(4), 376–382 (2003)

4. Marques, M., Elvas F., Nunes, I.L., Lobo, V., Correia, A.: Augmented reality in the context of naval operations. In: Ahram, T., Karwowski, W., Taiar, R. (eds.) IHSED 2018, AISC, vol. 876, pp. 307–313 (2019). https://doi.org/10.1007/978-3-030-02053-8_47
5. Nunes, I.L., Lucas, R., Simões-Marques, M., Correia, N.: Augmented reality in support of disaster response. In: Nunes, I.L. (ed.) Proceedings of Advances in Human Factors and System Interactions, AHFE 2017 Conference on Human Factors and System Interactions, Advances in Intelligent Systems and Computing, 17–21 July, AISC, vol. 592, pp. 155–167. Springer (2018). https://doi.org/10.1007/978-3-319-60366-7_15
6. Nunes, I.L., Lucas, R., Simões-Marques, M., Correia, N.: An augmented reality application to support deployed emergency teams. In: Bagnara, S., Tartaglia, R., Albolino, S., Alexander, T., Fujita, Y. (eds) (eds.) Proceedings of the 20th Congress of the International Ergonomics Association (IEA 2018), Advances in Intelligent Systems and Computing, 26–30 August, AISC, vol. 822, pp. 195–204. Springer (2019). https://doi.org/10.1007/978-3-319-96077-7_21
7. Simões-Marques, M., et al.: Empirical studies in user experience of an emergency management system. In: Nunes, I.L. (ed.) Proceedings of Advances in Human Factors and System Interactions, AHFE 2017 Conference on Human Factors and System Interactions, Advances in Intelligent Systems and Computing, 17–21 July 2017, pp. 97–108. Springer (2018). https://doi.org/10.1007/978-3-319-60366-7
8. Azevedo, S.G., Carvalho, H., Machado, V.C.: Agile index: automotive supply chain. World Acad. Sci. Eng. Technol. **79**, 784–790 (2011)
9. Azevedo, S.G., Govindan, K., Carvalho, H., Machado, V.C.: Ecosilient index to assess the grenness and resilience of the up stream automotive supply chain. J. Clean. Prod. **56**, 131–146 (2013). https://doi.org/10.1016/j.jclepro.2012.04.011
10. Fraga, M: A economia circular na indústria portuguesa de pasta, papel e cartão. Master Thesis, FCT, Universidade Nova de Lisboa, Almada (2017)
11. De Vet, E., Brug, J., De Nooijer, J., Dijkstra, A., De Vries, N.K.: Determinants of forward stage transitions: a Delphi study. Health Educ. Res. **20**(2), 195–205 (2005)
12. Gunaydin, H.M.: Impact of information technologies on project management functions. IIT Doktora Tezi, Chicago, USA (1999)
13. Adler, M., Ziglio, E.: Gazing Into the Oracle: The Delphi Method and its Application to Social Policy and Public Health. Kingsley Publishers, London (1996)
14. Wissema, J.G.: Trends in technology forecasting. R & D Manag. **12**(1), 27–36 (1982)

Automatic Clinic Measures and Comparison of Heads Using Point Clouds

Pedro Oliveira[1]([⊠]), Ângelo Pinto[1], António Vieira de Castro[1,2],
Fátima Rodrigues[1,3], João Vilaça[4], Pedro Morais[4],
and Fernando Veloso[4]

[1] Institute of Engineering, Polytechnic of Porto (ISEP/IPP), Porto, Portugal
`{pmdso, ajlpa, avc, mfc}@isep.ipp.pt`
[2] LAMU, Multimedia Laboratory, Porto, Portugal
[3] Interdisciplinary Studies Research Center (IRSC), Porto, Portugal
[4] 2AI - Instituto Superior do Cávado e Ave, Braga, Portugal
`{jvilaca, pmorais, fveloso}@ipca.pt`

Abstract. Nowadays the necessity to automate processes is increasing. One of the areas where automation can help is medicine. This document shows a process of how to analyze a point cloud of a head to define reference points and extract important measures. It also describes a head comparison process to compare a base head with a set of heads and find out the most similar one. The extracted measures will be applied for the diagnose and treatment of positional plagiocephaly disease that affects babies. A solution, integrating different technologies, has a graphical interface to present the measures retrieved from the point clouds of the heads and the results of heads comparison. The interface allows the visualization of the point clouds of the heads and helps to see the difference between heads. With the auxiliary of this solution it is possible to create a more adequate orthosis to treat each patient with positional plagiocephaly.

Keywords: Point clouds · Positional plagiocephaly · Automatic measures

1 Introduction

The technology to capture a 3D representation of an object is increasingly more accessible and reliable. This makes possible to easily capture a 3D representation of an object. Therefore it is possible to use that representation for analysis of the object, in some cases it can help to identify medical conditions. This paper explains the process used to automatize the extraction of anthropometric measures from 3D point clouds of heads. The measures will be used to help the creation of an orthosis to treat positional plagiocephaly.

Positional plagiocephaly is a condition characterized by an asymmetrical distortion of the cranium. A baby's skull is not strong, so it can suffer deformations, which cause plagiocephaly. In severe deformations, to fix the problem, the baby may need to use an orthosis, that should be used when the baby has between 3 and 18 months of age [1]. The orthosis will allow the head to grow gently back into a normal shape. This treatment must be done early, before the fontanels closes and the cranium fuses together.

A. M. Madureira et al. (Eds.): HIS 2018, AISC 923, pp. 560–569, 2020.
https://doi.org/10.1007/978-3-030-14347-3_55

2   Point Clouds

Point clouds are a set of data points in space, represented in a coordinate system. They are generally produced by 3D scanners, which measure a large number of points on the external surface of the real-world object [2]. Point clouds can be used to create 3D meshes to be used in many fields including medical imaging, architecture and 3D printing. When point clouds need to be aligned with other point clouds a process called Point Set Registration [3] is usually used. The process of aligning multiple point clouds into a globally consistent model includes the merge or comparison of the point clouds. In this case it was not necessary to perform this process. The alignment was made through the detection of anthropometric reference points in the head. Knowing the x, y and z coordinates of some reference points it is possible to apply maths to determine other points. The alignment of heads was made rotating the heads relatively to those reference points, this process is detailed in Sect. 3.

In this case study it was used a dataset with 11 point clouds. The point clouds were stored in PLY files, which represent 11 baby heads. PLY is a computer file format known as Polygon File Format. It's designed to store three-dimensional data from 3D scanners [4]. The graphical objects stored are described as a collection of vertices defined by the x, y and z coordinates. This format permits to define other properties like polygons, vertices, polygons' colours and normal vectors.

3   Point Recognition

In order to extract the cranial measures needed it was first necessary to identify some important reference points in the heads, such as the sellion, tragions and the top of the head. These points are shown in Fig. 1.

All the point clouds of the heads were initially in a different coordinate system, making it impossible to determine the skull measures. For that to be possible it was first indispensable to align the point clouds of the heads. This way heads are oriented in a well-defined coordinate system, which allows to know the orientation of the heads, to know, for instance, to which side is the head facing. Hence, the reference points are detected in an easier way, which helps to calculate head measurements.

To position the point clouds in the same place, it was calculated the mean point of all points in each point cloud. After that all heads were translated so that their center were positioned in the origin. After the translation process, the clouds had to be rotated to align them. To achieve that, the first step was to find the eye corners by finding the closest points to the mean point of each cloud. This was possible because the datasets provided had more definition in the face area, therefore the mean point was nearer the face, making the closest points to that mean point to be the eye corners.

To correctly align the point clouds, it was calculated the angle in x, y, and z axis from the vector between the center of the point cloud and eyes and the front vector (positive z axis). Centering and aligning the heads made it possible to detect other reference points more strictly. The point of the top of the head was detected as the point with maximum value in the y axis. The sellion was detected as the point in the middle of the eye corners [2].

In order to detect the tragions, it was first necessary to find the tragus points, also represented in Fig. 1. To calculate them it was found the local maximum in the x axis of the region of the ears, which corresponds to the most exterior points of the head in the ears area. With the tragus detected, the tragion was found as the local minimum in the x axis of the tragus region with a y coordinate greater than the tragus point. So, starting in the position of the tragus, it was searched in the above points for the most interior point of the head, the point with minimum x. Lastly, the back point of the head was detected by finding the point that was behind the sellion. The detected points are represented in Fig. 1.

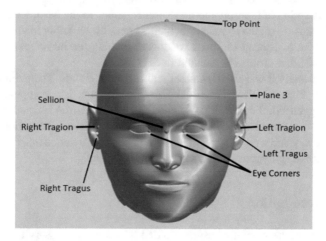

Fig. 1. Representation of the head showing the reference points and plane.

4   Measures

To calculate the measures it was necessary to define some planes in the heads, the plane formed by the tragions, sellion and back point was named as plane 0. The head was divided in 12 equal planes (10 planes from plane 0 to top of the head and 2 planes bellow plane 0). The circumference of the head was measured in the plane 3. To calculate the circumference the points in plane 3 were detected. The circumference is equal to the sum of the distance between all those points. The calculation was made according to Eq. (1).

$$Cranium\,Circumference = \sum \sqrt{(x_1 - x_2)^2 + (z_1 - z_2)^2} \qquad (1)$$

The cranial width and length were also calculated using the plane 3. The width was determined according to Eq. (2) and length according to Eq. (3).

$$Width = \max(x) - \min(x) \qquad (2)$$

$$Length = \max(z) - \min(z) \tag{3}$$

The cephalic index was calculated by the ratio of the width to the length, applying the formula of Eq. (4).

$$Cephalic\ index = \frac{cranialwidth}{craniallenght} \tag{4}$$

The diagonal difference was calculated by the difference of the diagonals of plane 3, using the Eq. (5).

$$Diagonals\ Difference = |diagonal_1 - diagonal_2|cm \tag{5}$$

The ear offset was calculated as the difference of values in z between left and right tragions, applying Eq. (6).

$$Ear\ Offset = |leftTragium_z - rightTragium_z|\ cm \tag{6}$$

The anterior and posterior cranial asymmetry index was measured using the following formula (7), where l is the larger quadrant volume and s is the smaller quadrant volume.

$$Cranial\ Aymetry\ index = \frac{(l - s) * 100}{s}\% \tag{7}$$

The volume of each quadrant was determined as the sum of the tetrahedron formed by a triangle and the center of the cloud [6]. The formula used is shown in Eq. (8), where a, b, c are the points of the triangle and d the center of the cloud.

$$Volume_{tetrahedron} = \frac{|(a - d) \cdot ((b - d) \times (c - d))|}{6} cm^3 \tag{8}$$

Table 1. First set of measures of the heads.

Heads	Cranial circumference (mm)	Cranial width (mm)	Cranial length (mm)	Cephalic index (%)
Normal	525.9	132.6	149.2	88.9
Head 1	503.4	133.1	150.1	88.6
Head 2	432.4	126.6	147.0	86.1
Head 3	460.4	128.8	143.2	90.0

Using the presented formulas it was possible to calculate the wanted cranial measures. The calculated measures for 5 of the heads are presented in Tables 1 and 2.

Table 2. Second set of measures of the heads.

Heads	Diagonal difference (mm)	Ear offset (mm)	Anterior cranial asymmetry index (%)	Posterior cranial asymmetry index (%)
Normal	0.1	0.4	5.3	0.5
Head 1	7.9	2.6	18.6	1.4
Head 2	19.3	113.8	15.2	11.7
Head 3	26.0	113.5	2.1	23.0

5   Head Comparison

Another important step in the process of building an orthosis for treating plagiocephaly is to compare heads. When building a model for the orthosis, it is necessary to compare the head with plagiocephaly with models of normal heads. This is done to determine which of the normal models is more similar to the abnormal head. Thus, the orthosis can be made using a model that perfectly fits the head, providing a good development of the head's growth to a normal shape.

In order to compare heads' shapes, it was necessary to compare point clouds. There are different algorithms to compare point clouds, such as nearest neighbor, nearest neighbor with local modelling, normal shooting and iterative closest point (ICP) [7]. There are also other methods to compare shapes of objects like Procrustes analysis. Procrustes can be used to calculate the Procrustes distances between two objects. The distance calculated is a measure of shape difference between the two objects [8]. Another similar measure with Procrustes is the Hausdorff distance. Hausdorff distance has been applied in many areas, such as object matching, image processing and face detection [9]. This technique is often used to determine similarity between shapes [10]. If two shapes have a small Hausdorff distance between them they are more similar than if they had a big distance [11]. So, as with Procrustes distance, the smaller the distance the bigger the resemblance between shapes.

In order to compare heads, both Procrustes and Hausdorff distances were used. To be able to calculate the correct distance between two objects using these algorithms, these objects must have been previously aligned. This was not a problem in this case because the alignment of the heads had already been made for the extraction of head measurements. In this case study, a dataset of 3D scans with 11 baby heads was used. The dataset had 1 normal head and 10 abnormal heads with deformities. In this dataset it was only known the real similarity of head number 1 with the other heads, so the comparison could only be made in this case. Procrustes and Hausdorff distances were used to assess which of the heads was more similar to head number 1. In order to compare the shapes of the heads, different approaches could be made. Comparing all the points of the heads, comparing 2D planes or comparing just sections of the heads. In this case study, 3 experiments were made. In Experiment 1, Procrustes distance was calculated considering all points of the head. In Experiment 2 and 3, the points from three 2D planes above the eye level were used. In Experiment number 2, Procrustes distance was used and in the 3rd it was used Hausdorff distance. The planes used were the planes 3, 4 and 5 represented in Fig. 2. These planes were used as they represent sections of the head where many deformities are present in cases of plagiocephaly.

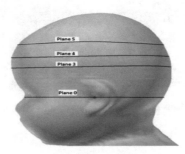

Fig. 2. Representation of the head planes. Adapted from [12].

To calculate the Procrustes distances between heads the software R [13] was used, along with "shapes" package [14], which has several functions for Procrustes analysis. Hausdorff distance was also determined using R software, using the "pracma" package [15]. To be able to compare heads, all heads must have the same number of points. Also, in Experiment 1, to use functions from "shapes" package, the number of points had to be reduced to 17000, more or less 30% of the original number. This had to be done because functions from "shapes" package are very demanding and require a lot of RAM to be executed. So, to execute these functions in most computers the number of points should not be higher than 17000. This creates a problem, as reducing the number of points in a point cloud can decrease the representation of the original model, especially if done randomly. To overcome this issue a technique called voxelization was used.

Voxelization algorithms are based on voxels. Voxels are the 3D equivalent of pixels [16], they are basically cubes that can contain points within them [17]. Voxelization can be described as "the process of producing a discrete 3D representation of an object" [18]. This technique is one of the simplest and easily understandable ways to transform 3D objects into regular structures, while keeping its original geometry and structure [19]. Voxelization of the baby heads was performed using "VoxR" package [20] of R software. After applying the voxelization algorithm it was possible to see that the head shape was the same, even though the point cloud had almost 70% less points. Figure 3 shows the normal head of the dataset, where the head on the left is the original head and the head on the right is the head after voxelization. Voxelization was also used in Experiments 2 and 3. As the number of points in each plane have to be equal to compare heads, voxelization was used to reduce the number of points of each plane to an equal number.

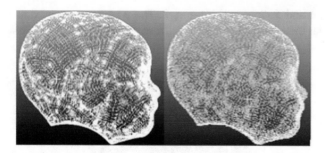

Fig. 3. Comparison of original head (left) and voxelized head (right).

6   Results

After the voxelization process it was possible to calculate the Procrustes and Hausdorff distance between the heads. The heads more similar to head number 1 were the normal head and head number 2, so it was expected that the least distance were in these heads. Results from Experiment 1 were close to expected. The results showed that the most similar heads with the head no. 1 were, by the following order: 2, 4, Normal, 8, 10, 7, 6, 3, 5, 9. Even though the normal head was only considered to be the 3rd most similar, the two most similar heads, head number 2 and the normal head were in the top 3 closest heads. Hence we can conclude that comparing using Procrustes distance might be a good approach for these type of cases.

In Experiment 2, as explained before, heads were only compared considering the points in 2D planes, planes 3, 4 and 5. Figure 4 shows an upper view of planes 3, 4 and 5 of the normal head, on the left, and head number 8, on the right. Planes are represented by different colors. Although these are only 2D planes of the head, deformities are clearly seen, as there is a prominence in the right head on the lower right corner. In order to compare heads, Procrustes distance was calculated by each plane individually and the final result was given by the mean of distances among the planes. In this case it was also necessary to perform voxelization, as the planes of each head had different number of points. Since the number of points to compare was much less than in Experiment 1 there were not performance issues, so the number of points used was the minimum number of points per plane among all heads. The results of Experiment 2 showed that the most similar head was head number 4, which is not true. Also, the normal head and head no. 2 were not in the top 3 of the closest heads. So, using Procrustes with 2D planes did not provide good results.

In Experiment 3, heads were compared considering the points in 2D planes, planes 3, 4 and 5. The distance used was Hausdorff. The distance was also calculated by each plane individually and the final result was given by the mean of distances among the planes. Voxelization was also needed, so that all planes of each head had the same number of points. The most similar heads with head number 1 are, by the following order: Normal, 2, 7, 8, 5, 3, 4, 10, 6, 9. Results from Experiment 3 showed that the most similar heads with head number 1 were, by the following order: Normal, 2, 7, 3, 4, 10, 9, 6, 5, 8. These results can be considered good, as the three most similar heads were correctly determined. From the 3 methods experimented it is possible to conclude

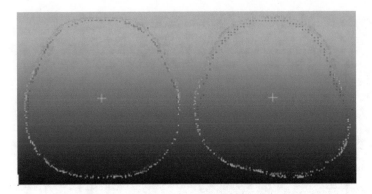

Fig. 4. Planes 3, 4 and 5 of normal head and head number 8.

that the best one was from Experiment 3, where heads were compared using Hausdorff distance and using three 2D planed of the heads.

7 Implementation

The solution was implemented using Unity, C# and R. The Unity was chosen by its capabilities in building interfaces that conjugate 2D and 3D elements and because Unity allows to see the results more easily and fast while the solution is developed. The R language was chosen because of the diversity of processing algorithms already developed in this language.

The graphical interface of the developed solution has two main areas. The first area is dedicated in the visualization and manipulation of point cloud. The second area is dedicated to view the results and make the necessary operations. In the Measures functionality, it's shown at left top corner the top view of the head. The top view of the head allows the user to see where the width and length was measured and also allows the user to see easily and more clearly the shape of the head.

The Export button is a utility button that lets the user export the point cloud of the head uniformly aligned and centered at origin. The Manual Alignment button has been so that the user can change the position of tragions and eye corners markers. This allows the user to be able to correct the position of initial markers for a more exact alignment and measures (Fig. 5).

The R language was used only in the comparison feature through the calling of a R script in the Unity. Two R scripts were made to handle two modes of comparison of the heads. The first handle the "Complete Head" mode, it uses the Procustes algorithm to calculate the head more similar to the selected one. The second handle the "Plane 3, 4, 5" mode, it uses the Hausdorff algorithm to calculate the head more similar to the selected one (Fig. 6).

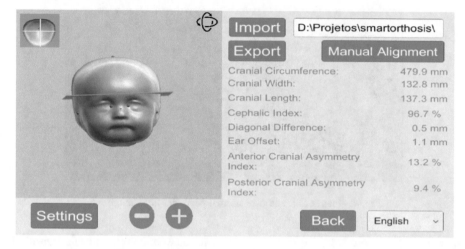

Fig. 5. Graphical interface of the feature measures

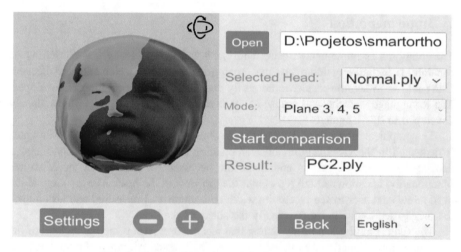

Fig. 6. Graphical interface of the feature comparation.

8 Conclusion

This paper showed several methods to be used for helping the creation of an orthosis for the treatment of plagiocephaly. It was shown how to extract cranial measures like the cranial circumference from point clouds of heads. It was also detailed a process for comparing point clouds of heads, which can be also used for comparing point clouds of other objects. The results of the comparisons seem promising, although it is still necessary to make more tests using larger datasets with different characteristics like variations in the density of points along the point cloud to assure this is a trustful method.

Acknowledgement. This work was funded by projects NORTE-01-0145-FEDER-024300 ("SmartOrtho-sis"), supported by Northern Portugal Regional Operational Programme (Norte2020), under the Portugal 2020 Partnership Agreement, through the European Regional Development Fund (FEDER), and has also been funded by FEDER funds, through Competitiveness Factors Operational Programme (COMPETE).

References

1. GREOP: Cranial remolding orthoses. Plagiocephaly (2018). https://www.greop.com/cranial-remolding-helmets. Accessed 03 Sept 2018
2. Slupchynskyj, O.: Regions of the nose. Rhinoplasty Surgeon New York. http://rhinoplastysurgeonnewyork.com/regions-of-the-nose/
3. Rouse, M.: Point cloud. https://whatis.techtarget.com/definition/point-cloud
4. Myronenko, A., Song, X.: Point-set registration: coherent point drift. IEEE Trans. Pattern Anal. Mach. Intell. **32**(12), 2262–2275 (2010)
5. Bourke, P.: PLY - Polygon File Format. http://paulbourke.net/dataformats/ply/. Accessed 04 May 2018
6. StackOverflow: Math - calculate volume of any tetrahedron given 4 points. Stack Overflow. https://stackoverflow.com/questions/9866452/calculate-volume-of-any-tetrahedron-given-4-points. Accessed 23 May 2018
7. Tsakiri, M., Anagnostopoulos, V.: Change detection in terrestrial laser scanner data via point cloud correspondence. ResearchGate. https://www.researchgate.net/publication/282001967_Change_Detection_in_Terrestrial_Laser_Scanner_Data_Via_Point_Cloud_Correspondence. Accessed 10 July 2018
8. Claude, J.: Morphometrics with R. Springer, New York (2008)
9. Kraft, D.: Computing the Hausdorff distance of two sets from their signed distance functions (2015)
10. Nutanong, S., Jacox, E.H., Samet, H.: An incremental Hausdorff distance calculation algorithm. Proc. VLDB Endow. **4**(8), 506–517 (2011)
11. Kim, I., McLean, W.: Computing the Hausdorff distance between two sets of parametric curves. Commun. Korean Math. Soc. **28**(4), 833–850 (2013)
12. Plank, L.H., Giavedoni, B., Lombardo, J.R., Geil, M.D., Reisner, A.: Comparison of infant head shape changes in deformational plagiocephaly following treatment with a cranial remolding orthosis using a noninvasive laser shape digitizer. J. Craniofac. Surg. **17**(6), 1084–1091 (2006)
13. R Core Team: R: a language and environment for statistical computing (2018)
14. Dryden, I.L.: Shapes - statistical shape analysis software in R. https://www.maths.nottingham.ac.uk/personal/ild/shapes/. Accessed 10 July 2018
15. Borchers, H.W.: Pracma: practical numerical math functions (2018)
16. Rouse, M.: What is voxel? - Definition from WhatIs.com. WhatIs.com. https://whatis.techtarget.com/definition/voxel. Accessed 10 July 2018
17. Kim, K., Cao, M., Rao, S., Xu, J., Medasani, S., Owechko, Y.: Multi-object detection and behavior recognition from motion 3D data. In: CVPR 2011 Workshops, pp. 37–42 (2011)
18. Laine, S.: A topological approach to voxelization. Comput. Graph. Forum **32**(4), 1 (2003)
19. Han, Z., Liu, Z., Han, J., Vong, C.M., Bu, S., Chen, C.L.P.: Unsupervised learning of 3-D local features from raw voxels based on a novel permutation voxelization strategy. IEEE Trans. Cybern. **99**, 1–14 (2017)
20. Lecigne, B., Delagrange, S., Messier, C.: VoxR: metrics extraction of trees from T-LiDAR data. R package. ResearchGate. https://cran.r-project.org/web/packages/VoxR/index.html. Accessed 10 July 2018

Structural and Statistical Feature Extraction Methodology for the Recognition of Handwritten Arabic Words

Marwa Amara[1]([✉]), Kamel Zidi[2], and Khaled Ghedira[1]

[1] SSOIE-COSMOS, Tunis, Tunisia
amara1marwa@gmail.com
[2] Community College, Tabuk University, Tabuk, Saudi Arabia

Abstract. Knowledge concerning the topography of Arabic letters, as well as the structural characteristics between background regions and character components is investigated as a novel approach for Arabic recognition. The suggested feature extraction method reduces the classifier input data to only the most significant and essential.

First, connected components consisting of more than one character are segmented into characters. Secondly, the primitives are extracted according to the knowledge of character structures and some statistical characteristics. Finally a hybrid model based on the combination of support vector machines (SVM) classifier and particle swarm optimization (PSO) is used to evaluate the performance of features extracted.

Keywords: Arabic handwritten recognition · Feature extraction · Support vector machines · Particle swarm optimization · IFN/ENIT-database

1 Introduction

Using a scanner, the input document is transferred into a digital file. For most documents with complex backgrounds, the image has to go through a preprocessing algorithm to remove non-significant information. Then, the lines then word/sub-words in the cleaned image are isolated or segmented to extract the character. Those characters which will be used in the feature extraction and the classification steps. OCR steps work together to achieve one goal which is converting an image into a text file that can be edited. Therefore, the output of each step in the OCR system propagates to the next one making the fail of one stage affects the whole system. In the classification stage, the extracted features are compared to the model; if the features are corresponding of one class of the model, the input character is classified into the appropriate class. In this stage, if segmentation algorithm fails to yield a good segmented character, the classifier will be enabled to affect the forms to the appropriate classes.

© Springer Nature Switzerland AG 2020
A. M. Madureira et al. (Eds.): HIS 2018, AISC 923, pp. 570–580, 2020.
https://doi.org/10.1007/978-3-030-14347-3_56

The poor understanding of the Arabic language is considered as one of the problems associated with character recognition. The calligraphic nature of the Arabic script is distinguished from other languages in several ways. Actually, Arabic is written cursively even when printed or handwritten. It has also right to left flow. The Arabic language consists of 28 letters, added to Hamza أ which is most often used as a complementary sign. The widths of letters in Arabic are variable. Its letters shapes change depending on their position in the word; a single character can have from one to four shapes start, middle, end and isolated. Some characters have a similar shape with only a change in the number of dots or diacritics as shown in Fig. 1.

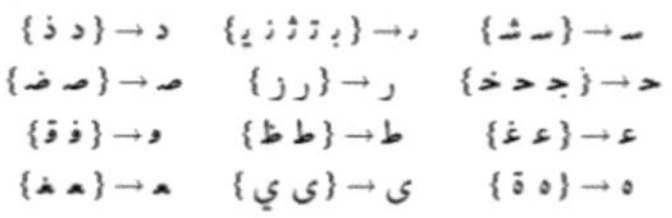

$$\{ \text{ذ د} \} \rightarrow \text{د} \quad \{ \text{ي ن ث ت ب} \} \rightarrow \text{ب} \quad \{ \text{ش س} \} \rightarrow \text{س}$$
$$\{ \text{ض ص} \} \rightarrow \text{ص} \quad \{ \text{ز ر} \} \rightarrow \text{ر} \quad \{ \text{خ ح ج} \} \rightarrow \text{ح}$$
$$\{ \text{ظ ط} \} \rightarrow \text{و} \quad \{ \text{ظ ط} \} \rightarrow \text{ط} \quad \{ \text{غ ع} \} \rightarrow \text{ع}$$
$$\{ \text{م م} \} \rightarrow \text{م} \quad \{ \text{ي ی} \} \rightarrow \text{ی} \quad \{ \text{ة ه} \} \rightarrow \text{ه}$$

Fig. 1. Similar shape with different diacritics number.

Besides, the use of ligature in Arabic is common (Fig. 2). There is one compulsory ligature (لا), that for ل and ا. Similarly, the vertical overlap may occur by the intersection of connected components for some characters combinations where characters occupy a shared horizontal space creating vertically overlapping connected or disconnected blocks of characters.

Fig. 2. Overlap and ligature.

In this paper, our goal is to identify the best features extraction method, especially, for Arabic words. We propose here a new evaluation methodology which avoids the features extraction in order to enhance the classification rate. The main contribution of this research is to enhance the features extraction process of an Arabic OCR system based on structural features to reduce the recognition error. The system was tested with IFN/ENIT-database.

The rest of this paper is organized as follows: Sect. 2, presents the state of the art of previous features extraction algorithms. Section 3 explains the introduced method used to extract features. In Sect. 4, experimental results are discussed. Finally, some concluding remarks are given in Sect. 5.

2   Related Works

The primitives extraction consists in transforming an image (character, grapheme, ...) into a vector of fixed size. The choice of primitives is important, which clearly differs from the result of the classification. The extracted information is then transmitted to the recognition process to construct models. This model will be used later for classification. The primitives extracting techniques proposed by researchers can be classified into two main types; structural features and statistical features.

Structural Features: are aspects totally related to the characteristics of writing such as loops, diacritic points, etc. For example, a number of Arabic letters share the same body but differ only in the number and location of the didactic points, such as the example of the three characters (ب), (ت), And (ث). In order to distinguish these letters during classification, it is necessary to have information on the number of diacritical points and their positions. The structural features create an adequate method to emphasize the characteristics specific to the Arabic language.

One of the first attempts to extract the structural features in an Arabic handwriting recognition system was proposed by Almuallim and Yamaguchi [1]. The researchers used the word skeleton and the structure to recognize the words. Clocksin and Fernando [2] have addressed the field of Syriac manuscripts (a Semitic language of the Near East, belonging to the Aramaic language) based character representation and invariant moments. In their paper, Mozaffari et al. [3] detected the end points and intersection points from the skeleton of the word. The vector of primitives used is composed of different structural characteristics as well as other statistics. Statistical characteristics: Are numerical measurements computed on images or regions of images.

Statistical Features: These features include pixel densities, Freeman code, invariant moments, Fourier descriptors, and so on. The Freeman codes have been used by Abdelazeem and El-Sherif [4] and Alaei et al. [5]. The gradient was used by Al Amri et al. [6], and Awaidah and Mahmoud [7]. In Alamri et al. [6], the direction of the gradient was used as characteristics of the character image. The invariant moments were used as a characteristic by Amara and Zidi [8] and Al-Khateeb et al. [9]. Chergui et al. [10] have proposed a system that uses features invariant to scale for the recognition of Arabic handwritten words. In addition, other features include Gabor wavelet transform [10], discrete cosine transform [11], wavlet transform [12], and hough transform [13] were used by the researchers.

However, it is possible to combine several features extraction methods in order to give a better description of the form to be classified later. Different types of structural primitives (loops, curves, ...) and statistical (invariant moments, projections, ect) have been used for the description of Arabic writing. The performance of a classifier can rely as much on the quality of the features as on the classifier itself. Therefore, an efficient set of features should represent characteristics that are particular for one class and be as invariant as possible to changes within this class. This may be a reason that structural features remain more common for the recognition of Arabic script. Our paper discusses the extraction of structural features for the recognition of handwritten Arabic words.

3   Feature Extraction

It is obvious that the recognition rate of an OCR system depends closely on the discriminating power of the extracted features. Indeed, the addition of a new characteristic or the reduction of the total number can influence the system reliability. According to the studies carried out, this choice depends mainly on the problem to be solved and used tools. In this work, we chose the statistical primitives in order to characterize the image. We have also proposed some structural features which take into account the special graphics of the Arabic letters. In the following, we discuss the selected characteristics.

3.1   Hu's Moment

Hu [14] proposed a function that allows the generation of geometric moments. These moments characterize the image of characters through coefficients called invariant moments. They are invariant to translation, rotation, reflection, and the combination of those different factors. In the case of document recognition, we reduce ourselves to binary images. The used image of character is defined by the function of the Eq. 1.

$$\begin{cases} f(x,y) = 1 \; If \; the \; pixel \; belongs \; to \; the \; image \\ f(x,y) = 0 \; If \; the \; pixel \; does \; not \; belongs \; to \; the \; image \\ does not \end{cases} \tag{1}$$

Hence, an image can be represented by a continuous function $f(x,y)$. The moment of order $p+q$ is defined by the Eq. 2.

$$\eta_{pq} = \int_{-\infty}^{+\infty} \int_{-\infty}^{+\infty} (x - \bar{x})^p (y - \bar{y})^q f(x,y) dx dy \; For \; p, q = 0, 1, 2, ... \tag{2}$$

$\bar{x}$ and $\bar{y}$ can be obtained by the formula 3.

$$\bar{x} = \int_x \int_y x f(x,y) dx dy \; and \; \bar{y} = \int_x \int_y y f(x,y) dx dy \tag{3}$$

A set of seven moments invariant to translations, rotations and homotheties is then computed. These moments are illustrated below:

$I_1 = \eta_{20} + \eta_{02}$
$I_2 = (\eta_{20} - \eta_{02})^2 + 4\eta_{11}^2$
$I_3 = (\eta_{30} - 3\eta_{12})^2 + (3\eta_{21} - \eta_{03})^2$
$I_4 = (\eta_{30} + \eta_{12})^2 + (\eta_{21} + \eta_{03})^2$
$I_5 = (\eta_{30} - 3\eta_{12})(\eta_{30} + \eta_{12})[(\eta_{30} + \eta_{12})^2 - 3(\eta_{21} + \eta_{03})^2] + (3\eta_{21}$

$$-\eta_{03})(\eta_{21} + \eta_{03})[3(\eta_{30} + \eta_{12})^2 - (\eta_{21} + \eta_{03})^2]$$
$$I_6 = (\eta_{20} - \eta_{02})[(\eta_{30} + \eta_{12})^2 - (\eta_{21} + \eta_{03})^2$$
$$+4\eta_{11}(\eta_{30} + \eta_{12})(\eta_{21} + \eta_{03})]$$
$$I_7 = (3\eta_{21} - \eta_{03})(\eta_{30} + \eta_{12})[(\eta_{30} + \eta_{12})^2 - 3(\eta_{21} + \eta_{03})^2] + (\eta_{30} - 3\eta_{12})(\eta_{21} +$$
$$\eta_{03})[3(\eta_{30}$$

3.2 Freeman Chain Code

The Freeman chain encoding is used to represent the boundary of a certain shape by a connected sequence of segments. This representation is based on 8-connectivity segments, in our case. The direction of each segment is encoded using a numbering scheme. Since each pixel of an image has separate 8-neighbors. Possible connectivity between pixels are illustrated in Fig. 3.

Fig. 3. The possible neighborhoods of a pixel.

The extraction of freeman coding proceeds through the following steps:

- Choose the initial pixel of the contour and the direction of browse;
- Code the direction that allows to pass from a pixel of the contour to its immediate neighbor;
- Continue until you return to the initial pixel.

Moving from one pixel to another will produce a direction number (code). For example, consider the segment of Arab characters ‬ as shown in Fig. 4.

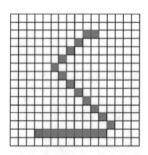

Fig. 4. Character ‬ representation.

In Fig. 4, the starting point is marked in red. The generated codes are defined in this case by tracking the white pixels in the contour boundary:

00000000133333331111045555777777544444444

3.3   Intersection with Zones

We defined during our paper [15] the baselines from the image of the word. The special lines include upper line (UL), baseline (BL), Baseline start (BLS) and lower lines (LL) as shown in Fig. 5.

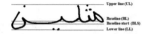

Fig. 5. Special lines detection.

The area covered by a word, can be horizontally divided into three sub-areas using these lines mentioned above. These areas are organized as follows:

- The upper zone (UZ) delimited by the baseline start and the upper line;
- The middle zone (MZ) located between the baseline and the baseline start;
- The lower zone (LZ) that lies between the baseline and the lower line.

Figure 6 describes the location of the different areas in the image of word.

(1)   (2)

Fig. 6. Location of zones; UZ, MZ, LZ.

The intersection of the segment (character) with the defined zones allows to describe its topography. In other words, if there is an intersection with one of the zones, the vector takes the value 1. Otherwise, it takes the value 0. Let us take the illustrative example of the character (Í). This character has a first intersection with the median zone and another with the upper zone. Hence the primitive vector will be described in the following way:

$$\begin{cases} UZ = 1 \\ MZ = 1 \\ LZ = 0 \end{cases}$$

3.4   Loops Detection

Arabic characters can be classified into two categories; Characters that include loops and others that do not include them. The existence of loops in certain characters also depends on the position of the character within a subword (initial, middle, final, isolated). For example, a "Meem" in its initial position (ﻣ) and

Fig. 7. Loops detection.

middle (ـهـ) includes a loop, while in its final (ﻢ - Mـ) does not include it (see Fig. 7).

Indeed, the detection loops reduces the possible ambiguities between characters.

3.5   Direction of the Background Region

We used the direction of the background region presented by [16] to detect the direction of the segment. This type of characteristic consists of two main categories: characters that have the "face up" (FU) and others that have the "face down" (FD) (see Fig. 8).

(1)   **(2)**

Fig. 8. Direction of region; (1) FU, (2) FD.

Let's take the example of the ﺑ character that will be described in the following way:

$$\begin{cases} FU = 1 \\ FD = 0 \end{cases}$$

In summary, we can say that the vector of primitives representing the character consists of a mixture of different primitives. Static primitives exploiting the Hu moments. Characteristics obtained from the Freemam code. Other characteristics that describe the intersection of the character segment with the zones. Finally primitives that represent the direction of region background and the existence of loops.

This mixture produces a better presentation of the characteristics of the Arabic letters. However, other criteria are also taken into account in the choice of primitives such as; Ease of calculation, non-redundancy and discriminating power. After a number of tests, we selected from a set that was extracted the primitives that presented the best rates.

4   Experimental Settings

The goal of the study is to evaluate the usefulness and the effectiveness of structural proposed features on the recognition of handwritten Arabic script. In all experiments, we use a support vector machine classifiers of type One Against one hybridized with the pso. In the OAO approach, an SVM is trained to discriminate each two classes (k, m). Thus, the number of SVMs used in this approach is $M(M - 1)/2$ [20, 21].

The PSO produces the particles of the initial population randomly through the evolutionary computation to find the optimal solution. In each evolution, the particle would change the individual search direction by two search memories. The first search is the optimal individual variable memory and the other is the optimal variable memory of the population. After the computation, the PSO would calculate the optimal solution according to the optimal variable memory.

The purpose of this stage is to find the best (C, γ) SVM parameters using PSO intelligence. The algorithm proposed in [19] describes the entire process of the hybrid PSO-SVM process.

4.1   Database Description

The experimentation is conducted using the IfN/ENIT database [16]. The IfN/ENIT database was created by the Institute of Communications Technology (IfN) at Technical University Braunschweig in Germany and the Ecole Nationale d'Ingenieurs de Tunis (ENIT) in Tunisia. The version 1.0 of this database consists of 26459 images of the 937 names of cities and towns in Tunisia, written by 411 different writers. The database contains 115585 pieces of Arabic words and 212211 characters. The images are partitioned into four sets. To this date, this database is the most used by many researchers of Arabic handwritten text recognition.

4.2   Experiments and Results

During all the experiments, we tested our approach on sample (a) of the IFN/ENIT database. This sample contains 569 city names written by 10 different writers. A summary of the recognition results obtained is shown in the Table 1.

By evaluating the histogram of the Fig. 9), we can notice that the recognition rates obtained using the two different models are the same in most cases. Nevertheless, the OAO-OEP hybrid model performs better in some cases.

The most noticeable improvement is that a 100% recognition rate is achieved in the training phases for some confused characters, which is better than those presented in the literature. We believe that our results are competitive and quite promising since we used only limited number of features. Otherwise, we conclude from the results that the achieved accuracy is due to the discriminatory power of features and the capabilities of OAO-PSO classifier.

Table 1. Handwriting character recognition rates.

Parameters	OAO classifier	OAO-OEP classifier
Meilleur (C, γ)	$(2^3, 2^{-1})$	$[1.5, 2]$
Noyau	RBF	RBF
Recognition rate	87.65	87.88
Learning time (s)	34.61	32.84
Recognition time (s)	26.01	24.69

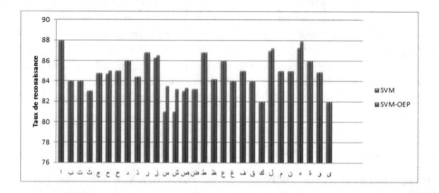

Fig. 9. Variation of the recognition rate per character with the IFN/ENIT database.

Based on experiments, we can conclude that the proposed features gives encouraging results for recognizing handwritten Arabic characters. In fact, the main advantage of the structural features is that it allows better description of the special characteristics of Arabic script. In general, we can conclude that the use of the structural features can be a good alternative for enhancing the performance of an AOCR system.

5   Conclusions and Future Works

The recognition of Arabic handwritten writing, with its morphology problems, imposes a cooperation of several types of features. Those features must cover to the variability of the word shapes. Among the different type of features, we proposed new structural features with some proposed static features to resolve the problem.

An efficient features vector should represent characteristics that are particular for one class and be as invariant as possible to changes within this class. In this paper, the following features are detailed: Hu's moment, Feeman chain code, Intersection with zones, loops and direction of the background region. A feature set made to feed a classifier can be a mixture of such features. In fact, the performance of a classier can rely as much on the quality of the features as on the classifier itself. For this prepose with have used an hybred OAO-PSO classifiers to recognize words. Extracted structural features will be tested on the database IFN-ENIT.

Acknowledgment. This research and innovation work is carried out within a MOBIDOC thesis funded by the EU under the PASRI project.

References

1. Almuallim, H., Yamaguchi, S.: A method of recognition of Arabic cursive handwriting. IEEE Trans. Pattern Anal. Mach. Intell. **5**, 715–722 (1987)
2. Clocksin, W.F.: Towards automatic transcription of Syriac handwriting. In: Proceedings of the 12th International Conference on Image Analysis and Processing, pp. 664–669. IEEE, September 2003
3. Mozaffari, S., Faez, K., Ziaratban, M.: Structural decomposition and statistical description of Farsi/Arabic handwritten numeric characters. In: Proceedings of the Eighth International Conference on Document Analysis and Recognition, pp. 237–241. IEEE, August 2005
4. Abdleazeem, S., El-Sherif, E.: Arabic handwritten digit recognition. Int. J. Doc. Anal. Recogn. **11**(3), 127–141 (2008)
5. Alaei, A., Nagabhushan, P., Pal, U.: Fine classification of unconstrained handwritten Persian/Arabic numerals by removing confusion amongst similar classes. In: 10th International Conference on Document Analysis and Recognition, ICDAR 2009, pp. 601–605. IEEE, July 2009
6. Alamri, H., He, C., Suen, C.: A new approach for segmentation and recognition of Arabic handwritten touching numeral pairs. In: Computer Analysis of Images and Patterns, pp. 165–172. Springer, Heidelberg (2009)
7. Awaidah, S.M., Mahmoud, S.A.: A multiple feature/resolution scheme to Arabic (Indian) numerals recognition using hidden Markov models. Sig. Process. **89**(6), 1176–1184 (2009)
8. Amara, M., Zidi, K.: New mechanisms to enhance the performances of arabic text recognition system: feature selection. In: Handbook of Research on Machine Learning Innovations and Trends, pp. 879–896. IGI Global (2017)
9. AlKhateeb, J.H., Jiang, J., Ren, J., Khelifi, F., Ipson, S.S.: Multiclass classification of unconstrained handwritten Arabic words using machine learning approaches. Open Sig. Process. J. **2**, 21–28 (2009)
10. Elzobi, M., Al-Hamadi, A., Saeed, A., Dings, L.: Arabic handwriting recognition using Gabor wavelet transform and SVM. In: 2012 IEEE 11th International Conference on Signal Processing (ICSP), vol. 3, pp. 2154–2158. IEEE, October 2012
11. Lawgali, A.: Handwritten digit recognition based on DWT and DCT. Int. J. Database Theory Appl. **8**(5), 215–222 (2015)
12. ElAdel, A., Ejbali, R., Zaied, M., Amar, C.B.: Dyadic multi-resolution analysis-based deep learning for Arabic handwritten character classification. In: 2015 IEEE 27th International Conference on Tools with Artificial Intelligence (ICTAI), pp. 807–812. IEEE, November 2015
13. Mukhopadhyay, P., Chaudhuri, B.B.: A survey of Hough Transform. Pattern Recogn. **48**(3), 993–1010 (2015)
14. Hu, A.K.: Pattern recognition by moment invariants. Proc. IRE **49**, 1428 (1961)

15. Amara, M., Zidi, K., Ghedira, K., Zidi, S.: New rules to enhance the performances of Histogram projection for segmenting small-sized Arabic words. In: Hybrid Intelligent Systems, pp. 167–176. Springer International Publishing (2016)
16. Xiao, X., Leedham, G.: Knowledge-based English cursive script segmentation. Pattern Recogn. Lett. **21**(10), 945–954 (2000)
17. Pechwitz, M., Maddouri, S.S., Märgner, V., Ellouze, N., Amiri, H.: IFN/ENIT-database of handwritten Arabic words. In: Proceedings of CIFED, vol. 2, pp. 127–136, October 2002
18. Amara, M., Zidi, K.: Feature selection using a neuro-genetic approach for Arabic text recognition. In: Metaheuristics and Nature Inspired Computing (2012)
19. Amara, M., Zidi, K., Ghedira, K.: Towards a generic M-SVM parameters estimation using overlapping swarm intelligence for handwritten characters recognition. In: International Conference on Advanced Concepts for Intelligent Vision Systems, pp. 498–509. Springer International Publishing, October 2016
20. Amara, M., Zidi, K., Zidi, S., Ghedira, K.: Arabic character recognition based M-SVM: review. In: Advanced Machine Learning Technologies and Applications, pp. 18–25. Springer International Publishing (2014)
21. Amara, M., Ghedira, K., Zidi, K., Zidi, S.: A comparative study of multi-class support vector machine methods for Arabic characters recognition. In: International Conference on Computer Systems and Applications (2015)

Correction to: Electrogastrogram Based Medical Applications an Overview and Processing Frame Work

Ahmed Al Taee and Adel Al-Jumaily

Correction to:
Chapter "Electrogastrogram Based Medical Applications
an Overview and Processing Frame Work"
in: A. M. Madureira et al. (Eds.): *Hybrid Intelligent Systems*,
AISC 923, https://doi.org/10.1007/978-3-030-14347-3_50

The original version of the book was inadvertently published with the author's name as 'Ahmad A. Al-Tae' in chapter 50, which has now been changed to 'Ahmed Al Taee'.

The updated version of this chapter can be found at
https://doi.org/10.1007/978-3-030-14347-3_50

Author Index

© Springer Nature Switzerland AG 2020
A. M. Madureira et al. (Eds.): HIS 2018, AISC 923, pp. 581–583, 2020.
https://doi.org/10.1007/978-3-030-14347-3